The polyurethanes book

The polyurethanes book

Editors: David Randall and Steve Lee

Distributed by

JOHN WILEY & SONS, LTD

Copyright © 2002 Huntsman International LLC, Polyurethanes business.

All rights reserved.

Reproduction, transmission, or storage of any part of this publication by any means is prohibited without the prior written consent of Huntsman International LLC via its business unit Huntsman Polyurethanes.

Library of Congress Cataloging-in-Publication Data:

The Huntsman polyurethanes book/edited by Stephen Lee.
    p. cm.
  Includes bibliographical references and index.
  ISBN 0-470-85041-8
  1. Polyurethanes. I. Lee, Stephen.

  TP 1180.P8 H86 2002
  668.4'239--dc21
          2002033074

British Library Cataloguing-in-Publication Data:

A catalogue record for this book is available from the British Library.

ISBN 0 470 85041 8

Design, layout and typesetting by Dunholm Publicity Ltd.
Jacket design by C^2 creative communications.
Printed and bound in The United Kingdom.

Foreword

The world of polyurethanes has matured so much in the past 20 years, reflecting the increasing and changing demands of a global market.

When we first published this book in 1987, polyurethanes were well-established materials that had found their way into an unusually wide variety of uses, from foam insulation to shoe soles, and from car seats to abrasion-resistant coatings.

When I look back at that time I am struck by how versatile these materials were. Today, they remain so versatile and so full of promise.

This publication first appeared as *The ICI Polyurethanes Book*, with a second edition in 1990. Today, under our owners since 1999, we publish *The Huntsman Polyurethanes Book*.

The original purpose was to reflect an understanding of the production, properties and potential of polyurethanes. Our parent has changed, but the purpose of this book is the same.

In this fully revised and updated edition, we look at the various sectors in polyurethanes, including flexible and rigid foam, elastomers (including thermoplastics), coatings, adhesives, sealants and encapsulants, as well as composites.

In addition to the raw materials and chemistry of polyurethanes, the important areas of product stewardship and environment, health and safety are considered in detail.

I trust, as before, this is a contribution that will be of value to all seeking a better understanding and broader knowledge of polyurethanes.

Patrick Thomas
President, Huntsman Polyurethanes
Everberg, Belgium
November 2002

Acknowledgements

The first idea to produce this new Huntsman Polyurethanes handbook was initially discussed with the publisher, John Wiley & Sons, in early 1999 with work beginning in earnest the following year. Just two years later the book has been published. The editors, literary and technical, who took on the task to design, structure, project manage and, in part, write this book are very happy that it has been completed and are also indebted to many people for their help and support.

Thanks are due to our joint editors Steve Lee and David Randall for whom this work became a full-time project.

Thanks are also due in particular to Kevin Dunn and his team of Mike and Janet Pattison at Dunholm Publicity who have been the consummate professionals and helped to deliver the final product. Internally within Huntsman I would like to thank Katrien Perremans and Marina Dehaes for their excellent administrative and secretarial support and the group of people who read and approved the book on behalf of Huntsman: Philip Baken, Bill Brooks, Erik Casselman, Judith Dobbs, Marnix Moens, John Nevard, Annemie Swinnen and Sarah Wijns. I would also like to thank all the authors who worked so closely with us on this project. I know how hard the struggle was and how many long hours were put into this project.

Not least, I would like to thank especially: David Sparrow for reading everything in the final typescript and for spotting all the things that others missed; Alan Hamilton for being able to find information, no matter where from or what about; and Martin Routh for ensuring the grammar and punctuation are correct.

Richard Northcote
Director of Communications

Editors and authors

Technical editor: David Randall

Editor: Steve Lee

Authors:
Tony AbiSaleh
Mike Anderson
Martyn Barker
Guy Biesmans
Joris Bosman
Diane Daems
Kristof Dedecker
Joris Deschaght
Judith Dobbs
Nick Duggan
Berend Eling
Brian Fogg
Alan Hamilton
Nick Hernandez
Mike Jeffs
Vikram Kapasi
Steve Lee
Nick Limerkens
Paul Mackey
Steve O'Nien
Alain Parfondry
Chris Phanopoulos
Wolfgang Pille-Wolf
David Randall
Sachchida N Singh
David Sparrow
David Thorpe
Gaby Verhelst
Jacquin Wilford-Brown
Robert L Zimmerman

While the information and recommendations in this publication are based on our general experience and are given in good faith to the best of our current knowledge, NOTHING HEREIN IS TO BE CONSTRUED AS A GUARANTY, WARRANTY OR REPRESENTATION, EXPRESS, IMPLIED OR OTHERWISE.
IN ALL CASES, IT IS THE RESPONSIBILITY OF THE USER TO DETERMINE AND VERIFY THE ACCURACY, SUFFICIENCY AND APPLICABILITY OF SUCH INFORMATION AND RECOMMENDATIONS, AS WELL AS THE SUITABILITY AND FITNESS OF ANY PRODUCT FOR ANY PARTICULAR USE OR PURPOSE.
THE MENTIONED PRODUCTS MAY PRESENT UNKNOWN HAZARDS AND SHOULD BE USED WITH CAUTION. WHILE CERTAIN HAZARDS ARE DESCRIBED IN THIS PUBLICATION, NO GUARANTEE IS MADE THAT THESE ARE THE ONLY HAZARDS THAT EXIST.
Hazards, toxicity and behavior of the products may differ when used with other materials and is dependent upon the manufacturing circumstances or other processes. Such hazards, toxicity and behavior should be determined by the user and made known to handlers, processors and end users. NOTHING IN THIS PUBLICATION IS TO BE VIEWED AS A LICENCE UNDER ANY INTELLECTUAL PROPERTY RIGHT. NOTHING IN THIS PUBLICATION IS TO BE CONSTRUED AS ALLOWING OR RECOMMENDING THE INFRINGEMENT OF ANY PATENT OR OTHER INTELLECTUAL PROPERTY RIGHT, NOR OF ANY APPLICABLE LAWS AND REGULATIONS.

Contents

Chapter	Page
1. **Introduction to polyurethanes**	1
Cost and processing advantages	3
Properties of polyurethanes	3
Types of polyurethanes	4
Applications of polyurethanes	6
Industry structure	7
Future trends	8
2. **The global polyurethanes market**	9
Drivers for growth	9
Technology growth factors	12
Supply and demand balance	13
Applications	16
Automotive	17
Coatings	18
Construction	18
Footwear	19
Furniture	20
Thermal insulation	20
Replacement challenges	21
3. **The life-cycle of polyurethanes**	23
Life-Cycle Assessment (LCA)	24
Polyurethane raw materials production phase	26
The product manufacture phase	28
Polyurethane factory emissions	28
Production waste and scrap	29
The use phase	30
Energy efficiency	30
Ecolabels	32
Environmental declarations	33
Indoor air	34
End-of-life phase	35
Waste management of polyurethanes	35
4. **Product Stewardship**	39
Regulatory framework	40
Product design and development	40

Chapter	Page
Raw materials used in polyurethanes	42
Isocyanates	42
Polyol blends	43
Pre-formulated mixtures	44
Thermoplastic polyurethanes	44
Handling of polyurethane chemicals	44
Occupational hygiene monitoring for isocyanates	46
Storage and transport of polyurethane chemicals	47
Drum stock	47
Intermediate bulk containers (IBCs)	48
Bulk storage	48
Manufacture and handling of polyurethanes in the factory	49
Flexible foam	49
Rigid foam	50
Thermoplastic polyurethanes	51
Use of polyurethanes	51
Polyurethanes and fire	51
Flexible foam and fire	53
Furniture and mattresses	53
Automotive	55
Flame retardancy	55
Test methods	56
Rigid foams, composite wood panels and fire	58
USA	59
China	60
EU and EFTA	60
Performance-based codes	61
Other polyurethane applications and fire	61
Effluents from a fire	61
5. **Isocyanates**	63
Reaction chains	63
Benzene to MDI	63
Toluene to MDI	64
Hexamethylene diamine to HDI (and derivatives)	64

Chapter	Page
Acetone to IPDI and derivatives including other uses for isophorone and IPDA	64
MDA (DADPM) to H_{12}MDI	69
Nitration process	70
Hydrogenation process	71
Liquid slurry phase process	72
Liquid-vapour slurry phase process (aniline)	72
Vapour phase fixed bed (aniline)	72
Vapour phase fluidised bed (aniline)	73
Aniline from phenol	73
Aniline-formaldehyde condensation	73
Phosgenation process	76
Purification	78
Isocyanate derivatives	79
Analysis	80
Isocyanate value	80
Hydrolysable chlorine	80
Ionisable chlorine	81
Acidity	81
Viscosity	81
Gas chromatography	82
Storage/transportation	82
Isocyanate product characteristics	82
MDI	83
Polymeric MDI	84
TDI	86
NDI	86
HDI	87
IPDI	87
H_{12}MDI	87
Other diisocyanates	87
Alternatives to phosgene	87
6. Polyols	**89**
Manufacture of propylene oxide	89
Chlorohydrin process	89
Hydroperoxidation	91
Propylene oxide polyether polyols	93
Anionic polymerisation with potassium hydroxide	94
Product purification	96
Side reactions	96
Alternative catalysts	97

Chapter	Page
Structure	98
Propylene oxide-ethylene oxide polyether polyols	99
Standard polyether polyols	101
Rigid foam applications	101
Flexible foam applications	103
Modified polyether polyols	104
Graft dispersions	104
Amine-terminated polyethers	106
Polyether polyols from tetrahydrofuran	106
Polyester polyols	107
Linear or lightly branched aliphatic polyester polyols	108
Polycaprolactones	110
Aromatic polyester polyols	110
Polycarbonate polyols	111
Miscellaneous polyols	111
Quality and analysis	112
7. Outline of polyurethane chemistry	**113**
Isocyanates	113
Isocyanate reactions with hydroxyl	114
Isocyanate reaction with water	115
Isocyanate reaction with amines	116
Isocyanate reaction with urea	116
Isocyanate reaction with urethanes	117
Isocyanate reactions with isocyanates	118
Other reactions of isocyanates	121
Polyurethane degradation reactions	123
Polyether polyol	123
Polyester polyol	125
Ultra-violet radiation	125
Physical chemistry	126
8. Blowing agents	**127**
Montreal Protocol and other regulations	128
Blowing agents	131
Hydrochlorofluorocarbons (HCFCs)	132
Hydrofluorocarbons (HFCs)	132
Hydrocarbons (HCs)	133
Liquid carbon dioxide and other physical blowing agents	134
Chemical blowing	134

Chapter	Page
Selection criteria	135
Environmental considerations	135
Feasibility	136
Performance	136
9. Catalysts	137
Preparation of amine catalysts	137
Preparation of organotin catalysts	139
Reaction mechanisms	141
Amine catalysts	141
Organotin catalysts	143
Isocyanurate catalysts	145
Reaction kinetics	145
Traditional experiments	147
Model systems	147
FTIR experiments	148
Ion viscosity	150
10. Additives	151
Blowing agents	151
Catalysts	151
Flexible foams	151
Rigid foams	152
Coatings, adhesives, sealants and encapsulants	153
Surfactants	156
Structural parameters of silicone surfactants	157
Flexible foams	158
Rigid foams	159
Elastomers and other applications	160
Fire retardants	160
Flexible foams	162
Rigid foams	162
Other applications	162
Cross-linking agents, chain-extending agents and their reactions	163
Other additives	163
Adhesion promoters	163
Anti-static	163
Anti-oxidants	165
Fillers	165
Hydrolysis	166
Lubricants	166
Anti-microbials	167
Pigments	167

Chapter	Page
Viscosity reducers	167
UV resistance	167
11. Introduction to flexible foams	169
Markets	169
Raw material history	169
Basic flexible foam chemistry	171
Foam and polymer morphology	174
Cellular structure	175
Polymer morphology	175
Functionality and performance tests for flexible foams	177
Basic properties	179
Durability cushioning/comfort tests	180
Ageing tests	187
Fire test methods	188
Environmental performance	188
12. Moulded foams for automotive seating, sound insulation and furniture	189
Applications	189
Automotive seating	190
Sound insulation foams	192
Moulded furniture foams	193
Raw materials and formulations	193
Cold-cure foams	194
Hot-cure foams	197
Processing of flexible foams	198
Cold-cure moulding	199
Hot-cure moulding	202
13. Slabstock foams	203
Foam technologies	203
Raw materials and formulations	204
Polyester foams	206
Conventional polyether foams	207
High-resilience foams	208
Processing technologies	211
Foam machinery	212
Foam curing	214
Carbon dioxide-assisted blowing	214
Variable pressure foaming	214
Batch-block	215
Scorch	215
Post-processing of slabstock foam	216

Chapter	Page
14. Technical foams and applications	217
Semi-rigid foams	217
Applications	217
Formulations	218
Foam properties	219
Viscoelastic foams	220
Packaging	221
Polyurethane gels	223
Hydrophilic foams	223
Textile laminated polyurethane foams	224
Post-treated polyurethane foams	225
Reticulated foams	225
Impregnated foams	226
Rebonded foams	226
Carpet backing foams	227
15. Introduction to rigid foams	229
Basic science	229
Chemical processes during foam formation	230
Physical processes during foam formation	230
Formulation design for optimum processing and end properties	231
Blowing agents	232
Key criteria for blowing agent choice	233
Heat transfer	233
Fundamental aspects of thermal conductivity	234
Radiative transfer ($\lambda_{radiative}$)	235
Solid conduction (λ_{solid})	235
Gaseous conduction (λ_{gas})	235
Thermal conductivity ageing	236
Basic theory	236
Thermal conductivity ageing under service conditions	238
Industry standards for thermal performance	239
Open-cell foam	239
Key testing methods	241
Thermal conductivity	241
Compression strength	242
Dimensional stability	243
Closed-cell content	243
Water vapour transmission	244

Chapter	Page
16. Appliances	245
Appliance design	246
Formulations and properties	250
Process technology	253
17. Construction	257
Insulation materials	257
Formulation technologies	258
Boardstock	259
Formulations	260
Blowing agents	261
Mixing	262
Facers	262
Lay-down	262
Conveyor	263
Line speed – board thickness	265
Cutting and stacking	265
Sandwich panels	266
Formulations	268
Continuous lamination	269
Discontinuous manufacture	271
18. Other construction applications	273
Insulated pipe	273
Discontinuous moulding	274
Continuous moulding	277
Continuous spray	277
Spray insulation	278
One-component foam (OCF)	281
Water heaters	282
Other niche applications	283
Buoyancy	283
Surfboards	283
Aircraft propellers/windsails	283
Floral foam applications	283
Wall, soil and mine stabilisation	284
Low-density semi-structural foams	284
High-density structural foams	284
19. Introduction to elastomers	285
Theory of polyurethane elastomers	286
Test methods	294
Standard mechanical properties	294
Dynamic property testing	297
Frictional properties	298
Thermal analysis	298
Environmental exposure testing	299

Chapter	Page
20. Elastomers for footwear applications	301
Two-component polyurethanes	303
Polyester technology	303
Polyether technology	306
Hybrid technology	307
Additives	308
Process technology for two-component polyurethanes	309
Machinery	309
Moulds	310
Mould release	310
Thermoplastic polyurethanes	311
21. Thermoplastic polyurethanes	315
Applications	316
Automotive	316
Engineering	317
Footwear	318
Medical	318
Pipe, hose and tube	319
Wire and cable	320
Film, sheet and calendared articles	320
Raw materials and properties	321
Production	324
Batch process	325
Band casting	325
Reactive extrusion	326
Processing	327
Compounding	327
Polymer blends	328
Powder	328
Drying	328
Injection moulding	328
Extrusion	329
Rotational moulding	330
22. Other two-component elastomers	331
Cast elastomers	331
Synthetic leathers	336
Elastomeric fibres	338
Integral skin foams	339
Reaction injection moulding elastomers	342
Polyurea elastomers	344

Chapter	Page
23. Introduction to coatings, adhesives, sealants and encapsulants	347
Raw materials	349
Isocyanates	349
Polyols	349
Other ingredients	350
Surface interactions	350
Reactions at surfaces and interfaces	352
Solidification and bonding	353
Test methods	353
Properties of basic materials	354
Sample preparation	358
Final product properties	358
Environmental issues and future trends	361
24. Coatings	363
Materials selection	364
Isocyanates	364
Polyols	364
Amines	365
Solvents	365
Technology of reactive coatings	366
Two-component polyurethane coatings	366
Oven-curing or stoving systems	372
One-component polyurethane coatings	374
Technology of non-reactive coatings	375
Solvent-borne lacquers	375
Polyurethane dispersions	375
Urethane oils and alkyds	377
Radiation curing	377
25. Adhesives	379
Types of adhesive technology	379
Materials selection	382
Non-reactive polyurethane adhesives	384
Solvent-borne adhesives	384
Water-borne adhesives	386
Hot-melt adhesives	387
Reactive polyurethane adhesives	388
One-component adhesives	388
Two-component adhesives	391
Reactive hot-melt	393
Hybrid systems	394

Chapter	Page
26 Wood adhesives	395
Mechanisms of adhesion	397
Types of wood composites	399
Oriented strand board (OSB)	399
Medium density fibreboard (MDF)	403
Engineered lumber (EL)	405
Other types of wood composites	408
27. Sealants and encapsulants	409
Sealants	409
Applications	410
Formulations and properties of sealants	410
Encapsulants	416
Applications	416
Formulations and properties of encapsulants	417
28. Introduction to polyurethane composites	419
General properties of composites	420
Design considerations when working with composites	421
Stress analysis	421
Detailed design	421
Raw materials	422
Polyurethane systems	422
Form and types of glass fibre	422
Mineral reinforcements	423
Carbon fibre	424
Natural fibre	424
Processing techniques	424
Test procedures	424
29. Polyurethane composite technology	427
SRIM, P4 and LD-SRIM	427
RRIM	429
Sprayed chopped fibre	429
Filament winding	430
Pultrusion	431
Mould technology	431
Formulations and properties	432
Choice of process	434

Chapter	Page
30. The use of polyurethane composites in automotive applications	437
Automotive interiors	437
Door panels	439
Sun-shades	440
Package trays	441
Headliners	441
Seatbacks	442
Load floors and floor pans	442
Automotive exteriors	442
Future trends	444
Appendix 1: Calculations	447
Appendix 2: Conversion factors	453
Appendix 3: Physical properties of isocyanates	455
Index	457

1. Introduction to polyurethanes

Although the reaction between isocyanate and hydroxyl compounds was originally identified in the 19th Century, the foundations of the polyurethanes industry were laid in the late 1930s with the discovery, by Otto Bayer, of the chemistry of the polyaddition reaction between diisocyanate and diols to form polyurethane. The first commercial applications of polyurethane polymers, for millable elastomers, coatings and adhesives, were developed between 1945 and 1947, followed by flexible foams in 1953 and rigid foams in 1957. Since that time they have been finding use in an ever-increasing number of applications and polyurethanes are now all around us, playing a vital role in many industries – from furniture to footwear, construction to cars. Polyurethanes appear in an astonishing variety of forms, making them the most versatile of any family of plastic materials.

Comfortable, durable mattresses and automotive and domestic seating are manufactured from flexible foam. Rigid polyurethane foam is one of the most effective practical thermal insulation materials, used in applications ranging from domestic refrigerators to large industrial buildings. Polyurethane adhesives are used to make a wide variety of composite wood products from load-bearing roof beams to decorative cladding panels. Items such as shoe soles, sports equipment, car bumpers and 'soft front ends' are produced from different forms of polyurethane elastomers.

Many of us are clothed in fabrics containing polyurethane fibres or high-performance breathable polyurethane membranes. Highly demanding medical applications use biocompatible polyurethanes for artificial joints and implant coatings. Polyurethane coatings protect floors and bridges from damage/corrosion and adhesives are used in the construction of items as small as an electronic circuit board and as large as an aircraft. Advanced glass and carbon fibre reinforced composites are being evaluated in the automotive and aerospace industries. Examples of typical applications are shown on page 2.

Commercially, polyurethanes are produced by the exothermic reaction of molecules containing two or more isocyanate groups with polyol molecules containing two or more hydroxyl groups. Relatively few basic isocyanates and a far broader range of polyols of different molecular weights and functionalities are used to produce the whole spectrum of polyurethane materials. Additionally, several other chemical reactions of isocyanates are used to modify or extend the range of isocyanate-based polymeric materials. The chemically efficient polymer reaction may be catalysed, allowing extremely fast cycle times and making high volume production viable.

Applications of polyurethanes

Cost and processing advantages

Although a unique advantage of polyurethanes lies in the very wide variety of high-performance materials that can be produced, they also differ from most other plastic materials because the processor is able to change and control the nature and the properties of the final product, even during the production process. This is possible because most polyurethanes are made using reactive processing machines, which mix together the polyurethane chemicals that then react to make the polymer required. Changes in the detailed chemical nature of the polyols, isocyanates or additives allow the user to produce different end polymers. Minor changes in the mixing conditions and ratios allow for a fine-tuning of the polymers produced. The polymer is usually formed into the final article during this polymerisation reaction and this accounts for much of the versatility of polyurethanes. They can be tailored with remarkable accuracy to meet the precise needs of a particular application.

Another important property of polyurethane reaction mixtures is that they are powerful adhesives. This enables the simple manufacture of strong composites such as building panels and laminates, complete housings for refrigerators and freezers, fully integrated instrument panels for vehicles and reinforced structures in boats and aircraft. It is, in part, this dual functionality that makes the material so valuable for manufacturing industries since it is possible to eliminate a number of complex and expensive assembly steps when using polyurethanes rather than alternative polymers.

It is this combination of high material performance coupled with processing versatility that has resulted in the spectacular growth and wide applicability of the polyurethane family of materials. The processing benefits enable polyurethanes to compete with lower cost polymers since raw material costs are not the only consideration in the total cost involved in producing an article. Factors of at least equal importance are cycle time, the cost of tooling and finishing as well as reject rates and opportunities for recycling. As polyurethane reaction moulding requires only comparatively low pressures, moulds can be made of less expensive materials such as aluminium or glass reinforced polyester rather than steel. This is of particular importance for low volume production, but also allows the simple and rapid production of inexpensive prototypes for the development of new products and processes or the refinement of established ones.

Properties of polyurethanes

Polyurethanes can be manufactured in an extremely wide range of grades, in densities from 6 to 1,220 kg/m^3 and polymer stiffness from flexible elastomers to rigid, hard plastics. Although an over simplification, the following chart, Figure 1-1, illustrates the broad range of polyurethanes, with reference to density and stiffness.

Figure 1-1   Property matrix of polyurethanes

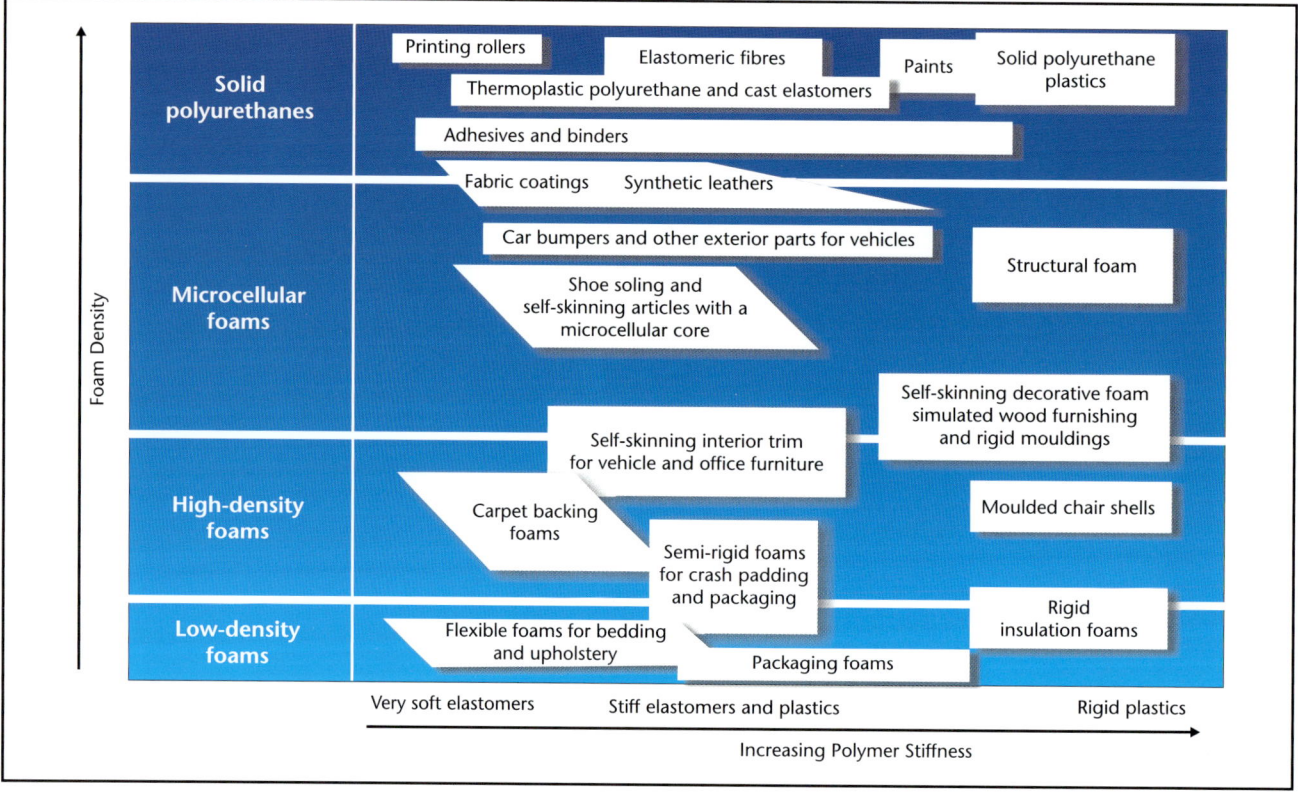

Types of polyurethanes

A consideration of particular properties of certain grades of polyurethanes and the way these are used serves to demonstrate their versatility.

Foamed

By itself the polymerisation reaction produces solid polyurethane and it is by forming gas bubbles in the polymerising mixture, often referred to as 'blowing', that a foam is made. Foam manufacture can be carried out continuously, to produce continuous laminates or slabstock, or discontinuously, to produce moulded items or free-rise blocks.

Flexible foams can be produced easily in a variety of shapes by cutting or moulding. They are used in most upholstered furniture and mattresses. Flexible foam moulding processes are used to make comfortable, durable seating cushions for many types of seats and chairs. The economy and cleanliness of flexible polyurethane foams are important in all upholstery and bedding applications. Strong, low-density rigid foams can be made that, when blown using the appropriate environmentally acceptable blowing agents, produce closed cell structures with low thermal conductivities. Their superb thermal insulation properties have led to their widespread use in buildings, refrigerated transport, refrigerators and freezers.

A fast, simple moulding process can be used to produce rigid and flexible foam articles, having an integral skin, that are both decorative and wear resistant. Fine surface detail can be reproduced in the integral skin of the foam allowing for the simple manufacture of simulated wood articles, 'leather-grain' padded steering wheels and textured surface coatings.

Three foam types are, in quantity terms, particularly significant: low-density flexible foams, low-density rigid foams and high-density flexible foams, commonly referred to as microcellular elastomers and integral skin foams.

Low-density flexible foams have densities in the range 10 to 80 kg/m^3, made from a lightly cross-linked polymer with an open cell macro structure. There are no barriers between adjacent cells, which results in a continuous path in the foam, allowing air to flow through it. These materials are used primarily as flexible and resilient padding material to provide a high level of comfort for the user. They are produced as slabstock, which is then cut to size, or as individually moulded cushions or pads. There are semi-rigid variants of this material, where the chemistry of the building blocks has been changed, and these are mainly used in energy management systems such as protective pads in cars. An example of the cellular structure is shown in Figure 1-2.

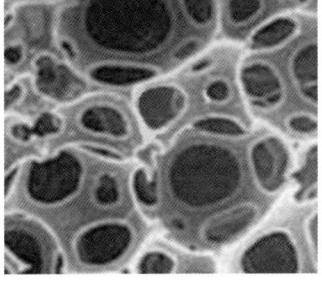

Figure 1-2
Scanning electron micrograph showing the open cells of flexible foam

Low-density rigid foams are highly cross-linked polymers with an essentially closed cell structure and a density range of 28 to 50 kg/m^3. The individual cells in the foam are isolated from each other by thin polymer walls, which effectively stop the flow of gas through the foam. These materials offer good structural strength in relation to their weight, combined with excellent thermal insulation properties. The cells usually contain a mixture of gases and depending on their nature and relative proportions the foams will have different thermal conductivities. In order to maintain long-term performance it is necessary for the low thermal conductivity gases to remain in the cells, consequently more than 90 per cent of the cells need to be closed. An example of the cellular structure is shown in Figure 1-3. Recently, fully open celled rigid foams specifically developed for vacuum panel applications have been developed.

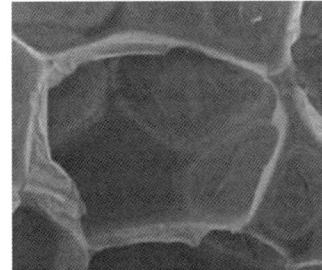

Figure 1-3
Scanning electron micrograph showing the closed cells of rigid foam

High-density flexible foams are defined as those having densities above 100 kg/m^3. This range includes moulded self-skinning foams and microcellular elastomers. Self-skinning or integral skin foam systems are used to make moulded parts having a cellular core and a relatively dense, decorative skin. There are two types: those with an open cell core and an overall density in the range up to about 450 kg/m^3 and those with a largely closed cell or microcellular core and an overall density above 500 kg/m^3. The microcellular elastomers have a much more uniform density in the range of 400 to 800 kg/m^3 and mostly closed cells, which are much smaller than those in the low-density applications. The biggest applications for integral skin and microcellular elastomers are in moulded parts for upholstery, vehicle trim and shoe soling. Another similar material is the microporous elastomer in which the porous structure is often created in ways other than by the expansion of gases. Often produced in thin films these materials have an open cell like structure, which allows movement of gases, but have the appearance and physical integrity of a solid film.

Solid
Although foamed materials account for a substantial proportion of the global polyurethanes market there is a wide range of solid polyurethanes used in many, diverse applications.

Cast polyurethane elastomers are simply made by mixing and pouring a degassed reactive liquid mixture into a mould. These materials have good resistance to attack by oil, petrol and many common non-polar solvents combined with excellent abrasion resistance. They are used amongst other things in the production of printing rollers and tyres, both low speed solid relatively small units and to fill very large, pneumatic off-road tyres.

Polyurethane elastomeric fibres are produced by spinning from a solvent, usually dimethylformamide (DMF), or by extrusion from an elastomer melt. The solvent process is the dominant one and has two forms, one in which the completed elastomer is dissolved and then a fibre spun as the solvent is removed and the other in which the isocyanate and polyol are mixed into a DMF solution and the fibre spun as the reaction occurs. The major applications are in clothing where these fibres have effectively replaced natural rubber.

Thermoplastic polyurethanes are supplied as granules or pellets for processing by well-established thermoplastic processing techniques such as injection moulding and extrusion. By these means elastomeric mouldings having an excellent combination of high strength with high abrasion and environmental resistance, can be mass-produced to precise dimensions. Applications include hose and cable sheathing, footwear components and high-wear engineering applications. Recent advances have shown the possibility to foam the polymer during injection moulding, extending even further the range of applications.

Polyurethanes are also used in flexible coatings for textiles and adhesives for film and fabric laminates. Paints and coatings give the highest wear resistance to surfaces such as floors and the outer skins of aircraft and for the automotive industry. Binders are used increasingly in the composite wood products market for oriented strand board and laminated beams for high performance applications.

Applications of polyurethanes

A detailed breakdown of the markets for polyurethanes is given in Chapter 2, but the versatility of this material can be demonstrated by looking at the applications in five major areas.

Automotive
The use of polyurethanes in this area is now well established to the benefit of both the manufacturer and the end consumer. Applications include seating, interior padding, such as steering wheels and dashboards, complete soft front-ends, components for instrument assemblies and accessories such as mirror

surrounds and spoilers. Door panels, parcel shelves, sun roofs, truck beds, headliners, components mounted in the engine space and even structural chassis components are now made from polyurethanes.

Furniture
The market for cushioning materials is mainly supplied by polyurethane flexible foam, which competes with rubber latex foam, cotton, horse hair, polyester fibre, metal springs, wood, expanded polystyrene, propylene and PVC. Polyurethanes are also ideal where strong, tough, but decorative integral-skinned flexible or rigid foam structures are needed.

Construction
When sandwiched between metal, paper, plastics or wood, polyurethane rigid foam plays an important role in the construction industry. Such composites can replace conventional structures of brick, concrete, wood or metal, particularly when these later materials are used in combination with other insulating materials such as polystyrene foam, glass fibre or mineral wool. Technically advanced wood composites can be produced for use in load-bearing applications and wood construction boards for flooring and roofing.

Thermal insulation
Rigid polyurethane foam offers unrivalled technical advantages in the thermal insulation of buildings, refrigerators and other domestic appliances and refrigerated transport. Competitive materials include cork, glass fibre, mineral wool, foamed expanded and extruded polystyrene and phenol formaldehyde.

Footwear
Soles, some synthetic uppers and high performance components for many types of footwear are produced from polyurethanes. These compete with traditional leather and rubber, PVC, thermoplastic rubber and EVA. Polyurethane adhesives are widely used in shoe manufacture and coatings are used to improve the appearance and wear resistance of shoe uppers made from both real and synthetic leather.

Industry structure

The industrial base for polyurethanes that has evolved over the last half century is driven by two key factors. The first is the economies of scale associated with the raw material manufacture. The process to make isocyanates, starting from chemicals such as benzene or toluene, is complex and expensive. As with all chemical processes there are many safety and environmental issues to be considered and it is essential to have the lowest possible cost base for production of these large volume chemicals. Consequently, the manufacture resides in the hands of a few major global chemical companies. The situation for the polyol side is more complex since the production of propylene oxide (PO), ethylene oxide (EO) and polyesters is more wide spread, in part due to the wider variety of processes available and the suitability to run economically, at a relatively smaller scale. In addition, both PO and EO can be purchased on the open market and then used to manufacture polyols whereas the isocyanates are produced and sold by the major companies. It is, of course, still beneficial to have world-scale production economics for the polyol components.

The second factor is the diversity of the market and the routes to serve that market. Although all the major isocyanate manufacturers are also basic in PO and EO and make and sell polyols, there are strategic choices made about the way that the chemical 'package' is delivered to the customer. This choice is governed principally by economics and supply chain logistics. In addition, the customer is not always the end producer. An illustrative example of this is the automotive industry where there are a number of 'tiers' of manufacturers all creating more complex product offers as they move towards the final specifiers/designers: the car companies, often referred to as OEMs (original equipment manufacturers) who assemble and sell cars under their own name and brands. Thus, a chemical manufacturer could be providing a fully-formulated system, isocyanate and polyol with a full additive package to a flexible foam seat moulder, who will then provide the cushions to another company to finish the seat assembly. This, in turn, goes to the car company, which will fit the seat into the car.

To some customers the chemical manufacturer will provide only the basic isocyanate and polyol and the customer, using its own, in-house expertise, will formulate a system and then manufacture the end article itself. An alternative approach is the use of a 'system house', either independent or aligned, which buys in the base chemicals from a number of sources, designs and manufactures a formulated system and then sells it on to the end polyurethane producer.

Future trends

The future of the polyurethanes industry will be driven by the continued innovation in both the chemistry and the polymer physics of this highly versatile material and research and development will continue to provide new capabilities. The integration of raw material supplies, such as propylene to propylene oxide to polyols, will have an effect on the cost base of the major chemical manufacturers. At the same time, as the economics of production improve and market development into previously unexplored end uses identifies new targets, so the growth of the industry will be fuelled not only by economic growth, but also by the replacement of other materials in existing markets.

An example of this is the progress being made in the area of composite materials, where metals and other engineering plastics are being displaced by polyurethane composites offering superior properties and processing advantages.

2. The global polyurethanes market

Guy Biesmans

The global market for polyurethane products is diverse and complex with a myriad of end-uses, which makes it difficult to determine the specific global consumption for each application. Production details are reasonably easy to acquire, providing a global volume for the amount of polyurethane raw materials produced each year. But these raw materials then have to be transferred to processors, through a variety of routes including intermediate traders, which is why the precise consumption volumes broken down by end-use are so difficult to determine. However, it is the versatility of polyurethane chemistry, with its capability to both tailor the raw material formulations and the processing of them to match the innumerable end-uses that accounts for the major growth decade-after-decade. The polyurethanes market has a dynamic and evolving structure driven not only by the increase in global population and wealth, but also by advances in technology.

Drivers for growth

The polyurethanes market has been growing at an average rate of over seven per cent per annum for the last 15 years and even in times of economic recession has maintained its global sales. This can be seen in Figure 2-1 together with the extrapolation of growth to 2005.

Figure 2-1   Volume growth of the global polyurethanes market

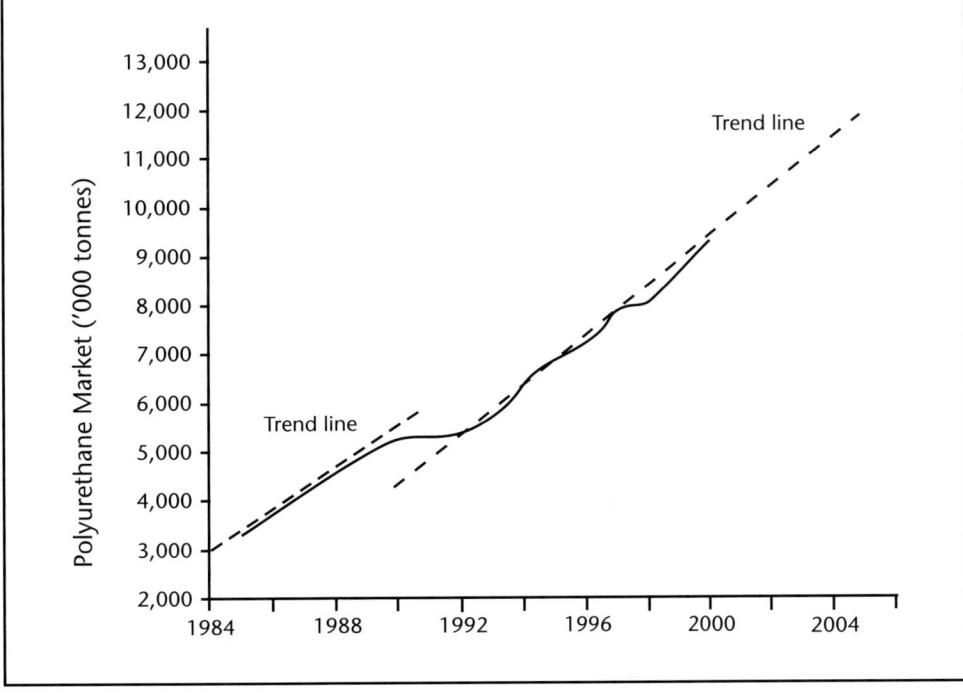

Four factors have contributed to this growth:

- Population.
- Economic development.
- Social pressures.
- Technology.

The world population has been growing, over the same period, at an average rate of 1.5 per cent per annum. This is a global figure and individual regions and countries have vastly different rates; in some north European countries the population has actually been decreasing whilst in the Peoples Republic of China it has increased by 17 per cent. The global trend is shown in Figure 2-2 and for specific regions and countries in Figures 2-3 and 2-4 (1985 to 2000).

Population increase alone, whilst important, cannot be responsible for all the polyurethane growth, and a second significant consideration is wealth. With increasing wealth comes an increase in demand for many products containing polyurethane, such as refrigerators, shoes, and eventually cars. Although income levels are directly related to purchasing power, a correction is required for local manufacturing factors and local economic considerations – an identical global brand refrigerator can be sold for a different absolute amount and a different proportion of an individual's disposable income dependent on the country.

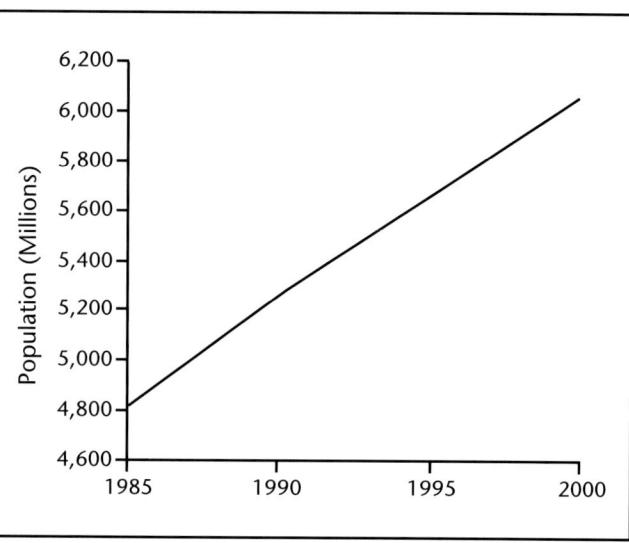

Figure 2-2   World population increase

Figure 2-3
Population increase for China, India, Rest of Asia and Africa/Middle East

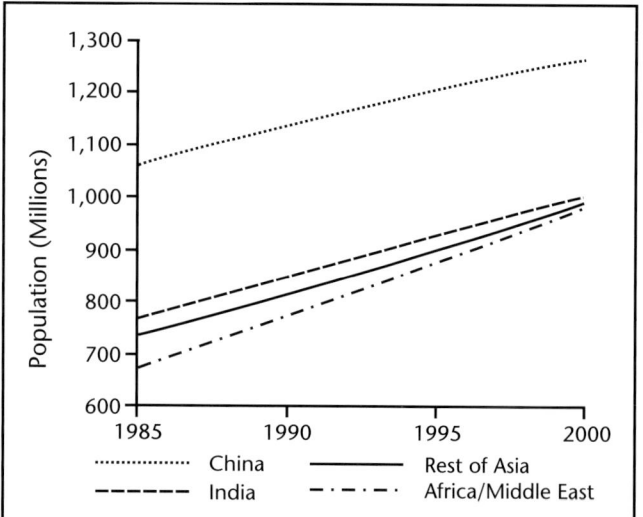

Figure 2-4
Population increase for Japan, Latin America, North America, Eastern Europe and the European Union

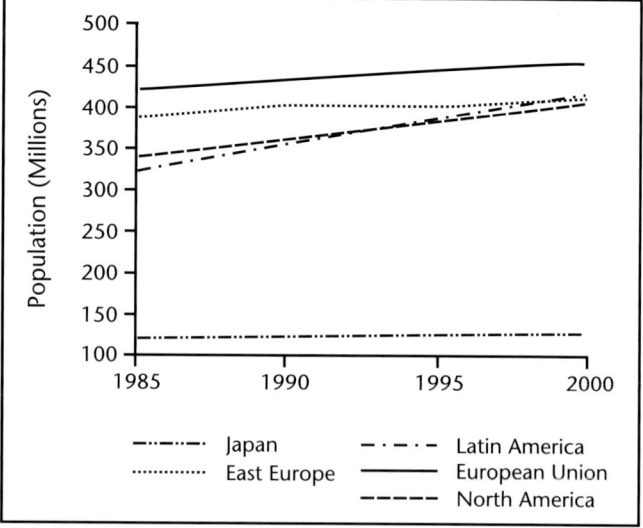

Thus, the purchasing parity power (PPP) is the correct measure to compare the wealth of different countries or regions and the development of PPP over the last 25 years has shifted the consumption balance around the world. The PPP of regions and some individual countries (1985 to 2000) is shown in Figures 2-5, 2-6 and 2-7, with the PPP quoted in 1997 US$. Although almost all areas showed a slowing down in the growth of PPP in the last five years of the 20th Century, it is important to note that Eastern Europe, Africa and Asia, excluding Japan and China, showed a substantial decrease whilst China has grown at such a rate that it doubled its PPP every seven to eight years.

An additional influence in the growth of the polyurethanes market, ahead of PPP and population growth, can be social pressure. Many of the items produced using polyurethane are considered to be 'luxury' goods and owning them is a mark of success or demonstrates belonging to social groups with refrigerators and shoes as prime examples. In these cases a disproportionate amount of an individual's wealth can be used to purchase these items increasing polyurethane consumption ahead of PPP.

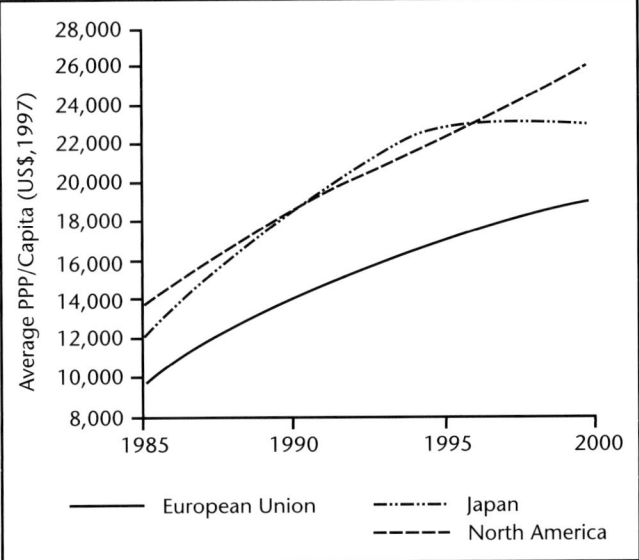

Figure 2-5
Average PPP/capita for European Union, Japan and North America

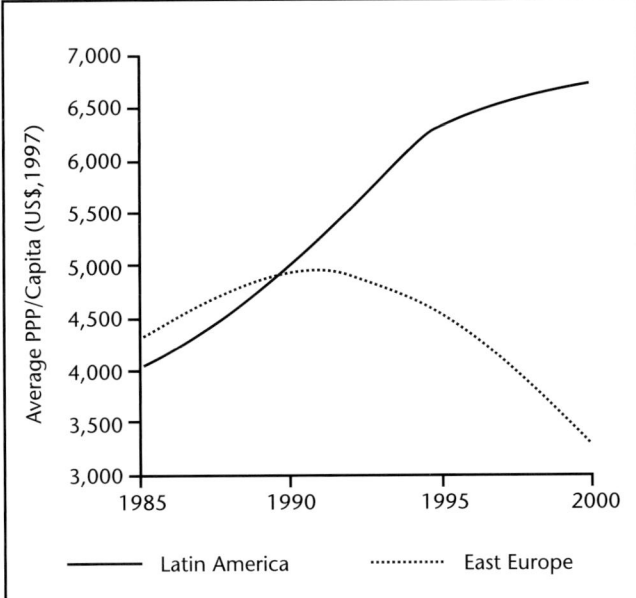

Figure 2-6
Average PPP/capita for Eastern Europe and Latin America

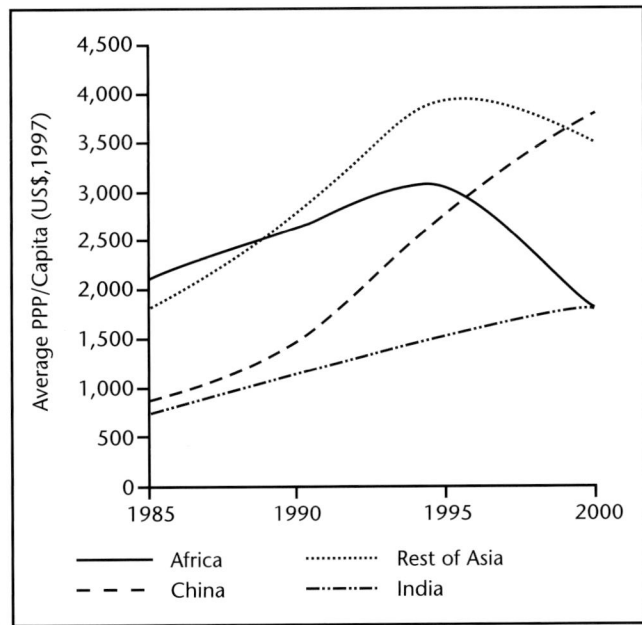

Figure 2-7
Average PPP/capita for China, India, Africa and the Rest of Asia

Technology growth factors

There has been a technology shift in the polyurethanes industry, which can be seen in the relative balance between products based on MDI and TDI, Figure 2-8. In the period 1985 to 2000 TDI has had a reasonably consistent average annual growth rate of 5.7 per cent, around twice the growth rate of global GDP/Capita at 2.9 per cent (GDP/Capita – the Gross Domestic Product – a measure of the economic activity, which includes the total amount of goods and services produced in a year, divided by population). Prior to 1994 TDI-based polyurethanes had the larger share of the market, with both TDI and MDI showing similar growth rates. However, after 1994 this situation changed as the growth of MDI accelerated. Overall from 1985 to 2000 MDI has had an average annual growth rate of 8.5 per cent, about three times global GDP/Capita. The reasons for this shift are:

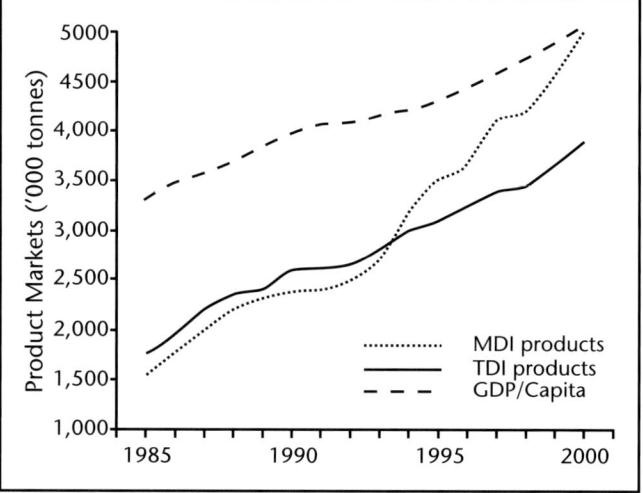

Figure 2-8
Comparison of the volume growth of MDI- and TDI-based products and GDP/Capita

- Faster growth of the traditional market segments served by MDI compared with those for TDI.
- Substitution of MDI into markets previously dedicated to TDI, especially in the moulded flexible foam area.
- Development of new markets for MDI, such as: partial replacement of the traditional phenol/formaldehyde resin in wood binding, the use of polyurethane products for artificial leather applications, increased use of polyurethane in the footwear industry.

Approximately 87 per cent of TDI is consumed today by the manufacture of foams, compared to a figure of 64 per cent for MDI. The TDI-based foams are predominantly flexible, whereas the MDI-based foams were traditionally rigid or semi-rigid. However, the extension of MDI-based foam capabilities has resulted in some substitution for TDI in flexible foams, particularly for moulding applications. Although TDI has had limited use in rigid foams in the past, its use in this market sector has almost completely disappeared. Aliphatic isocyanates are used primarily in lacquers, paints and coatings and account for a much lower volume than TDI or MDI. The percentage shares of the technologies for the polyurethanes market are given in Figure 2-9, comparing 1985 and 2000.

All sectors of the polyurethane industry have shown significant growth over this period, but this has been particularly rapid in three areas: adhesives, sealants and binders; elastomers; and coatings. The increase in the elastomer market has been due to both a shift in the relative contributions of the products within this area and the development of elastomers for a wide range of different and often new applications, where they have replaced other materials. Cast elastomers and especially thermoplastic polyurethanes have grown the fastest, whilst reaction injection moulding (RIM) products have shown little growth or even contraction.

Figure 2-9   Split of polyurethane market by technology for 1985 and 2000

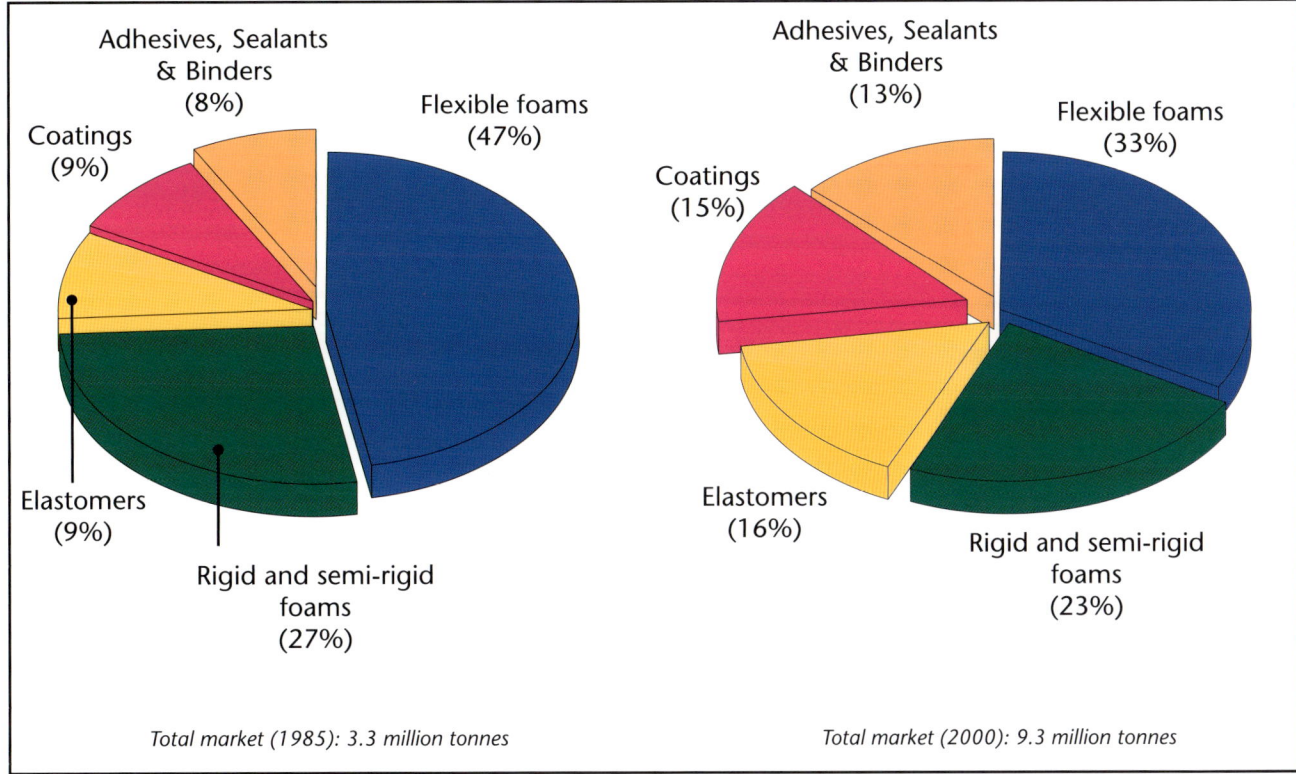

The versatility of polyurethanes in being able to meet market requirements has resulted in a change in the ratio of the different application sectors, and whilst flexible foam is still the largest sector of the polyurethanes industry, its market share has declined from a dominant 47 per cent in 1985 to 33 per cent in 2000.

Supply and demand balance

The supply and demand balance of polyurethane chemicals has seen a change over the period 1985 to 2000, which reflects the movements in PPP as described earlier. The strong growth of purchasing power of the large Chinese population combined with the decline in purchasing power of the East European and African countries strongly contributed to the shift observed in the relative market importance of polyurethanes chemicals indicated in Figure 2-10.

In 1985, West and East Europe, Africa and the Middle East, the EAME region, consumed about 44 per cent of the global polyurethane volume closely followed by the American region, North, Central and Latin America, with 36 per cent, and the Asian region accounting for the other 20 per cent. By the year 2000, consumption in the Asian region had grown to 28 per cent, the Americas stayed constant at 35 per cent with the EAME region dropping to 37 per cent, comparable in size to the Americas.

Figure 2-10   Global split of polyurethanes chemical consumption for 1985 and 2000

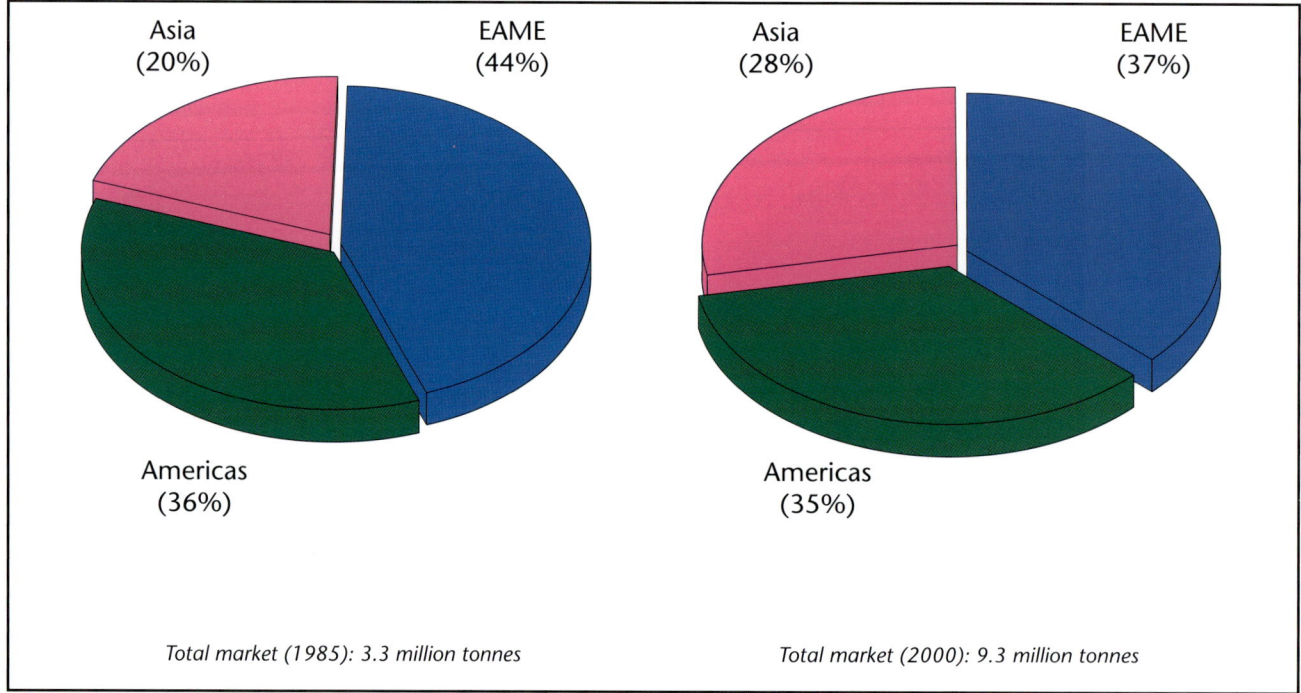

Total market (1985): 3.3 million tonnes   Total market (2000): 9.3 million tonnes

The consumption of polyurethane chemicals, as defined above, needs to be compared to the production and more specifically to the location of production. On a global basis, the capacity for the production of polyurethanes chemicals – isocyanates, polyols and other components, with the latter accounting for about eight per cent of the total volume – is greater than the amount consumed. This has been the case since 1985 and is currently predicted to continue to 2005 and beyond as shown in Figure 2-11.

These are average annual figures and there will be times when one or more of the polyurethane components might be in short supply. The global utilisation rate of raw material capacities has fluctuated in the range of 80 to 90 per cent and is higher for isocyanate than polyol plants. The world-scale isocyanate plants built from 1985 to 2000 have almost doubled in size and propylene oxide plants, providing the base chemical for polyether polyol production, have also increased substantially in size during the same period. The occupacity for polyol plants tends to be lower as the capital cost of these plants is much less than for an isocyanate plant and there is more temptation to over invest. It is relatively easy to cope with the wide range of polyol grades required by using more storage tanks. New polyol plants now being built are significantly larger than in the past.

Figure 2-11   Global consumption versus nameplate capacity for polyurethanes chemicals

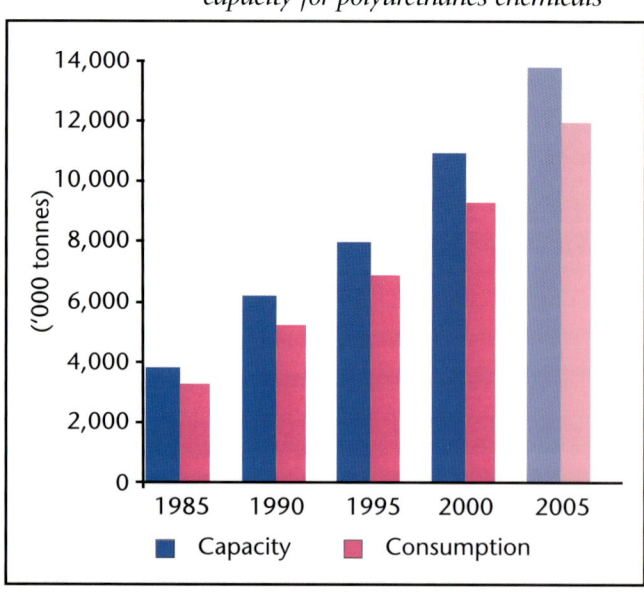

The location of this capacity, although sufficient globally, can be under- or over-supplied locally, presenting a supply challenge for the industry. This is illustrated by comparing the Japanese industry with the rest of Asia. The evolution of capacity and consumption for Japan is shown in Figure 2-12 and for the rest of Asia in Figure 2-13. Japanese consumption has remained more or less constant since 1990, indicative of the lack of growth in the economy and the move of manufacturing 'off-shore' to lower labour cost countries, such as China and the rest of South East Asia. The figures for the rest of Asia reflect not only the rapid growth of local markets such as China, but also the increase in production for export to the rest of the world.

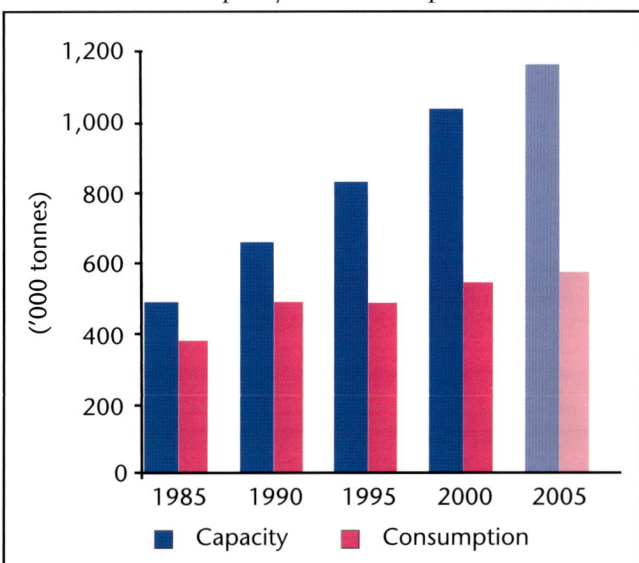

Figure 2-12　Japanese polyurethanes production capacity and consumption

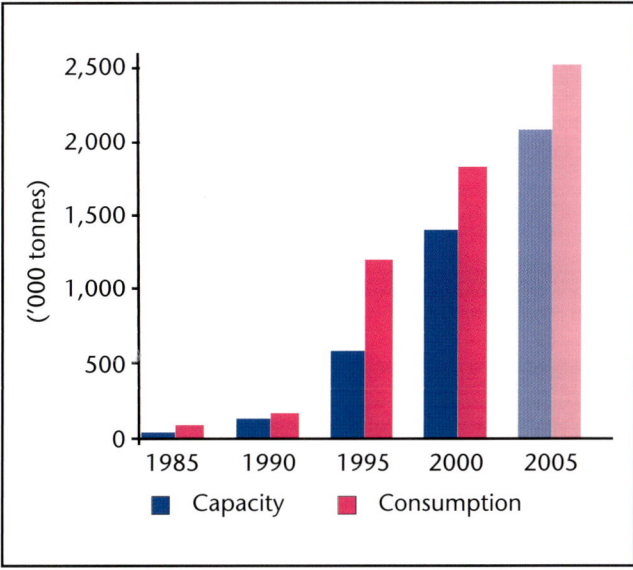

Figure 2-13　Rest of Asia polyurethanes production capacity and consumption

In the second half of the 1990s a consolidation of the major raw material suppliers took place and the Americas region currently has four major suppliers with a small number of other marginal players providing speciality products or additives to the polyurethanes industry. Further consolidation is going on, focussing mainly on the supply of isocyanates and polyols.

Six major suppliers exist in Western Europe alongside a number of smaller players situated in Eastern Europe and the Middle East. Further consolidation is occurring in these areas, again focussed on rationalisation of isocyanate and polyol production. Alliances and mergers have been the major trend over the past few years.

The situation in the Asian area is different, as here the four major global players are establishing their positions in the emerging markets while the Japanese suppliers are consolidating to face up to the internal competition in the region. A large number of small plants serving local needs is scattered over the region. These plants are mostly based on imported Western or Japanese technology.

The total polyurethanes base chemicals capacity in North America, Western Europe and Japan exceeds the local demand and these regions tend to be net exporters. Western Europe mainly exports to Eastern Europe and Asia whilst Japanese exports go to the Asian region with North America exporting to Latin America and Asia. Both Latin America and the rest of Asia have excess polyol capacity.

Applications

The polyurethanes market can also be described in terms of end-use applications, which is of particular use in defining market strategies since each sector requires its own approach. The shift in relative size of the different application areas between 1985 and 2000 is shown in Figure 2-14.

Figure 2-14   Split of polyurethanes market by end-use application

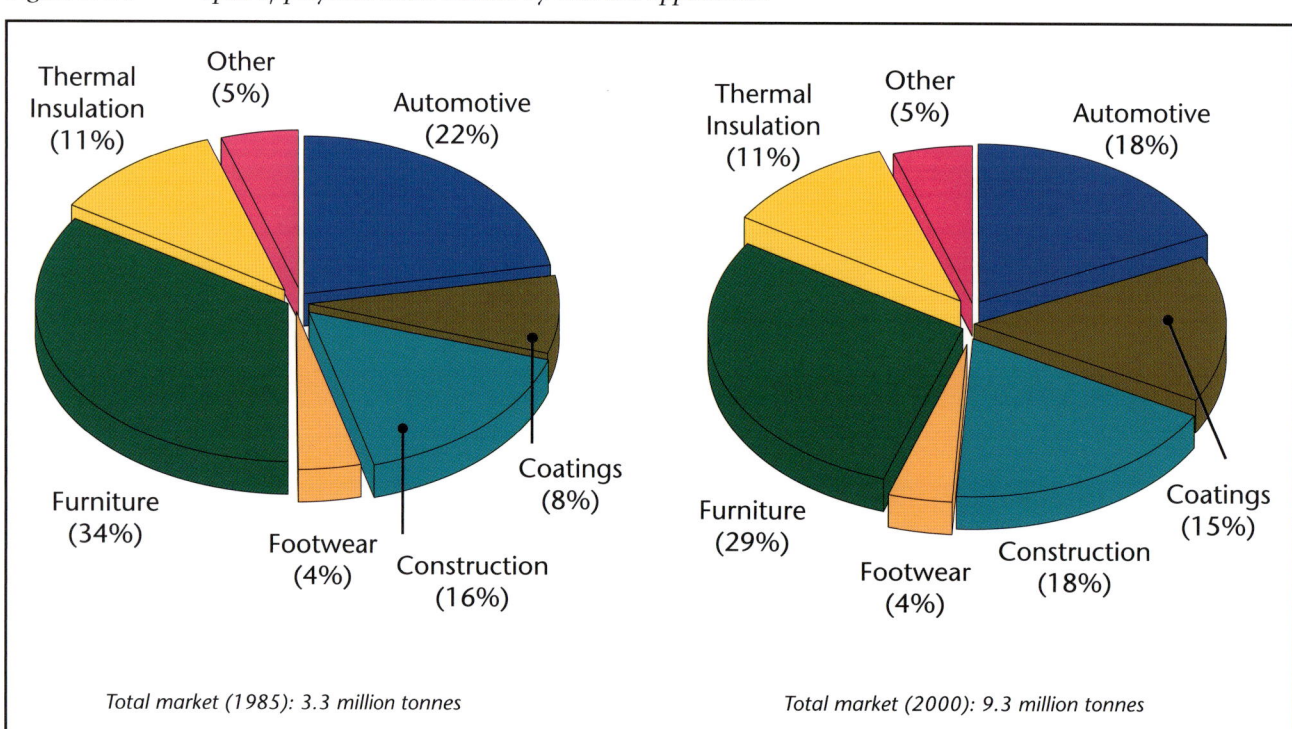

Total market (1985): 3.3 million tonnes   Total market (2000): 9.3 million tonnes

Coatings and construction applications have grown in relative importance at the expense of the automotive and furniture markets whilst all the other segments have more or less maintained their relative share. The furniture sector, at 29 per cent, is still the largest market for polyurethanes followed by automotive and construction, both at 18 per cent, with thermal insulation at 11 per cent and footwear at 4 per cent.

Automotive

The transport industry uses the widest variety of polyurethane products spanning almost the entire product range and physical properties achievable. Polyurethane products are used to make car seats, head rests, liners, internal body parts, dashboards, fascias, bumpers, energy absorption parts, sound insulation, clear top coats, powder coatings, lacquers and refinishes.

Flexible foams are used in car seats, cut and shaped from slabstock or moulded, with the major driver being weight reduction, through using lower density materials and/or thinner seats. The second driver is an enhanced comfort level, which often conflicts with the drive for reduced weight. Other uses for flexible foams are: improved sound insulation, carpet underlay and headrests both sometimes made from bonded scrap foam.

Semi-rigid foams are used to make liners, some internal body parts and energy absorbing structures whilst elastomers are used to produce gaskets to seal doors and windows. Integral skin foams, based on microcellular polyurethane elastomers, are used for steering wheels, armrests and dashboards.

The use of polyurethane-based coatings, especially clear topcoats, is growing due to the trend for enhanced corrosion and impact resistance of lacquers. Demand for water-borne and/or powder coatings is rising as they meet the more stringent environmental regulations being applied. The water-borne systems have found wider acceptance in Europe than in the USA. Polyurethane coatings for automotive refinish have grown considerably in the past 15 years, but the rate is expected to slow down due to the high penetration already achieved.

The use of high-performance structural polyurethane adhesives is also increasing as it enables automotive designers to produce more aesthetic designs.

RIM or reinforced RIM (RRIM) products are used to produce fascias, body panels and other components for cars and trucks, but some of these products are threatened by the substitution pressure of thermoplastic polyolefins. However, glass fibre-reinforced polyurethane panels are in direct competition with the polyolefins and the low tooling cost of the polyurethane RRIM technology provides a cost advantage in the trend towards more customised car trims.

Overall, the transport applications for polyurethane are expected to grow above GDP level, despite the sluggish progress of the automotive industry, due to the continuing growth of plastic materials used in cars, required to achieve the continuing weight reductions demanded by environmental legislation. Polyurethane products will be in direct competition with other engineering plastics, but are expected to maintain or slightly increase their proportion of the total amount of plastics used in a vehicle.

Coatings

The polyurethane coatings industry consists of a broad range of products that are used across all application areas and the market growth has significantly exceeded the rise in GDP for all regions. Two-component systems, water-borne and powder coatings have shown the highest growth rates and increased awareness and acceptance of these products, in combination with environmental pressure, will be the key drivers for future growth, which is expected to be 8 to 10 per cent per year.

Coatings for wood products have a major share of the market segment and are primarily two-component systems applied to furniture with future growth expected to be at GDP level with a tendency for increased use of water-borne systems. Architectural coatings, used mainly for interior clear wood finishes, represents another significant market segment and growth in this area has been driven by the enhanced use of water-borne systems supported by increasing legislative pressure on solvent-based systems.

Polyurethane anti-corrosion coatings are increasingly used instead of traditional coatings because their superior properties and longer lifetimes greatly lengthen the intervals between refurbishment. Growth of the anti-corrosion market segment is expected to follow GDP, but polyurethane coatings may see a faster growth due to their ability to meet the increasingly more stringent environmental demands.

High-performance light-resistant coatings are dominated by products based on aliphatic diisocyanates, which are mainly used in the more demanding application areas such as automotive refinishing. This is mainly due to the penetration of water-based coating technology replacing the traditional solvent-based coatings, increased used of powder coating technology and higher penetration of clear topcoats in the automotive industry.

Construction

The growth of the polyurethane construction market segment has mainly been driven in the past 10 years by the penetration of polyurethanes into the wood binding area. The use of composite wood products is widespread in the North American construction industry and MDI-based resins are replacing the phenolic resins, traditionally used to produce bonded wood composites. About 27 per cent of the North American oriented strand board (OSB) market now uses MDI as the bonding material, Figure 2-15.

The use of MDI-based resins has only recently entered the European market, but new mills are now starting to make panels that are immediately manufactured using polyurethane technology for the core of the material. It is clear that the continuing growth in the wood composite market will mainly be captured by polyurethane-based materials as they more efficiently use fast growth timber, natural resources and support the spreading trend for wood driven construction, as currently used in the USA. Polyurethanes will continue to replace phenolic resins through a blend of lower resin usage and shorter production cycles.

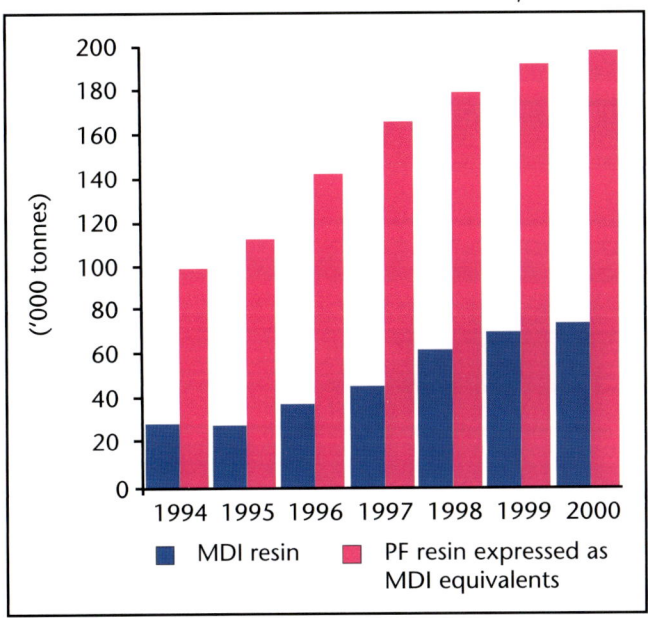

Figure 2-15 Market penetration of MDI resins in the North American OSB industry

The use of polyurethane insulation panels is expected to grow due to the need for higher energy efficiencies and the drive towards reduced emission of green house gasses. Polyurethane panels offer the best insulation value per unit thickness and are the material of choice when space is restricted. The market is expected to grow, as more stringent legislation will come into effect in the European Union and the United States, regarding building energy efficiency. Renovation of existing buildings will be another main driver for growth with a focus on a major rise in demand for spray and cavity filling technologies.

New buildings will be required to be built with less construction time, as the average wealth of the population increases and the cost of labour becomes more expensive. This favours the use of prefabricated panels for rapid construction on site, leading to the increased demand for panels, either rigid or flexible faced, that use the structural strength of rigid polyurethane foams in combination with its excellent thermal insulation.

Footwear

The footwear segment is declining in Western Europe, due to the relocation of the industry to the Asian region with China now dominating the global production of footwear and there is a rapid growth in demand for both microcellular polyurethane elastomers, for soling, coatings for the production of artificial leather and polyurethane adhesives. The major production of the polyurethane raw material is still mainly in Western Europe, the USA and Japan which all export to China, but increasingly production units are starting up in China.

Other key footwear producing areas besides China are Latin America, Eastern Europe, the Middle East and specific countries such as Italy, which is still seen as the design trend setter, Thailand and Vietnam.

Growth in the footwear industry is driven strongly by population growth and societal pressures – most people need basic footwear for protection and increasing numbers buy shoes to make a statement about their position in society. The dynamics of the footwear market are driven by import-export with shoes specified in developed countries made in low-cost labour countries, often using components and raw materials from the developed countries, before being transported back to the developed markets.

Furniture

Replacement sales will be the main driver for growth in developed countries due to the demand for enhanced cushioning and comfort that is achieved with higher density materials. This comfort premium will become more important as the average wealth per capita increases and has already been seen in the USA and Western Europe. As the wealth of other regions increases the need for comfort seating will grow proportionally. This trend for enhanced foam performance and flexible foam consumption is illustrated in Figure 2-16, which shows a linear correlation between the amounts of flexible foam consumed per capita versus the average purchasing power per capita. Local shifts in production location towards the lowest cost producing country provide a balanced scatter around the trend.

Figure 2-16   Flexible foam consumption versus wealth

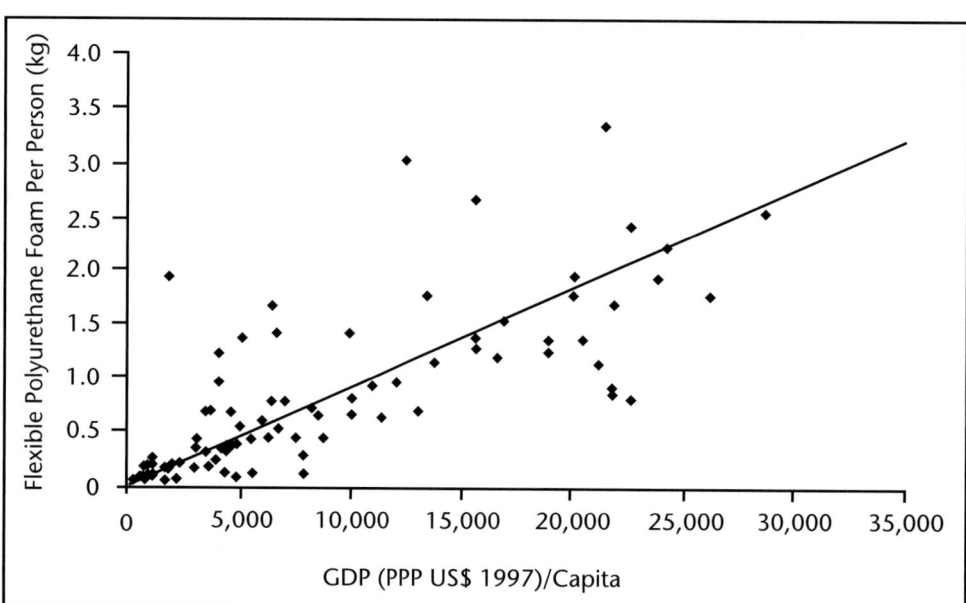

Large growth can be expected from regions with fast growing disposable incomes such as in Asia where many populations are increasing their standard of living very rapidly.

Thermal insulation

An increasing world population will require more efficient use of food, with the emphasis on better food preservation to avoid waste. This can only be done through more efficient insulation during transport and storage. Developed regions will see growth through replacement of existing or old units by more energy-efficient units driven by increasing environmental pressure and legislation.

Larger units, having thicker walls, will increase the demand for polyurethane insulation, which provides both insulation and structural support. The fast-growing population in the developing regions will need more refrigerated transport and storage. As the population keeps growing, food will need to be transported in larger quantities over longer distances. Government investments in infrastructure will enhance the sales of units in developing regions. Increasing personal wealth will further drive the demand for polyurethane insulated appliances.

Replacement challenges

The fact that polyurethane is used in many forms and finds its way into many end applications is confirmation of its versatility as a polymer. The research and development being undertaken by both the raw material suppliers and the systems formulators continues to extend the areas of application and, therefore, the usefulness of polyurethane. At the same time, there is an ever-present and growing desire from the end-users for more cost-effective solutions. Polyurethanes will deliver some of these, but potentially there will be alternative, perhaps cheaper materials, which will displace polyurethanes from some of the market segments that it currently occupies. Similarly, technical requirements are also increasing, although mainly as another facet of 'cost-down', for instance, making a stronger rigid foam would allow appliance manufacturers to use lower density, and hence lower cost, foam. The challenge to the polyurethane industry is to continue to adapt its technology to meet the challenges.

In flexible foam applications, the main alternative 'comfort' material used in seating or bedding has traditionally been latex rubber, but there are now other new potential products such as foamed elastomeric polypropylene and polyethylene or bonded polyester fibre mats. These radically new materials are most likely to first appear in automotive applications. So far, however, there are still performance and processing requirements which they cannot achieve. Consequently, polyurethane flexible foam has a relatively secure market position for the near future, but these new technological advances, if successfully implemented, could dramatically change the position.

Rigid foam should be examined as two separate areas: construction and refrigeration appliances. In construction, there have always been many alternative materials and polyurethane only has a relatively small proportion of the market. Although being one of the best insulating materials and offering the capability to produce integral building composites there will be pressure from inorganic insulants, with superior fire performance, via legislation. The main pressure in appliances comes from energy consumption regulations. Currently, the whole assembly process for an appliance is dependant upon polyurethane foam as a structural member and adhesive as well as an insulant.

However, there are rapid advances taking place in design and in the area of vacuum insulation technology. Polyurethanes may have a role in this, but there are also alternative technologies which can be used. Vacuum technology can also be used in construction and it has been shown that it can form part of a total energy management approach to housing. Another factor for appliances comes from outside the polymer industry; genetic engineering has already been able to produce vegetables with a much longer shelf life not requiring chilling. Similarly, gamma radiation is used to sterilise certain foods for long-term storage and reduce the low temperature requirements for others. Although still a long way from widespread implementation further developments may well remove the necessity for cold storage.

The market segments for elastomeric materials and for coatings, adhesives, sealants, and encapsulants are varied and growing. Many of these are niche and polyurethanes offer a unique set of properties and processes, which no other material can currently approach. Additionally, because of their individual and varied nature, no concerted effort by any single replacement material is likely to be either successful or financially attractive. There will continue to be a flux of applications moving in and out of the polyurethane market segment, but the long-term sustainability and the potential for growth is high.

Composite technology is the newest of the polyurethane application areas and is growing by the replacement of other established technologies. Either metal or wood constructions, or other less versatile polymers, can and are being substituted in a range of application areas, most notably the automotive industry. As the range of polyurethane composites grows and the processing capabilities increase, so more and more substitution will take place.

3. The life-cycle of polyurethanes

Mike Jeffs

The Bruntland Commission of the United Nations Commission on Sustainable Development defined sustainable development in 1987 as:

> "A continuing process of economic and social development, in both developing and industrialised nations, that meets the needs of the present without compromising the ability of future generations to meet their own needs."

Sustainable Development is commonly said to have three pillars – economic, social and environmental development. This cannot be applied to a material as such, but can be to an industry. In order to determine whether an industry is following the principles, the economic and social as well as the environmental factors have to be examined.

For the polyurethanes industry more technical effect can be obtained from less material with the prime example being the insulating foams used extensively in the food chain. Their use preserves food along the chain from production to preservation in the home. Because of this there is less food wasted and less energy used in the preservation.

Polyurethanes are also very durable and their long-lasting properties find many applications such as shoe soles and protective coatings. They are also excellent adhesives and isocyanates are extensively used as binders to make high-quality boards for interior and exterior use from waste materials such as weed trees and rice hulls as well as wood.

Renewable materials, such as sucrose and starch, have been used for many years as starters for polyols and new materials are being researched. However, care must be taken in sourcing these materials to ensure a sustainable development approach.

The impact of a material on the Earth's eco-system has become an important factor, which influences all aspects of its design, manufacture and use. This is certainly the case for polyurethanes where environmental issues are critically important in how it is made, including which additives are used in the formulations, the benefits and drawbacks in its use and what happens at the end of its useful life. This is termed a life-cycle approach and the overall elements are shown in Figure 3-1.

The importance of the environment was first highlighted for the polyurethanes industry by the ozone issue, which reached a critical point in the mid-to-late 1980s. Because chlorofluorocarbon (CFC) blowing agents were implicated as a major cause of the depletion of the ozone layer, the industry was made aware of

Figure 3-1   *Generic life-cycle issues for the polyurethane industry*

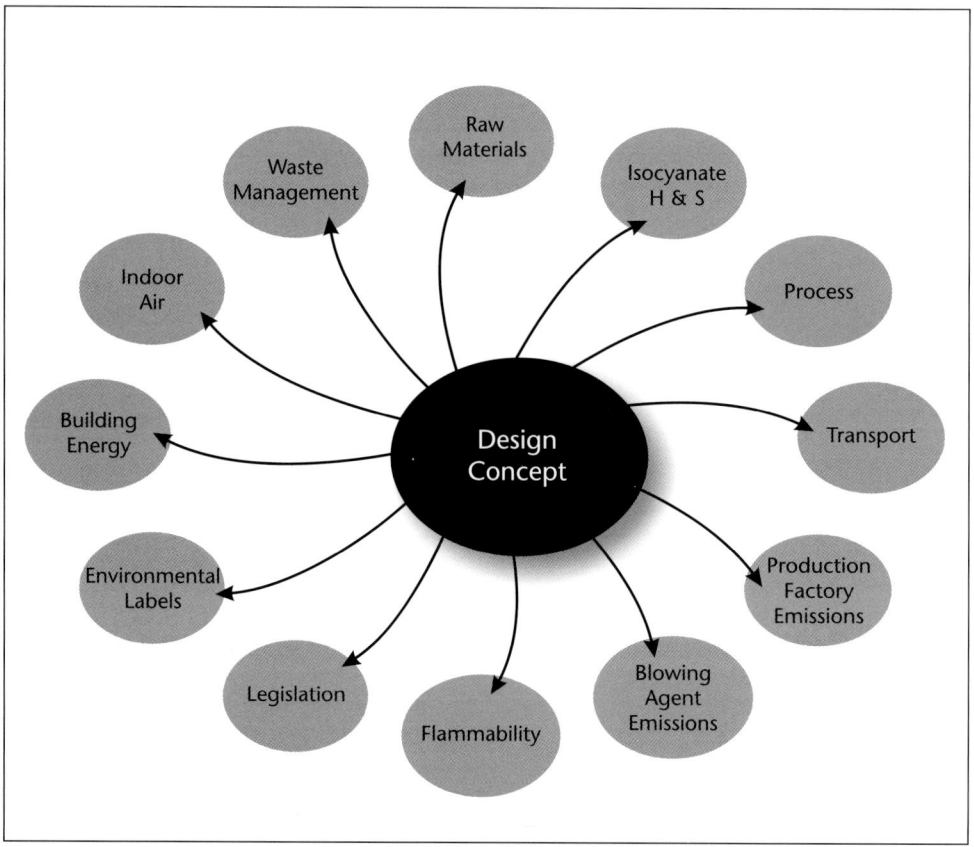

the fact that environmental issues can force rapid change. Since then, the polyurethanes industry's awareness of the importance of such issues has increased significantly and so has its ability to manage them.

Part of this ability is due to the establishment of effective trade associations where the combined resources of members can be deployed on common environmental, health and safety issues. For instance, the European Isocyanate Producers' Association, ISOPA, was set up in 1987 specifically to deal with the CFC issue.

The environmental impact has become an important factor in choosing materials and one that is increasingly used by materials engineers, specifiers, designers and architects. The provision of accurate and, it must be emphasised, objective information for polyurethanes is now essential.

Life-Cycle Assessment (LCA)

Life-cycle inventory (LCI) studies, providing representative and good quality data, are essential to establish detailed life-cycle assessments in order to assess a product's environmental impact from cradle-to-grave. Although comparative assessment between alternative materials is commonly practised, life-cycle assessment is ideally suited to support environmental improvement strategies: identifying the factors, such as emissions and energy use, which can then lead to process and end-use improvement strategies.

The technique of life-cycle assessment has become well established and the polyurethanes industry has commissioned several studies to develop its own data. One example is the study commissioned by ISOPA, which describes the life-cycle inventory for the production of TDI, MDI and polyols. Further, complementary, studies have been commissioned to extend the inventories to the manufacture of products such as rigid foams. These studies have been prepared using average data to ensure that the information is representative of the industry as a whole.

The ISOPA study uses the boundary limits starting with raw material extraction from the earth and terminating at the factory gate of the isocyanate or polyol manufacturer. An important consideration is the energy used and its source. The type of the fuel, such as gas oil, natural gas or coal has to be taken into consideration. Further, the source of electricity has also to be known because different countries in Europe use a different mix of fuels, ranging over gas, coal, oil, nuclear and, increasingly, renewable sources to generate their electricity.

Such studies generate huge amounts of data: the quantities of energy, primary fuels, water and raw materials required to produce each unit, such as one kilogram, of the product (input data); the air, water and solid emissions associated with the production of each unit of product (output data). Typical illustrative data are summarised in Table 3-1.

Table 3-1 *Production factors for 1 kg MDI, TDI and polyether polyol*

		Fuel production	Fuel use	Transport	Feedstock/ Process	Total
MDI	Energy, *MJ/kg*	14	47	0	34	95
	Air emissions, *mg/kg*	5,000	8,700	240	1,300	15,000
	Water emissions, *mg/kg*	16	<1	–	23	39
	Solid waste, *mg/kg*	14,000	5,300		3,100	23,000
TDI	Energy, *MJ/kg*	18	59	1	32	110
	Air emissions, *mg/kg*	6,100	13,000	330	1,500	21,000
	Water emissions, *mg/kg*	22	<1	–	170	190
	Solid waste, *mg/kg*	18,000	9,400		1,500	29,000
Polyol	Energy, *MJ/kg*	16	41	0	36	94
	Air emissions, *mg/kg*	5,800	8,100	260	250	14,000
	Water emissions, *mg/kg*	12	<1	–	23	36
	Solid waste, *mg/kg*	19,000	2,300		2,000	23,000

Energy	*Gross energy required to produce 1 kg of product*
Air emissions	*Gross air emissions of NO_x associated with the production of 1 kg of product*
Water emissions	*Gross water emissions of hydrocarbons associated with the production of 1 kg of product*
Solid waste	*Gross solid waste, slag/ash, associated with the production of 1 kg of product*

The information from the ISOPA studies can be added to additional data to determine the environmental factors of polyurethane products used in a range of applications. The European insulation board association, BING, has carried out an exercise to determine the input elements for a polyurethane roofing board. The example was for one square metre of a board with core density of 32 kg/m³, thickness of 60 mm and covered with aluminium foil facings. The total energy to produce one square metre of board is 284 MJ. Table 3-2 shows that the relative contributions to the gross energy of the board are dominated by the production of the polyurethane raw materials and the aluminium facings. The contribution of the foaming stage is small because this is an energy efficient process involving an exothermic reaction.

The calculations can be extended to cover the use phase of the board, but this has to be done using a model building plus assumptions about weather patterns and other factors. The calculations show that the energy used per square metre of insulation has a very rapid payback in terms of energy saved of less than one year compared to a board lifetime of 50 years or more.

Table 3-2 Relative contribution to the production of 1m² of foam board

Total energy for production, MJ	284
Relative contribution, %	
Production of Al facing	20
Production of polyurethane raw materials	66
Delivery of polyurethane raw materials	1
Foaming	4
Packaging	6
Final transport of PUR foam boards	3

Polyurethane raw materials production phase

To rank emissions against their potential for impact and to prioritise improvement plans, site emissions to air, land and water are reported as environmental burdens. For all environmental impact categories the environmental burden is calculated by multiplying the weight of emissions in that category by a conversion factor reflecting its potency. This provides a more meaningful picture of the potential impact of emissions, compared with the customary practice of reporting the weights of substances discharged.

All the major polyurethane raw material manufacturers have active programmes which apply the life-cycle assessment philosophy. Particular attention is being paid to the following environmental burdens:

- Acidity to air.
- Aquatic ecotoxicity.
- Aquatic oxygen demand.
- Hazardous air emissions.
- Solid wastes.
- Energy efficiency.

Acidity to air concerns the release of gases with the potential to form acid rain.

Aquatic ecotoxicity covers emissions that may have adverse effects on aquatic plants and animals.

Aquatic oxygen demand relates to emissions of substances to water which remove dissolved oxygen that could otherwise support fish and other aquatic life.

Hazardous air emissions are specific airborne emissions of substances that are potentially hazardous to human health.

All of the emissions described above can be reduced by improving the treatment facilities of the plants, such as more efficient gas scrubbers or water purifiers. As new technology becomes available so it is being introduced.

Energy consumption is another important area for improvement plans, but its use is linked to total production so the total energy use can increase when total production levels go up. More important is the energy efficiency of the process. This can be improved by the use of new technology especially the use of waste heat.

The approach described above can be exemplified by the process that Huntsman Polyurethanes followed from 1995 to 2000. During this period all of the issues mentioned above were addressed on a global basis. The improvements achieved were:

- Acidity to air – reduced by 47 per cent.
- Aquatic ecotoxicity – reduced by 77 per cent.
- Aquatic oxygen demand – reduced by 77 per cent.
- Hazardous air emissions – reduced by 87 per cent.
- Solid wastes – reduced by 13 per cent.
- Energy efficiency – increased by 10 per cent.

The energy efficiency of the plants was achieved against the background of an increase in plant production and a global increase of 26 per cent in energy consumption. In the cases where new plants are being constructed, high energy efficiency can be designed in, such as at the Rozenburg 2 MDI plant in Holland. In this plant the energy efficiency was improved by 35 per cent due mainly to the installation of a combined heat and power plant.

Figure 3-2   Rozenburg 2 MDI plant, Holland

Generally, there will be a constant improvement in these burdens as new modern plants are installed and older ones closed down. Environmental performance is a key part of modern design philosophy.

The general reduction in all environmental burdens should be seen in the context of a significant increase in the production of isocyanates and polyols from 1985 to 2000, as described in Chapter 2.

The product manufacture phase

The conversion of isocyanates and polyols to a wide range of polyurethane products introduces several additional environmental issues, but the amount of energy required is low compared to that used in producing isocyanates and polyols. Most of the additional environmental issues introduced at this stage are linked to the addition to the polyurethane formulation of additives, such as blowing agents and fire retardants. For example, blowing agents can have ozone, climate and volatile organic compound (VOC) impacts during the production, use and end-of-life phases of the life-cycle. The issues around blowing agents are discussed in Chapter 8.

Emissions from the product manufacturing process have also to be considered. As there is a vast array of different processes and products, the relative importance of these environmental impacts will also vary.

Polyurethane factory emissions

Prevention of emission to the atmosphere of chemicals used in the manufacture of polyurethane products is important not only to preserve the local environment around the factory, but also in terms of global issues such as ozone depletion and climate change and there is an increasing public concern about the health issues from chemicals in general.

The International Isocyanate Institute, III, is running a project, which started in 1990, to accurately determine the gaseous emissions of TDI and MDI and other chemicals and additives from a wide range of production processes.

The project measured the emissions from factories producing flexible slabstock foam (both methylene chloride and liquid carbon dioxide (LCD) based processes), flexible moulded foams (cold- and hot-cure), rigid foam boards, blocks and panels, appliances and shoe soles. Measurements were made in the emission stacks of the factories and the efficiency of stack emission abatement equipment, or scrubbers, was also examined by the project. So far, all the measurements have been carried out in European factories.

The results show, in all cases, that the level of isocyanates being emitted by factories is extremely low and well within the allowable limits. To illustrate these results, the case of the rigid foam production of boards and blocks can be considered, Table 3-3, which confirms that the amount of MDI escaping to the atmosphere is very low.

However, when the emissions of other compounds, such as blowing agents, were measured a different picture emerged. The emission levels, particularly those from factories using pentanes, may pose problems with regard to levels being proposed in developing legislation to control smog generating substances known as volatile organic compounds or VOCs. The technology to abate these

Table 3-3   Emissions from rigid foam factories

	Board 1	Board 2	Board 3	Block 1	Block 2
MDI, mg/m^3	0.0015	0.0013	0.0004	0.0037	0.0016
Mass loss/year, *grammes*	76	8	<1	23	18
Annual PU, *tonnes*	3,000	6,800	1,200	3,500	6,000
MDI loss, %	4×10^{-6}	2×10^{-7}	$<1 \times 10^{-7}$	1×10^{-6}	6×10^{-7}
Blowing agent	n-pentane	iso-pentane	n-pentane	HCFC-141b	HCFC-141b
Mass loss/year, *tonnes*	9.7	11.7	1.5	0.94	0.33
Blowing agent loss, %	5.6	3.6	2.9	0.4	0.4

emissions is known and some factories have already installed equipment such as thermal oxidisers.

Emissions of MDI and TDI from flexible foam moulding factories were, in general, low, Table 3-4, with the exception of the hot-cure TDI process where the addition of an abatement system would be necessary to comply with legislation currently in force in a number of European Union (EU) member states. Emissions of volatile organic compounds from flexible foam moulding factories are associated with secondary operations, such as application of release agent, mould cleaning and solvent flushing, and amounted to 26 tonnes/year for car seating (Factory 7), 7.1 tonnes/year for car seating (Factory 8) and two tonnes/year for moulding (Factory 9).

Table 3-4   Emissions of MDI and TDI from flexible foam moulding factories

Factory	Operation	Concentration (mg/m^3)		Mass loss per year (g)		Chemicals processed (tonnes)	Percentage loss (%)	
		MDI	TDI	MDI	TDI		MDI	TDI
Car seating 7	Production & trimming	0.0002	0.019	26	1,760	1,400	2×10^{-5}	6×10^{-4}
Car seating 8	Moulding	0.0125	n/a	794	n/a	420	6×10^{-4}	n/a
Car seating 8	Moulding headrests	0.0009	n/a	75	n/a	250	9×10^{-5}	n/a
Moulding 9	Hot-cure	n/a	0.38	n/a	13,520	510	n/a	9×10^{-3}

Production waste and scrap

There is a certain level of scrap from all production processes, which can be as high as 40 per cent for some rigid and flexible foam block production processes where the foam is cut into useable forms. In continuous insulation board production the waste, of the order of 5 per cent, is from trimming the edges of boards.

In moulding processes there will be reject mouldings, which is especially the case for integral skin foam mouldings where skin quality is a premium property. In all processes there may be scrap arising from the start of a production process or when the grade of product is being changed.

Some years ago such waste from production processes was considered to be just an economic issue. However, the challenge of waste disposal is growing as the pressures on landfill as a cheap means of disposal is increasing and alternatives to landfill will be discussed in the section *End-of-life phase*.

The use phase

The main factors in the use phase are:

- Energy efficiency.
- Ecolabels.
- Environmental declarations.
- Indoor air.

Energy efficiency

There is a consensus that the combustion of fossil fuels needs to be decreased and there is a growing concern in many countries, both developed and developing, about the security of energy supply. The continued use of fossil fuels can be supported as long as more efficient energy power delivery systems are used, demand side management is improved and there is a programme to replace fossil fuel by renewable energy sources. The polyurethanes industry is already making significant contributions to energy conservation.

Demand side management provides the largest opportunity for polyurethanes. In both North America and in Europe the heating and cooling of buildings generates about 45 per cent of the carbon dioxide emissions, Figure 3-4. This opportunity has been emphasised by industry groups and is now on the regulatory agenda in both regions. Polyurethanes offer significant advantages in energy efficiency improvements through their ability to control the cold chain. Examples are the insulation of buildings – controlling temperature, insulation of refrigerators – conservation of food.

Combined heat and power (CHP) coupled with district central heating (DCH) is an efficient means of supplying energy for heating buildings in urban areas and is used in many cold climate areas in North America, Northern Europe and in China. A key element in these systems is the insulation of the pipe network that transports the hot water and polyurethane rigid foam now dominates this market.

The excellent thermal insulation capability of polyurethane is the most obvious benefit, but it also

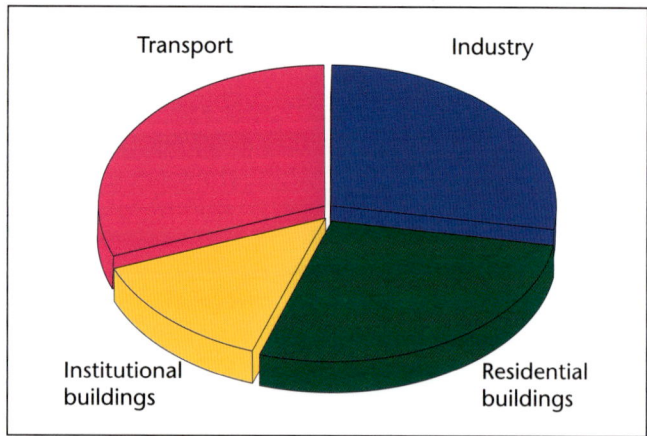

Figure 3-4   Importance of buildings in terms of emissions of carbon dioxide

has a role in transportation through its ability to produce lighter parts, saving weight, which leads to improvements in fuel efficiency.

The focus on improving energy efficiency by the use of polyurethanes is mainly related to the concern about climate change, which is another global atmospheric issue. Whereas ozone depletion resulted in a technology change and its impact was borne primarily by industry, the impact of climate change will be considerably more far reaching.

The concern over climate change led to the Rio Conference in 1992 with agreement reached to reduce the emissions of carbon dioxide (and other gases measured as carbon dioxide equivalents) to 1990 levels by 2000. Little was achieved and growing concern led to the Kyoto Protocol in 1997, in which it was agreed that industrialised countries would reduce the emissions of the gases by 5.2 per cent to 1990 levels by 2008 to 2012. Individual countries and the EU have varying targets and within the EU the reduction burden is shared between the individual member states according to their circumstances.

The science of climate change is far more complex than that of ozone depletion. There are a large number of interacting effects and even ozone depletion affects climate change because ozone is a greenhouse gas. Thus, the depletion of the ozone layer has reduced the climate effect. Other parameters include the cyclical behaviour of the output of radiation from the Sun, the cycles of climate on Earth and the effects of oceans and forests in addition to a wide range of anthropogenic effects, which have to be factored into the 'equation'. It is not surprising that many non-scientists and even some scientists are sceptical about climate change. Another contributing factor is the long time scale of the effect. The atmospheric lifetime of carbon dioxide, the main greenhouse gas, is several hundred years and legislators and even scientists have difficulty in taking actions on behalf of future generations.

The Inter Governmental Panel on Climate Change, IPCC, which is a global panel of over 2,000 atmospheric scientists, assesses the science of the issue. The IPCC issued its Third Assessment Report in 2001. This report showed the increasing confidence of the contributors in the interpretation of observations of effects and in predicting the impact of the issue over the next several hundred years. For the future, the IPCC predicts an acceleration of climate effects with the atmospheric concentration of carbon dioxide increasing by 250 per cent by 2100. The average temperature will increase by up to 5.8°C and the sea level will rise by 88 cm. Beyond 2100 there will be a further acceleration of effects. The impact of climate change will also be manifest in changing and extreme weather patterns with effects on food production and bio-diversity.

The report records that the global average surface temperature has increased by 0.6°C since the pre-industrial era, with the 1990s being the warmest decade and 1998 the warmest year for the past 1,000 years. The retreat of glaciers is very pronounced, as is the thinning of Arctic sea ice and it is estimated that the sea level has risen by 20 cm. The changes in climate are believed to have been caused by increases in the atmospheric concentrations of several gases, namely carbon dioxide, methane (CH_4) and nitrous oxide (N_2O), Figure 3-5. The radiative forcing factor (Wm^{-2}) of the three gases varies considerably in the ratio $CO_2/CH_4/N_2O$ of 10:3:1. (Radiative forcing is a measure of the ability of a gas to 'capture' infra red radiation and convert it into heat.)

Energy labels

In the USA the EPA implemented the Energy Star programme, which is a voluntary scheme involving certification of residential and commercial buildings' energy efficiency. In the EU a directive is being put into place, which will require both new and existing buildings to be more energy efficient and again certification is included. Both of these initiatives are major opportunities for polyurethane rigid insulating foams in the construction industry and particularly for upgrading existing buildings. This is because a highly efficient insulation material such as polyurethane will achieve a given insulation factor for a thinner section thus conserving more usable space. In addition, the climate change issue boosts the revision and upgrading of building regulations.

The use of energy/electricity in domestic appliances and particularly in domestic refrigerators and freezers is also a significant contributor to emissions of carbon dioxide. This sector has been regulated for several years in the USA and the EU and the practice to add energy consumption labels for consumer information, and then to limit and reduce consumption per unit volume of cooled space is spreading in developed countries.

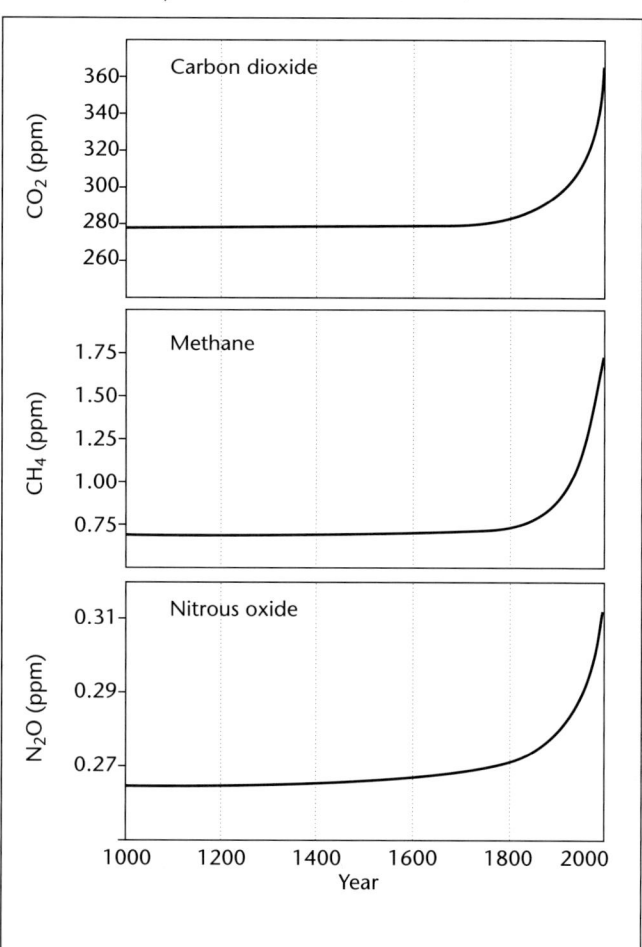

Figure 3-5
Increase in the atmospheric concentrations of carbon dioxide, methane and nitrous oxide

Ecolabels

The intention in developing ecolabels is to inform consumers and help them select products with improved environmental performance. The labels, represented by a logo attached to the product, are awarded to the top few per cent of a product category and normally are only held for a limited time, such as three years. Then the criteria for selection are adjusted, as the intention is to keep up with and reflect technological progress. This is an important area for polyurethanes because of their use in a very wide range of consumer products.

Ecolabels are a developing area and are already applied widely to appliances, refrigerators and freezers, and are being introduced for composite wood panels, footwear, mattresses, coatings and adhesives. The criteria used differ widely between countries and product categories. For those articles, which consume electricity, the energy efficiency is an important criterion. For articles based on foam, the absence of ozone depleting substances or those with a significant Global Warming Potential (GWP) are normally included.

Typical examples of energy labels and ecolabels are shown in Figure 3-6.

Figure 3-6   *Typical examples of energy labels and ecolabels*

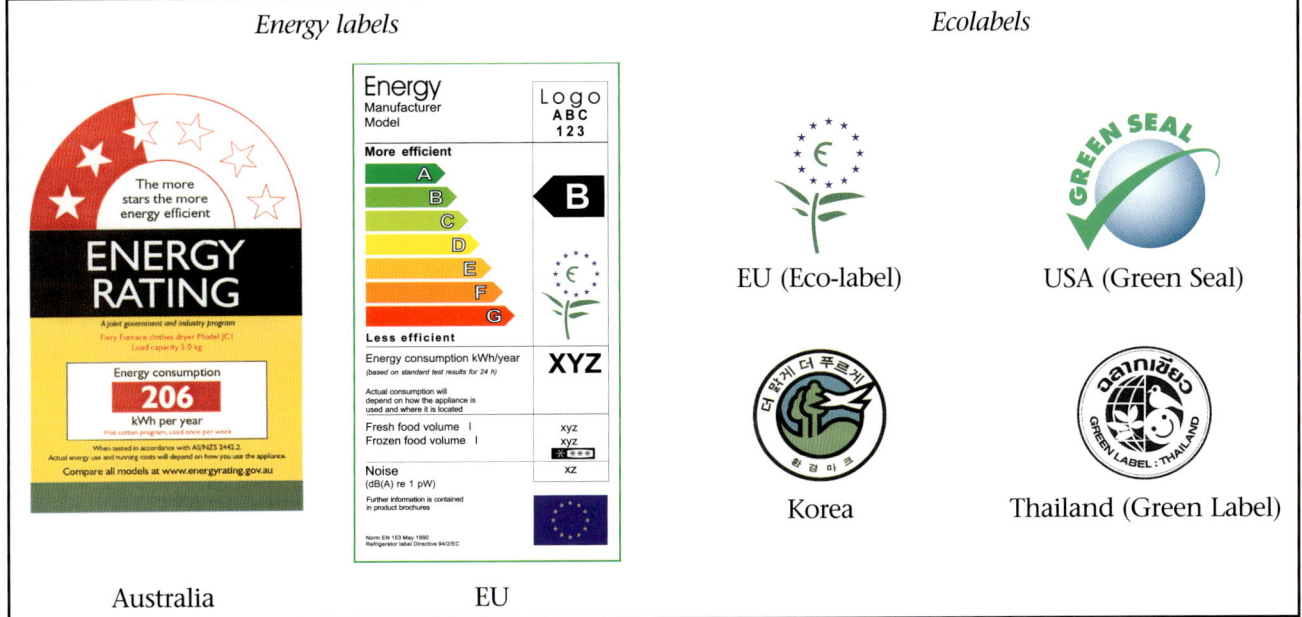

Environmental declarations

Environmental declarations are a recent development. They provide information on environmental performance using detailed life-cycle assessment data and differ from ecolabels as they do not set criteria for selection. That process is left to the designer or specifier who, unlike a consumer, has expert knowledge.

Environmental declarations are being developed for materials in the construction, transportation and electric and electronic equipment sectors. The area receiving most attention is the former because of the large range of alternative materials and a wide audience of architects and specifiers. Environmental declarations on building materials are being developed in several EU countries and a harmonised pan-European scheme is being considered. Computer software is also being developed to compile the data on materials, to the level of the whole building or construction, to support material and design choices.

But, environmental declarations need to be used carefully, especially when comparing materials. A recent report from the EU Commission stated:

> "Construction products cannot be assessed on a stand-alone basis since construction works with the highest 'green' credentials may use products which might have relatively high environmental loads but which will significantly contribute to reducing a building's environmental impact throughout its lifetime. Construction products need to be viewed in terms of functional units, how they perform throughout the life-time of the construction works in which they are installed then when deconstruction or demolition takes place."

This is particularly true for materials such as polyurethane insulation foam where its low density and high thermal resistance have a key role to play in the overall energy efficiency of a building throughout its long lifetime. Polyurethane has a higher environmental load during the manufacturing phase than many competitive materials, but its performance during the use phase far outweighs this factor. A holistic approach to evaluating a situation is required.

Indoor air

The quality of the air within a room or a car is dependent on two factors. The first is the emissions from the contents and construction itself and the second is the throughput of ventilating air. Concerns about health effects from insulation materials dates back to the 1970s when improperly installed urea formaldehyde foam insulation caused high levels of formaldehyde emissions in homes. This concern regarding emissions has been exacerbated by the development of ultra sensitive analytical techniques, but the science of dose/response relationships to interpret the data has not kept pace. So far, it has rarely been possible to prove a causal relationship between an adverse health effect and the concentration levels of a pollutant encountered in non-industrial indoor air.

The issue is relevant for all materials, both natural and synthetic. Legislation is still in development with the Scandinavian nations, Germany, Canada and the USA being the most active in developing indoor air pollution policies. The essential requirements of the EU's Construction Product Directive will include hygiene, health and safety aspects regarding the use of the whole building structure.

The implications for polyurethanes are that additional measures will be required as the legislation on low volatility raw materials and additives is introduced.

End-of-life phase

The volumes of commercial, industrial and domestic wastes are growing year-by-year, Figure 3-7, and the control and management of wastes presents a general challenge to society. The waste of material resources is also becoming apparent and the necessity to recycle if waste cannot be avoided is a key point. However, recycling is another 'industrial' process and life-cycle assessment in comparison with other routes should be undertaken to determine the best environmental option.

Figure 3-7   The growing waste mountain, Western Europe

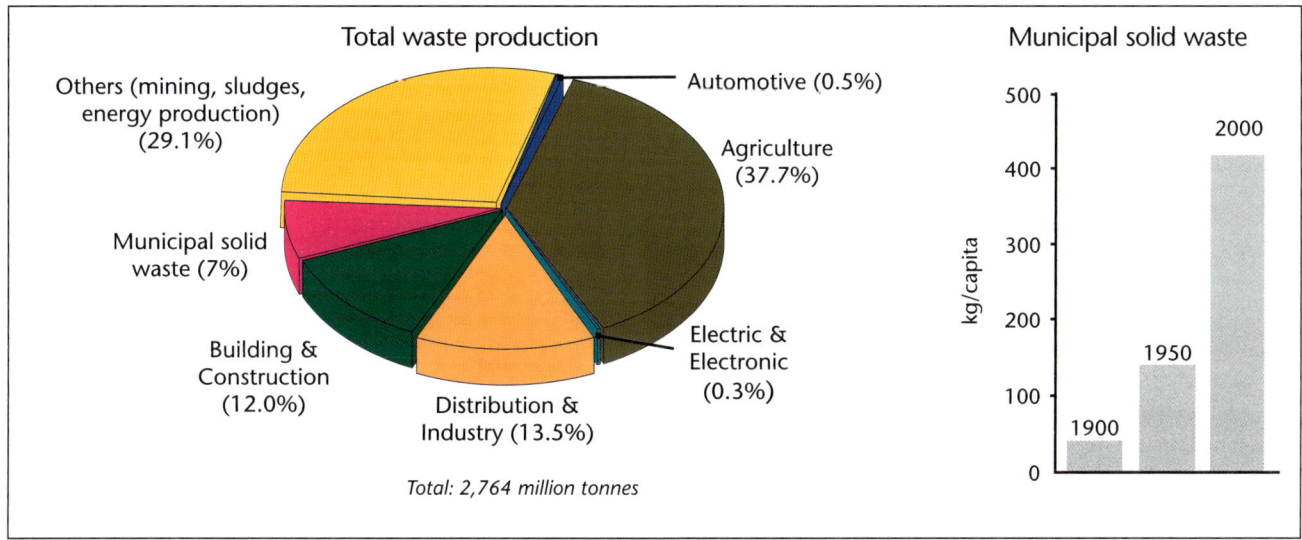

There is a developing consensus that the waste hierarchy is, in descending order: avoid, re-use, recover, incinerate or landfill. At the same time, an on-going debate is being conducted as to whether, within recovery, material recycling should be favoured or not over energy recovery. Increasing use of life-cycle assessment to develop waste management strategies has demonstrated that the choice between the various recovery technologies depends on several technical factors such as volume, qualities and consistency of waste streams, market capacity for recyclates and logistics. Local and regional conditions will be key in defining the most environmentally sensible and technically and economically feasible balance between the various waste management technologies.

Waste management of polyurethanes

The same technology used to deal with manufacturing and post consumer waste will be discussed together. However, the scale of the problem with post consumer waste is orders of magnitude greater because of four issues:

Volume
The volume of polyurethane to be treated is much greater. The average level of production waste and scrap is of the order of one to three per cent whilst the volume of post-consumer waste is an order of magnitude higher.

Collection
Post-consumer waste is widely dispersed around the world in every home and car and in all commercial, institutional and industrial locations. Before it can be economically treated it has to be collected to central points where the chosen process can be applied.

Separation
Polyurethane in practical applications is either adhered to one or more other materials, for example in a refrigerator or a shoe, or is used in conjunction with one or more materials, as in a mattress or a car seat. Thus, the feasibility, both technical and economic, of the separation of the polyurethane away from co-materials is another factor in determining how post-consumer waste can be treated.

Composition
The fourth point is the composition of the product. Closed-cell rigid foams will contain a blowing agent and the release of the blowing agent to the atmosphere, whether because of the ozone, climate change or smog issue, should be avoided. The longevity and durability of polyurethane and products made from it means that products made 50 or more years ago will have to be treated. At that time, the choice of additives was not governed by the stringent environmental, health and safety requirements of today and, for example, the heat treatment during a recycling process of a mix of (unknown) fire retardants should be avoided.

There are many technical options for the management of waste polyurethanes, Table 3-5.

Table 3-5   Technical options for waste management for polyurethanes

Energy recovery	Physical recycling	Feedstock recycling
Co-combustion in Municipal Solid Waste	Powdering/Fillers	Single Phase Glycosis
Fluidised bed	Particle/flake/granulate rebonding	Split Phase Glycosis
Rotary kilns	Compression moulding	Hydrolysis
Specialist boilers (co-combustion fuel)	Injection moulding	Pyrolysis
Cement kilns	Adhesive pressing	Hydrogenation
		Gasification
		Blast furnace iron ore reduction

Because of the wide range of applications and articles, energy recovery is a very attractive option for many cases. It does not require separation from many substrates and can treat mixed plastics and mixtures with municipal solid wastes (MSW). Furthermore, it is capable of dealing with 'old' polyurethanes containing CFCs since incineration is an approved United Nations Environmental Protection Agency (UNEP) method and it has been shown to be more than 99.99 per cent effective in destroying these substances. Its effectiveness as a fuel has been shown in practical trials in cement kilns, power generation units and in municipal solid waste incinerators.

Incineration is a strictly controlled option as the emissions are controlled, as are the incineration temperatures and other conditions. An example is the EU Directive, which would on implementation subject products containing halogenated components to temperatures of a minimum of 1,100°C.

Physical recycling can be a very attractive option for some production and post-consumer waste. A first option is to grind waste polyurethane to a particle size of less than 100 microns and incorporate it into polyol formulations for re-use. The limit on viscosity increase is reached before there is an effect on physical and thermal properties. Another option, widely practised in the carpet industry, is to cut post-consumer and virgin flexible foam production waste into small flakes and then bond it together to make carpet underlays, Figure 3-8.

Figure 3-8
Rebound carpet underlay

A third area of interest is chemical recycling and feedstock recovery. It has been demonstrated that a range of polyols of exceedingly high quality can be obtained by glycolysis. Feedstock recycling processes such as gasification and iron reduction in blast furnace are capable of dealing with mixed plastic streams. This is of increasing relevance in view of the requirements to collect and recycle end-of-life cars and electric and electronic equipment containing many different plastic components difficult to separate.

These options are all available to polyurethanes and demonstrate that this, mostly thermosetting, material can fit into a wide range of practical waste management streams. There should not be a competitive material disadvantage from this issue.

The regulatory field for the recovery and recycling of consumer waste streams is developing in several parts of the world, particularly in densely populated countries and regions where landfill is becoming a very unattractive and expensive option. However, the effect will be global because of the trade of goods into these regions from those with lower pressures on non-landfill routes.

The EU is a leader in this area of regulatory development. It has 'priority waste streams' for treatment and these are packaging, vehicles and electrical and electronic goods. Construction and demolition waste, furniture and mattresses will sooner or later arrive on the agenda.

EU Directive (94/62/EC) on packaging and packaging waste has relatively minor impact on the polyurethane industry. In terms of chemical packages, such as drums, these are already being collected for re-use in many areas. The use of polyurethane as a packaging material is confined to industrial and commercial applications where it is feasible to tackle the logistical problem.

EU Directive (2000/53/EC) on 'end-of-life vehicles' has set many precedents in this legislative field. It stipulates that producers are responsible for the end-of-life aspect after their customers, the consumer, has dispensed with the vehicle and for the consumer there is free take-back of the car, Figure 3-9.

Figure 3-9   Scrap cars waiting for treatment

A particularly important section is that which sets targets for re-use, recycling and recovery. By 2006 it will be necessary to recover 85 wt-% of the vehicle and 80 wt-% has to be re-used or recycled. By 2015 the targets increase to 95 wt-% and 85 wt-% respectively.

The EU Directive on electrical and electronic goods is under development and the Council and European Parliament are considering a Proposal (EC2000/0158) from the European Commission. This proposal, covering the category of large household equipment, which includes refrigerators, Figure 3-10, has a recovery target of 80 wt-% with a re-use/recycling target of 75 wt-% from 2006 and these targets will apply to old refrigerators too. There is also a Proposal (EC2000/0159) on the prohibition of the use of cadmium, chromium, lead and certain brominated fire retardants in new equipment as from 2008.

Japan introduced a home appliance recycling law in place in 2001, which was again applicable to old equipment. It requires that 50 wt-% of a refrigerator is recycled. The equipment has to be taken back to the manufacturer or importer and a fee on consumers will finance this scheme.

Figure 3-10
Old domestic refrigerators waiting for treatment

Taiwan also has legislation, enacted in 1997, which requires collection of old equipment, including refrigerators. It has not set material recycling targets.

Some European member states, particularly Germany, Italy and The Netherlands, also have collection and disposal schemes, which are financed by a levy on the new, replacement appliance.

The development of EU regulations for Construction and Demolition waste may result in a recommendation rather than a directive. Its aim will be to increase the level of re-use and recycling and energy recovery for mixed or contaminated waste. Polyurethane foam could be a minor constituent of the latter.

4. Product Stewardship

Judith Dobbs
Diane Daems

Product Stewardship is one of the six elements of the Responsible Care™ programme widely used by companies in many countries; the elements are:

- Community Awareness and Emergency Response (CAER).
- Pollution Prevention.
- Process Safety.
- Distribution.
- Employee Health and Safety.
- Product Stewardship.

The Community Awareness and Emergency Response (CAER) code promotes emergency response planning and calls for on-going dialogue with local communities whilst the Pollution Prevention code commits industry to safe management and reduction of wastes. The Process Safety code is designed to prevent fires, explosions and accidental chemical releases. The Distribution code focuses on reducing employee and public risks from the shipment of chemicals and applies to activities involved in the transportation, storage, handling, transfer and repackaging of chemicals. The Employee Health and Safety code is designed to protect employees and visitors to chemical sites.

A Product Stewardship programme demonstrates commitment to Responsible Care and ensures health, safety and environmental protection through active assessment of products throughout their life-cycle; from design and development, raw material selection and sourcing, through manufacture, transport, storage, handling, processing and use to eventual disposal.

Such an assessment involves evaluation of potential health, safety and environmental issues at all stages providing any necessary risk management advice and technology, developing analytical and measurement techniques as appropriate, providing tools such as Safety Data Sheets and adequate training. To meet the increasing demands for safer and more environmentally sustainable products everyone in the supply chain needs to fully understand these issues.

The development of regulations involving chemicals has grown over the past 45 years, but there are definite regional differences. The USA, Canada, the European Union (EU), Australia and Japan have had chemical control regulations for a number of years and many countries in Asia, Eastern Europe and South America are at various stages of introducing regulations. The EU is undertaking a major revision to its chemicals policy, which will have an impact on all companies making and using chemicals.

These regulations concern the manufacture, marketing and use of chemicals through registration and approval processes. There is, however, still considerable concern expressed both by governmental and non-governmental organisations that there is insufficient information known about the potential health effects and environmental impacts of many chemical substances that have been in use for a number of years.

The Product Stewardship process, through minimising the possibility of marketing defective products, minimises the risk of harm both to mankind and the environment. It is, therefore, considered an essential and added value part of any product development strategy.

Regulatory framework

Most of the regulatory systems require the provision of hazard data for new chemical substances through notification and/or registration schemes and for several years there have been drives for chemicals that have been available for many years to be reviewed with regard to hazard data and risk assessment. All control regulations have hazard classification systems relating to the toxicity and physico-chemical properties of the chemicals and some include environmental properties. These systems consider acute, chronic and sub-chronic effects. There is a major on-going initiative to harmonise the various systems, known as the Global Harmonised chemical classification and labelling System (GHS), which would simplify the situation and lead to greater transparency.

The United Nations classifications and recommendations for the transport of dangerous chemicals are published in its 'Orange Book', which the various authorities responsible for regulating air, sea, road and rail transport around the world have adopted. Classification for transport purposes considers only acute hazards, regarding toxicological, environmental and physico-chemical properties, as the prime rationale for classification is that exposure in a transport emergency situation is short term.

In addition to chemical control regulations many countries have introduced specific marketing and use regulations designed to safeguard the consumer. There are a number of application areas, such as medical devices, where approvals for the use of polyurethane are directly governed by regulations. In other areas such as food packaging, adhesives for food packaging and toys, regulations exist, but they are usually covered by lists defining permissible residual levels of monomers in the final product.

On top of this complex regulatory environment many companies have developed their own lists of chemical substances that they do not want in the final products they make and sell to the consumer, giving them an alleged 'green' marketing position.

Product design and development

Before looking more deeply into the product life-cycle it is useful to consider the meaning of risk. Risk of injury or disease is a natural part of everyday life in all that we do through work, at rest and in leisure activities. Safe working practices are designed to avoid, or at least minimise, any risk associated with work, over the normal events of everyday life. Risks need to be assessed at all different stages in a product life-cycle and the key areas are:

- The risks associated with the manufacture and use of chemicals for the production of polyurethanes as they affect workers inside a factory.
- The risks affecting the safety of people outside the factory such as those associated with the transport of chemicals, emissions and disposal of waste.
- The risks that can potentially affect the customer using the final product.

To comprehend risk it is necessary to understand the potential hazards involved and also to know the exposure levels and risk is equal to hazard times exposure. Therefore, to determine the risk in using a chemical, requires a detailed knowledge of both hazard and exposure.

The hazards of chemicals can be physical, health related or environmental and examples are respectively the flammability or explosiveness of chemicals, their toxicological properties and the effect they or waste products have on the environment. It is important to understand the reactions that have taken place to produce the final product, the amount of residual monomers that may be present and also the need to recognise the use to which a product may be put so that any further treatment, potentially causing degradation, can be evaluated.

To evaluate exposure requires consideration of the entry route of a material to the body, the level and quantity involved and the duration, whether it is a single or multiple event. Chemicals can enter the body in a number of different ways:

- Inhalation – dusts, particulates or aerosols small enough to be considered inhalable, less than 10 microns, vapours or gases can enter the body in the course of breathing.
- Ingestion – chemicals swallowed alone or through contaminated food are carried into the digestive system where they can do damage, or can be absorbed into the body, or can pass through and be excreted without effect.
- Absorption – although the skin is an excellent protective barrier, some chemicals have the potential to pass through the skin and be absorbed into the body. Sometimes, however, irritation can occur, chemicals can cause visible changes to the skin or eye without being absorbed.

To evaluate exposure and its impact on the environment and/or humans, consideration needs to be given to emissions from production processes to air or groundwater and to the disposal of final polyurethane products. Disposal of waste chemicals is covered by legislation in many parts of the world, see Chapter 3, and therefore, apart from accidental release, the potential for large-scale ground contamination is small.

A positive approach to Product Stewardship is, therefore, critical at the design and development stage of a new product, because environmental benefits and positive health and safety features can be more easily 'designed in' to the product than introduced as add-ons later. Getting things right at the start makes the stewardship task throughout the rest of the life-cycle more straightforward. Product designers and development technologists should be tasked with developing lower risk products than those already available on the market and take the necessary steps to avoid the misuse of a product in an inappropriate application. The key issues to be considered when developing a new product are:

- The suitability of the product for the application being considered.
- If it is intended to replace something else then an analysis of benefits against disadvantages.
- If it is novel then a detailed knowledge not only of the product, but also what it is made from, will be required.
- Are all the ways known about how it can be used or misused.
- Whether the product or its proposed application requires regulatory clearance.
- The hazard profile of the product and any issues that may arise from further processing.
- The development of new advice and training materials.

Consideration of these elements leads to effective risk management, which is an essential element in new product introduction, whether the product is a new polyurethane material, a new application or a new chemical or formulation for making polyurethanes.

Raw materials used in polyurethanes

The way products are used in the polyurethanes industry can be divided into the following generic systems in order to discuss the hazard elements of the raw materials:

- Isocyanates.
- Isocyanates plus polyol blends.
- Pre-formulated mixtures.
- Thermoplastic polyurethanes.

Isocyanates

Many publications have been written on the health and environmental properties of diisocyanates and only a brief overview is given here. The main hazard associated with diisocyanates arises from the inhalation of vapour, spray or dust, which can give rise to respiratory problems. All isocyanates are respiratory irritants and potential sensitisers the effect of which can vary from slight irritation to the eye, nose and throat, to, in severe cases, acute bronchial irritation and difficulty in breathing. Individuals who have developed sensitivity to an isocyanate may experience wheezing, tightness of the chest and shortness of breath and these symptoms of both irritation and sensitisation can be delayed for some hours after exposure. The inhalation hazard of isocyanates is related to their vapour pressure, which depends upon the nature of the isocyanate and increases with temperature, as shown in Table 4-1.

Table 4-1   *Vapour pressure of diisocyanates as a function of temperature*

Temperature (°C)	Vapour pressure (Pa)				
	MDI	TDI	HDI	H_{12}MDI	IPDI
20	0.0004	2.1	7	0.13	0.04
40	0.0025	11.1	–	–	0.93
100	0.02	492	–	–	–

Diisocyanates can also cause moderate irritation to the skin and repeated or prolonged contact can cause skin sensitisation; protective clothing including long-sleeved overalls and gloves should be worn at all times when handling these chemicals or in maintenance work. There is some evidence to suggest that skin contact with isocyanates can influence the onset of respiratory sensitisation. Diisocyanates, as liquid or dust, can irritate the eyes causing watering and discomfort and this is also experienced from exposure to aerosols or vapours above the occupational exposure limit. There is no evidence, after 50 years of industrial handling, to suggest that diisocyanates have a carcinogenic effect in humans.

Not all diisocyanates have been subjected to long-term animal studies. Life-time (two-year) inhalation studies in animals of HDI vapour showed no increase in tumour incidences at any concentration (0 to 1.2 mg/m^3). Inhalation studies of TDI vapour in animals have shown that it is not a carcinogen at exposure levels well above the maximum permissible Occupational Exposure Limit (OEL) in the workplace. Massive doses of TDI in corn oil, given directly into the stomach of animals through tubes, showed some carcinogenic effect but this route of exposure has no relevance to the normal handling and use of the chemical. Similar inhalation studies in animals with MDI, in the form of a respirable aerosol, resulted in chronic pulmonary irritation at high concentrations. Only at the top dose, about 300 times the commonly adopted OEL of 0.02 mg/m^3, was there a significant incidence of a benign tumour and one malignant tumour. No effects were seen at 0.2 mg/m^3. The increased incidence of tumours is associated with prolonged irritation and the concurrent accumulation of yellow material in the lung, which occurred throughout the study

Polyol blends

In comparison to the above risks for isocyanates, the polyol blends based on polyester and polyether polyols are normally of very low toxicity. Polyols are usually pre-blended, dependent on the application, with additives such as catalysts, fire retardants, blowing agents, chain extenders, colouring agents and fillers. Whilst these additives individually have different, and some even hazardous properties, when incorporated into a pre-blended polyol the levels are sufficiently low that the toxicity of the polyol blend itself remains low to moderate. Material safety data sheets should always be consulted before using any chemical.

Pre-formulated mixtures

Adhesives and coatings are complex mixtures where adhesives can be one- or two-component systems with or without solvents, heat-activated thermoplastic polyurethanes, dispersions and even mixtures of polyurethanes or isocyanates with other polymers. Generally, either the unreacted isocyanate or the solvent will be the prime hazardous component.

Thermoplastic polyurethanes

Thermoplastic polyurethanes are solids available in a variety of forms presenting a low hazard, but they require heating during processing, which can give rise to some isocyanate vapour.

Handling of polyurethane chemicals

The key elements in developing a safe handling procedure for the use and processing of polyurethane chemicals are:

- Medical surveillance programme.
- Ventilation, especially local extract ventilation.
- Workplace monitoring programme.
- Training.
- Good housekeeping.
- Provision of safety equipment, decontaminants and breathing apparatus.
- Waste minimisation.
- Personal protective equipment.

It is recommended and increasingly regulated, that a health or medical surveillance programme should be implemented for all those handling and using diisocyanates as problems can arise due to inhalation of vapour, aerosol or dust, leading to respiratory sensitisation. The programme, accounting for local regulations, should start with a pre-employment medical to include a health/respiratory questionnaire and a lung function test, followed by repeat tests at six weeks and thereafter at six-monthly intervals. Lung function tests should be carried out on individuals who may have been exposed to diisocyanates as a result of an accident or incident.

Substances from which a health hazard can result, if the airborne concentration is not controlled, are assigned Occupational Exposure Limits (OELs) and many countries publish lists, which include the diisocyanates and many of the additives such as catalysts, or solvents used in adhesives. OELs are often set by reference to the Threshold Limit Value (TLV), of a substance. TLVs are recommendations issued by the American Conference of Governmental Industrial Hygienists (ACGIH) as guidelines for good practice in workplaces.

The TLV relates to the airborne concentration of a substance and represents conditions under which it is believed that nearly all workers can be daily exposed without adverse effect. It must be noted that the TLV limit relates only to airborne concentrations and takes no account of exposure by ingestion or skin absorption. The TLV-TWA is the time-weighted average concentration for a normal eight-hour workday and a 40-hour workweek. The TLV-C is the maximum value, which must not be exceeded. TLVs are usually expressed as mg/m^3 or ppm. In some countries a different phrase, such as maximum allowable concentration (MAC or MAK), can be used. Values can be revised from time to time and so the current list should always be consulted.

To ensure that the OEL is not exceeded for any process using diisocyanates, wherever there is potential for worker exposure, then good local extract ventilation (LEV) is required to ensure that all vapours or aerosols arising from a process are controlled. Other chemical products, such as catalysts, blowing agents, release agents, solvents or cleaning agents can also have OELs and the LEV needs to be designed to ensure all potential emissions are captured. The ventilation system should be checked regularly to ensure there is no build-up of reacted material such as foam, which could reduce the efficiency of the system. Whilst ensuring a safe working environment through good ventilation, it is equally important to ensure that the extracted air is not vented directly to the atmosphere, but is treated by scrubbing or an equivalent process to prevent contamination. In many parts of the world regulatory authorities have set emission standards and checks against these should routinely be made.

Monitoring of both the workplace environment and individual workers, by determining levels and exposure in relation to OEL values, should be part of a regular industrial hygiene programme. Some countries set specific requirements on the frequency of testing. Details of the procedures used are given in the next section.

Training is a vital component in the safe handling of chemicals as it is important that employees fully understand the hazards of chemicals, how to handle them and the processes in which they are used. On-the-job-training and refresher sessions are essential to ensure safe operation, as is training in the use of emergency equipment in case of accident or incident.

Good housekeeping is also of prime importance, as is availability of both solid and liquid decontaminants, Table 4-2, emergency safety equipment, safety showers and eyewash equipment all of which should be clearly labelled. Regular inspections of equipment especially hoses and valves, placing drip trays containing solid decontaminant beneath valves to capture any potential leakages, keeping waste to a minimum and the routine testing of the emergency safety equipment are all essential elements.

Table 4-2   Decontaminants

	Liquid 1	Liquid 2	Liquid 3**	Solid
Application	Preferred		For equipment	For spills
Concentrated ammonia*, %	–	3 – 8	5	Non-flammable, absorbent carrier, such as sand or a proprietary adsorbent wetted with decontaminant liquid 1 or 2
Sodium carbonate, %	5 – 10	–	–	
Liquid detergent, %	0.2 – 2	0.2 – 2	–	
Industrial alcohol, %	–	–	50	
Water, %		To make 100	45	

* Concentrated. Ammonia solution is corrosive, hazardous to health and the aquatic environment and should be used with care. Consult supplier's Safety Data Sheet

** Only use in 'protected areas' as the mixture has a fire point of 46°C

The wearing of personal protective equipment (PPE) is necessary whenever chemicals are being handled or used. For normal operations where the LEV ensures exposure is below the defined OELs the standard PPE is overalls, eye protection, gloves and safety shoes. This equipment should be regularly inspected to ensure it is maintained in good condition and to ensure the correct level of protection it is important that the right materials for overalls and gloves are selected with respect to the chemicals being used. It is important that long sleeved overalls are used since some chemicals can be absorbed through the skin.

Eye protection should be worn at all times when handling chemicals or operating machines, not only as a protection from chemicals, but also as a protection from reacting polyurethane. Safety data sheets of the chemicals being handled should be consulted to determine the best advice on which PPE to select. In the case of an operation with no LEV, such as in roof spraying, maintenance or a major spillage, fresh air-fed breathing apparatus should be worn along with a splash suit, eye protection, gloves, a safety helmet and if necessary PVC boots and the area should be marked-off to prevent unprotected access.

Occupational hygiene monitoring for isocyanates

To assess workplace exposures to isocyanates a combination of both personal and area samples should be taken. Personnel from a range of disciplines and carrying out different tasks should be assessed such as supervisors and operators on process lines, electricians, fork-lift truck drivers, maintenance, quality control and warehouse staff and operators of saw lines and grinding/finishing.

Area samples need to be collected at key locations along production lines to evaluate the risk of exposure to personnel who may be required to work in specific locations for long periods. In addition, area sampling is used to highlight any potential 'hot spots' and areas that may require special attention, for example, to assess the efficiency or the need to introduce additional local exhaust ventilation and the use of personal protective equipment.

There are a number of analytical methods and techniques available for sampling and analysing isocyanates in workplace atmospheres. However, environmental analysts and occupational hygienists need to take great care to ensure that a representative sample is collected and need to consider the regulatory requirements when selecting a method. For instance: is a method capable of measuring 'total isocyanate' and what is the physical state of the isocyanate in the atmosphere being sampled, for example, will the isocyanate be present as a vapour and/or condensation aerosol, or will the isocyanate be coated on another medium such as dust?

There are two general methods for monitoring isocyanates, divided into direct or indirect reading instruments. The direct reading instruments or devices: pumped tubes and badges, passive badges and paper tapes, are generally only suitable for monitoring vapours and some of the methods are listed in Table 4-3.

Table 4-3   Monitoring of isocyanates by direct reading instruments or devices

Method	Sampling technique	Isocyanate	Range	Notes
Drager Tube 0.02/A	Sample pump	TDI	0.02 – 0.2 ppm	Not recommended for assessing personal exposures
SureSpot (Scott-Bacharach GMD)	Sample pump		1 ppb	Measured using a colour comparator
K&M SafeAir	Passive badge	TDI	5 – 700 ppb/h	Measured using a colour comparator
MDA Model 7100	Paper tape		1 – 200 ppb	
MDA SPM	Paper tape		2 – 60 ppb	Single point monitor
GMD AutoStep Plusr	Paper tape	MDI & TDI	1 – 200 ppb	Data logging/manipulation
GMD AutoStep Plusr	Paper tape	HDI	1 – 500 ppb	Data logging/manipulation
GMD Remote Intelligent Sensor	Paper tape	TDI	1 – 2,000 ppb	Continuous monitoring system

The indirect methods collect samples using a pumped impinger and/or filter system with the former requiring standard chemicals and analytical facilities for final analysis. There are two basic measuring systems. The colorimetric system uses a modified marcali method (UK MDHS 49) with the sample collected by impinger pumping and the colour reaction, isocyanate and agent, measured using a spectrophotometer. The more common method, sampling by impinger pumping and/or impregnated filter sampling is followed by in-situ derivatisation of the isocyanate and then analysis by high performance liquid chromatography (HPLC). There are several methods available, which use different chemicals, for instance: 1-(2-methoxyphenyl)piperazine or 1-(2-pyridyl)piperazine coated onto glass fibre filter for total isocyanate.

Storage and transport of polyurethane chemicals

Storage areas for drum stock, intermediate bulk containers (IBCs) or bulk storage should be separated from the factory working area by a fire-resistant wall and for preference, be roofed to protect the materials from water ingress, frost and direct heat from the sun. Neither diisocyanates nor polyols are considered flammable, but as with all organic materials, they will burn if involved in a fire so it is advisable to segregate them from flammable materials. Diisocyanates and polyols can be stored in the same area, but should be segregated by a bund wall.

Drum stock

Drums should be off-loaded and handled on pallets, using a fork-lift truck, in preference to grabbing or rolling techniques that can cause damage, and they should be inspected for damage before being placed in storage. Stocks need to be used in date order and only stacked three high, if on pallets and banded.

The storage area should be well ventilated and regularly inspected for leaking or pressurised drums. If bulging of a drum is observed such that the lid has bowed, but is still below the top outer rim then the bung can be slowly unscrewed, by a trained operative wearing normal PPE, to release pressure. If the drum is bulging above the top rim then it is considered to be under high pressure and expert assistance is required. Isocyanate drums that have become contaminated with moisture should not be re-sealed as a hazardous increase in pressure can result.

Drums of isocyanates may need to be heated in order to melt the contents or to raise the temperature to that required for a given process. Drums should be carefully inspected, to ensure that they are in good condition and heating, which can be in a hot air oven, ideally with air circulation, or with a steam chest or in a hot water bath, needs to be carried out with caution and under responsible supervision. The heating period should be kept to a minimum for technical and safety considerations and direct heating such as a hot plate or naked flames should not be used. Materials should be transferred from drums to machine day tanks using immersion transfer pumps or a vacuum system.

Isocyanate drums need to be decontaminated immediately after emptying by the addition of a liquid decontaminant solution, see Table 4-2, ensuring that the walls are well rinsed and the drums should be left unsealed for 24 hours. The decontaminant solution can then be poured into a storage vessel or another drum since it can be re-used several times. Final disposal of the decontaminant solution should be according to prevailing legislation. The decontaminated drums can then be sent to an approved reconditioner for recycling, or crushed and shredded for scrap metal. On no account should drums, which have contained isocyanates, be used for other products before an approved reconditioner has recycled them. Other drums do not require decontamination before reconditioning.

Intermediate bulk containers (IBCs)

IBCs should be inspected on arrival for any damage that may have been sustained in transit and to ensure the integrity of outlet fittings, vents and seals. The storage area for IBCs should be bunded such that the area is capable of retaining 110 per cent of the contents of the largest IBC. Discharge of material from an IBC should only be done by transfer pump.

Bulk storage

Tanker delivery should always be preceded by discussions and inspections to ensure a safe transfer of product to the customer's bulk storage tanks. Hoses and connections should be different sizes, clearly marked and ideally colour-coded for the isocyanates, polyols, polyol blends and activator solutions. As it is vital that materials are not transferred to the wrong tank then having selected sizes and colour coding for the transfer point, it is good safety practice to continue this on all pipe work and valves, plus the day tanks of the machines, throughout the production facility.

Isocyanates should be stored in tanks, fitted with relief valves, bursting discs and pressure and level gauges, blanketed with dry air or nitrogen and transfer should be by means of pumps. Routine inspection and maintenance programmes are essential to ensure that the gauges have not become blocked and are capable of measuring pressure rises due to the formation of polyureas. Isocyanate bulk storage tanks should be surrounded by an imperviously coated bund capable of containing 110 per cent of the largest storage tank. If the storage area containing isocyanates is totally enclosed it is recommended that an isocyanate monitor is installed with an alarm fitted outside which is automatically triggered if there is a major release. At least one set of emergency PPE should be stored in a secure, dirt-free container outside the entrance of the storage area for use in the event of the alarm being triggered.

Manufacture and handling of polyurethanes in the factory

The properties of raw materials have already been discussed, but emphasis also needs to be placed on working practices, equipment design, factory layout and regular maintenance schedules. Common generic mistakes that have been identified are listed below, followed by sections on specific applications for polyurethane:

- Lack of management awareness of environmental health and safety (EHS) responsibilities.
- No EHS policy/standards in place.
- Inappropriate or insufficient personal protective equipment.
- Inadequate or imbalanced ventilation or local exhaust systems.
- Poor 'housekeeping'.
- Poor waste control.
- Decontaminants either not available, not visible or not clearly labelled.
- Poor availability and positioning of breathing apparatus for emergency situations.
- Safety posters not displayed, either general safety or product specific.
- Poor markings identifying areas where hazardous materials are being used.

Flexible foam

When fully cured, flexible polyurethane foam is a non-irritant material that does not present a toxic hazard. However, as the reaction is exothermic, freshly manufactured foam will be warm and there will be residual amounts of catalysts, volatile organic compounds and other additives, which could give rise to skin irritation, emitted or remaining on the surface, so protective gloves should be worn when handling it. These materials will dissipate during the curing and cooling of the foam, a process that can be accelerated by forced extraction of slabstock or by crushing moulded pieces of foam.

Auto-ignition or scorch only occurs during the curing of freshly-made foam and although it is principally related to formulation, high component temperatures or an imbalance in mix ratios can exacerbate it. A temperature of greater than 165°C, which can be reached at the centre of foam blocks, leads to auto-ignition. It is, therefore, critical to control the reactions by using metering equipment fitted with suitable automatic monitoring devices, which signal if the material flows are out of balance or the temperature is too high. The potential for auto-ignition to occur in freshly made foam makes it essential that they are stored in an area separated from the main factory by a fire wall and fitted with an automatic sprinkler system. It is advisable not to remove foam blocks until their temperature has fallen by at least 20°C.

Flexible polyurethane foam will burn if ignited or involved in a fire and so factories converting it into bedding, cushions, upholstery and other products should also have policies of good-housekeeping, keeping waste to a minimum, regularly removing all scraps and trimmings and adopt a no smoking policy to minimise the risk of fire.

Rigid foam

The manufacturing of rigid foam boards or blocks involves exothermic processes and care needs to be taken in the storage of freshly made products.

Spraying of rigid polyurethane foams in a factory should be done inside a specially-designed spray booth having extraction and side baffles to ensure the air is drawn away from the operator. Maintenance is key in these processes as spray can build up inside the ducts thus rendering them less efficient. If the spray process is used outside, for example to insulate a roof or to coat a component, then full PPE must be worn including fresh air-fed breathing apparatus. Spray foam should not be left uncoated, but should be coated with a flame-retardant paint or clad with a material that does not sustain fire.

Rigid polyurethane foam is often cut with saws having fine cutting teeth, a process producing fine dust, so extraction is needed to immediately remove this as it is generated. The dust should be collected in dust-proof sacks for disposal or recycling. In a boardstock factory this part of the process is usually done in an enclosed space, which is well extracted. If rigid foam dust is not removed at source it will settle on surfaces and then become an explosion hazard, if dust clouds are formed, or increase the fire risk as fire will quickly spread across dust layers.

All dusts, including those that are chemically and physiologically inert, create nuisance and discomfort problems in the working environment. For this reason, it is generally recommended that dust concentrations do not exceed a level of 10 mg/m^3, 8-hour time-weighted average (TWA) of total inhalable dust, 4 mg/m^3, 8-hour TWA of respirable dust. However, results from animal experiments with polyurethane foam dusts have indicated that they cannot be regarded as completely inert and so it is recommended that the concentration should be kept below 5 mg/m³. Polyurethane dust can also cause irritation to the skin, eyes and mucous membranes. Adequate PPE should be worn whenever rigid foam is cut, including installation on construction sites.

Thermoplastic polyurethanes

Thermoplastic polyurethanes require melting out when processed. It is important to ensure that the processes (injection, extrusion and moulding) are well ventilated and protection is taken with regards to handling hot materials.

Use of polyurethanes

Polyurethanes can be further processed before becoming a final product used by consumers and the techniques used for such processes need careful consideration in the way they are carried out.

If heat is involved, attention should be given to the temperature reached as decomposition can occur, which whilst not impacting on the performance of the product, can result in some evolution of residual components. Sufficient time for complete curing needs to be given in such processes, especially when the final product is used in the consumer market, such as in food packaging.

Polyurethanes and fire

Polyurethanes, like all organic materials, are combustible with the ignitability and rate of burning dependent on its chemical constitution and physical form. Flaming combustion is a gas phase reaction between volatile combustible materials and oxygen. Flames from solid materials are the result of thermal decomposition of the solid to produce flammable vapours, which then burn. The flames become self-sustaining when the heat produced keeps the surface of the solid above its decomposition temperature so maintaining the supply of combustible vapour. The maintenance of a high surface temperature and the generation of combustible gas are aided by a high surface to volume ratio.

The way fire develops in buildings is shown in Figure 4-1 and there are now many tests available to assess the performance of materials, composites, finished articles or even complete full-scale room systems, using a variety of ignition sources from smouldering cigarettes to kerosene burners.

Each method is designed to address a specific aspect of fire hazard and the tests can be categorised as follows:

- Reaction to fire.
- Resistance to fire.
- External fire.

Figure 4-1   Stages of development of a fire in a building

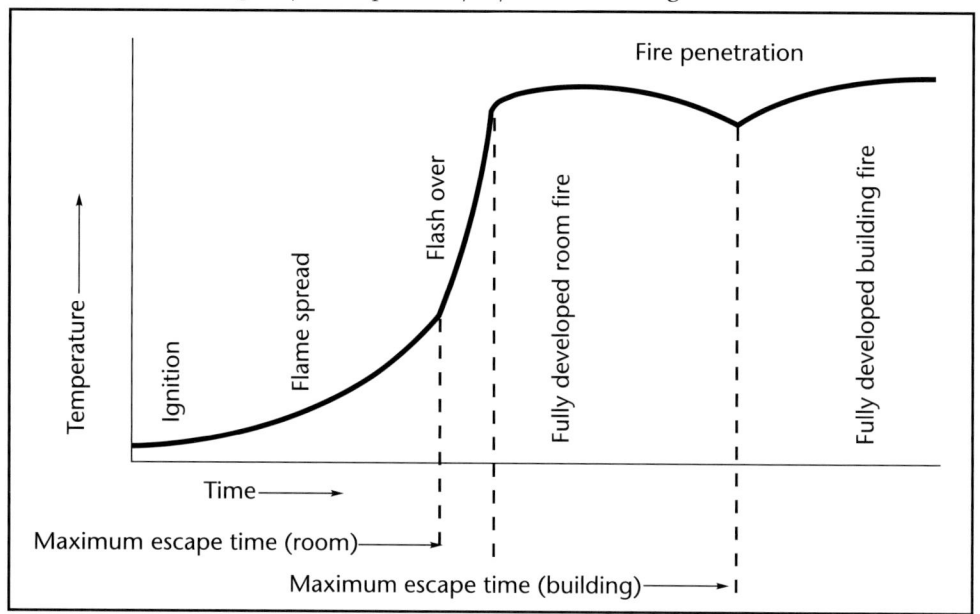

Reaction to fire tests addresses internal fires, looking at ignitability, fire propagation and smoke/toxic gas formation. Ignitability is assessed using low energy sources such as cigarettes or a small flame whilst the rate of burning, fire spread, emission of smoke and toxic gases are measured using larger sources, such as a gas burner, a wood crib or a large calibrated heat source. These tests can be used for research purposes, quality control or to comply with regulatory, insurance or industry standards. In the latter cases, the results are related to pass fail criteria and the tests are only valid if performed by approved laboratories.

Resistance to fire tests, which assess whether structural boundaries can contain an internal fire, once it is developed beyond flash-over stage, and external fire tests, which assess the impact on structural components of external fires, are always performed on structural components, such as part of a building, a railway carriage, a bus or the surrounding of a ship. Resistance to fire tests are carried out in furnaces fitted with kerosene burners that achieve temperatures above 1,000°C after 30 minutes. The heat sources for external fire tests are smaller and can include flying brands, simulate radiation from an adjacent building or a burning waste container near a wall. Resistance and external fire tests are always developed for technical specifications, either set by regulation or by insurance companies.

Fire regulations are changing from prescriptive to performance-based codes and standards, which means that traditional small-scale tests are being replaced by tests correlated with behaviour in large-scale, or real-life scenarios. This change represents both threats and opportunities as, for many applications, prescriptive codes only approved materials with low calorific content. In future, the fire performance of the whole article in an appropriate large-scale context, will be evaluated. Flammability issues emerge as a result of changing fire standards and regulations, which lead to different product performance requirements and all polyurethane developments need to take this into account so that products can be adapted where necessary and guidance given to customers.

Flexible foam and fire

Because of the open-celled nature of flexible polyurethane foam, oxygen is available from within the foam, to feed a fire, which can lead to smouldering, and a chimney effect can occur that speeds up propagation. Dependent on the formulation and additives, flexible foams can either char or melt, with the former giving better results in horizontal and the latter in vertical small-scale tests. Foams can be formulated to meet regulatory standards, but this can have a significant effect on cost and physical properties, such as comfort with aircraft seats, which have to meet stringent regulations. Upholstered furniture and bedding is regulated in some countries and, for example, California, Hong Kong, Singapore and the UK have already implemented ignitability and post-ignition criteria whilst France and the USA are in the process of doing so and Australia has ignitability criteria.

Furniture and mattresses

The carelessly dropped cigarette is the most common source of ignition for furniture, mattresses and bedding with other small open flame sources being matches, cigarette lighters and candles. US residential fire loss estimates for 1998, from the Consumer Products Safety Committee (CPSC), are given in Table 4-4.

Table 4-4   *US residential fire loss estimates 1998*

	Fires	Deaths	Injuries	Property loss ($m)
Total Residential	332,300	2,660	15,260	3,563
Ignition source				
Cigarette	22,200	800	1,990	285
Match	6,500	90	610	72
Lighter	6,000	140	920	99
Candle	12,800	170	1,200	175
Sub Total	*47,500*	*1,200*	*4,720*	*631*
Percentage of Total Residential	*14*	*45*	*31*	*18*
Material first ignited				
Upholstered furniture	10,200	520	1,420	207
Per cent of Total Residential	*3*	*20*	*9*	*6*
Mattress, bedding	18,900	410	2,260	255
Percentage of Total Residential	*6*	*15*	*15*	*7*

Data from US Fire Administration.
Total deaths equivalent to 9.8 per million US inhabitants.

These data show that whilst smoker materials or candles caused 14 per cent of all residential fires in the USA, the first ignited item was upholstered furniture, mattresses or bedding only 9 per cent of the time. However, the number of deaths related to these fires was 35 per cent of all residential fire casualties and for smoker materials and candles, this rose to 45 per cent with cigarettes being the prime cause. The number of fires started in furniture, mattresses and bedding was much lower than those in kitchens, as a result of cooking, but the consequence in terms of casualties was quite the reverse. The reason is that when fires start in kitchens people are awake, so the fire is detected quickly and intervention or escape can be rapid. But fires ignited by a smouldering cigarette usually start at night, when people are at rest or asleep in a different room and – in the absence of a smoke detector – it can be too late for escape before the fire is detected.

There has been a significant decline in the total number of residential and, particularly of upholstery fires, in both the UK and the USA between 1980 and 1998, as shown in Table 4-5. The introduction of flammability regulations in the UK and California is only one of the explanations for the decline. An important reason for the USA is that since 1980 the Upholstered Furniture Action Council (UFAC) has introduced a voluntary industry standard to ensure that all upholstered foams have a cigarette resistance and more than 80 per cent of the furniture manufacturers in the USA now comply with the standard. In Europe, a similar voluntary industry standard has been introduced by the European Upholstered Furniture Action Council (EUFAC).

Other reasons for the decline are the reduction in smoking and the introduction of smoke detectors, with almost all homes in the USA now having smoke detectors fitted. There are significant differences in Europe, which varies from a few per cent in southern countries such as France, to as high as 90 per cent in the north, such as Sweden and the UK. To put these numbers in perspective, the association of European slabstock foam producers (EUROPUR) has estimated that about 2.1 deaths per million are due to careless disposal of smoking materials or candles in Europe.

Table 4-5   US residential fire loss estimates 1980-1998

Fire Deaths	1980	1994	1996	1998	Change 1980-1998 (%)
Total	*4,560*	*2,980*	*3,440*	*2,660*	*-42*
Ignition source					
Cigarette	1,940	860	1,100	800	-59
Match	220	140	100	90	-59
Lighter	270	260	150	140	-48
Candle	20	80	130	170	+750
Sub-total %	*54*	*45*	*43*	*45*	
Material first ignited					
Upholstered furniture	1,350	680	650	520	-61
Mattresses, bedding	850	470	660	410	-52
Sub-total %	*48*	*39*	*38*	*35*	

Once ignited, upholstered foam can burn rapidly with the generation of a high exotherm flash-over within 10 minutes of ignition in a living room with open doors. However, with all doors and windows closed, the fire may not lead to propagation, but be extinguished because of lack of oxygen, dependent on the size of the space. Investigations into the effect of many different foam and fabric cover combinations, using full-scale burning tests of upholstered furniture, have shown that the ignitability and rate of burning can be reduced considerably by a suitable choice of foam filling, cover material and the use of an inter-liner.

Mattresses are seldom the first item ignited. Rather, it is the bedding, from a small ignition source such as smokers' materials or, in some cases, children playing with matches or a cigarette lighter. Untenable conditions in a small room can be created from the bedding fire alone and at some stage the mattress can become involved. If the foam is ignited, the fire may propagate quickly to generate a high exotherm and development work has focused on providing resistance to large flaming sources using a combination of special fabrics, inter-liners and padding.

Automotive

In the transport sector, an extensive range of polyurethane products is used for seating, headliners, sunshades, dashboards (instrument panels), steering wheels and as insulation for refrigerated container lorries. The risk of fire in cars is small and, in most cases, is related to fuel. But there is a universal voluntary agreement regarding fire performance for the interior products in cars: MVSS 302, set by the federal motor vehicle safety standard unit in the USA. This has been adopted globally by the automotive industry. However, for public transport, more stringent fire performance is required by national regulations.

Flame retardancy

The major method to improve fire performance of flexible foams is by the addition of flame-retardant additives to the polyurethane formulation. The introduction of more stringent fire regulations for flexible foams by both California and the UK has led to a much higher usage of flame retardants. For details of the flame retardants available, see Chapter 10. Despite the advantage of using flame retardant and the pressure from lobbying groups to increase fire regulations there is a counter environmental pressure regarding the compounds used and several of the flame retardants are being scrutinised by new regulations in the USA and Europe. For instance: pentabromodiphenylether will be prohibited in the EU from 2003 onwards and is also being investigated in the USA, because of its bio-accumulative characteristic. This counter-pressure increases the complexity of finding the best solution for the issue of flammability in flexible foams.

Test methods

All tests for flexible foams are reaction to fire tests and most methods focus on the resistance to ignition and fire propagation. They are now mainly being used for the generation of data for fire safety engineering (FSE) and computer modelling. Care should be taken in relating laboratory-scale tests to real fire situations as in many cases data cannot simply be extrapolated due to limitations of both the test and choice of sample. Also, an acceptable result on a component test might not lead to a good performance for the total article.

Ignition tests

A smouldering cigarette is the smallest ignition source used, especially for composites since cover materials are recognised as having less resistance than foams. US standards include 16CFR302 part 1632 for mattresses and Cal 117D for furniture composites, which uses an L-shaped cushion assembly, see Figure 4-2. The European standards are respectively ISO 1021-1 and ISO 597-1. Most conventional and high-resilience (HR) flexible foams pass these standard tests without additives.

Figure 4-2 California L-shaped cushion sample

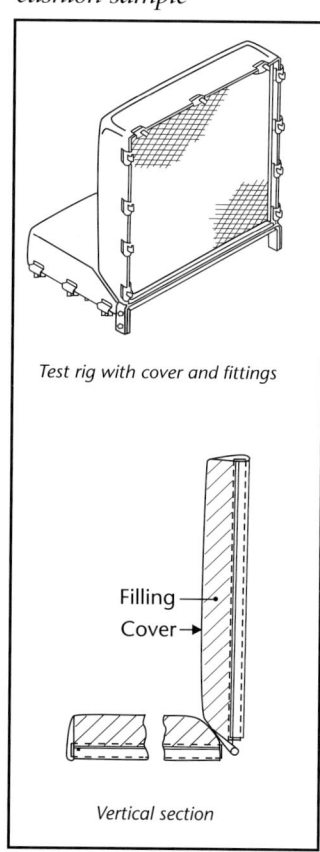

The most widely used ignitability test is ISO 3582, in which a small flame is applied for 60 seconds to the end of a horizontal foam sheet supported by a wire grid, shown in Figure 4-3. The rate and extent of burning plus the extinction time are recorded with a maximum allowable burning rate of 100 millimetres per minute, for any of the five samples tested. A similar test is ISO 3795, which is based on the MVSS 302 test originally developed as a US Federal Motor safety Standard and now used throughout the automotive industry, with the variation that the flame is only applied for 15 seconds.

HR foams pass these tests more easily than conventional foams, even without additives. When testing cover/foam composites, the fabric materials have a decisive influence on test results, cotton and wool are prone to smouldering, but synthetic fibre fabrics tend to melt leaving the foam exposed to the flame.

Fire propagation tests

Testing for the post-ignition behaviour of an article requires energy sources larger than the size for ignition and examples are:

- Large gas burner, such as the Belfagor type (19 kW) used in the California 133 test and the NIST or CBUF furniture calorimeters.
- Wooden cribs, BS 5852 test series, sources 4 to 7, in order of increasing size.

These large ignition sources are mainly used for research purposes and are only included in fire regulations for special cases or higher risk areas, with the UK and California being the only authorities to have them as part of their domestic furniture and mattress standards.

The UK upholstered furniture regulations (1988), require that all upholstered furniture and mattresses meet BS 5852: Part 1 which has two ignition sources, a

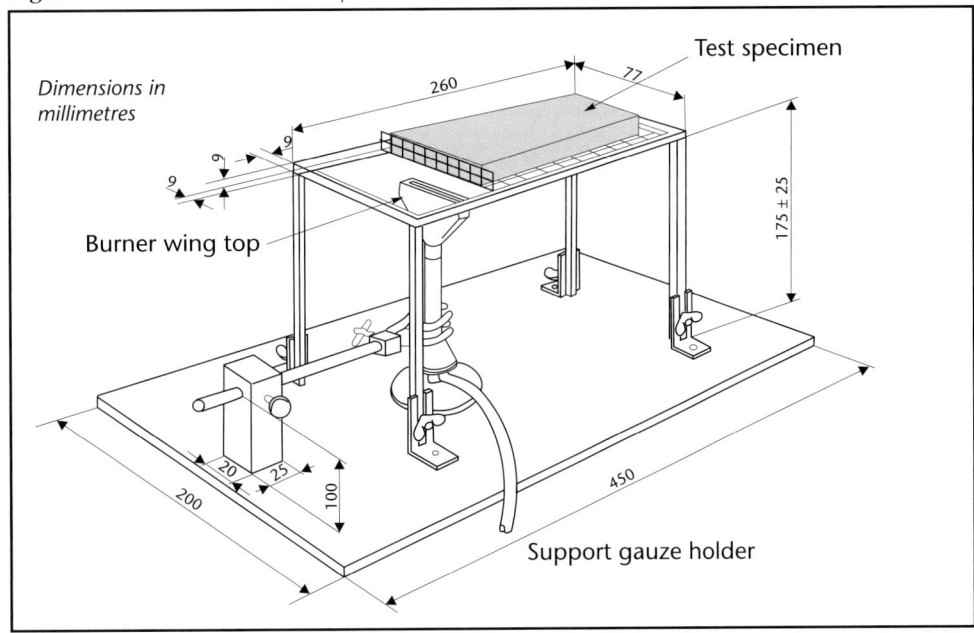

Figure 4-3   ISO 3582 2 flame test

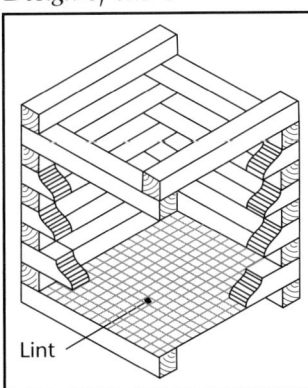

Figure 4-4
Design of crib 5

freshly lit untipped cigarette which smoulders at a defined standard rate (source 0) and a butane flame equivalent to a burning match (source 1). Both tests are carried out on a two-cushion cover/foam combination on a metal frame positioned at right angles to simulate a real seat and back assembly. Part 2 of BS 5852 defines sources 2 and 3, larger flame sources, and 4 to 7, wooden cribs of increasing size; the crib 5 design is shown in Figure 4-4.

The crib 5 test is specific for flexible polyurethane foams and must be passed by all upholstered furniture and mattress foams intended for domestic use in the UK, while larger cribs are reserved for higher hazard locations. The 17-gramme wooden crib 5 is lit with a small amount of isopropanol and allowed to burn on a seat assembly similar to that described for sources 0 and 1. To pass the test, the rig must not lose more than 60 grammes and a standard fire-retarded polyester fabric is used when comparing the performance of different foams.

Conventional foams need high levels of halogenated flame-retardants to pass this test and combustion-modified foams, with flame-retardants are better. Combustion-modified high-resilience (CMHR) foams based on special polyether polyols, additives such as melamine or exfoliating graphite, which lead to char formation, and other fire-retardants are preferred.

The UK regulation is currently under revision since, whilst polyurethane foam must pass a crib 5 test, other filler materials are only required to pass the flame 2 test. It is accepted that this is a discriminative rule with no sound justification, since research has shown that the performance of the fabric is key. The performance of a piece of furniture, filled with foam that passes crib 5, can be either good or poor dependent on the choice of fabric. Therefore, the performance of the whole item must be considered.

Higher levels of fire protection may be required. To satisfy aerospace standards for seat cushions, FAR 25.853c kerosene burner test, it is necessary to impregnate foams with slurries containing high levels of fire-retardant or intumescent resins. Upholstery designed for public buildings such as hotels, theatres or prisons may have to pass crib 6 or 7 fire tests and special non-burning inter-liners may be required. For international trains, changes to performance-based standards and new requirements are being introduced in the USA and Europe.

Other tests

The oxygen index test, ASTM D2813 or ISO 4589, is used to determine the minimal concentration of oxygen required to sustain a small flame. A small square-sectioned bar test specimen is placed in a standard vertical glass chimney and ignited in an atmosphere of a known nitrogen/oxygen mixture with a fixed flow-rate. The level of oxygen required to ensure that at least 50 millimetres of the sample burns for a minimum of 180 seconds is then determined to an accuracy of 0.3 per cent. The correlation between oxygen index data and other ignition fire tests is difficult, especially when dealing with open-cell foams, because variations in air flow through the test specimen can influence the results.

A fundamental assessment of fire propagation, especially when dealing with fabric/foam combinations, is possible using a cone calorimeter, test method ISO 5660. The sample is exposed to a cone-shaped calibrated heat source under defined ventilation conditions and the concentration of gases in the ventilation duct is monitored, along with the weight loss of the test sample. The heat released is derived by calculation from the oxygen depletion as a function of time.

Larger scale versions of the cone calorimeter have been developed for a single item of furniture, Nordtest NT Fire 032 and for a fully furnished room, ISO 9705, but are mainly used for fire modelling and engineering purposes. In the mid-1990s a multinational consortium, Combustion Behaviour of Upholstered Furniture (CBUF), developed models to correlate different scale calorimetry data with the post-ignition behaviour of complete articles in room fires and it has been claimed that the results can be used to predict safety-related parameters, such as the time to escape in case of a room fire.

Rigid foams, composite wood panels and fire

Rigid polyurethane foams are mainly used in construction for insulation, such as boards for walls and roofs, spray for pipes, one-component froth (OCF) as sealants and as prefabricated composite panels. Polyurethane adhesives and MDI-bonded composite wood panels are also extensively used. Whilst well-insulated buildings are environmentally better due to energy conservation, they create a problem in that fires can develop more quickly because the heat is kept inside the building and temperatures rise rapidly.

The chimney effect does not occur in closed-celled insulation foams and oxygen is not available so unlike open-celled foams, they do not smoulder, but their insulating properties can lead to rapid surface spread of flame. The blowing agent in them also affects the fire performance, especially the use of the highly flammable hydrocarbons, such as iso-pentane. An explosive mixture with air is not formed in the closed cells and it has been possible to develop formulations that meet fire regulations.

Building fire regulations are initially focused on the resistance-to-fire performance of construction units such as wall sections, doors or windows and tests are not specific to foam or product. On the other hand, the regulations for reaction to fire (RTF) performance include tests for ignitability, propagation and combustibility. RTF requirements can be directly related to foam, a composite product or an assembly, hence the polyurethane industry has spent most of its fire research efforts in this area.

Most countries distinguish between products with high or low calorific content using ISO 1182 or other non-combustibility tests. Low calorific products are stone, steel, concrete or composites containing only a small amount of organic binder or adhesive. All plastics, wood and products with higher than 10 per cent organic material are classified as combustible. ISO 1182 is not a fire test, but provides material characteristics that are used in fire regulations to define, for instance, permissible materials for escape ways or for heat protection.

Whilst fire is a global issue for the construction sector, building techniques differ significantly between countries and regions, leading to variations in regulations.

USA

Since the early 1970s the regulations in the USA have been based on the use of a thermal barrier, internal wall and ceiling linings, separating the interior of the building from foam insulation, which must survive for 15 minutes to comply with fire resistance test ASTM E 119. In addition, building products have to meet the criteria in ASTM E 84, the Steiner Tunnel test, which is the main method for classification of products. This involves applying a 320-kilowatt gas flame at one end of an eight metre long tunnel fitted with a ceiling of un-clad foam samples and measuring the flame spread and smoke density.

Several exceptions to the thermal barrier rule are allowed, for instance when the construction can meet an intermediate or large-scale test accepted by the building code or when certain regulatory rules are fulfilled. An example is steel roofs, which alone do not provide a thermal barrier. However, when covered with polyisocyanurate rigid foam (PIR) they meet the requirements of FM 4450. This has now become a major application, with roofs installed in warehouses, shopping malls and industrial buildings. The first FM 4450 approvals for PIR-insulated steel deck roofs were obtained around 1985 and the market has grown rapidly; this is now the most popular roof construction in the USA.

China

Until 2000, the GBJ 16-87 code required walls to have at least a 30-minute fire resistance time and lightweight constructions, such as those based on polyurethane foams, were unable to comply, but the rules have been changed allowing non-resistant walls and roofs to be used for low risk industrial buildings and warehouses. The new regulation is based on German tests, but China continues to monitor all European fire classification systems and test methods. Ignitability and degree of combustibility are determined in a similar way to DIN 4102 Part 1, the small flame test, and part 15, 16, the Brandschacht test.

EU and EFTA

The European Union (EU) classification of building products regarding reaction, resistance and external fire performance is being restructured between 2002 and 2006 under the umbrella of the Construction Products Directive (CPD) and eventually all building products across the EU and the European Free Trade Association (EFTA) area will be tested and classified using the same methods and criteria, which are summarised in Table 4-6.

Table 4-6   New European classification system for reaction to fire

Fire situation	Euroclasses	Methods
Fully developed fire in a room	A1	Bomb calorimeter and furnace test and list of non-combustible products
	A2	Bomb calorimeter and/or furnace test and SBI
	B	SBI and Kleinbrenner (30 s)
Single burning item in a room	C	SBI and Kleinbrenner (30 s)
	D	SBI and Kleinbrenner (30 s)
Small fire attack on a limited area	E	Kleinbrenner (15 s)
	F	No performance determined

It is expected that, although initially national regulations will continue to apply, there will be a gradual harmonisation of regulations, standards and tests throughout the EU and EFTA. The new classification system will have a major impact on the construction market as the EU is moving, at the same time as the harmonisation process, to performance-based standards and these developments will extend well past 2006.

The main EU tests for fire classification are EN 13823 (the SBI test) and the ENISO 11925-2 (the small flame test) for ignitability determination. The SBI test is a new intermediate-scale fire propagation test in which two product samples, 0.5 by 1.5 metres and 1.0 by 1.5 metres, are mounted in a corner configuration and subjected to a 30-kilowatt output burner for 20 minutes. The main fire classification is determined from measuring fire growth, total heat release and lateral flame spread. Smoke and burning brands classification is determined by measuring smoke rate, total smoke and the occurrence of burning droplets or particles.

Performance-based codes

Several countries, Canada, Sweden, New Zealand and Japan, have changed their fire regulations to performance-based codes, with other countries such as Australia about to follow suite, whilst in the USA the National Fire Protection Association (NFPA) has produced a new performance-based code as an alternative to the unified International Building Code (IBC). In performance-based codes, test results are related to how a product performs in the end-use application and it is possible to deviate from stringent prescriptive rules for resistance to fire to allow cost-effective, lightweight constructions, provided that fire safety engineering principles are applied in the design stage. Performance-based codes make use of cone calorimeter data as input to modelling and large-scale reference scenarios. Small- and intermediate-scale tests can address critical parameters, but correlation must exist with the reference scenario and the end-use condition and the result of the analysis is only as good as the reliability of the correlations.

Other polyurethane applications and fire

Another major application for polyurethanes in construction is as an adhesive to bond concrete-to-concrete, wood-to-wood, steel to foam or as a gap-filling material. For all steel-faced sandwich panels, with cores other than polyurethane, the core material, rock wool, glass fibre, polystyrene and phenolic foam, is glued to the steel facing using polyurethane adhesive. The adhesive is considered a 'non substantial' component of the composite and it does not significantly affect performance in the SBI test. However, the total composite may exceed the requirements set for ISO 1182. Polyurethane floor coatings can also fall within the building regulations and be subject to fire criteria. Fire regulations may exist or be introduced for clothing and coatings for clothing.

Effluents from a fire

Heat, smoke and toxic gases are secondary fire effects and their extent is governed by the likelihood of ignition, speed of flame spread, heat release, rate of build-up, oxygen depletion and the nature of the materials. Fire effluents can reduce the ability of occupants to escape or even prevent it and the critical factors that affect human behaviour and restrict escape are:

- Smoke obscuration – a visibility of less than two metres can lead to disorientation requiring longer times for escape even from a small room.

- Sensory and upper respiratory effects from irritant gases – these are concentration-dependent and instantaneously occur once a critical level has been exceeded. Typical irritants formed in fires are hydrogen chloride, bromide and fluoride, acrolein, sulphur dioxide, nitrogen dioxide and formaldehyde.
- Asphyxiation from depletion of oxygen or inhalation of asphyxiants – these are dose-related and after a certain exposure the victim will feel dizzy and escape is affected whilst higher doses can lead to incapacitation and death. Examples of such gases are carbon monoxide and hydrogen cyanide.
- Heat inhalation.
- Skin burns.

In addition to sensory effects, most irritants penetrate deep into the lungs, causing pulmonary irritation, but it normally occurs a few hours after exposure and does not usually affect escape.

The toxic potency, defined as the potential to produce a harmful effect, of materials in fires has been characterised by laboratory-scale evaluations, but the toxic hazard from a fire is more dependent on the rate and conditions of burning than on the chemistry of the materials that burn.

When most synthetic and natural materials burn their acute toxicity is similar rather than different and carbon monoxide is the most abundant toxicant in fires involving polyurethanes and natural materials under most combustion conditions.

Because of this, it is unwise to restrict organic materials with significant amount of hetero-atoms on the basis of the potential to produce more toxic atmospheres since it is often the presence of these hetero-atoms that results in better resistance to ignition and flame spread.

Historically, building regulations have focused on fire testing and classification with only a few countries regulating for smoke and even fewer for toxicity. The reason is that it has not been possible to define an adequate way to quantify for smoke and toxic hazard in a fire on the basis of laboratory potency results. During the 1970s and 1980s most toxicity work concentrated on toxic potency measurement with later work looking at the toxic effects of effluents from specific fire situations, defined in the International Standards technical report ISO TR 9122.

Tools for the estimation of the time available for escape, based on toxicity, visual obscuration and heat exposure, are offered in ISO TS 13571. Fire safety engineers use input from small-scale tests to estimate protection measures for a building and whilst FSE is worthwhile, it has the weakness that the model's predictions are highly dependent on the assumptions used and the quality of input data. Until now, a small-scale tube furnace, which correlates with results obtained in real fires, has not been defined, but work in the area continues.

5. Isocyanates

David Thorpe

Methylene diphenyl diisocyanate (MDI), with a 61.3 per cent share and toluene diisocyanate (TDI) at 34.1 per cent dominated the global isocyanate market of 4.4 million tonnes in 2000, Figure 5-1. In comparison to these aromatic diisocyanates, the two major aliphatic diisocyanates, hexamethylene diisocyanate (HDI) and isophorone diisocyanate (IPDI) had a 3.4 per cent share whilst all other commercial isocyanates such as 4,4'-diisocyanatodicyclohexylmethane (H_{12}MDI), 1,5-naphthalenediisocyanate (NDI), tetramethylxylenediisocyanate (TMXDI), *p*-phenylenediisocyanate (PPDI), 1,4-cyclohexanediisocyanate (CDI) and tolidine diisocyanate (TODI) together only amounted to 1.2 per cent.

Figure 5-1   Global split of isocyanate market in 2000

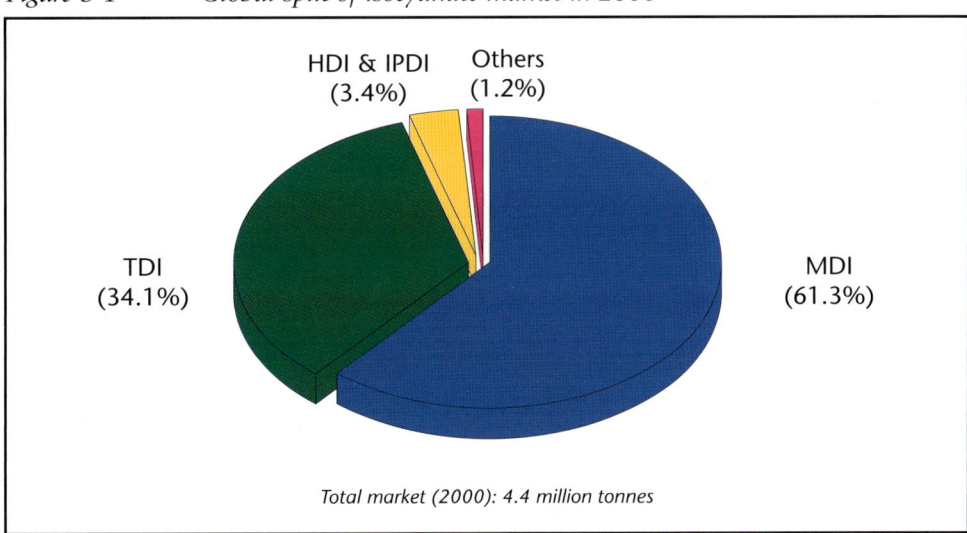

The production of isocyanates is based 99.9 per cent on first synthesising the required molecular structure, but with primary amine instead of isocyanate groups, followed by phosgenation of the amine groups to produce high yields of isocyanates.

Reaction chains

Benzene to MDI

Benzene is nitrated to nitrobenzene, a reaction with a high yield, much less than 1 wt-% by-products and a small, less than 10 wt-%, benzene recycle. The nitrobenzene is then catalytically hydrogenated to aniline, which after purification is reacted with formaldehyde to generate a complex polyamine mixture known as methylene dianiline (MDA) or diaminodiphenylmethane (DADPM). After removal and recycle of excess aniline, MDA is phosgenated to convert the amine groups to isocyanates. After removal and recycle of solvent and excess phosgene, the complex polyisocyanate mixture is separated by distillation into a crude volatile mixed diisocyanate stream and a residue which is a complex mixture of oligomeric polyisocyanates, known as polymeric MDI. The crude mixed diisocyanate stream is fractionated into pure MDI and a mixed isomer stream. The complete process is shown in Figure 5-2.

Toluene to TDI

Toluene is nitrated to dinitrotoluene (DNT) producing an isomer mix with an approximate ratio of 77/19/4 for the 2,4/2,6/(2,3 + 3,4) isomers that is catalytically hydrogenated to crude toluene diamine (TDA). This is purified by fractional distillation to remove the mixed 2,3 and 3,4 isomers, known in the industry as 'ortho isomers'. The purified TDA is then phosgenated to convert the amine groups to isocyanate groups and after removal and recycling of solvent and excess phosgene, the isocyanate mixture is distilled to generate a liquid TDI product (80/20 2,4/2,6 isomer mix) and a residue. The liquid TDI can be further processed, by bulk crystallisation, to separate out a pure 2,4-TDI isomer leaving a liquid 65/35 2,4/2,6 isomer mix. The complete process is shown in Figure 5-3.

Hexamethylene diamine to HDI (and derivatives)

Hexamethylene diamine (HDA) is the amine produced in large volume for the nylon 6,6 industry by a range of commercial processes. A small proportion is phosgenated to convert the amine groups to isocyanate groups and after removal and recycling of solvent and excess phosgene the crude diisocyanate is distilled to generate pure HDI and a residue. Because of the high volatility of the product much of it is converted into isocyanurate or biuret derivatives prior to sale. The complete process is shown in Figure 5-4.

Acetone to IPDI and derivatives including other uses for isophorone and IPDA

The driving force for the production of IPDI, the most recent diisocyanate to penetrate the market, came from the phenol/acetone co-production process and the need to produce value-added products from acetone. Acetone is first converted to isophorone, a solvent used in the paints and coatings industry, which is then reacted with hydrogen cyanide to form isophorone nitrile. This is reductively aminated to the product commonly known as IPDA. About 50 per cent of the IPDA produced is used as an epoxy-curing agent whilst the remainder is phosgenated to produce a crude IPDI stream, which is distilled to yield pure IPDI and a residue. Because of the high volatility of IPDI a large percentage is converted into isocyanurate or biuret derivatives prior to sale. The complete process is shown in Figure 5-5.

Figure 5-2   Benzene to MDI

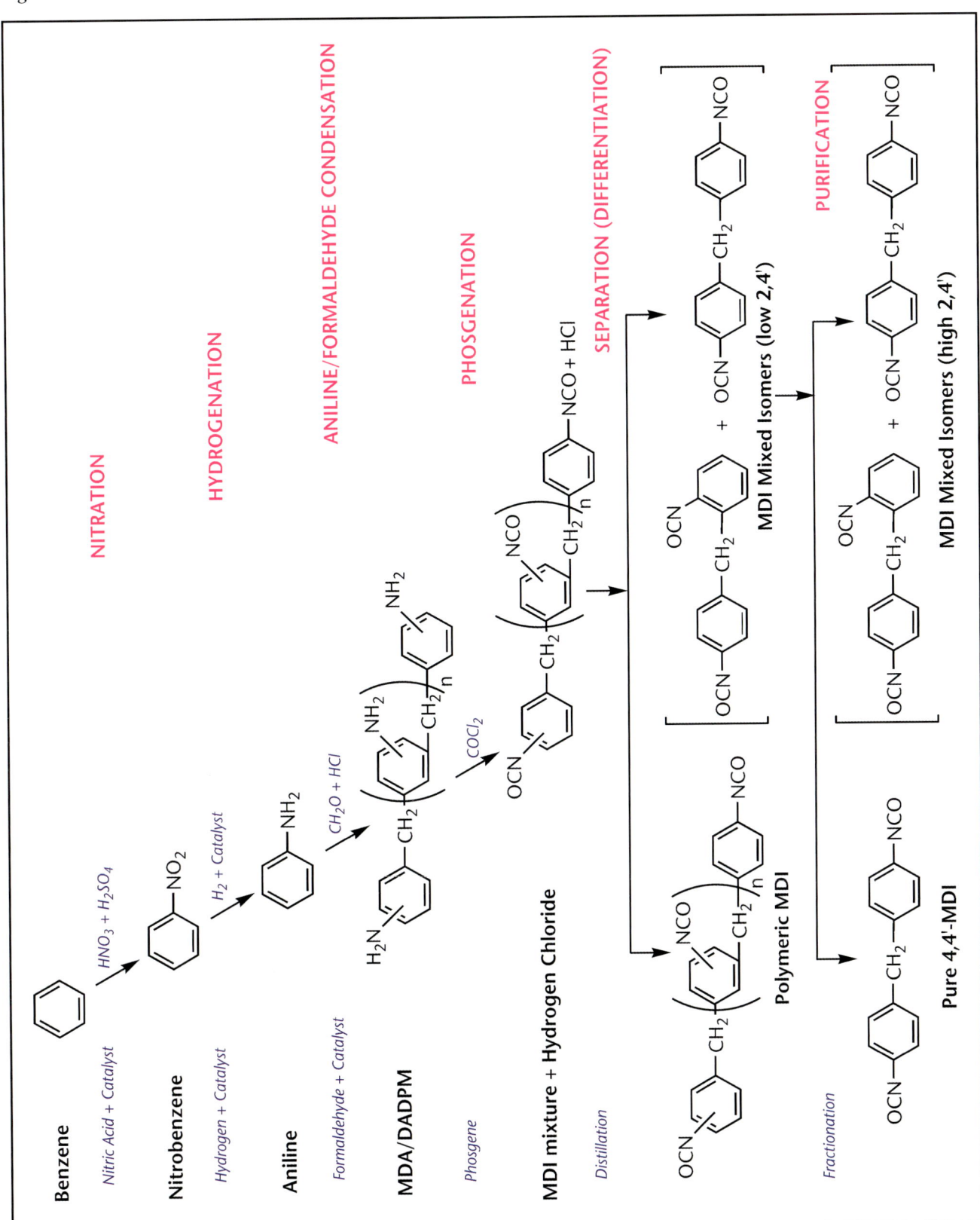

Figure 5-3   Toluene to TDI

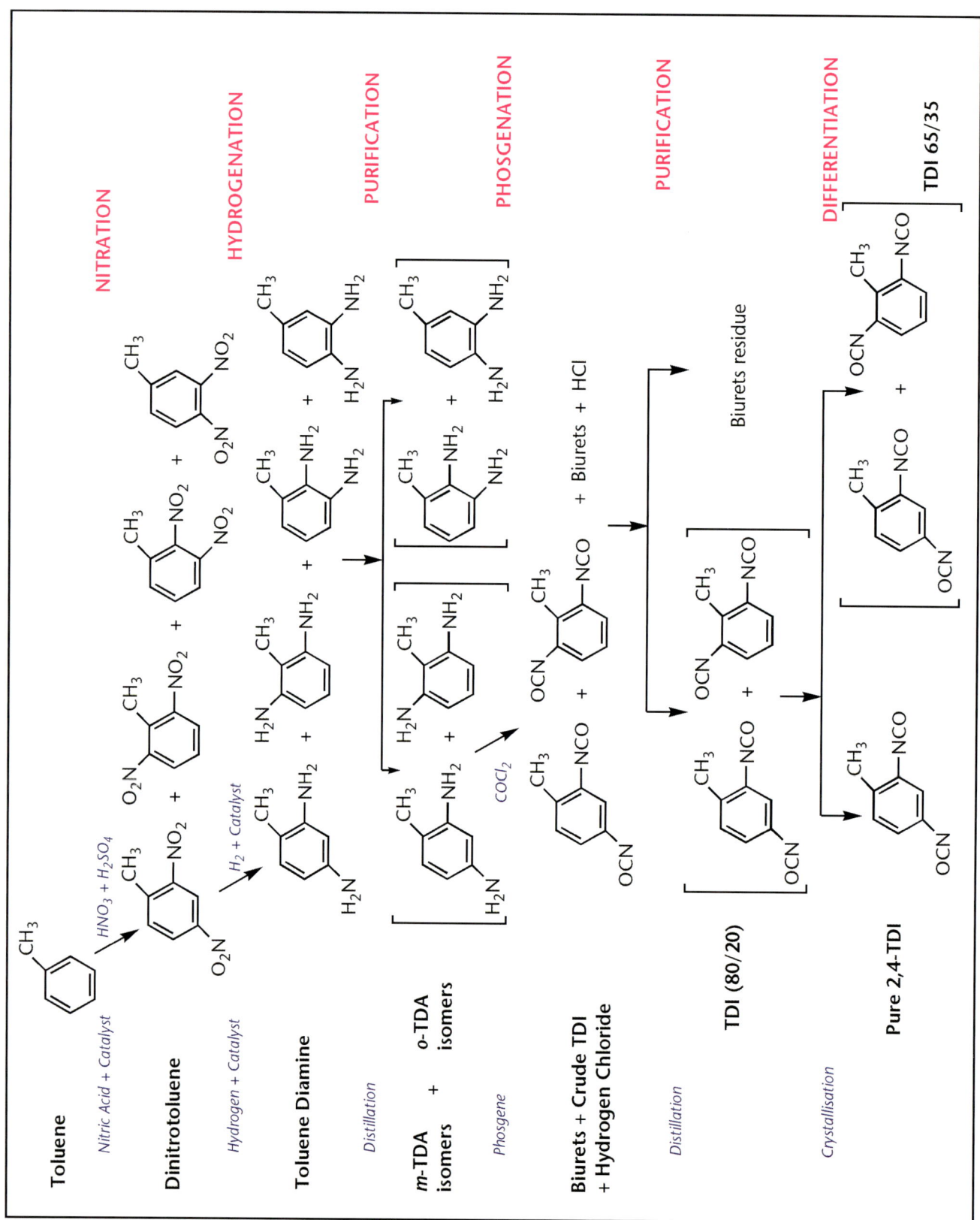

Figure 5-4   Hexamethylene diamine to HDI

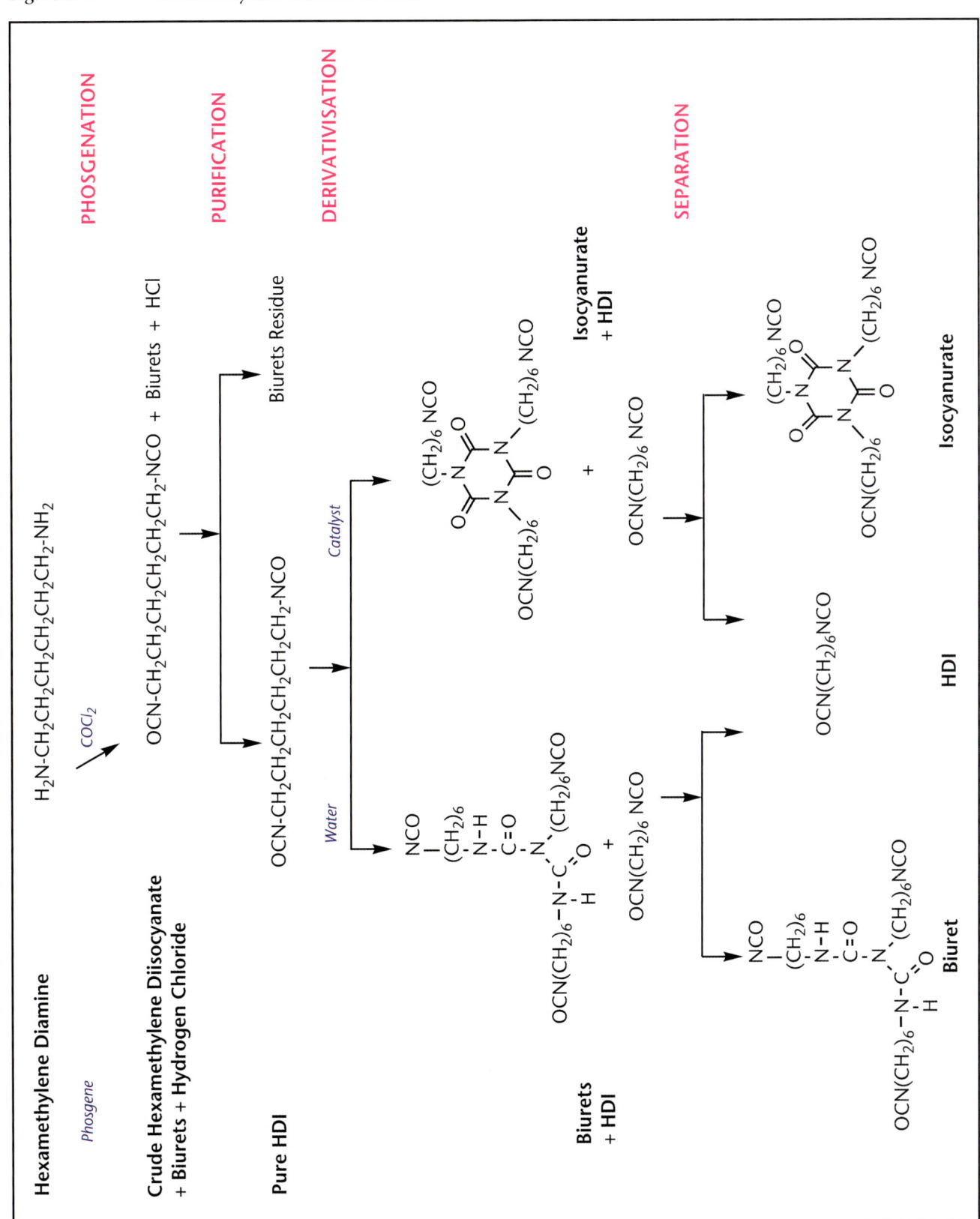

Figure 5-5   Acetone to IPDI

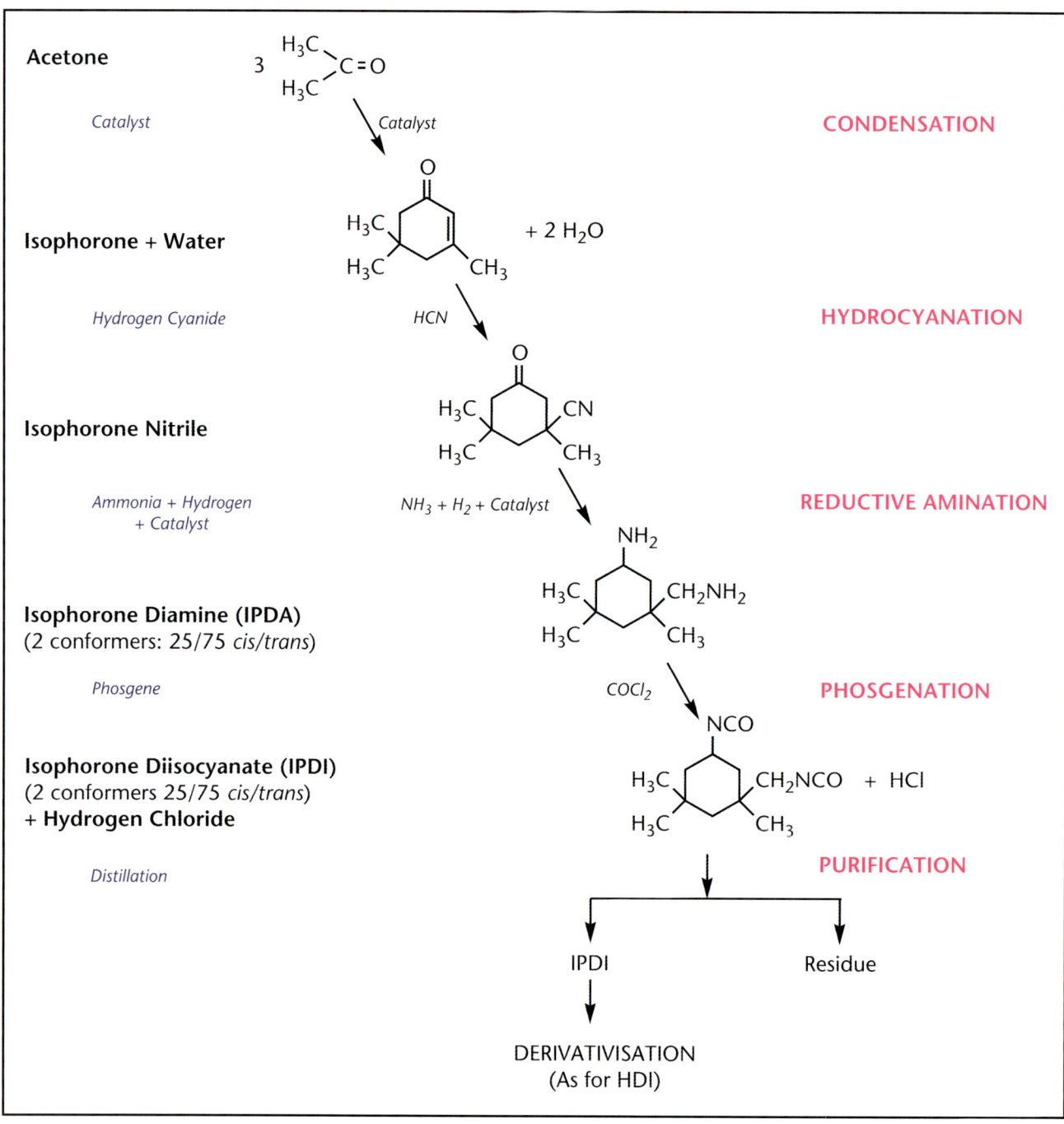

MDA (DADPM) to H₁₂MDI

H$_{12}$MDI is a fully ring-reduced pure MDI made as a side stream from the MDI chain. A pure diamine stream, 90 per cent 4,4' and 10 per cent 2,4', is obtained by either distilling part of the MDI intermediate amine (MDA) or a special diamine-rich grade is made by operating a campaign at a very high aniline/formaldehyde ratio. The diamine is catalytically hydrogenated to convert benzene rings to cyclohexyl rings. The resulting cycloaliphatic amine is distilled to yield a pure H$_{12}$MDA, that is then phosgenated to a produce a crude isocyanate, plus a residue used as an epoxy-curing agent. The crude isocyanate mix is distilled to yield a pure diisocyanate and a residue. The complete process is shown in Figure 5-6.

Figure 5-6   MDA to H$_{12}$MDI

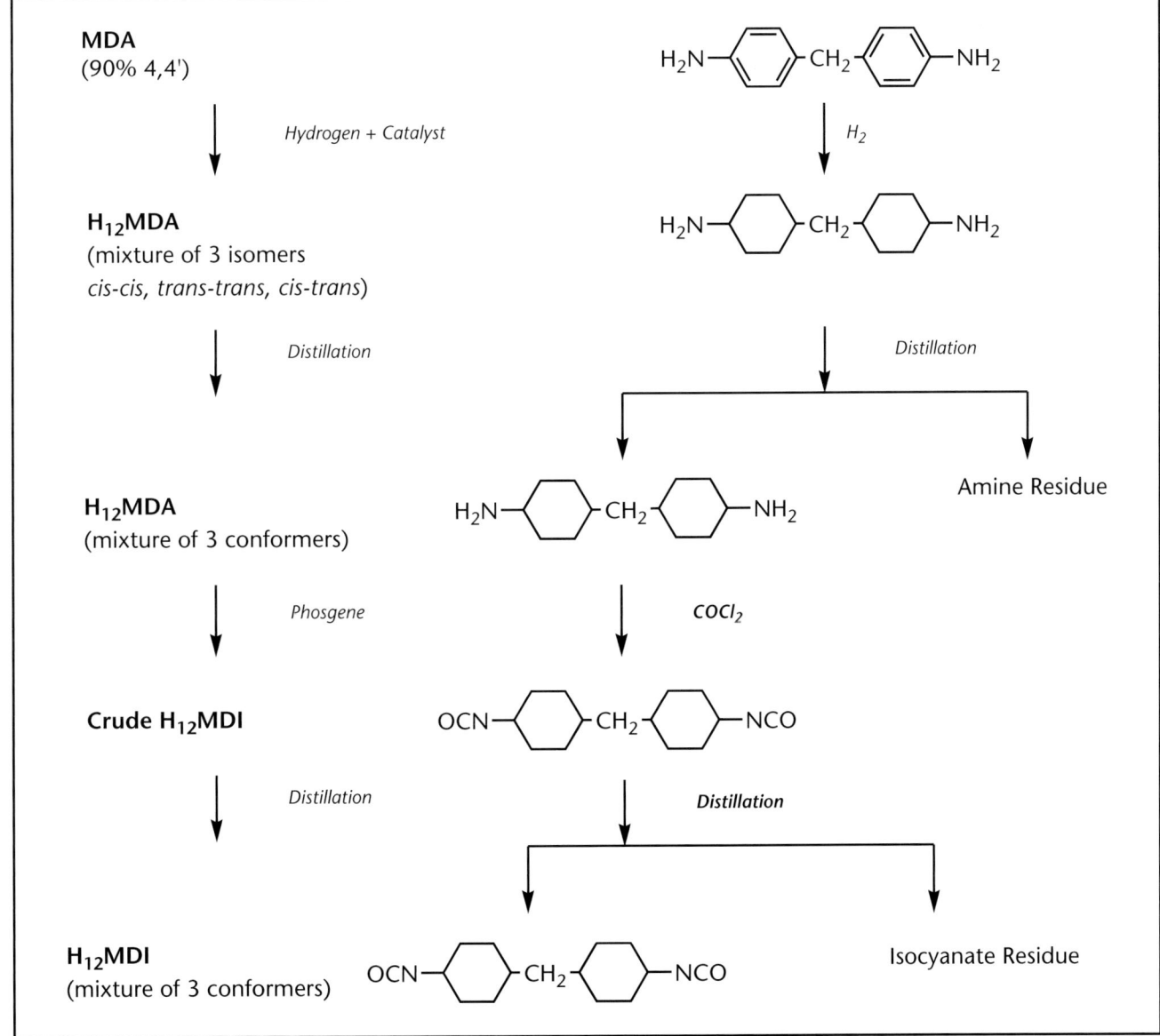

Nitration process

More than 4.5 million tonnes of nitrobenzene and dinitrotoluene were produced worldwide in 2000 using a nitration reaction discovered in the 19th Century that has now been developed into a sophisticated and extremely efficient process, Figure 5-7.

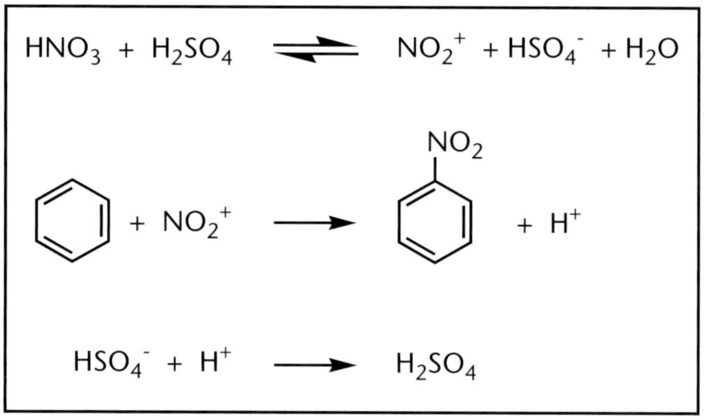

To produce nitrobenzene, a mixture of concentrated nitric and sulphuric acids plus immiscible benzene is intimately blended using high intensity mixing. The resulting exothermic reaction inserts a nitro group into the aromatic ring via a nitronium ion and generates a mole of water. The reaction is carried out adiabatically, the final temperature reaching 120 to 130°C. The mixture is allowed to separate into a dilute sulphuric acid stream and the immiscible nitrobenzene, which contains less than 20 mole-% of unreacted benzene. Water is removed from the dilute sulphuric acid by flash distillation and used to wash the crude nitrobenzene, which is then stripped to remove benzene for recycle. The concentrated sulphuric acid is also recycled and the excess reaction heat recovered as steam. The economics of the process is influenced by the amount of water that needs to be recycled. This produces nitrobenzene of better than 99.7 per cent purity.

Figure 5-7   Nitration process

$$HNO_3 + H_2SO_4 \rightleftharpoons NO_2^+ + HSO_4^- + H_2O$$

$$C_6H_6 + NO_2^+ \longrightarrow C_6H_5NO_2 + H^+$$

$$HSO_4^- + H^+ \longrightarrow H_2SO_4$$

The main impurities are:

- 1,3 dinitrobenzene (less than 300 ppm) resulting from the further nitration of nitrobenzene. This can be minimised by maintaining a small excess of benzene.
- Mixed mono nitrotoluenes that result from the nitration of toluene impurities in benzene with the level dependent on the original benzene specification.
- Mixed polynitrophenols (less than 3,000 ppm). These result from an oxidation reaction between benzene and nitric acid to generate phenol. The polynitrophenols are easily removed from the crude nitrobenzene by adding alkali to the wash-water, which transfers them completely to the aqueous phase.

The (di)nitration of toluene is a two-stage process with the first stage similar to the above. However, chemical differences arise because of the presence of the methyl group, which activates the ring to electrophilic attack in the ortho and para positions. The initial nitration, therefore, results in a mixture of the 2- and 4-nitrotoluenes in the ratio 57/43. This first stage has to be pushed further than with benzene so that only a small amount of unreacted toluene, less than five mole-% is left and about 15 mole-% dinitrotoluenes are formed. This crude mixture is separated from the sulphuric acid, which goes for flash distillation and recycle, passing to a second identical nitration stage to introduce a second nitro group into the ring.

The desirable second substitution on toluene is easier than the undesirable one on benzene because of the activating effect of the methyl group, which partially counteracts the deactivating effect of the nitro group. The 2-nitrotoluene converts approximately 65/35 to 2,4/2,6 dinitrotoluene, while the 4-nitrotoluene converts almost 100 per cent to 2,4 dinitrotoluene. The final product, after water washing, is thus a mix of approximately 80/20 2,4/2,6 dinitrotoluene isomers of about 95 per cent purity.

The main impurities in the crude product are:

- About four per cent of a mixture of the 2,3 and 3,4 dinitrotoluene isomers.
- Up to one per cent of mixed nitrocresols, which result from the same oxidation and nitration sequence that leads to polynitrophenols in the nitrobenzene process. The nitrocresol impurities are easily removed by adding alkali at the washing stage to transfer them into the aqueous phase.
- Less than one per cent mono nitrotoluene isomers.

Hydrogenation process

The hydrogenation of the nitro group to an amine group was first exploited commercially in the 19th Century and the original 'Bechamp' process is still operated today to produce more than 100,000 tonnes a year of pure aniline by dissolving iron in hydrochloric acid in the presence of nitrobenzene. The hydrogen generated from the reaction of the metal with the acid reacts with the nitro group to generate aniline. The aqueous acidic product mix is neutralised then separated to provide pure aniline and a range of iron oxide pigments. Aniline production by this process is still commercially successful, but is limited by the size of the market for the co-produced iron oxide pigments.

The main process currently used to convert nitro to amine groups is based on catalytic hydrogenation using hydrogen gas, Figure 5-8.

The process is highly exothermic and it is the control of this exotherm on the large scale that leads to various different, but successfully competing processes:

Figure 5-8   Hydrogenation process

$$R-NO_2 + 3H_2 \xrightarrow{\text{Catalyst}} R-NH_2 + 2H_2O$$

$$(\Delta H = -540 \text{ kJ/mole})$$

- Liquid slurry phase process.
- Liquid-vapour slurry phase process (aniline).
- Vapour phase fixed bed (aniline).
- Vapour phase fluidised bed (aniline).
- Aniline from phenol.

Liquid slurry phase process

The amine product forms the liquid phase and finely divided catalyst, usually nickel on an inert support, is stirred into this under hydrogen pressure at a temperature of 130°C. The nitro aromatic is pumped into this stirred liquid where it rapidly converts to the amine plus water. The heat is removed through internal cooling coils and the very rapid reaction rate maintains a low-ppm-level of the nitro compound.

The reaction mix is continuously decanted from the reactor leaving the catalyst behind and the amine product is distilled following separation from the water. The conversion of nitro to amine is better than 99.5 per cent.

In the case of TDA, the refining stage is designed to remove the approximately four per cent 'ortho isomers', the 2,3 and 3,4 toluene diamines, which have a large negative effect on TDI yield and quality. Thus the yield of TDA is approximately 95 per cent and this process was used to produce one million tonnes of TDA in 2000, virtually all of the global demand. The yield for aniline by this process is greater than 99 per cent and approximately 200,000 tonnes were produced in 2000.

The following processes are also used, but only for aniline because all exploit the vapour phase, which is unsuitable for the high boiling TDA.

Liquid-vapour slurry phase process (aniline)

This is a modification of the liquid slurry phase process in which the product is continuously removed as a vapour. Most of the heat of reaction is removed from the reactor as latent heat, recovered downstream in the product condenser. The heat balance is maintained by recycling either crude liquid product aniline or liquid product water to the reactor. The reactor temperature is typically 200 to 230°C.

There are two variations to this process; one in which the reactor is operated at two bar hydrogen pressure using nickel catalyst with intensive agitation plus aniline recycle and the other at 15 bar hydrogen pressure using precious metal catalysts in a column reactor with water recycle. The yield of aniline is greater than 99 per cent for both and approximately 900,000 tonnes of aniline was produced by these methods in 2000.

Vapour phase fixed bed (aniline)

This process is commonly carried out using a copper-based catalyst in a fixed bed at a temperature of 400 to 450°C. The nitrobenzene is vaporised using some of the high-pressure steam generated by the high temperature exothermic reaction. This vapour is fed, along with an approximately 100-fold excess of hydrogen, to control the exotherm, at a pressure of one to two bars through a hot tube bundle containing the catalyst.

The aniline and water are condensed in a steam-generating condenser whilst the hydrogen is recycled via powerful compressors. As the catalyst loses its activity it is regenerated by oxidation, to burn off organic tars, so a continuous plant would normally have two or more reactors, one regenerating whilst the others are in production. The yield for aniline is greater than 99 per cent and this process produced approximately 700,000 tonnes of aniline in 2000.

Vapour phase fluidised bed (aniline)

The fluidised bed process operates at similar temperature and pressure to the fixed bed process, but overcomes the 'excess hydrogen' problem because the heat can be efficiently removed by means of steam registers within the mobile bed of copper-based catalyst. However, the catalyst needs to be re-generated much more frequently. The yield for aniline is greater than 99 per cent and this process produced approximately 600,000 tonnes of aniline in 2000.

Aniline from phenol

About 200,000 tonnes of aniline was produced in 2000 by a vapour phase, fixed bed amination of phenol using ammonia. This process is commercially viable because it co-produces the higher priced diphenylamine.

Aniline-formaldehyde condensation

With all polyisocyanates, other than MDI, the functionality (usually di-) has been built in by the time the amine groups are generated, but with MDI the 'functionality building' step is based on a separate aniline-formaldehyde condensation reaction, Figure 5-9.

The process consists of reacting aniline and formalin (aqueous formaldehyde) in the presence of an acid catalyst, usually hydrochloric acid. The reaction is exothermic and predominantly homogeneous. After a series of chemical transformations a complex mixture of polyfunctional primary aromatic amines is obtained still containing the acid catalyst. This mix is neutralised with aqueous sodium hydroxide, producing a two-phase system of organic amines and an aqueous sodium chloride solution. After separation and water washing the organic phase is stripped to remove and recycle excess aniline. The product is a very complex mixture of diamine, triamine, tetramine, pentamine, hexamine and other higher oligomers, Figure 5-10. The relative amount of each species depends on the process conditions used and typical ranges are given in Table 5-1.

Figure 5-9 Aniline-formaldehyde condensation process

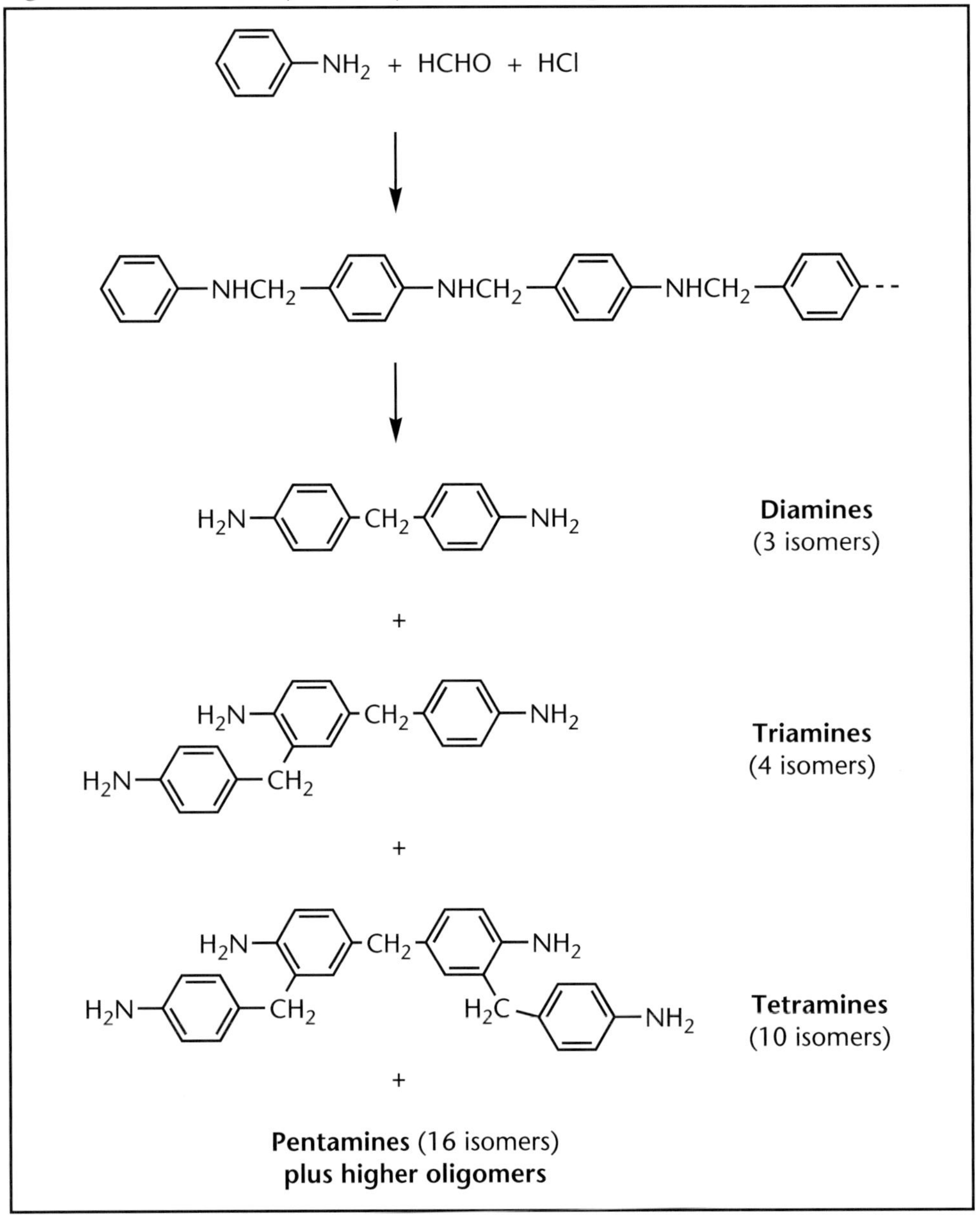
NH2 + HCHO + HCl
NHCH2 NHCH2 NHCH2
H2N CH2 NH2
Diamines
(3 isomers)
+
H2N CH2 NH2
H2N CH2
Triamines
(4 isomers)
+
H2N CH2 NH2
H2N CH2 H2C NH2
Tetramines
(10 isomers)
+
Pentamines (16 isomers)
plus higher oligomers

Figure 5-10 Molecular structure of generic amine products

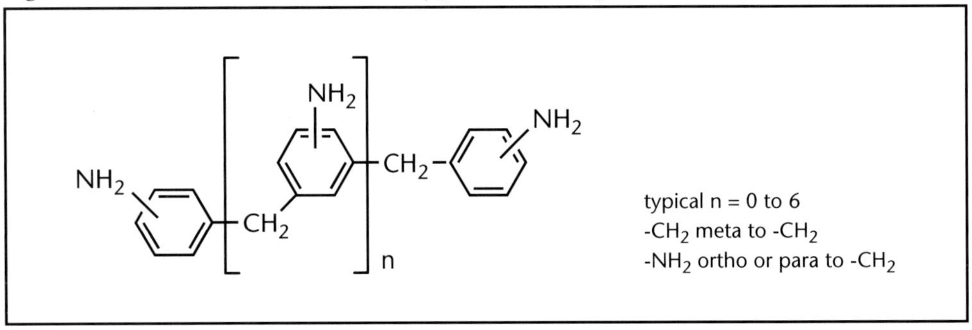
NH2
NH2
CH2
NH2
CH2
n
typical n = 0 to 6
-CH2 meta to -CH2
-NH2 ortho or para to -CH2

Table 5-1   Typical concentrations of amine species

Species	Range (%)	Example 1 (%)	Example 2 (%)
Di	50 – 80	50	80
Tri	10 – 20	20	10
Tetra	5 – 12	12	5
Penta & higher	5 – 12	18	5

Each oligomer can exist in a number of different isomeric forms. Thus the diamine has three possible isomers, Figure 5-11.

The triamine has up to seven theoretically possible isomers, but three of these are so unlikely on steric grounds that they are not detectable in most products, Figure 5-12.

Taking this same steric limitation into consideration, it can be shown that the number of isomers which exists for any oligomer is: 2^n when n is even and $2^n + 2^{(n-1)/2}$ when n is odd, where n is the number of CH_2 bridges in the oligomer.

Figure 5-11   Molecular structure of diamine isomers

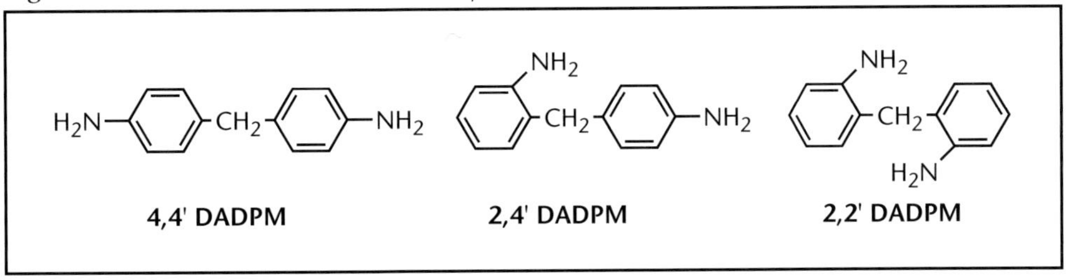

4,4' DADPM   2,4' DADPM   2,2' DADPM

Figure 5-12   Molecular structure of triamine isomers

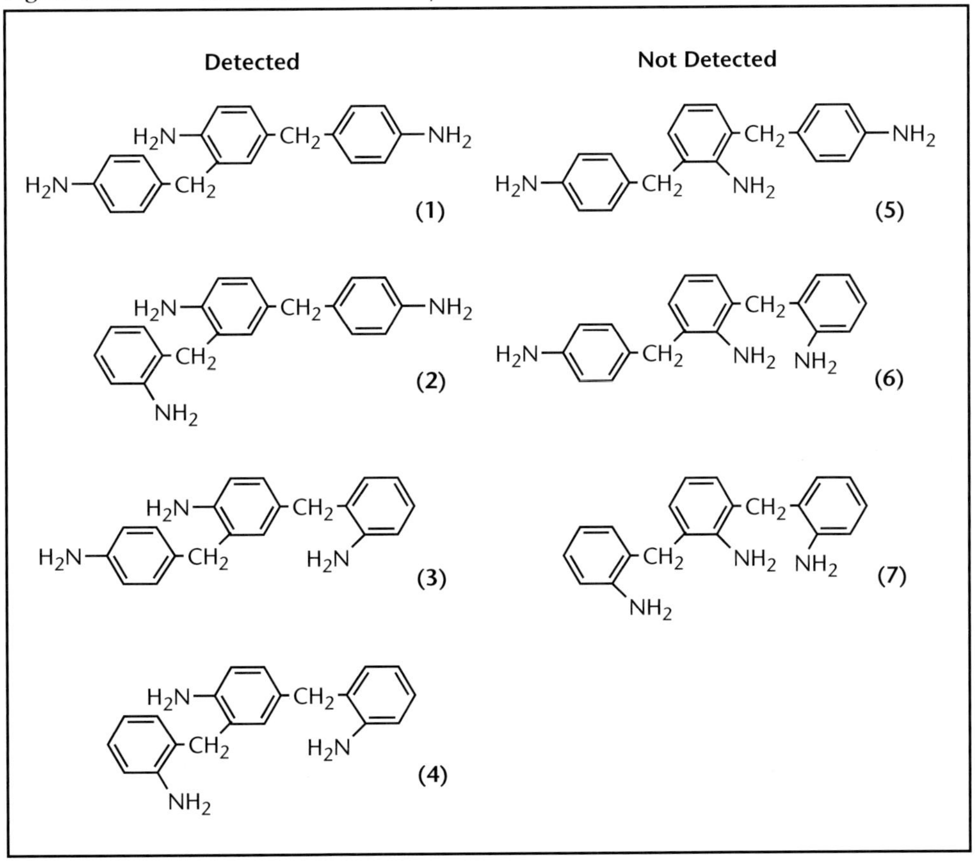

Detected: (1), (2), (3), (4)

Not Detected: (5), (6), (7)

Thus the tetramine (n = 3) exists in 10 isomeric forms, the pentamine (n = 4) in 16 and the hexamine in 36. Mass spectrometry has detected traces of oligomers containing up to 15 aromatic rings and amine groups (n = 14) in commercial products or 16,384 isomers.

The chemistry of the process is complex, but extremely versatile and by selection of process conditions it is possible to adjust the distribution of the different functionality amines in the mixture, by varying the aniline/formaldehyde ratio, or adjusting the different isomer ratios, by varying the temperature.

For these reasons it is not possible to define here a single product mixture. In reality, every producer makes a range of products, depending on their production strategy and the eventual requirement of the marketplace. Although the chemistry is versatile, it does not allow all mixtures to be made. For instance, it is not possible to change the isomer mixture of the diamines independently of the isomer mixtures in the tri and higher polyamines.

Generally, a plant will be designed to make a particular product mix and to dramatically change that product mix the producer would need to accept a cut in production rate or invest in major plant changes. Thus, once a plant has been designed and built it can usually only operate economically within a narrow range of processing conditions.

One important side-reaction in this process is that formaldehyde molecules (less than 0.5 per cent) can act as a methylating agent. This produces a few N-methyl groups distributed through all the polyamine oligomers and isomers. Since these modified molecules remain in the complex amine mixture they have a significant effect on the eventual MDI products.

Phosgenation process

The phosgenation process is the final chemical stage in the production of all the aromatic isocyanates and 90 per cent of the aliphatic isocyanates, Figure 5-13.

In principle, the process is simple as the highly reactive phosgene molecule reacts very rapidly with the amine group to produce the carbamoyl chloride group and hydrogen chloride gas, which is easily removed from the relatively non-volatile product. The carbamoyl chloride group decomposes at elevated temperature and/or low pressure to isocyanate and a second molecule of hydrogen chloride gas. The overall process is approximately thermo-neutral.

Figure 5-13   Phosgenation process: main reaction

$$R-NH_2 + COCl_2 \longrightarrow R-NHCOCl + HCl$$

$$R-NHCOCl \rightleftharpoons R-NCO + HCl$$

Although the main reaction is extremely fast, less than one millisecond, there are two exothermic side reactions that occur in less than five milliseconds, shown in Figure 5-14.

Figure 5-14   Phosgenation process: side reactions

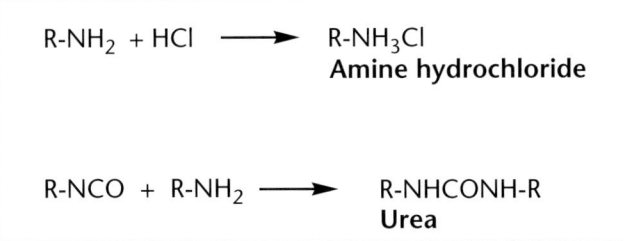

The first leads to an amine hydrochloride precipitate that is highly insoluble in the reaction medium. This precipitate reacts slowly and endothermically with phosgene to produce the desired isocyanate and three molecules of hydrogen chloride. As the base strength of the amine increases more of the insoluble amine hydrochloride is generated and the resulting amine hydrochlorides are less soluble in the typical phosgenation solvents. The other reactions, producing ureas, are undesirable because they result in yield loss since the formation of one urea group effectively leads to the loss of two molecules of diisocyanate.

Another reaction, which only involves MDA, converts N-methyl group impurities to stable N-methyl carbamoyl chlorides during the phosgenation stage.

Phosgene is produced as required in a closed system, by a simple gas phase reaction of carbon monoxide with chlorine through a fixed bed of carbon catalyst and is immediately absorbed in solvent. It is then intimately mixed, in excess, with a solution of the amine in solvent. The most commonly used solvents are monochlorobenzene (MCB), ortho dichlorobenzene (ODCB) and toluene (T). These are utilised for the production of MDI, TDI and aliphatics as shown in Table 5-2.

Table 5-2
Solvent utilisation

Isocyanate	MCB	ODCB	T
MDI	x	x	
TDI	x	x	x
Aliphatics	x	x	

The urea reaction is minimised by using excess phosgene. Various proprietary processes exist in which the temperature and pressure vary widely. In principle, low-pressure processes allow more of the slow reacting amine hydrochloride to form, while high-pressure/high-temperature processes prevent this. However, the higher pressure processes need to overcome the carbamoyl chloride equilibrium.

After the reaction stage, hydrogen chloride and excess phosgene are boiled off and separated either by fractionation or by absorption of phosgene in solvent. The phosgene is recycled and the anhydrous hydrogen chloride is used as a raw material in other processes which is why isocyanate producers are the world's largest suppliers of hydrogen chloride gas. The solution of crude isocyanate in solvent is distilled, to remove the solvent for recycle, leaving the crude isocyanate product that requires purification.

Figure 5-15
Cyclic urea from a 1,3 aliphatic diamine

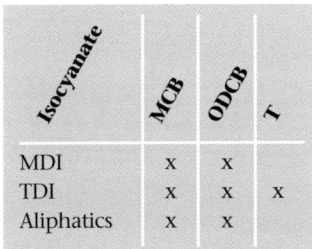

A vapour phase phosgenation process has recently been commercialised specifically for the production of HDI, in which gaseous phosgene and gaseous hexamethylene diamine are intimately mixed and then quenched into solvent. This same process is also claimed in a recent patent application to solve the low yield problem associated with conventional phosgenation of *cis*-1,3 cycloaliphatic diamines, which can easily produce six-membered cyclic ureas, Figure 5-15.

Purification

After removal of phosgene, hydrogen chloride and solvent the crude isocyanates are normally purified by fractional distillation. Because the isocyanates are very reactive and tend to be unstable at elevated temperatures, thin film evaporators are used to minimise heating times. The higher boiling and more reactive aromatic isocyanates are particularly sensitive.

All the aliphatic isocyanates are obtained as single isomer pure compounds, ignoring conformers, whilst TDI is obtained predominantly as an 80/20 isomer mix (2,4/2,6) but MDI consists of a complex mixture of isomers and oligomers.

When crude isocyanates are distilled any urea groups present can react with a further diisocyanate molecule to form biuret and again to form triuret and again to form tetra-uret. These can cyclise to form isocyanurate and re-generate the urea to start the cycle again. Ureas can 'dehydrate' to carbodiimide, which can further react with a diisocyanate molecule to form uretonimine. (See Chapter 7 for details of these reactions.) Therefore, a small number of urea groups, formed at the phosgenation stage, can result in a large yield loss during purification.

The distillation step generates a polymeric residue that increases in average molecular weight as pure diisocyanate is removed, so the residue needs to be kept mobile, or special equipment added to extract the maximum yield from the increasingly viscous, even solid, residue. The yield is dependent on many factors; the relative reactivity and purity of the amine feeding into the phosgenation, the efficiency of phosgenation, the temperature and time of distillation, the relative reactivity of the isocyanate and the complexity of the residue processing equipment. In recent years some TDI plants have added a hydrolysis step to the distillation residue to recycle the resulting amine, rather than burning the residue.

If the 'ortho isomers' of TDA are not removed at the hydrogenation stage they react with phosgene to produce five-membered cyclic ureas, which subsequently form biurets and other compounds with the TDI isomers during the phosgenation and distillation stages leading to large losses in yield.

Figure 5-16 shows the formation of a typical cyclic urea and biuret.

A small amount of refined 80/20 TDI is treated via a bulk crystallisation process to generate essentially the pure solid 2,4 isomer and a second liquid product of approximately 65/35 2,4/2,6 isomer mix.

In the case of MDI the unpurified stream can be sold as polymeric MDI, though to provide a stable clear liquid it needs to be heated to about 200°C to degas and de-dimerise it before crash-cooling to 'freeze' the dimer equilibrium at a low level. Alternatively, it can be passed through a distillation step to remove a part of the diisocyanate isomers as a vapour stream. In this case the non-volatile distillation residue is crash-cooled and is sold as another polymeric MDI grade.

Figure 5-16   Biuret side reaction of phosgene and diamine

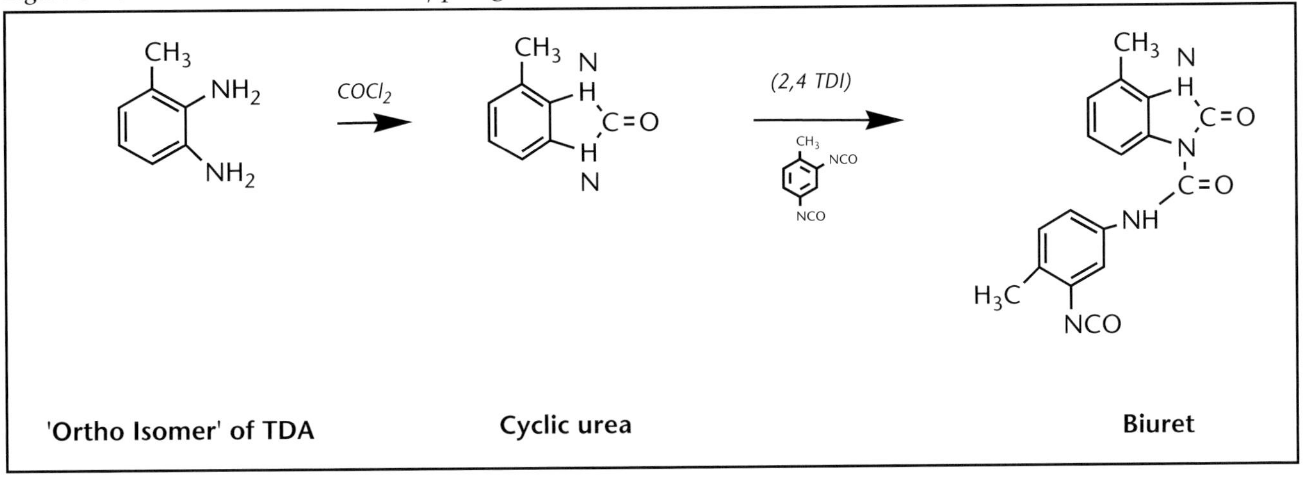

'Ortho Isomer' of TDA   Cyclic urea   Biuret

The volatile diisocyanate isomer stream is fractionated to produce a pure 4,4'-MDI stream and a mixed isomer stream (MI) of up to 60 per cent 2,4' isomer mixed with 4,4' isomer and a small amount of 2,2' isomer.

Isocyanate derivatives

Prior to end-use many isocyanates are modified to form derivatives, with the method used dependent on the product type required and the end application. There are many different reasons why an isocyanate needs to be modified, some of which are:

- To achieve liquid stability in storage (pure MDI).
- To match stream ratios and/or viscosities in polyurethane mixing equipment (polyurethane shoe soling systems).
- To remove some of the exotherm prior to mixing with polyol blend.
- To affect the miscibility of the isocyanate with the polyol blend.
- To decrease the hazard of a volatile isocyanate (isocyanurate or biuret derivatives of HDI and IPDI).
- To make the isocyanate emulsifiable in water.
- To modify the chemical reactivity of the isocyanate groups (chemical blocking in powder coatings).
- To add cross-linking (isocyanurate or biuret derivatives of HDI and IPDI).
- To ensure that a slower reacting component of a system reacts first (or completely) with the isocyanate.
- To affect the hard block molecular weight distribution in the final polyurethane forming.

The nomenclature which has grown up in the different application areas means that derivatives may be known as prepolymers, quasi-prepolymers, semi-prepolymers, partial prepolymers, variants or blocked isocyanates.

Because a large variety of these derivatives are required they are commonly made in a multi-product batch plant where the isocyanate is stirred, optionally heated and the reactive component, normally a polyol tightly quality controlled to minimise water and undesirable catalytic species, is added. The reaction is then adjusted to achieve the final isocyanate content and viscosity for the particular derivative.

A special case is where an isocyanate derivative is required with virtually none of the starting isocyanate monomer remaining in the product, as for use in coating applications, such as the trifunctional isocyanurates, Figure 5-17. To achieve low monomer content whilst producing a consistent low molecular weight oligomer only a small amount of product should react, so the reaction needs to take place in a large excess of monomer which is then removed for recycle. This may be done by distillation, using thin film evaporation for products, which are thermally stable such as the isocyanurates of TDI, HDI or IPDI, or by solvent extraction for thermally labile products like low monomer prepolymers.

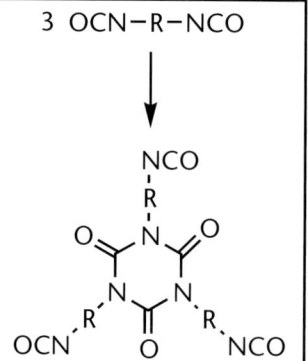

Figure 5-17
Trifunctional isocyanurate

Analysis

Isocyanates are analysed in order to quantify the parameters that will affect either the processing characteristics, the physical properties, or the environmental impact of the polyurethane.

Isocyanate value

This is required for calculation of the correct ratio of isocyanate to polyol. The standard method is to react a sample of the isocyanate with an excess of a secondary aliphatic amine to convert isocyanate groups to urea groups. The excess amine is titrated with standard nitric acid, to determine how much amine has reacted and a simple calculation determines the isocyanate content. However, most isocyanates contain acidic or 'hydrolysable chloride' groups, which neutralise some of the amine, leading to an over-estimate of the isocyanate value. Therefore, the isocyanate value is more accurate if it is corrected for hydrolysable chlorine and/or acidity.

Hydrolysable chlorine

The level of hydrolysable chlorine in an isocyanate can influence the rates of the various chemical reactions occurring during the polyurethane-forming reaction.

The normal method is to take the titrated sample from the isocyanate value determination and determine the chloride content by titrating with standard silver nitrate solution to a potentiometric end-point. The silver nitrate test measures the chloride ion, which has been released from the isocyanate by reaction with the strongly basic amine.

However, in real polyurethane processes there may not be a strongly basic secondary amine to release the chloride. In a polyurethane coating system the reaction takes place in a non-basic environment and at a low temperature compared to polyurethane rigid block foam, which contains tertiary amine catalysts and the temperature, may rise to 170°C. In the first application the isocyanate may not release all its chlorine whilst in the second it almost certainly will. Therefore, for the first application it is better to measure the chloride released by reacting the cold isocyanate with a simple alcohol rather than an amine. It is possible to 'hydrolyse' varying levels of chlorine by changing the chemical reactant, the reaction temperature or time.

Many tests have grown up, due to the variety of application areas for isocyanates, but often the term 'hydrolysable chlorine' is used without qualification, so care is needed when comparing hydrolysable chlorine figures to ensure that all figures have been obtained from the same test method. Generally the method using secondary amine reaction produces the highest hydrolysable chlorine figures.

Ionisable chlorine

The term ionisable chlorine normally refers to the chlorine released by cold reaction of the isocyanate with isopropanol. So, it could be considered as the mildest hydrolysable chlorine test, producing the lowest figures.

Acidity

Acidity is determined by reacting an isocyanate at room temperature with isopropanol then titrating the mix to a pH end-point with sodium hydroxide. Generally, the acidity and ionisable chlorine figures, both expressed as HCl, will be in close agreement with each other, because both are measuring the concentration of hydrogen chloride released when the isocyanate is reacted with isopropanol under the same mild conditions. These two figures would not be in agreement if the isocyanate was stabilised by adding non-chlorine based acid species.

Viscosity

Viscosity is an important parameter in defining the ease with which the isocyanate can be pumped and mixed with the polyol formulation. Most purified isocyanates have a consistent viscosity range and for polymeric MDI viscosity is used as a primary means of differentiating the different grades. For quality control purposes viscosity is normally measured by a simple U-tube method.

Gas chromatography

Gas chromatography is used to determine isomer ratios for TDI, MDI diisocyanates, MDI triisocyanates and to measure the low (ppm) levels of volatile impurities in isocyanates, such as residual phosgenation solvent and traces of mono-isocyanates. These impurities are maintained below agreed industry standard levels.

Storage/transportation

All isocyanates are reactive chemicals that should be stored in sealed containers and contact with water must be avoided. If water gets into a closed isocyanate container it can produce a dangerous situation because the reaction of the isocyanate with water produces urea and carbon dioxide. Water is immiscible with, and lower density than, the isocyanates so the urea reaction takes place slowly at the interface near the top of the container. Once this reaction has started the pressure will increase and even opening the top of the container may not prevent a violent rupture since a thick layer of foamed urea, on the top of the isocyanate, can block the top again.

High temperature storage of isocyanates should also be avoided. Chapter 7 describes how isocyanates can self-react exothermically to form isocyanurates, or thermo-neutrally to form carbodiimides and carbon dioxide. Prolonged heating in sealed containers can thus be dangerous.

Provided the correct precautions are taken, most isocyanates have storage lives of six months to two years. One notable exception is pure MDI, which is more stable at room temperature as the uretidine dione (dimer) than it is as the free isocyanate. For this reason pure MDI is transported either as a liquid at 40 to 45°C, storage life of only three weeks, or as a deep-frozen solid (minus 20°C), storage life of two years. It is important to avoid prolonged heating of pure MDI since it promotes dimerisation and a second concern is oxidation, as the bridge CH_2 can be oxidised slowly to a ketone.

Isocyanate product characteristics

A table with the physical properties of all the commercially produced polyisocyanates can be found in Appendix 3.

With the notable exception of polymeric MDI, all the other isocyanates are manufactured as either pure, single isomer compounds, 2,4-TDI, HDI, NDI, TMXDI; easily defined isomer mixtures, TDI and pure MDI; or as conformer mixtures, IPDI and H_{12}MDI.

Why are all these different molecules available? Why should one be chosen over another for a particular application? The answer, of course, is not simple, but some of the reasons are:

- The shape of the molecule can influence the properties of the polymer derived from it. Thus a linear diisocyanate molecule, 4,4'-MDI, HDI, H_{12}MDI, TMXDI, is ideal for elastomeric and thermoplastic properties by producing two-phase block copolymers and creating physical links between the polymer chains whilst an asymmetric molecule, 2,4'-MDI, TDI, IPDI, tends to produce clear and less stiff polymers.
- The number of isocyanate groups per molecule, functionality, influences the properties of the polymer produced and also the cure rate. Higher functionality results in hard, glassy and brittle polymers, which are quicker to cure.
- Aliphatic isocyanates lead to light-stable polyurethanes, but those made with aromatic isocyanates rapidly yellow on exposure to light. The major applications for aliphatic isocyanates are thus for applications that are exposed to a lot of light such as floor coatings and shatterproof laminated windscreens. Conversely aromatic isocyanates are either pigmented with dark colours or used in applications with little or no exposure to light.
- The reactivity of the isocyanate group influences how rapidly a polymer cures, which determines the length of time between mixing and gelation. Aromatic isocyanate groups are several hundred times more reactive than aliphatic isocyanate groups thus making the aliphatic isocyanates more suited to the slow curing requirements of the coating applications in which they are predominantly used.
- The differential reactivity between the two or three groups of an isocyanate molecule influences the way in which a polyurethane formulation builds up viscosity and this can modify the flow characteristics of the gelling polymer. This can be important in such widely diverse applications as rigid foam filling of a refrigerator cavity, or the coating of a large floor.

MDI

Pure 4,4'-MDI is a symmetrical molecule with two aromatic isocyanate groups of equal reactivity. Commercial products normally contain one to two per cent of the 2,4' isomer and have hydrolysable chlorine levels below five ppm.

2,4'-MDI is an asymmetrical molecule with two aromatic isocyanates of different reactivity. The 4-position is approximately four times more reactive than the 2-position and is of similar reactivity to the two groups in 4,4'-MDI. It is normally commercially available as a mixture with the 4,4' isomer (2,4'/4,4', 55/45). Typical hydrolysable chlorine levels are less than 50 ppm.

Polymeric MDI

This is a complex mixture of molecules. It contains both 4,4'- and 2,4'-MDI and different grades may have different ratios of these two. In addition there are the three functional, four functional and higher oligomers, each of which has multiples of isomers. Some of the molecules contain, randomly distributed, N-methyl carbamoyl chloride groups and also other potentially acidic impurity groups. Uretonimine, biuret or isocyanurate species are also present.

Polymeric MDI cannot easily be described in molecular terms because there are so many different molecules present that they cannot be analytically separated and individually characterised. Therefore, it can only be described overall in 'average' molecular terms. Gas chromatography can be used to determine the isomer ratios within the di and tri isomers, Figure 5-18. Tri iso one to four are derived from the tri amine one to four shown in Figure 5-12.

Gel permeation chromatography is a technique ideally suited to qualitatively characterising polymeric MDI and with the right conditions, base-line separation of the di, tri and tetra oligomers is possible, Figure 5-19.

However, isomer separation within these oligomer peaks is almost non-existent, so calibration is not possible because the detectors commonly used with this technique are not isomer sensitive. Beyond the 'tetra' oligomers, the chromatogram is complicated by the presence of the more complex molecules of uretonimine, biuret and isocyanurate derivatives, which, with the higher oligomers, effectively constitute a continuum of molecular volumes, so peak separation is not possible.

Figure 5-18   Gas chromatography trace of di and tri isomers of MDI

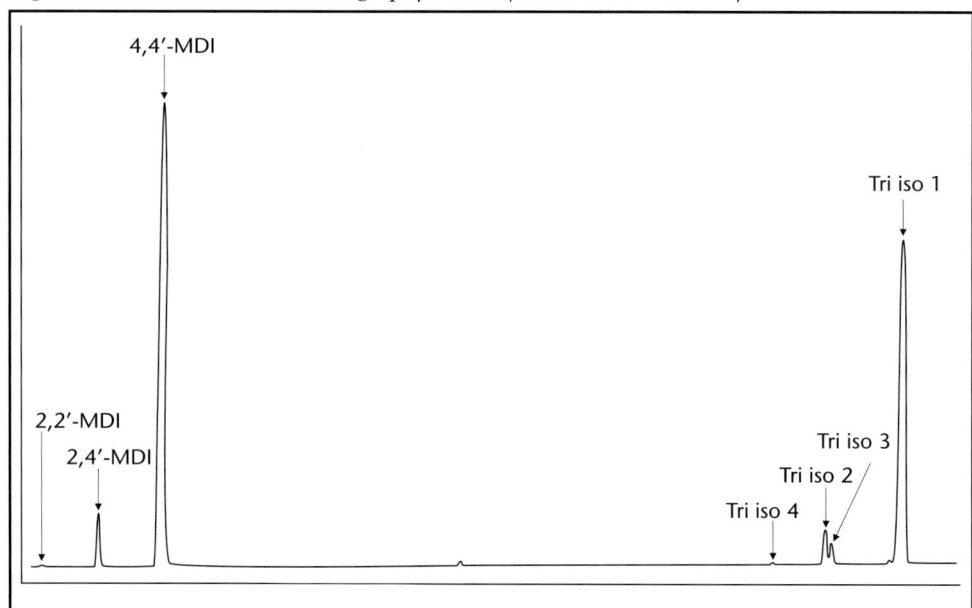

Figure 5-19   Gel permeation trace of polymeric MDI

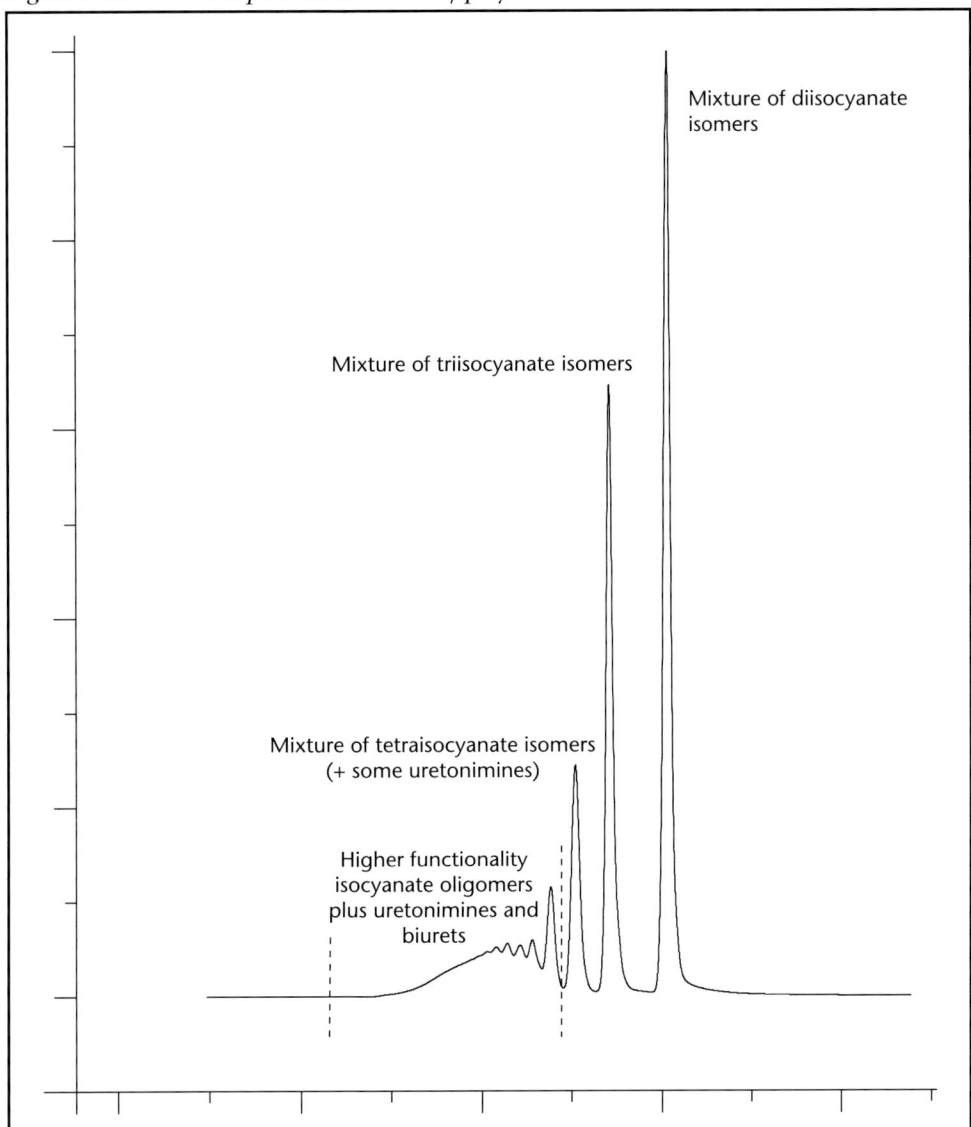

In practice, a range of different grades of polymeric MDI is supplied, normally specified and differentiated in terms of viscosity, hydrolysable chlorine, acidity and colour. There may also be more selective differentiation based on, for instance a reactivity test. Sometimes manufacturers quote 'functionality' for their polymeric MDI grades. These are normally very approximate numbers, because functionality is not easily measured with any degree of accuracy and to calculate functionality an accurate analysis of all species present is needed.

In polymeric MDI the isocyanate groups have three different reactivities: the 4-position at the end of a chain, with reactivity equivalent to 4,4'-MDI, the 2-position at the end of a chain, approximately 25 per cent of the 4,4'-MDI reactivity and those on phenyl rings in a chain, about 15 to 20 per cent the reactivity of 4,4'-MDI. In a typical polymeric MDI about 20 per cent of the isocyanate groups will always be of the third type. The split of the other two types will depend on the process conditions at the aniline-formaldehyde stage, but normally will be approximately 75 per cent of the first and five per cent of the second.

Of all the commercially available isocyanates, MDI is the one that produces polyurethanes which yellow fastest on exposure to light. This is because the urethanes derived from it can easily oxidise through a radical mechanism to produce a quinone-imide, conjugated structure, which is a yellow chromophore, Figure 5-20.

Figure 5-20   Formation of quinone-imide

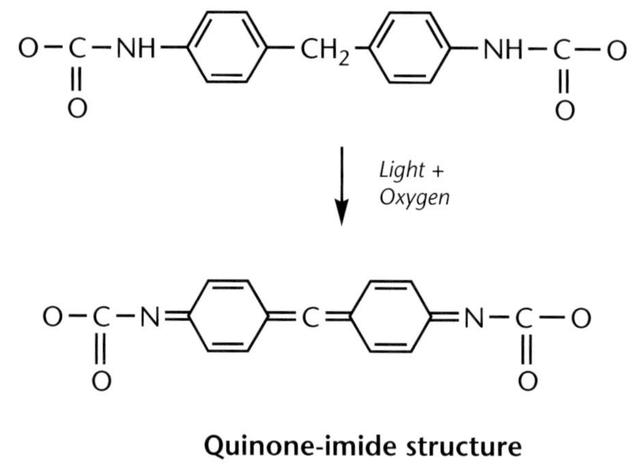

Quinone-imide structure

TDI

The isocyanate groups on 2,4-TDI have different reactivities with the 4-position approximately four times the reactivity of the 2-position and about 50 per cent more reactive than the 4-position group in MDI, whilst for the 2,6 isomer the groups have equal reactivity that is approximately the same as that of the 2-position in 2,4-TDI.

Because the two isocyanate groups in each TDI molecule are on the same aromatic ring, the reaction of one of the groups tends to cause a change to the reactivity of the second group.

TDI is normally available as pure 2,4, as an 80/20 (2,4/2,6) mixture or as a 65/35 (2,4/2,6) mixture.

NDI

NDI (1,5-naphthalenediisocyanate) is a very bulky and symmetrical molecule with two aromatic isocyanate groups of equal reactivity, and similar to that in the 4-position in MDI. It is normally supplied in flake form and requires melting at 130°C or dissolving in solvent for processing.

HDI

HDI is a flexible, linear, symmetrical molecule with two primary aliphatic isocyanate groups of equal reactivity. Their reactivity is at least two orders of magnitude lower than that in the 4-position of MDI. Of all the commercially available polyisocyanates, it has the highest isocyanate content. Because it is totally aliphatic, it gives rise to light-stable polyurethanes.

IPDI

IPDI is a bulky and very asymmetric molecule. In fact, of all the commercially available polyisocyanates, it is the only one with no degree of symmetry. It is totally aliphatic, therefore giving rise to light-stable polyurethanes. It is commercially available as a mixture of two isomeric forms (25/75 *cis/trans*). Because of this, it has effectively four different isocyanate groups. Two are secondary aliphatic groups with similar reactivity, about half that in HDI. The other two are primary groups, but both are sterically hindered, rendering them even slower, by a factor of about five than MDI. Thus, IPDI has the slowest reactivity of all the commercially available polyisocyanates.

$H_{12}MDI$

$H_{12}MDI$ is commercially available as a 90/10 blend of the 4,4'/2,4' isomers. The predominant 4,4'-diisocyanatodicyclohexylmethane consists of three conformational isomers, *cis-cis*, *cis-trans* and *trans-trans*. The two different isocyanate groups, either of which can be axial or equatorial, are secondary and are of similar reactivity to the secondary isocyanate groups in IPDI. The 10 per cent of 2,4'-diisocyanatodicyclohexylmethane, derived from the 2,4 isomer in the MDA, is made up of four conformational isomers, *cis-cis*, *cis-trans*, *trans-cis* and *trans-trans*. Because it is totally aliphatic it gives rise to light-stable polyurethanes.

Other diisocyanates

All other diisocyanates are only commercially available in limited or developmental quantities, so only have niche and specialised applications.

Alternatives to phosgene

The manufacture of isocyanates is currently carried out by a small number of large chemical companies, all of whom have very high engineering and safety standards for the handling of phosgene. Phosgene is environmentally clean because it results in high organic yields, but there is a potential safety issue with its use that is controlled by only producing it on an as required basis.

A search of patent and chemical literature over the last 30 to 40 years reveals that an enormous effort has been spent on seeking new ways to make isocyanates without using phosgene. However, despite all the research that has been undertaken, the reality is that no viable process has been found for the production of *all* isocyanates.

There are certain fundamental chemical and physical reasons for this, which are best described by examining the most successful of these alternative processes. This is the conversion of amine to urethane, followed by thermal decomposition of the urethane to isocyanate and alcohol, which are then separated by distillation and selective condensation. The chemistry required for the conversion of amine to urethane works quite well with the strongly basic aliphatic amines, but gives poor yields and many side reactions with the aromatic amines.

The thermal decomposition of urethanes to isocyanate and alcohol is also much easier with aliphatic urethanes than with aromatics. Aliphatic isocyanates react with alcohols much more slowly than aromatics, therefore the separation after thermal cleavage is easier. Finally, the aliphatic isocyanates are easier to distil, and have lower boiling points, than the aromatics. Therefore, we can conclude that via this 'non-phosgene route' it is easier at every stage to make and purify an aliphatic isocyanate than an aromatic. By comparison the aromatic amines are easier to phosgenate than the aliphatics, the strongly basic nature of the aliphatic amines tending to produce more ureas and amine hydrochlorides, so a much larger excess of phosgene is required, or longer reaction times or lower yields are achieved.

Therefore, if such an alternative process is to be viable then it needs to exploit its advantages and phosgenation's disadvantages which is seen in the manufacture of the aliphatic isocyanates, HDI or IPDI. Two small-scale plants were constructed in the early 1990s (Hüls in USA and BASF in Germany) but in 2000 they made only about 10 per cent of the world's aliphatic isocyanate with about half of that 10 per cent for captive use.

The chemistry and process will require much improvement to replace phosgene in just the rest of the aliphatic isocyanate production. The step to TDI would then need a major chemistry improvement, which is nowhere in sight, and the final step to MDI would require a further giant leap in the chemistry.

6. Polyols

David Sparrow

Although the highly reactive isocyanate group is the unique feature of polyurethane technology, it is the polyol that in large part determines the properties of the final polyurethane polymer. The extensive ranges of polyol types that are available to the industry explain why polyurethanes have become the most versatile family of plastic materials.

The term 'polyol' describes compounds with hydroxyl groups that react with isocyanates to produce polyurethane polymers. Typically 'polyols' contain two to eight reactive hydroxyl groups and have average molecular weights from 200 to 8,000. The two key classes of product are polyethers and polyesters.

The initial polyols used were predominantly polyesters until, in the late 1950s, it was realised that polyether polyols were particularly well suited for the manufacture of flexible slabstock foam. They quickly became the dominant class of polyol, and now account for 80 per cent of total consumption. Total polyol use had grown from 1.75 million tonnes in 1985 to 4.5 million tonnes in 2000.

A major factor in the choice of polyol for a polyurethane application, apart from its technical effect, is cost. A selected polyol must be competitive with other polyols and also enable the final polyurethane product to be cost competitive with other materials in the end application. The types of polyols used are shown in Figure 6-1.

Three cyclic ethers are used commercially in the manufacture of polyether polyols: ethylene oxide, propylene oxide and tetrahydrofuran. Of these, propylene oxide is the most significant whilst the use of ethylene oxide is limited to the production of copolymers with propylene oxide. Tetrahydrofuran is used for the manufacture of a range of speciality polyols.

Manufacture of propylene oxide

Global propylene oxide (PO) production was around five million tonnes in 2000, with a growth rate of 3.3 per cent per annum. Approximately 70 per cent of propylene oxide is used as an intermediate in polyether polyol production, Figure 6-2.

Propylene oxide is manufactured from propylene by two basic routes – chlorohydrin and hydroperoxidation – with the former being the earlier and simpler method.

Chlorohydrin process

Propylene is reacted with chlorine in water to produce a chlorohydrin, which is then dehydrochlorinated with ring closure to propylene oxide, Figure 6-3.

Figure 6-1   Polyol types used in polyurethane

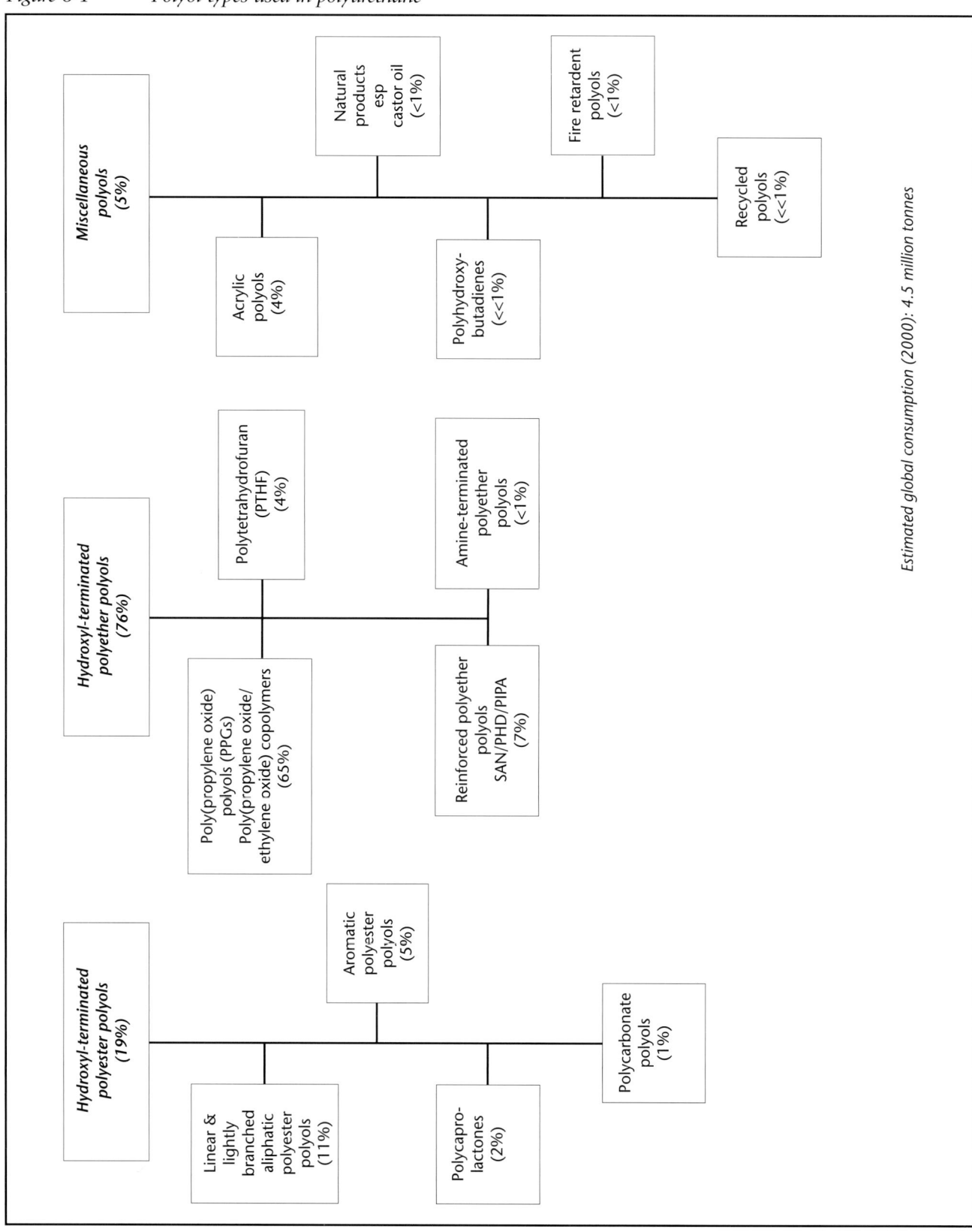

Estimated global consumption (2000): 4.5 million tonnes

Figure 6-2 *End-use of propylene oxide (2000)*

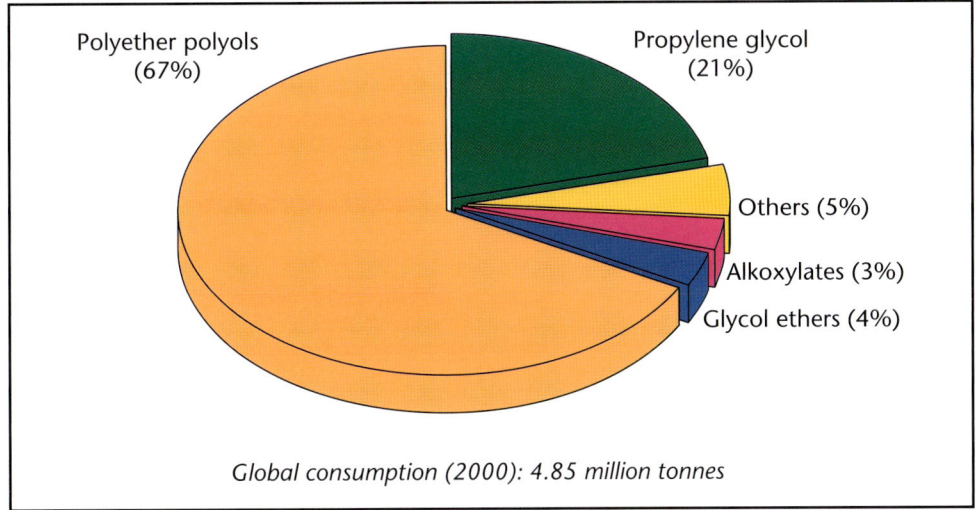

Polyether polyols (67%)
Propylene glycol (21%)
Others (5%)
Alkoxylates (3%)
Glycol ethers (4%)
Global consumption (2000): 4.85 million tonnes

Figure 6-3 *Chlorohydrin route to propylene oxide*

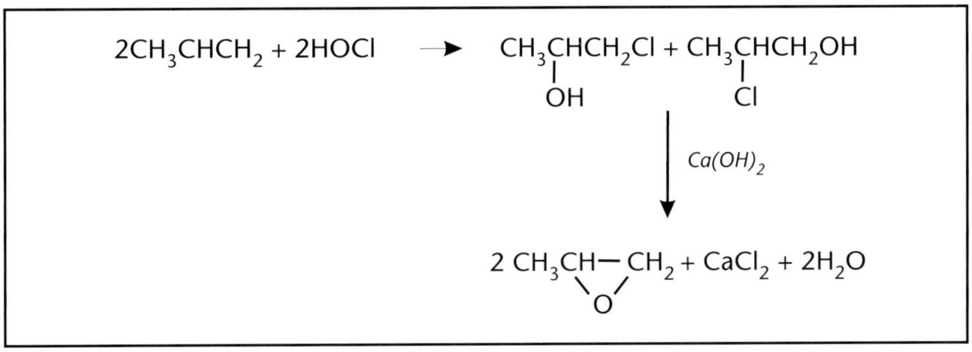

2CH3CHCH2 + 2HOCl
CH3CHCH2Cl + CH3CHCH2OH
OH
Cl
Ca(OH)2
2 CH3CH—CH2 + CaCl2 + 2H2O
O

Significant quantities of calcium chloride (or sodium chloride) are produced as a by-product, which is recycled, used for low-grade outlets or most frequently disposed of. This is the simpler of the two routes and is feasible for small-scale plants. It is, therefore, the process of choice in either fully integrated chloroalkali complexes or in developing markets.

Hydroperoxidation

Propylene is reacted with an organic hydroperoxide to give propylene oxide plus a co-product. Two hydroperoxides are used industrially, Figure 6-4.

- Tertiary-butyl hydroperoxide, obtained from isobutane, is reacted with propylene to produce propylene oxide plus tertiary-butyl alcohol (TBA). The tertiary-butyl alcohol is then reacted with methanol to give methyl tertiary-butyl ether (MTBE), which is used as a petrol additive.
- Ethyl benzene hydroperoxide, obtained from ethyl benzene, is reacted with propylene to produce propylene oxide and (1-hydroxyethyl)benzene. The latter is then dehydrated to produce styrene monomer (SM).

Figure 6-4   Co-oxidation routes to propylene oxide

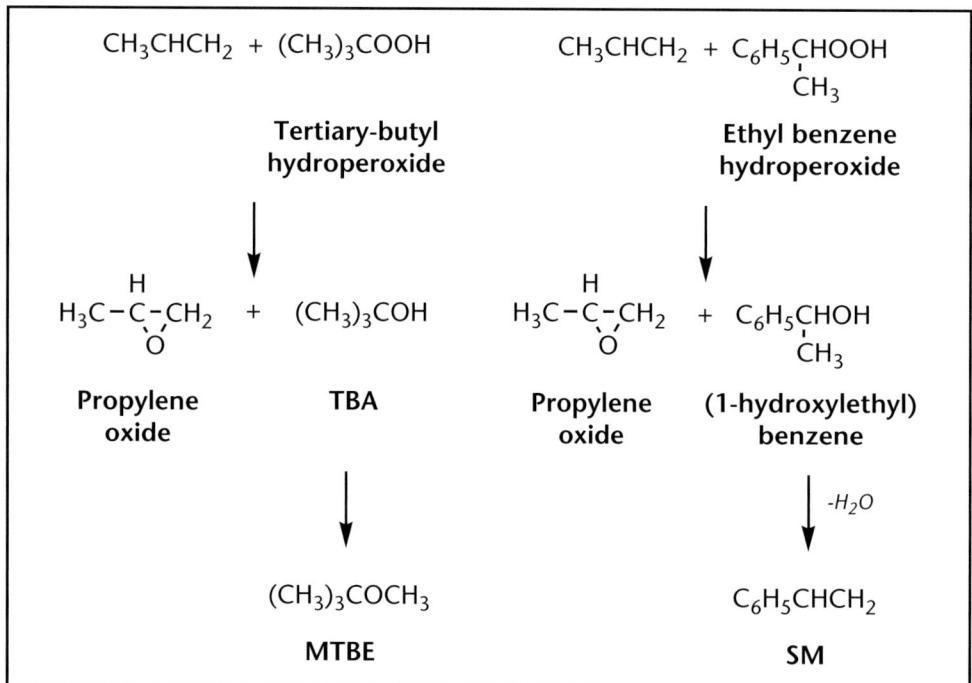

The production of propylene oxide by co-oxidation was first operated commercially in the early 1970s and is becoming the dominant process as new capacity is installed, Figure 6-5.

Figure 6-5   Propylene oxide production process trends

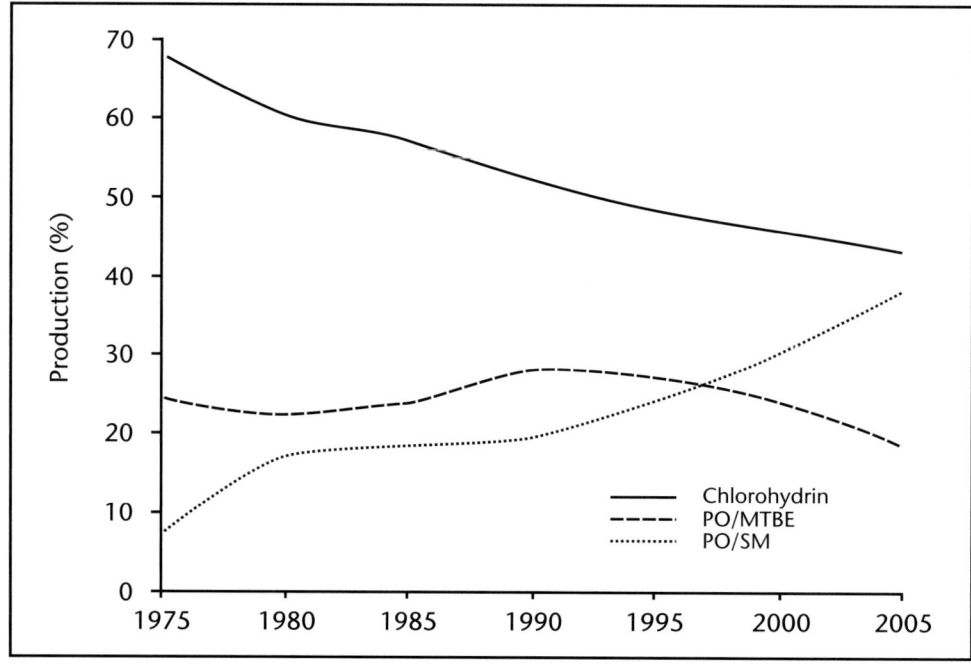

In both processes the by-products, TBA and SM monomer, are produced respectively in amounts between 3 to 3.5 and around 2.5 times that of propylene oxide. Co-oxidation plants need to be large, producing 200 to 250 thousand tonnes of propylene oxide a year. They are capital intensive and consequently are only operated by a limited number of producers.

The relative economics between PO/SM and PO/MTBE processes is dependent on the value of the co-product, whereas the chlorohydrin process is sensitive to energy costs. In general both co-oxidation routes are more competitive than the chlorohydrin process. Alternative routes, using hydrogen peroxide, cumene hydroperoxide or hydro oxidation which have the advantage of not producing a co-product, have been described in the patent literature and may become more significant in the future.

Propylene oxide polyether polyols

The manufacture of polyether polyols from propylene oxide involves the polyaddition of propylene oxide to an initiator in the presence of a catalyst followed by a catalyst treatment stage. The manufacturing chain is shown in Figure 6-6.

Figure 6-6   Polyether polyol manufacturing process

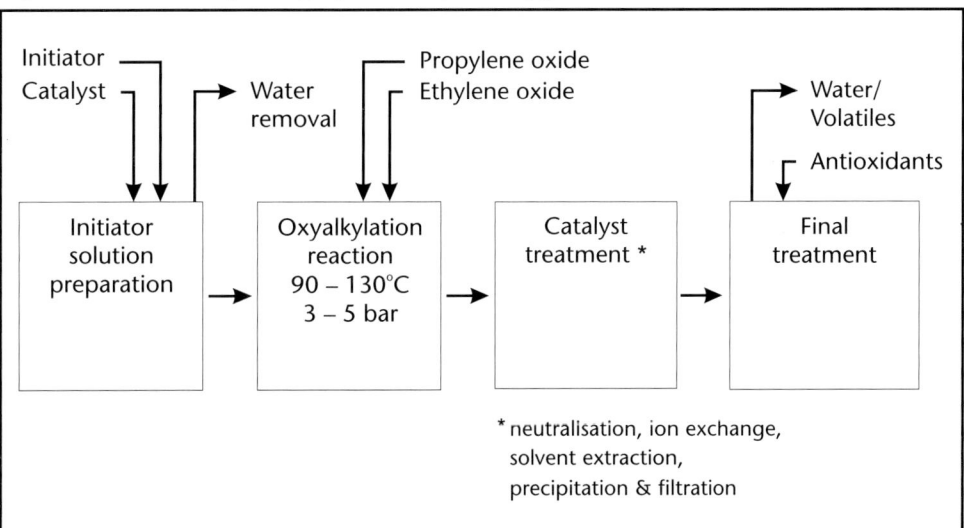

The reaction offers considerable scope to design polyols to meet the specific needs of end applications. Options available include:

- Wide choice of initiators (alcohols and amines).
- Control of functionality (two to eight).
- Polymerisation level (molecular weights up to 10,000).
- Reactivity of the hydroxyl group (primary or secondary).
- Process conditions.

The common initiators for polyether polyol manufacture are summarised in Table 6-1.

Table 6-1   Initiators for polyether polyol production

Polyol starter	Functionality	Polyol starter	Functionality
Carbohydrate sources		**Amine starters**	
Sucrose	8	Alkanolamines (e.g.	
Sorbitol	6	mono-, di-, triethanolamine)	3
Methyl glucoside	4	Ethylene diamine	4
		Diethylene triamine	5
Aliphatic starters		Toluene diamine	4
Glycols	2	Diaminodiphenylmethane	4 – 5.5
Glycerol	3	Mannich bases	3 – 7
Trimethylolpropane	3		
Pentaerythritol	4		

Three groups of catalysts will polymerise propylene oxide, with different reaction mechanisms and end products. The mechanisms are anionic (base catalysis), cationic (acid catalysis) and co-ordination rearrangement. Traditionally, the anionic catalyst, potassium hydroxide (used at 0.2 to 1.0 per cent on the final weight of the polyol) has been used for the production of polyether polyols.

Recently, caesium hydroxide and double metal cyanide catalysts have been commercialised for polyether polyol production whilst some rigid polyols are manufactured using tertiary amine catalysts.

Anionic polymerisation with potassium hydroxide

Mixing the selected initiator with a concentrated aqueous solution of potassium hydroxide, usually one potassium ion to 10 to 50 hydroxyl groups, produces a mixture (the initiator solution) containing the potassium salt of the initiator.

Since water is a di-functional initiator, its level in the initiator solution needs to be controlled. For many polyols the water is stripped out prior to propylene oxide addition to minimise propylene glycol formation.

Propylene oxide is added to the initiator solution, in the absence of oxygen, under pressure (three to five atmospheres) at reaction temperatures in excess of 90°C (typically 105 to 120°C). The reaction is exothermic, requiring heat removal.

The reaction is of the SN_2 type, with nucleophilic attack of the alkoxylate group on either of the two carbons of the oxirane ring, followed by ring opening, Figure 6-7.

The reaction occurs preferentially (95 per cent) at the less sterically hindered and more electrophilic primary carbon atom, and in consequence the commercial alkali-catalysed process results in propylene oxide polyols containing almost

Figure 6-7   Ring opening of propylene oxide

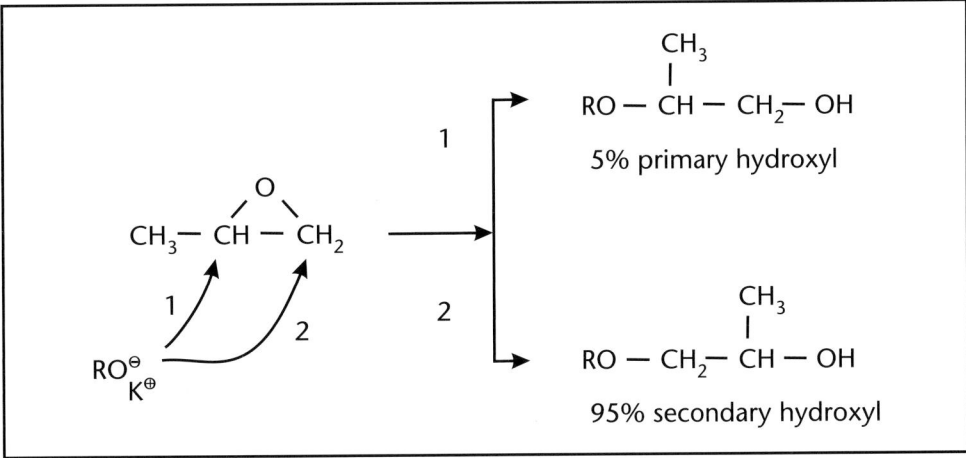

exclusively secondary hydroxyl groups which are much less reactive with isocyanates than the less hindered primary hydroxyl groups. It also means that the polymer backbone consists of repeating 'head-to-tail' units.

Proton transfer between the hydroxyl and alkoxide groups is very rapid and much faster than the rate of propylene oxide addition. This results in propylene oxide addition taking place with equal probability over all the end groups, producing a polymer with a much narrower molecular weight distribution compared to polyesters.

Addition of propylene oxide is continued until the desired molecular weight is reached. During polymerisation, the volume of product in the reactor increases and for a typical low molecular weight polyol, used for rigid foams, the build up ratio (final volume of polyol to volume of initiator) is between 2:1 and 4:1 and the polymerisation can be completed in a single step. For higher molecular weight polyols, used in flexible foam applications, the build up ratio is significantly higher, between 30:1 and 85:1 and polymerisation is carried out in at least two stages, initially preparing an intermediate of molecular weight around 500 to 1,000, and then reacting this in a second step to give the final polyol.

At the end of propylene oxide addition the reactor is under pressure and contains un-reacted propylene oxide dissolved in the polyol phase. As this reacts away the pressure in the reactor gradually decreases – a step referred to as 'cook-down'. In some processes a stripping stage may be included during or at the end of 'cook-down' either to reduce production time or to remove residual levels of alkylene oxide.

Reaction times vary considerably dependent on the polyol type, with low molecular weight ethylene diamine initiated polyols being completed in two to three hours, whilst for more complex high molecular weight polyols the reaction may take 12 to 24 hours.

Product purification

Once the desired molecular weight is obtained the polyol has to be treated to either neutralise or remove the catalyst.

For polyols that are used in fast reacting rigid systems the catalyst may be neutralised with an acid. Acetic acid is commonly used and leaves residual potassium acetate in the polyol, which is a strong catalyst in the subsequent reaction with isocyanates to form polyurethane.

In other rigid polyols, polyols used for prepolymer manufacture and all polyols for flexible foams and elastomers, the polyol is treated to remove the catalyst using one or more of the following techniques:

- Neutralisation with an acid producing an insoluble salt which is removed by filtration.
- Passing the polyol through an anionic ion exchange column.
- Adsorption, most commonly with magnesium silicate.
- Extraction with water combined with techniques to aid the separation of the polyol and aqueous layer such as the addition of toluene or hexane.

Polyols for rigid foams can accept 50 to 100 ppm of residual potassium whilst flexible foams require much lower levels, typically below five ppm.

In many processes, volatile species, a potential source of odour, are removed and an antioxidant is added to the final polyol.

Side reactions

There is a highly undesirable side reaction that occurs leading to the production of terminal unsaturated, monofunctional alcohols (mono-ols). This results from the rearrangement of propylene oxide to form allyl alcohol, Figure 6-8. This then reacts further with propylene oxide, giving chain extension to form a mono-ol. The presence of such mono-functional species is undesirable in polyurethane reactions since they result in chain termination, which interferes with the build-up of the polymer network.

The level of unsaturation is not very significant in low molecular weight polyols. However as the molecular weight rises the level grows and also the rate of formation, relative to chain extension, increases. Certain reaction conditions, such as higher temperatures, lead to a faster formation of unsaturation compared to polyol chain extension. Thus, the selection of the reaction conditions involves a balance between the speed of production and achieving the lowest possible unsaturation. This side reaction limits the molecular weight of polyols that can be produced by potassium hydroxide catalysis to around 8,000.

Figure 6-8   Formation of unsaturation in polyether polyols

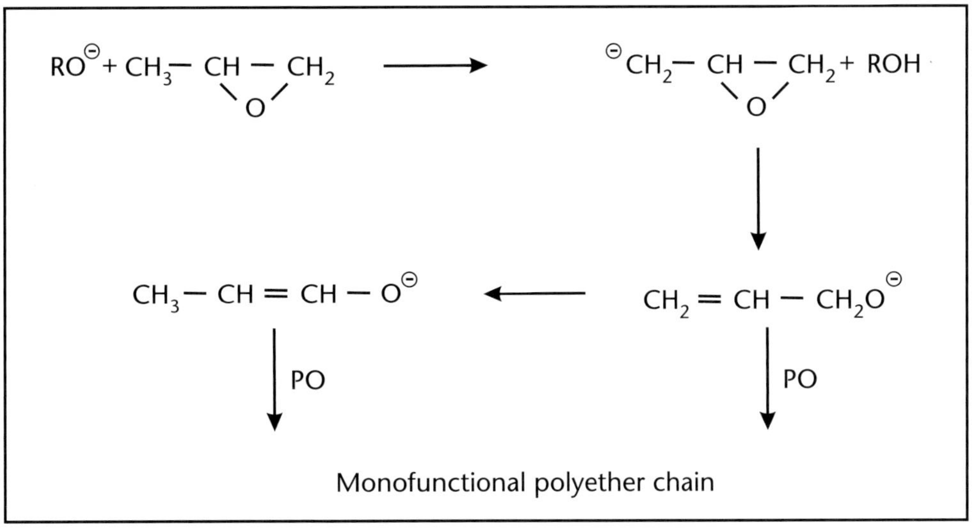

$$RO^\ominus + CH_3-CH-CH_2 \longrightarrow {}^\ominus CH_2-CH-CH_2 + ROH$$
$$\diagdown O \diagup \phantom{\longrightarrow {}^\ominus CH_2-CH}\diagdown O \diagup$$

$$CH_3-CH=CH-O^\ominus \longleftarrow CH_2=CH-CH_2O^\ominus$$

$$\downarrow PO \qquad\qquad\qquad\qquad \downarrow PO$$

Monofunctional polyether chain

The allyl group can rearrange to a propenyl group, which is less stable and under acidic conditions that may occur in the purification step can be cleaved to yield a difunctional polyol and propenal. This may also contribute to the formation of odorous by-products such as unsaturated aldehydes, cyclic acetals, substituted dioxanes and others.

Alternative catalysts

Caesium hydroxide
Caesium hydroxide is a more powerful catalyst than potassium hydroxide and gives less side reaction during polymerisation, providing polyether polyols with significantly lower unsaturation. However, the high cost, about 100 times that of potassium hydroxide, has severely limited the use of caesium hydroxide and restricted its use to the manufacture of polyols for specialist applications in the coatings and adhesives market and areas where polymers with improved dynamic performance characteristics are required such as in automotive seating.

Acid (cationic) catalysis
Propylene oxide polymerisation can also be catalysed at low temperatures by strong acids, typically Lewis acids such as boron trifluoride, but the reaction is much less regiospecific than the base catalysed process leading to polyols with equal numbers of secondary and primary hydroxyl groups. These catalysts also promote side reactions, producing high levels of impurities, such as dioxane and dimethyl dioxanes. Consequently, acid catalysis tends to be confined to situations where a base catalyst is precluded, such as the polymerisation of tetrahydrofuran, epichlorohydrin and trichlorobutylene oxide.

Double metal cyanide
Polyols derived from double metal cyanides, which are co-ordination rearrangement catalysts, were first developed in the 1960s, but the technology

has only recently been exploited commercially. Double metal cyanide catalysts can provide reaction rates around 1,000 times greater than are obtained with potassium hydroxide and fewer by-products are produced since the formation of unsaturation is greatly suppressed. The catalyst is used at very low levels (20 to 250 ppm) and consequently does not need to be removed from the final polyol. However, the catalyst is highly susceptible to contamination and cannot be used effectively for ethylene oxide tipping. It is currently used for the production of propylene oxide homopolymers, particularly for the coatings and adhesives market, and for polyols containing some ethylene oxide units for the conventional slabstock market, where it is able to provide significantly increased polyol output.

Structure

The average structure of a polyether polyol can be determined from the initial formulation plus measurement of the hydroxyl value, Figure 6-9, and the level of unsaturation of the final polymer, Figure 6-10. The relationship between the hydroxyl value and molecular weight and equivalent weight and functionality is shown in Figures 6-12 and 6-13. Another key parameter is acid value, Figure 6-11.

For low molecular weight polyols the addition of propylene oxide to a water stripped, base catalysed initiator provides a polyol close to the theoretical structure.

However, as the molecular weight is increased, significant deviations occur. This can be illustrated by looking at the composition of a typical commercial 16 per cent ethylene oxide tipped, oxypropylated glycerol polyol of hydroxyl value 35 mg KOH/g, used for the manufacture of high resilience foam. The polyol will typically have an unsaturation value of 0.05 meq/g and would have been produced with propylene oxide containing 50 to 100 ppm of water.

Although this polyol is assumed to be a triol with a molecular weight of 4,800, it will contain mono-ol species, resulting from the allyl chain transfer reaction and diols, formed from the reaction with water. So, in reality the polyol will comprise about 72 mole-% of the triol species, around 20 mole-% mono-ols and 8 mole-% diols – and have an actual (number average) functionality of 2.5 and actual number average molecular weight of 4,000. Figure 6-14 shows a calculated molecular weight distribution of such a polyol generated by modelling.

Figure 6-9

Hydroxyl value
A measure of the concentration of hydroxyl groups in a polyol, expressed as the milligrams of KOH equivalent to the hydroxyl groups in one gram of the polyol.
Units: *mg KOH/g*

Figure 6-10

Unsaturation value
A measure of the concentration of unsaturated end groups (allyl and propenyl) in a polyol, expressed as milli-equivalents of the unsaturated species in one gram of the polyol.
Units: *meq/g*

Figure 6-11

Acid value
A measure of the residual activitiy in a polyol, expressed as the milligrams of KOH needed to neutralize the acidity in one gram of the polyol.
Units: *mg KOH/g*

Figure 6-12

Relationship between hydroxyl value and equivalent / molecular weight

Equivalent weight = 56.1 x 1,000 ÷ Hydroxyl value
(number average)

Molecular weight = Equivalent weight x functionality
(number average)   (number average)   (number average)

Figure 6-13

Polyol functionality
The average number of hydroxyl groups per molecule in the polyol. The functionality of polyols for polyurethanes applications is in the range of two to eight.

Figure 6-14   Modelled molecular weight distribution of a polyether polyol

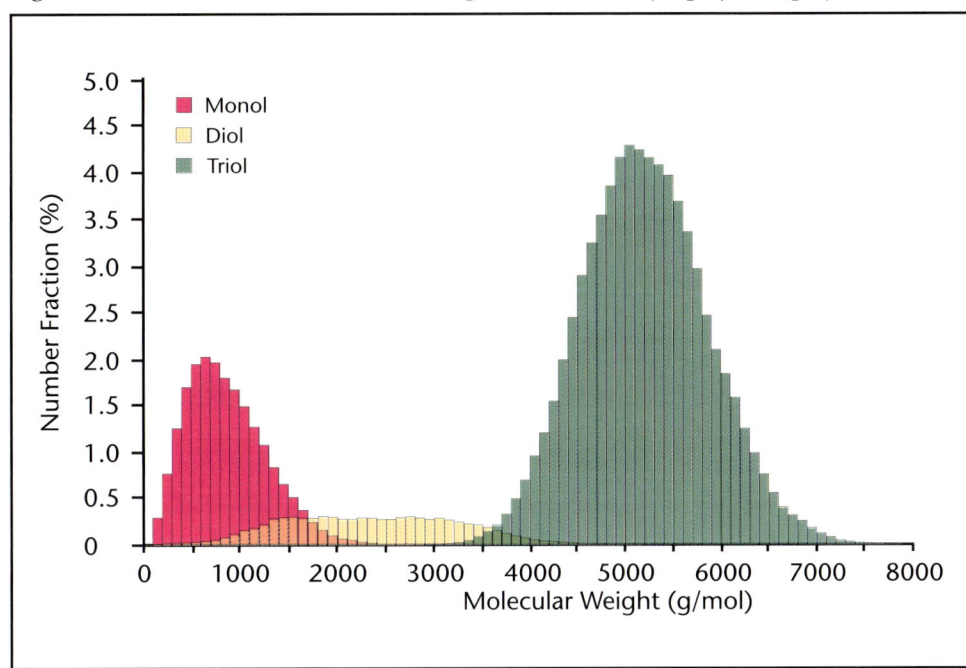

Such polyols continue to meet the requirements of the industry – most critically providing cost effective solutions. However, the need for improved performance, particularly lower density and greater durability, will continue to drive development towards optimum polyol structures.

Propylene oxide–ethylene oxide polyether polyols

Ethylene oxide can be used in conjunction with propylene oxide to produce a variety of copolymer polyol structures:

- As an end-cap (or tip).
- As a block in the polymer chain.
- As a 'random co-polymer', formed by polymerising ethylene oxide and propylene oxide together.
- As a combination of two or more of the above structures.

Ethylene oxide is incorporated into a polyol for a variety of reasons:

- To reduce its viscosity.
- To increase its reactivity through the introduction of primary hydroxyl groups.
- To adjust its miscibility in the subsequent polyurethane formation reaction.

Using ethylene oxide to reduce viscosity is most commonly exploited in rigid foam polyols. A polyol produced from four moles of propylene oxide, initiated with ethylene diamine, has a viscosity of 30,000 mPa.s at 25°C, whilst replacing one mole of propylene oxide with ethylene oxide reduces the viscosity to 17,000 mPa.s.

Ethylene oxide is symmetrical and will always ring open to give a primary hydroxyl group, which is more reactive with isocyanate than a secondary hydroxyl, and where higher reactivity is required, polyols are usually tipped with ethylene oxide. Prime examples are the polyols used to make high resilience flexible foam. These are 4,500 to 6,000 molecular weight polyols, manufactured in two stages; first propylene oxide is polymerised to an intermediate homopolymer, which is then reacted in a second stage with ethylene oxide at a level of between 10 and 20 per cent of the total alkylene oxide.

It is not possible to prepare a polyol with 100 per cent primary groups by ethylene oxide tipping, since each addition of an ethoxy unit generates a primary hydroxyl group that is more reactive to further ethylene oxide addition than the secondary hydroxyl groups of the initial intermediate. Consequently, the ethoxy units tend to build up longer chains on a few secondary hydroxyl groups rather than distributing evenly over all the polymer chains.

Reaction conditions influence the build up of primary hydroxyl content, with faster rates of addition resulting from a high catalyst level or increased polymerisation temperature leading to higher primary hydroxyl contents. The highest level of primary hydroxyl content that can be achieved is around 90 per cent, and that is only reached after the addition of seven to eight moles of ethylene oxide per secondary hydroxyl group in the intermediate. Figure 6-15 shows the build up of primary hydroxyl content in polyols with increasing ethylene oxide levels under typical reaction conditions.

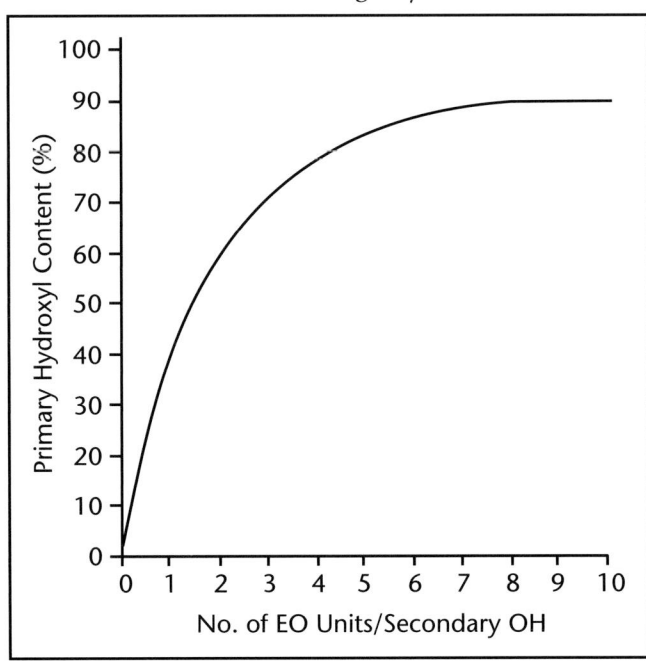

Figure 6-15   Build up of primary hydroxyl content with increasing ethylene oxide addition

Ethylene oxide can also be incorporated as a random comonomer into the backbone of a polyol to alter the miscibility characteristics. This can be critically important for certain polyurethane processes, such as conventional flexible slabstock production, that require the mixing of several components that differ considerably in their amount, physical and chemical properties, and mutual solubility. These polyols are made by polymerising ethylene oxide and propylene oxide simultaneously. The higher reactivity of the primary versus secondary hydroxyl group plus the higher reactivity of ethylene oxide versus propylene oxide results in ethoxy units in the polyol being intermingled with propoxy units in the polyol chain, but with the end group predominantly (95 per cent) being a propoxy, secondary hydroxyl, unit.

Standard polyether polyols

Rigid foam applications

There are significant regional differences in the type of rigid polyol selected. In Europe and Asia, almost all rigid foam is produced from polyether polyols. However, in the USA, whilst polyether polyols are used, the preferred polyols for the large rigid foam boardstock sector are low-cost aromatic polyester polyols.

Rigid foams have a highly cross-linked network, and the polyether polyols used are generally short chain oligomers with functionalities between three and eight. A wide range of initiators, both alcohols and amines, are used with sucrose being the commonest, but other significant products are: sorbitol, glycerol, diethylene glycol, aromatic amines (ortho-toluene diamine, diaminodiphenylmethane), aliphatic amines (ethylene diamine), alkanolamines (di- and tri-ethanolamine). Sucrose, with its low cost, ready availability and high functionality (eight) is an ideal candidate as a rigid polyol initiator.

However, sucrose is a solid, which discolours at temperatures above 130°C, presenting reaction problems for propylene oxide addition. The preferred methods are to oxypropylate sucrose in solution/suspension with a second initiator, such as glycerol/diethanolamine, or to use a 'heel' of polyol from the previous batch. The co-initiator reduces the functionality and the viscosity of the polyol and allows the polyol to be tailored to meet application requirements. For applications requiring high functionality, sucrose can be first partially oxypropylated in aqueous solution suspension, the water and low molecular weight glycols removed, and the oxypropylation then completed.

In Europe, sorbitol polyols are extensively used in preference to sucrose polyols for high temperature applications, such as district heating pipe insulation, for their superior thermal properties.

Aromatic amine-based polyols are used mainly in blends with sucrose polyols. The most common initiator for this group is ortho-toluene diamine, a by-product of the intermediate amine in the production of toluene diisocyanate. Diaminodiphenylmethane, the amine precursor in the production of MDI, is another initiator. The main benefit of these polyols is that they improve the miscibility of polyol blend and isocyanate, significantly improving processability during foaming.

Mannich polyols are a unique range of polyols produced firstly by the condensation of phenol or nonylphenol with formaldehyde and diethanolamine to produce a Mannich base (the Mannich reaction, Figure 6-16), followed by alkoxylation with propylene oxide and/or ethylene oxide.

Theoretically, a seven functional hydroxyl molecule can be produced from phenol and a five functional molecule from nonylphenol, but in practice commercial products have functionalities in the range of three to five. The Mannich polyols have high reactivities and are particularly suitable for cold climate spray applications.

Figure 6-16   Process for the production of a Mannich base

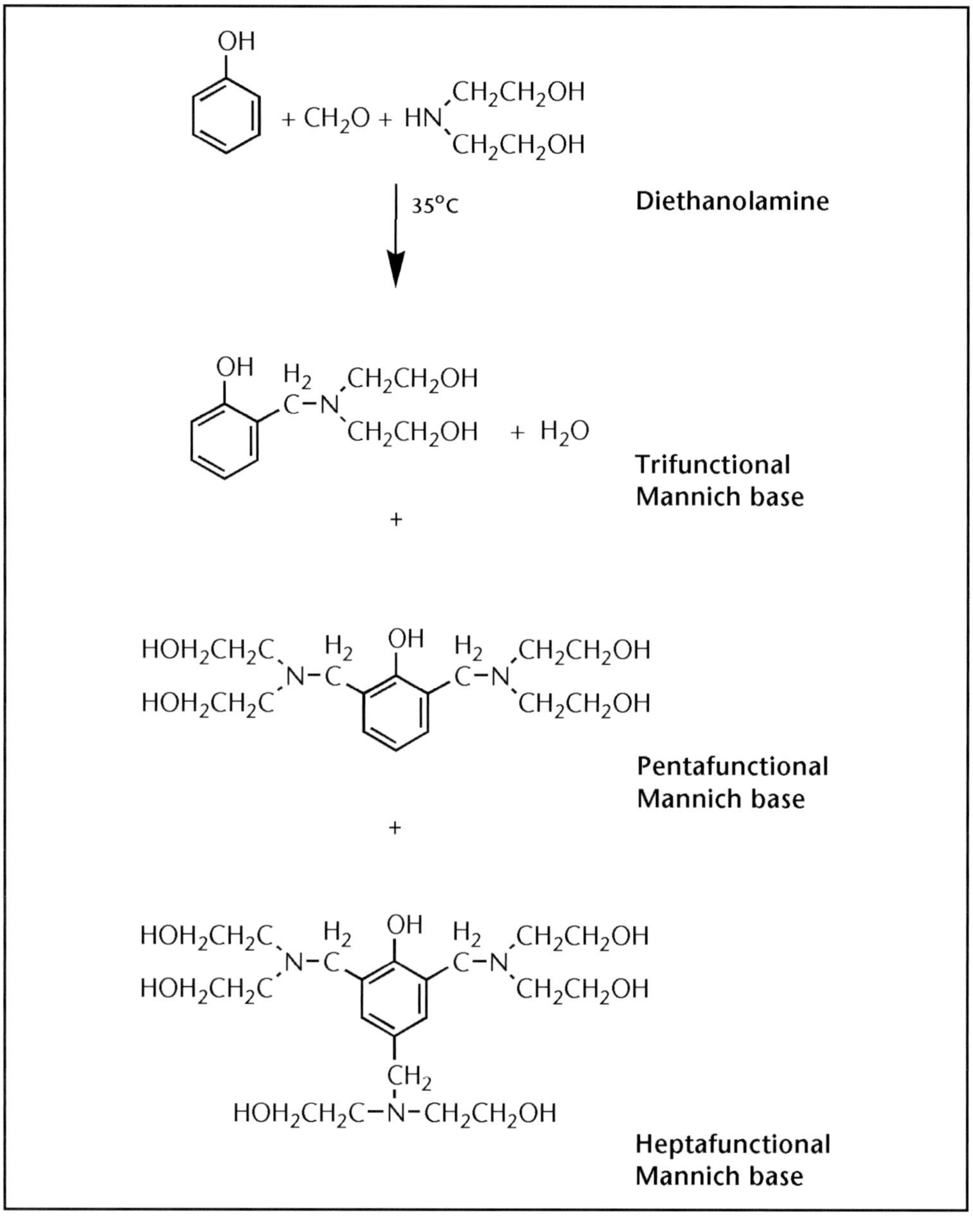

Flexible foam applications

Flexible foams are by far the largest outlets for polyether polyols.

Conventional slabstock applications

The first polyether flexible slabstock foams were produced by a two-stage 'isocyanate prepolymer' process, using a polyether polyol, typically a 2,000 molecular weight polypropylene glycol. Following the development of silicone surfactants and organo-tin catalysts, in the late 1950s, prepolymer systems were rapidly superseded by the 'one-shot' process, in which all ingredients (polyols, isocyanates, blowing agents and other additives) are mixed together simultaneously.

The first polyols used in the 'one-shot' process were 3,000 molecular weight oxypropylated triols, mainly based on a glycerol initiator and such polyols are still used in the industry today. By 1965 the benefits of incorporating some ethoxy units into the backbone of the polyol in non-terminal positions had been recognised as it improved miscibility. Today, the most commonly-used polyols for conventional slabstock foam are 3,000 to 3,500 molecular weight poly(ethylene/propylene oxide) copolymer glycerol-based triols containing 8 to 15 per cent ethoxy units with about 95 per cent secondary hydroxyl end groups.

Hot-cure moulding

'Hot-cure moulding' was the first process developed for the production of moulded flexible foam articles, such as car seat backs.

Unlike the slabstock process, where the exotherm of the reaction needs to be controlled, heat loss is a problem in hot cure moulding and therefore the polyols need to have higher reactivity than those used for slabstock foam. Typical products are glycerol initiated 3,000 molecular weight ethylene oxide tipped polyols, with a primary hydroxyl content of around 40 to 50 per cent. Some polyols also have ethylene oxide included in the polymer backbone to improve the miscibility of the polyol in the foaming formulation.

High-resilience flexible foam applications

For high-resilience flexible foam applications, both for slabstock and moulding processes, higher molecular weight triols are employed, typically in the range 4,500 to 6,000 nominal molecular weight. The products are tipped with between 13 and 20 per cent ethylene oxide, to give a polyol with primary hydroxyl content in the range 80 to 90 per cent.

High ethylene oxide containing polyols

Polyols produced from very high levels of ethylene oxide are used as cell openers in high resilience flexible foam formulations and for the production of special foams, such as super soft grades. The products are 4,000 to 5,000 molecular weight and usually contain up to 75 per cent ethylene oxide, which is the limit to ensure liquidity at ambient temperature.

Modified polyether polyols

Graft dispersions

Polyols containing organic fillers are extensively used to provide foams with increased hardness or load-bearing properties, that also provide improved processing by aiding cell opening at the end of rise in flexible foam production. There are three classes of polymer modified polyether polyols available commercially, each having a different dispersion:

- Polymer polyols – polyacrylonitrile/polystyrene (SAN) dispersions.
- 'Polyharnstoff dispersions' (PHD) polyols – polyurea dispersions.
- Polyisocyanate polyaddition (PIPA) polyols – polyurethane dispersions.

All are free-flowing liquids containing a stable, dispersed phase of finely-divided organic solids.

Polymer polyols

Polymer polyols with solids levels up to 45 per cent are commonly used in combination with a standard flexible foam polyether polyol, so that the overall solids content of the formulation is typically 5 to 15 per cent. For some heavy-duty applications such as low density moulded automotive seating and carpet underlay, higher levels are used.

Polymer polyols are produced, using batch and continuous processes, by the *in-situ* polymerisation of vinyl monomers, usually either acrylonitrile or mixtures of acrylonitrile and styrene, in a carrier polyol in the presence of a free radical catalyst, such as azobisisobutyronitrile or benzyl peroxide at temperatures around 80 to 90°C. The choice of carrier polyol is determined by the end application and is normally similar to the standard polyether polyol used. Apart from free polyol and dispersed polymer, the polymer polyol contains a third and very important species – graft co-polymer, which is made separately, for example by reacting a vinyl monomer onto the polyether backbone. Careful control of the reaction conditions is required to obtain stable dispersions with the required particle size distribution, in the range of 0.1 to 5 micron.

The introduction of styrene resolved the problems of polyol and foam discolouration. However, it is difficult to make stable polymer polyols as the level of styrene is increased, and 80 per cent styrene (based on the total vinyl monomers) is the maximum found in commercial products. Particular attention has to be paid to ensuring low levels (typically less than five ppm) of residual monomer, in polymer polyols to avoid odour and emissions problems, both in foaming operations and finished products.

It can be difficult to meet some fire standards with SAN polymer polyols, and better performance can be achieved with PIPA and PHD polyols.

PHD polyols

The second group of filled products has become known as the PHD polyols (an abbreviation of the German 'Polyharnstoff Dispersion'). These products are formed by the *in situ* reaction of a diisocyanate (usually TDI) with a diamine in a carrier polyether polyol, to give a stable dispersion of polyurea particles.

The stability of the dispersion results from the presence of an adduct, which is formed by the reaction of the diisocyanate with one hydroxyl group of the carrier polyol, followed by polymerisation of the diamine and isocyanate to form a polyol-urethane linked polyurea block copolymer, Figure 6-17.

Figure 6-17   Stabilisation of PHD polyols

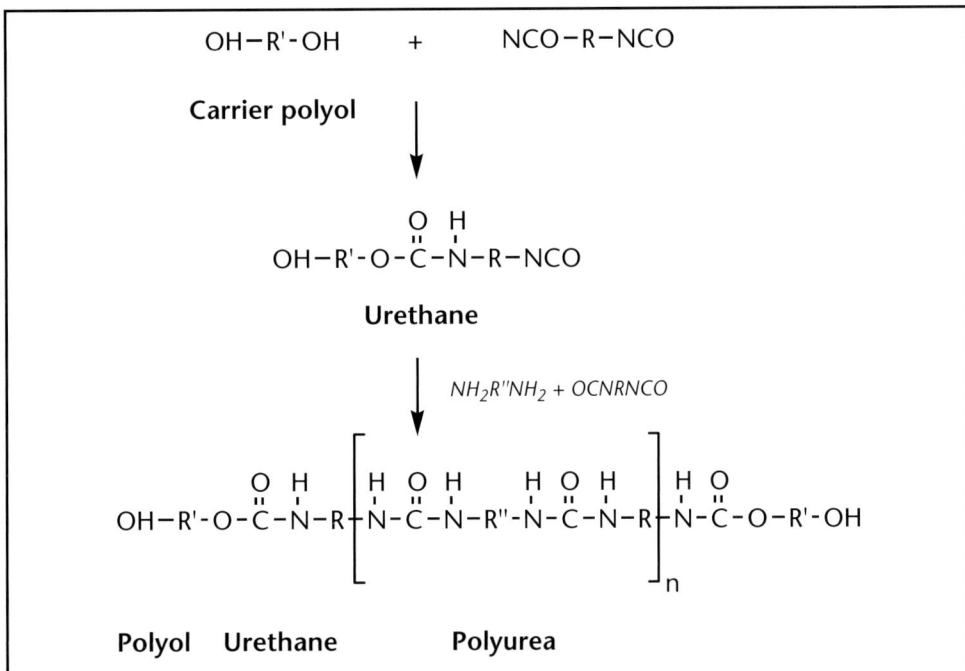

Tightly controlled process conditions are essential to balance the two reactions to form the required amount of the block copolymer (typically two to five per cent), to ensure good storage stability of the dispersion. However, the block copolymer increases the viscosity and this limits the manufacture of PHD polyols to a solids content of 30 per cent in order to maintain a reasonable viscosity.

PIPA polyols

PIPA (polyisocyanate polyaddition) polyols, are very similar to the PHD polyols except the dispersion is formed by the in situ reaction with isocyanate of an alkanolamine, instead of a diamine, to give a polyurethane dispersion in carrier polyol. The preferred amine is triethanolamine, and PIPA polyols based on both MDI and TDI are commercially available.

Amine-terminated polyethers

The hydroxyl groups of a polyether polyol can be converted in high yield (around 98 per cent) to amine groups by reductive amination. The polyether polyol is reacted with a mixture of ammonia and hydrogen in the presence of a nickel catalyst at temperatures in the range 190 to 230°C at pressures between 70 and 140 bar.

The resulting products are highly reactive to isocyanates and in the late 1980s there was significant interest in their use in RIM elastomers, which have been used to produce automotive body panels. This application has not developed significantly and the current main applications for amine-terminated polyethers are in epoxy curing agents (particularly in the E-coat for automotives), polyurea spray systems, and as fuel additives (for keeping the combustion chamber and valves clean). The most common amine-ended polyethers are based on oxypropylated diols of molecular weight 1,000 to 4,000.

Polyether polyols from tetrahydrofuran

Polytetrahydrofuran (PTHF), also called polytetramethylene ether glycol (PTMEG), is manufactured by the cationic polymerisation of tetrahydrofuran. The five-membered tetrahydrofuran ring is more stable than the three-membered rings of ethylene or propylene oxide, and cannot be polymerised using alkali catalysts. Acid catalysts, such as fluorosulphonic acid, are effective and at the completion of polymerisation the resulting polymer is hydrolysed to replace the SO_3F with hydroxyl end groups, Figure 6-18.

Figure 6-18   Manufacture of polytetrahydrofuran

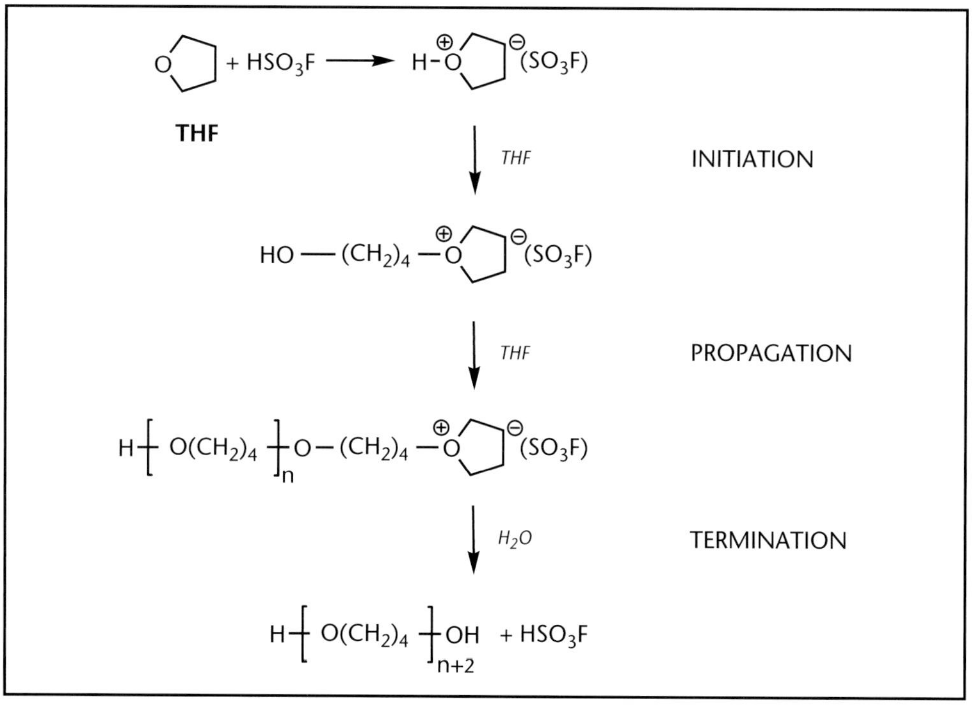

PTHF is available commercially with molecular weights in the range 650 to 3,000 and the polyols are waxy solids that melt into clear liquids at temperatures in the range 28 to 40°C.

PTHF is a premium product and is used in applications where the end effect justifies the higher cost. The particular benefits that PTHF confers on polyurethane products are:

- Excellent hydrolysis and microbial resistance (particularly compared to polyester polyols).
- Excellent dynamic properties, including resilience.
- Retention of elasticity at low temperatures.

Consequently, the most important applications of PTHF are:

- Polyurethane fibres, which take advantage of the resistance to hydrolysis.
- Cast elastomers requiring superior mechanical properties, such as conveyor belts, bearings etc.
- Thermoplastic polyurethanes (TPU) for low temperature.

Polyester polyols

There are four main classes of polyester polyols:

- Linear or lightly branched aliphatic polyester polyols (mainly adipates) with terminal hydroxyl groups.
- Low molecular weight aromatic polyesters for rigid foam applications.
- Polycaprolactones.
- Polycarbonate polyols.

The worldwide demand for polyester polyols in the polyurethane industry is estimated at around 850,000 tonnes, growing at four to five per cent a year and is broken down as follows, Figure 6-19.

Applications include the manufacture of flexible foam for textile lining, where superior resistance to dry cleaning solvents, flame bonding performance, elongation and tensile properties make polyester polyols the product of choice. The outstanding abrasion resistance of polyester polyol-based polyurethanes has led to their extensive use in surface coating and footwear applications, and the superior thermal and oxidative stability of the aromatic polyesters is exploited in the manufacture of rigid isocyanurate foams.

Figure 6-19   Worldwide demand for polyester polyols by application

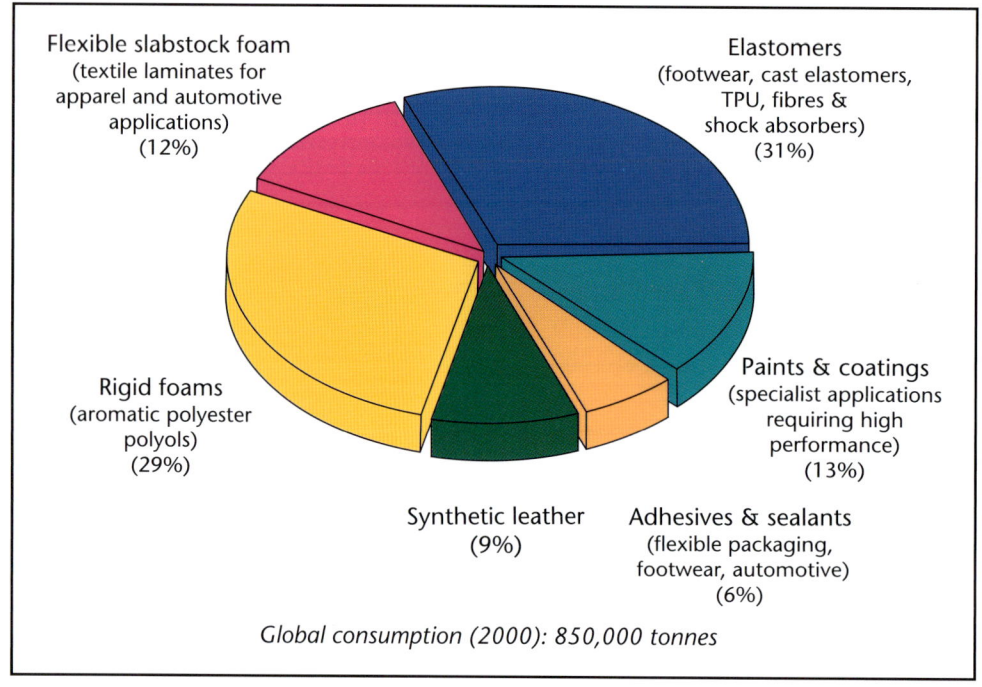

Linear or lightly branched aliphatic polyester polyols

The raw materials, Table 6-2, for the manufacture of polyester polyols are produced predominately for other applications, with only a small proportion being used in the polyurethane industry.

Table 6-2   Raw materials for the production of polyester polyols

Dibasic acids	Glycols	Branching agents
Adipic acid	Ethylene glycol	Glycerol
AGS mixed acids	Diethylene glycol	Trimethylolpropane
	Propylene glycol	Pentaerythritol
	Dipropylene glycol	
	1,4-butane diol	
	1,6-hexane diol	

Aliphatic polyester polyols are produced by direct esterification in a condensation reaction. This is a reversible equilibrium reaction, with water being removed during reaction to drive the process. As the reaction proceeds transesterification reactions also occur on the forming polymer backbone, giving rise to a relatively wide molecular weight distribution in the final polyester polyol (especially when compared to polyether polyols and polycaprolactones). Further, when polyesters are made from two or more glycols, they will be incorporated into the polymer chain in a statistical distribution irrespective of their sequence of addition. Careful control of the ratio of the ingredients is needed to ensure the product has the required hydroxyl, and not acid, end groups.

The yield is typically around 85 per cent due to the water removed during reaction whilst, by contrast, polyether polyols have yields close to 100 per cent. This lower yield, combined with the higher cost of several of the raw materials, explains the higher cost of many polyesters compared to polyethers.

The molecular weight of a polyester polyol is determined by the molar ratio of the glycols and adipic acid – the closer this is to 1:1 the higher will be the degree of polymerisation. The functionality of a polyester polyol can be increased by the introduction of triols such as glycerol or trimethylolpropane, leading to branching of the polyester backbone.

Polyester polyols are normally manufactured in a stirred batch reactor with solid adipic acid being added to the appropriate amount of glycol. The raw materials are heated in an inert atmosphere, and water distilled off through a column at reactor temperatures in the range 150 to 220°C.

Water initially distils off very rapidly, but the rate of polymerisation decreases as the concentration of acid groups decreases. Various techniques are applied to increase the end-phase reaction. These include azeotropic distillation, the application of vacuum and/or sparging with nitrogen. The reaction is closely monitored through measurement of viscosity, acid value, and hydroxyl value and continued until the acid value is typically below one mg KOH/g. Organometallic catalysts, such as tetrabutyltitanate and stannous octoate can be used to reduce reaction time.

Polyesters that are prepared from a single glycol are usually waxy solids with melting points up to 60°C. The exception is diethylene glycol that usually leads to liquid polyols. The hydrolysis resistance of polyester derived polyurethanes is increased as the length of the chains between the polyester linkages increases, the residual acidity of the polyester is reduced, the catalyst level is decreased and the branching of the polyester chains is increased. The latter factor also leads to improved resistance to swelling by solvents and oils

Polyester polyols for elastomeric applications are usually linear or lightly branched, with the latter used mainly for fast-curing shoe soling applications. The glycol selection is a compromise between cost, ease of processing and the required level of physical properties. For thermoplastic polyurethane (TPU) applications, adipates based on ethylene glycol, 1,4-butane diol and 1,6-hexane diol are most commonly used.

Polyester polyols are used for the manufacture of flexible foams in specialised applications such as fabric lining and semi-rigid foams. The complex equipment used to make continuous slabstock foams places practical limits on the viscosity of the raw materials. The most widely used polyesters are 2,000 to 3,000 molecular weight diethylene glycol adipates, usually branched with glycerol, trimethylolpropane or pentaerythritol. These products are stable liquids at ambient temperatures with viscosities around 15,000 to 25,000 mPa.s.

Polycaprolactones

Polycaprolactones are produced by the ring opening of caprolactone in the presence of an initiator and catalyst.

This gives rise to a polymer backbone structure of repeating head-to-tail units (as with polyether polyols and PTHF), in contrast to the repeating 'head-to-head, tail-to-tail' units of adipate-based polyester polyols.

The functionality of the polycaprolactone is controlled by the choice of initiator and products with functionalities between two and three with molecular weights from 250 to 4,000 are commercially available. Polycaprolactones have a much narrower molecular weight distribution than adipate polyester polyols and lower viscosities. They are relatively expensive and their use is restricted to high performance elastomer applications requiring outstanding low temperature flexibility or superior hydrolytic stability.

Aromatic polyester polyols

The use of polyesters in rigid foams was traditionally very limited, with polyether polyols being preferred. Following their introduction in the early 1980s, it was discovered that aromatic polyester polyols offered significant advantages in polyisocyanurate (PIR) rigid foam systems, where the highly cross-linked trimer structure can compensate for low functionality of the polyester polyol. Based on recycled or by-product streams, the aromatic polyester polyols are lower cost than polyether polyols and give superior performance in fire tests. They quickly became the polyols of choice in North America for the production of boardstock for building insulation; in combination with polyether polyols (typically Mannich-based) in spray systems; and in other applications, such as appliances and pour-in-place foam, as a diluent to cheapen the formulation.

Their impact has been so significant in North America, that aromatic polyester polyols now account for around 55 per cent of the polyols used for rigid foam manufacture. In Europe and Asia, aromatic polyester polyols have not yet gained the same market acceptance as in the USA, although capacity for these products has been installed in anticipation of a significant future growth in demand. Around 220,000 tonnes of aromatic polyester polyols were consumed in 2000, with North America accounting for 80 per cent of the total.

There are three types of aromatic polyester polyol used today:

- Products derived from the process residues of dimethyl terephthalate (DMT) production, commonly referred to as DMT still bottoms. They are typically transesterified at 180 to 230°C with at least one mole of diethylene or dipropylene glycol per equivalent of acid to produce a simple hydroxyl-ended, glycol-capped aromatic polyester polyol.

- Products derived from the glycolysis of recycled poly(ethyleneterephthalate) (PET) bottles or magnetic tape with subsequent re-esterification with di-acids or reaction with alkylene oxides.
- Products derived by the direct esterification of phthalic anhydride.

The polyesters have functionalities between two and three, typically closer to two, and hydroxyl values in the range 200 to 330 mg KOH/g. Compatabilisers and surfactants are often added during manufacture to reduce viscosity and to improve miscibility with blowing agents, other polyols and isocyanates.

Polycarbonate polyols

Polycarbonate polyols are a special class of polyester polyol – derived from carbonic acid – that can be produced through the polycondensation of diols with phosgene, although transesterification of diols, most commonly hexane diol, with a carbonic acid ester, such as diphenylcarbonate gives a smoother, preferred reaction.

The main advantage of polycarbonate polyols over conventional polyester polyols is that they provide significantly improved hydrolysis resistance. This is because reaction of the polycarbonate group with water produces an unstable carbonic acid derivative that decomposes to non-acidic species, thereby avoiding the further acid catalysed degradation that occurs when a conventional polyester group is hydrolysed.

Polycarbonate polyols are consequently used for the manufacture of adhesives for the shoe and packaging industries and in coating applications.

Miscellaneous polyols

The most significant of this miscellaneous group are the acrylic polyols that are prepared from hydroxyl-containing acrylate monomers such as hydroxy-ethylmethylacrylate (HEMA) by a free-radical reaction with comonomers, such as methylmethacrylate and styrene. The products, with functionalities of two to eight, are typically sold in solution or as dispersions and are used in coating applications.

The other miscellaneous polyols only have a small share of the market and consist of:

- Castor oil – the only polyol, of the numerous natural products, still in significant use. It is a three functional fatty ester of glycerol used in cast elastomers, coatings and adhesives.
- Polyols containing halogen and phosphorous – used to help meet fire regulations for some applications, mainly in rigid foams. An early example

was a propylene oxide adduct of phosphoric acid whilst epichlorohydrin can be used to introduce chlorine into a polyol.
- Hydroxyl-terminated polybutadienes – non-polar and hydrophobic polyols leading to a polyurethane with excellent hydrolytic stability and electrical resistance.
- Polyols regenerated through recycling of used flexible and rigid foam and shoe soles, described in Chapters 3 and 7.

Quality and analysis

Polyols used in the polyurethane industry are required to have an extremely high degree of reproducibility and consistency. This is achieved through tight specifications of the raw materials used for polyol manufacture, exact reproducibility of the manufacturing process conditions and vigorous analytical monitoring.

The properties that define the specification of a polyol are: hydroxyl value, water content, acid value, unsaturation value (for polyether polyols), colour, sodium and potassium content (from residual levels of catalyst in polyether polyol manufacture) and viscosity. Knowledge of the precise value of the first two of these, i.e. hydroxyl value and water content, is important in determining the appropriate ratio of polyol to isocyanate that is required for polyurethane preparation.

A comprehensive range of analytical techniques are also available, such as ^{13}C nmr spectroscopy and gel permation chromatography, that allow very detailed structural analysis, which is a valuable tool in the development and optimisation of polyols.

7. Outline of polyurethane chemistry

Joris Bosman

The formation of a urethane group from the reaction between isocyanate and hydroxyl compounds was originally identified in the 19th Century. However, it was only in the late 1930s that Otto Bayer discovered the chemistry of the polyaddition reaction between diisocyanates and diols developing urethane polymers, initially producing a range of elastomers and rigid foams.

Polyurethane formulations are generally designed to exploit several different isocyanate reactions simultaneously. Almost all the reactions are exothermic which usually provides sufficient energy to cure the polymer and to drive processing operations such as mould filling and foaming. Catalysts are used to balance the rates of the various reactions and to control the overall rate of cure during processing.

Isocyanates

The high reactivity of the isocyanate group is key to the chemistry of polyurethanes. The electronic structure of the isocyanate group makes it clear that the charge density is least on the carbon and the resonance forms are illustrated in Figure 7-1.

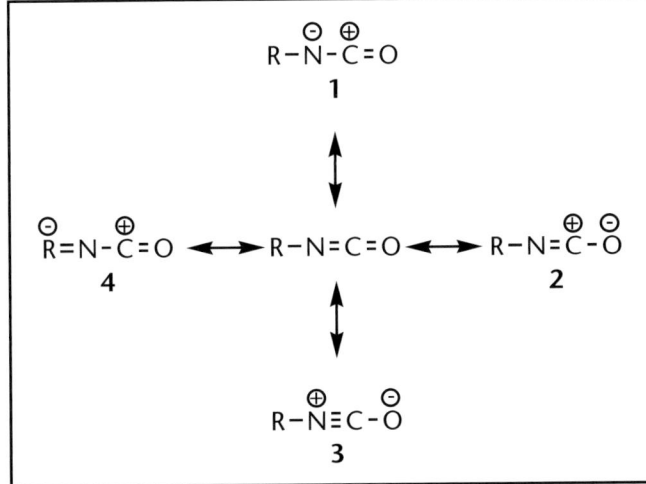

Figure 7-1   Resonance forms of isocyanates

Structure one is more important than structure two whilst the contribution of structure three is slight. The electron deficiency on the carbon explains the reactivity of isocyanates towards nucleophilic attack and since structure one contributes more than structure two, most reactions take place across the C=N bond. Structure four becomes important if R is aromatic, in which case the negative charge on the nitrogen will be distributed throughout the benzene ring, reducing further the electron charge on the central carbon of the isocyanate. This is the reason why aromatic isocyanates such as MDI and TDI are more reactive than aliphatic isocyanates like HDI and IPDI.

As a general principle, any electron-withdrawing group linked with R will increase the positive charge on carbon, thereby increasing the reactivity of the isocyanate group towards nucleophilic attack. Conversely, electron donating groups will reduce the reactivity of isocyanate groups. Consequently, TDI will be more reactive than MDI since the second isocyanate in the TDI molecule will activate the other isocyanate. However, once the first isocyanate has reacted with a nucleophile, the effect will reverse and the second one will be deactivated.

As an example, when TDI and MDI are reacted with butanol the isocyanate groups react at different speeds, Table 7-1. The first isocyanate of TDI is twice as reactive as that on MDI, but the second isocyanate of TDI is then half the reactivity of the second isocyanate on MDI.

Steric hindrance also plays a role in the reactivity of isocyanates and ortho substituents on aromatic isocyanates lower their reactivity. Examples are that 4-tolylisocyanate reacts faster than 2-tolylisocyanate and for MDI an isocyanate in the ortho 2-position is less reactive than an isocyanate in the para 4-position. With aliphatic isocyanates, for both electronic and steric reasons, primary isocyanates react faster than secondary isocyanates consequently, in IPDI, the primary isocyanate will react first.

Table 7-1
Relative reactivity of isocyanate groups in MDI and TDI with butanol

Position	TDI	MDI
2	0.08	
4	1.00	0.50
4'		0.16

In polyurethane chemistry the major focus is on the reactions of isocyanates with compounds that contain active hydrogen groups such as hydroxyl, water, amines, urea and urethane, but also the reactions of isocyanates with other isocyanates needs to be considered.

Isocyanate reactions with hydroxyl

The most important reaction in the manufacture of polyurethanes is between isocyanate and hydroxyl groups, Figure 7-2. The reaction product is a carbamate, which is called a urethane in the case of high molecular weight polymers. The reaction is exothermic and reversible going back to the isocyanate and alcohol.

Figure 7-2   Urethane reaction

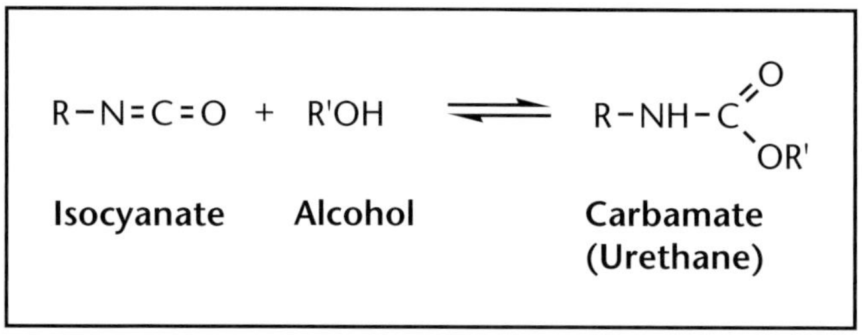

$$R-N=C=O + R'OH \rightleftharpoons R-NH-C(=O)-OR'$$

Isocyanate   Alcohol   Carbamate (Urethane)

Aliphatic primary alcohols are the most reactive and react much faster than secondary and tertiary alcohols due to steric reasons, but urethanes made from tertiary alcohols do not regenerate free isocyanate instead dissociating to yield the corresponding amine, alkene and carbon dioxide. The urethane back reaction starts at 250°C for aliphatic isocyanates, but is closer to 200°C for aromatic isocyanates.

Phenols react with isocyanate more slowly than alcohols and the resulting urethane groups are readily broken to yield the original isocyanate and phenol. A temperature of about 180°C is required for aliphatic isocyanates and 120°C for aromatic isocyanates. This reversible reaction with phenol is used in the manufacture of 'blocked' or hindered isocyanates, which are sometimes used in polyurethane coating applications. These blocked isocyanates are more moisture resistant and storage stable in comparison with free isocyanates.

The reaction of a blocked isocyanate with an active hydrogen-containing compound could involve either a two-step process in which there is an initial thermal dissociation to free isocyanate, heat initiated, or a direct bimolecular displacement reaction. Thermal dissociation generally requires heating to 120 to

180°C for 10 to 30 minutes. In an example concerning displacement, urethanes derived from aromatic diisocyanates and phenols will react with aliphatic amines at room temperature, but not with alcohols. Urethanes from aromatic diisocyanates and aliphatic alcohols are stable towards aliphatic amines under the same conditions. Other blocking agents than alcohol are known, such as beta-dicarbonyl compounds, oximes, caprolactam or dimethylpyrazole.

The reaction between isocyanates and alcohols is accelerated by the addition of catalysts such as acids, bases (mostly aliphatic tertiary amines) and metal complexes (organo tin compounds). Catalysts also promote the dissociation of urethanes and so the deblocking of blocked isocyanates can occur at lower temperature.

Isocyanate reaction with water

The reaction of isocyanates with water to produce an amine and carbon dioxide is highly exothermic. The initial reaction product is a carbamic acid, which breaks down into carbon dioxide and a primary amine, Figure 7-3. The amine will then react immediately with another isocyanate to form a symmetric urea. Due to the formation of carbon dioxide the water reaction is often used as a blowing agent as the level of blow can be tailored, simply by adjusting the amount of water in the formulation.

Figure 7-3   Water reaction

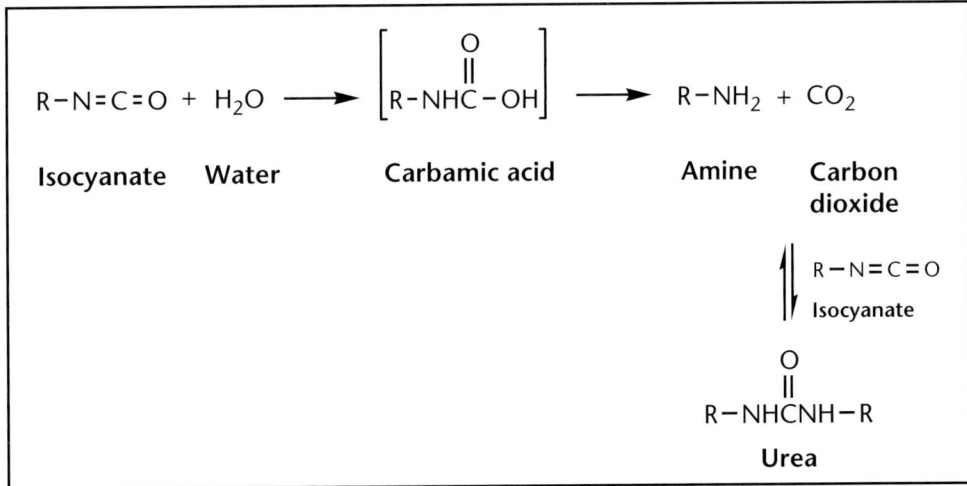

Diisocyanates having isocyanate groups of similar reactivity such as MDI, tend to chain extend to give crystalline polymeric urea. On the other hand, 2,4-TDI has an isocyanate group in the 2-position far less reactive than the one in the 4-position. Consequently, urea will be formed rapidly between TDI molecules in the 4-position leaving the 2-position unaffected, below 50°C.

Despite the high reaction exotherm the water reaction is generally slow in the absence of catalyst and one of the main reasons is that water is not very soluble in isocyanates such as MDI and TDI.

Isocyanate reaction with amines

Isocyanates react with primary and secondary amines to produce di- and tri-substituted urea respectively whilst tertiary amines form labile 1:1 adducts, but generally do not react with isocyanates, Figure 7-4.

Figure 7-4   Urea formation

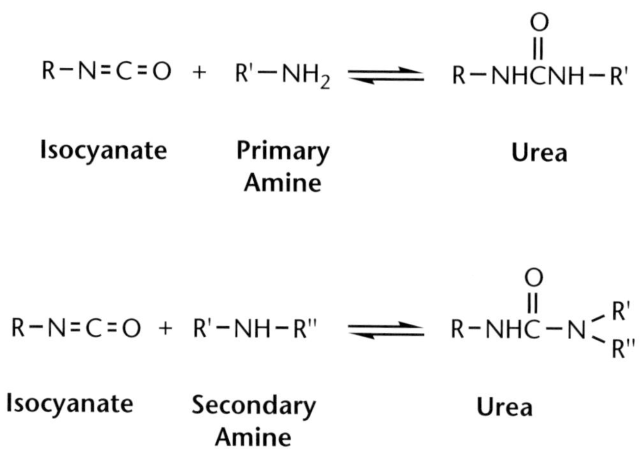

$$R-N=C=O + R'-NH_2 \rightleftharpoons R-NHCNH-R' \text{ (with C=O)}$$

Isocyanate   Primary Amine   Urea

$$R-N=C=O + R'-NH-R'' \rightleftharpoons R-NHC(=O)-N(R')(R'')$$

Isocyanate   Secondary Amine   Urea

These conversions are exothermic and diamines are used as chain extending and curing agents in polyurethane manufacture. The resulting polyurea segments increase the potential for cross-linking.

The reaction of unhindered isocyanates with primary amines at room temperature and in the absence of catalyst is 100 to 1,000 times faster than the reaction with primary alcohols. The reactivity of an amine increases with its basicity and consequently, aliphatic amines are much more reactive than aromatic amines. The reactivity of amines can be slowed down by the presence of electron withdrawing groups, another way is to increase the steric hindrance by branching on the carbon next to the nitrogen or introducing substituents in the ortho position of an aromatic amine.

The kinetics of the reaction of amines with isocyanates is complicated by strong product catalysis. Since the product urea is a much weaker base and more hindered than the amine, its catalysis is bi-functional and based on hydrogen bonds between urea and both amine and isocyanate.

Isocyanate reaction with urea

Biurets are formed from the exothermic reaction of an isocyanate with a urea. With di-substituted urea, mostly formed by the water reaction (see above) a biuret is formed through the active hydrogen, Figure 7-5.

Figure 7-5   Biuret formation and equilibria

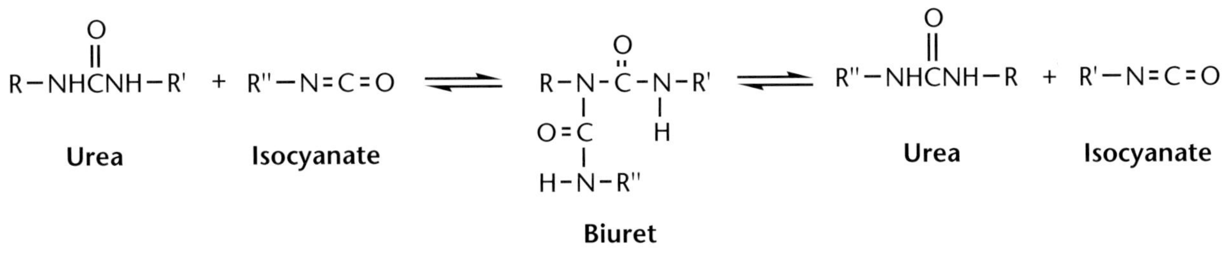

$$\text{R-NHCNH-R'} + \text{R''-N=C=O} \rightleftharpoons \underset{\text{Biuret}}{\text{R-N(C(=O)NHR'')-C(=O)-N(H)-R'}} \rightleftharpoons \text{R''-NHCNH-R} + \text{R'-N=C=O}$$

Urea   Isocyanate   Biuret   Urea   Isocyanate

This reaction is significantly faster than the allophonate reaction and occurs at lower temperature, about 100°C compared to 120 to 140°C. In polyurethane systems this reaction, that is reversible upon heating, is often a source for additional cross-linking.

Another important feature of this urea-biuret equilibrium is the potential for redistribution of the biuret across the spectrum of isocyanate species. For instance, if polymeric MDI and a diisocyanate prepolymer are mixed together then the molecules of the di, tri, tetra and higher species are not initially smoothly distributed across the spectrum of derivatives – biuret, allophonate, uretonimine. However, they slowly re-distribute through the various reversible reactions. This redistribution will be faster at higher temperatures, resulting ultimately in a product stable in composition and viscosity.

Isocyanate reaction with urethanes

An allophonate group is the result of the exothermic reaction of isocyanate with the active hydrogen on a urethane group, Figure 7-6.

Figure 7-6   Allophonate formation

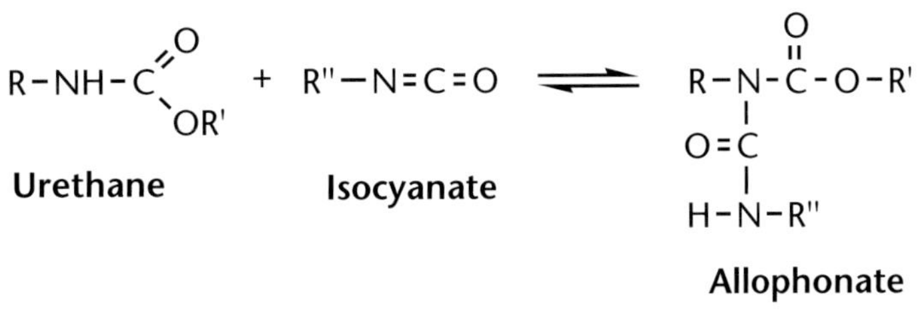

$$\underset{\text{Urethane}}{\text{R-NH-C(=O)-OR'}} + \underset{\text{Isocyanate}}{\text{R''-N=C=O}} \rightleftharpoons \underset{\text{Allophonate}}{\text{R-N(C(=O)NHR'')-C(=O)-O-R'}}$$

This reaction is slow compared to biuret formation and usually takes place uncatalysed at about 120 to 140°C. The reaction is reversible at temperatures above 150°C so, as with biurets, the reaction increases cross-linking in polyurethane systems. This reverse reaction takes place at lower temperature than with biurets so that the interchange of isocyanate homologues is faster. If the allophonates are heated with a third equivalent of isocyanate, the cyclic triisocyanurate or trimers can be obtained.

Isocyanate reactions with isocyanates

Dimerisation

Isocyanates undergo a [2+2] mildly exothermic cyclo-addition reaction across two C=N bonds resulting in a four-membered ring called a dimer or uretidinedione, Figure 7-7. With diisocyanates, polymeric dimers can be formed and these poly(diazetidiones) are insoluble crystalline polymers. The reaction can be catalysed and the most selective catalysts are the trialkylphosphines. Other base catalysts are known such as pyridine and 1,2-dimethyl imidazole and a particularly active one is dimethyl amino pyridine. The use of catalysts results in the formation of dimers at low temperature, generally as an intermediate in the formation of more stable trimers.

Figure 7-7   Dimer formation

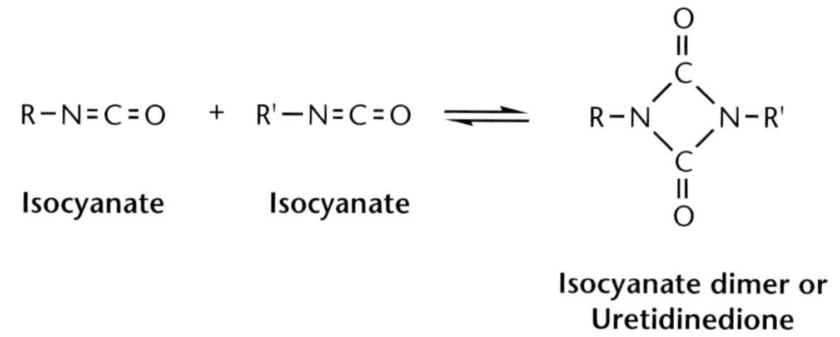

$$R-N=C=O \; + \; R'-N=C=O \; \rightleftharpoons \; R-N\underset{\underset{O}{\overset{\|}{C}}}{\overset{\overset{O}{\overset{\|}{C}}}{\diamond}}N-R'$$

Isocyanate dimer or Uretidinedione

Dimer formation mainly arises with aromatic isocyanates, but the reaction is slowed down by ortho substituents. Consequently, dimer formation is much more important for MDI than TDI. Thus 2,4- and 2,6-TDI do not form dimers at normal temperatures but 4,4'-MDI dimerises slowly when left standing at room temperature. Dimer precipitates as a fine solid as it only has a low solubility in MDI: 0.3 wt-% at 45°C and gives a turbidity and haziness to the liquid pure MDI that can only be stored for two weeks at this temperature before dimer comes out of solution.

Dimerisation can occur in prepolymers and isocyanate derivatives like uretonimines with the reaction rate and uretidione solubility dependent on product composition. The addition of water decreases the solubility of dimer in pure MDI so the dimer will come out of solution at an earlier stage. On the other hand, dimerisation is often suppressed in prepolymers and uretonimine derivatised isocyanate because of a reduced isocyanate concentration. Overall, haziness in prepolymers or isocyanate derivatives is often related to dimer content.

Even in the solid crystalline state, dimerisation of MDI occurs. Although the rate of dimerisation, Figure 7-8, increases with temperature below and above the melting point, the slope of the rate curve is higher in the solid. Presumably this is because the crystal structure is such that two NCO groups are placed in the correct conformation for formation of uretidinedione ring.

Figure 7-8   Rate of dimer formation and solubility

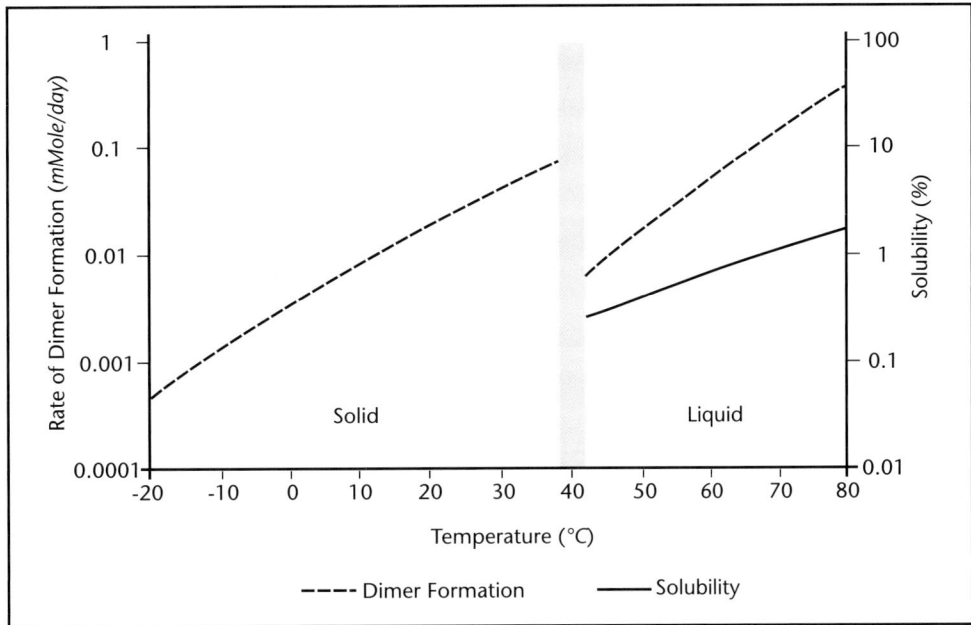

Because of the low rate of dimer formation at -20°C, pure MDI is best stored for a long period of time at this temperature and only for immediate consumption at 45°C whilst frozen MDI should be melted in the fastest possible way.

The reaction is reversible and it is the dimer, which is thermodynamically stable at room temperature although its rate of formation is extremely slow. The dimerisation rate rapidly increases with temperature, but eventually the equilibrium is shifted to isocyanate. For MDI the equilibrium mixture contains 6.3 per cent dimer at 120°C and dissociation is almost complete above 190°C, so dimer can be removed by heating MDI to 200°C followed by crash cooling to about 80°C. Dissociation of poly(diazetidinones) is much more difficult and will only take place at 270°C.

Aliphatic isocyanates do not dimerise readily although dimers have been obtained using trialkylphosphines or 1,2-dimethyl imidazole as catalyst.

Trimerisation

Three isocyanates can undergo a cyclisation reaction across the C=N bond resulting in a six-membered ring called a trimer or isocyanurate, Figure 7-9. Trimerisation is highly exothermic and continues until all NCO groups have reacted.

Both aliphatic and aromatic isocyanates readily form trimers. Under more severe conditions, polytrimerisation can occur. In general, trimerisation occurs mainly under the influence of basic catalysts, such as alkali metal alkoxides and carboxylic acid salts in particular. Other catalysts are acetates, carbonates, octoates and N,N-dimethylformamide. Organometallic compounds based on silicon and tin and transition metal elements also act as trimerisation catalysts

Figure 7-9   Trimer formation

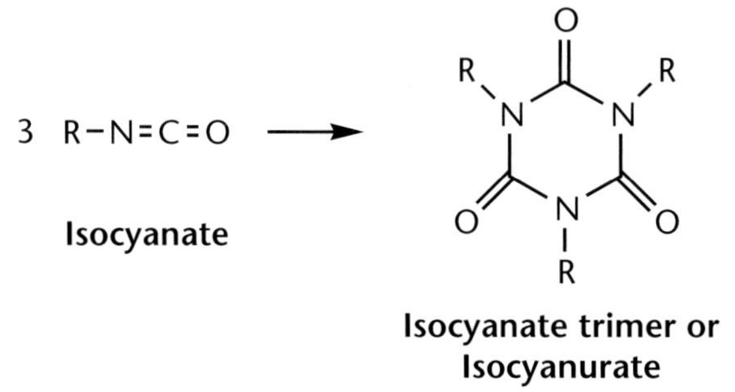

3 R-N=C=O → Isocyanate trimer or Isocyanurate

Isocyanate

with an insertion reaction being involved in the mechanism. Trimerisation can be an undesirable side reaction or a reaction used to introduce branching and cross-linking. It is used in the manufacture of rigid foams, where the cross-linked structure contributes to fire resistance.

When making prepolymers, undesirable trimerisation can occur if there is a high residual level of potassium salts present, in the form of acetates or alkoxides from the manufacture of the polyols and a temperature runaway is possible with the consequence that the reactor content could solidify.

Isocyanate trimers are thermally stable and not reversible like dimers, allophonates, biurets and uretonimines and isocyanurate foams of MDI only show significant degradation above 270°C. This is why the polyisocyanurate rigid foams (PIR) more easily meet national fire standards. They are used extensively in building insulation and roofing applications.

Carbodiimides and uretonimines
Isocyanates can react at high temperature (above 180°C) to form carbodiimides, but the mechanism is not readily understood with the most straightforward theory being a [2+2] cyclo-addition reaction across the C=N bond and the C=O bond of two isocyanate groups. The reaction product is unstable, potentially only exists as a transition species and loses carbon dioxide resulting in carbodiimides. This loss of carbon dioxide makes the reaction irreversible. Carbodiimide can then react with another isocyanate in a [2+2] cyclo-addition reaction across the C=N bonds forming a uretonimine, Figure 7-10.

The latter reaction is reversible producing carbodiimide at high temperature and slowly going to uretonimine at lower temperature. This equilibrium results in potential interchange of isocyanate homologues as already described for biurets and allophonates.

In the presence of special catalysts like 3-methyl-1-phenyl-pholene-1-oxide or triethyl phosphate, carbodiimide formation is much faster. The first catalyst is very active and is used commercially at ppm level at temperatures of 90 to 110°C. The catalyst is still active at room temperature so needs to be deactivated to stabilise the end product and typical products used are strong acids, often

Figure 7-10   Carbodiimides and uretonimines

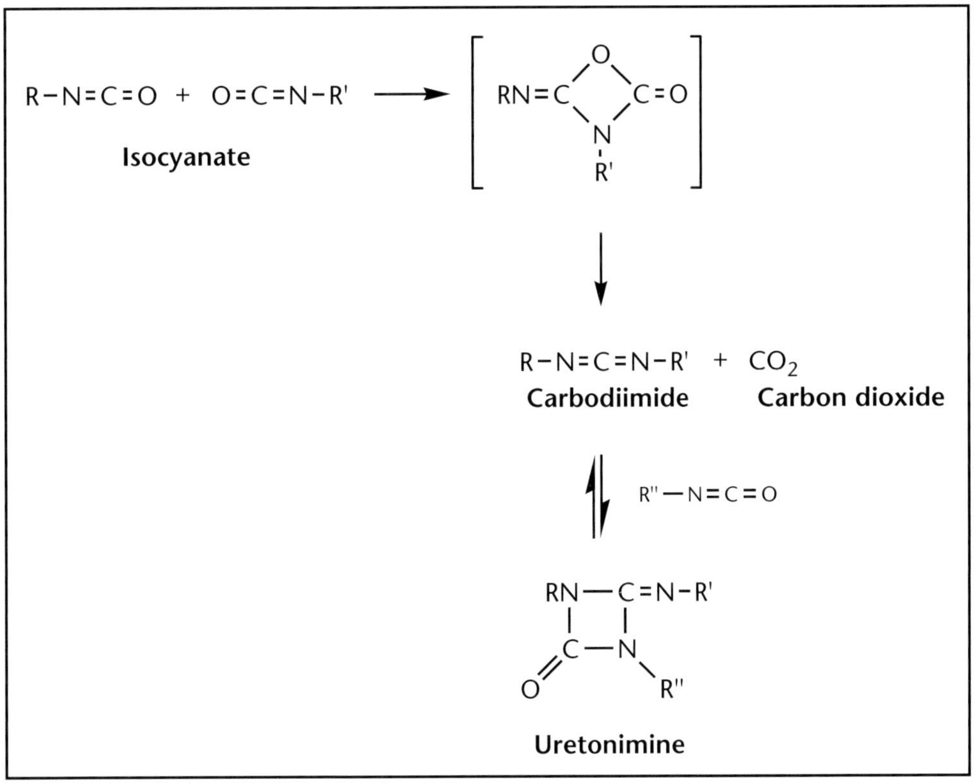

$$R-N=C=O + O=C=N-R' \longrightarrow \left[\begin{array}{c} RN=C\underset{\underset{R'}{N}}{\overset{O}{\diamond}}C=O \end{array} \right]$$

$$\downarrow$$

$$R-N=C=N-R' + CO_2$$
Carbodiimide   **Carbon dioxide**

$$\updownarrow \quad R''-N=C=O$$

Uretonimine structure:
RN—C=N-R'
with C=O and N-R'' forming a four-membered ring

Uretonimine

containing labile chlorine bonds. Therefore, chlorinated impurities present in the isocyanate can influence the catalyst activity. Triethyl phosphate is much less active and is used at temperatures above 200°C and in concentrations higher than one per cent, but the advantage is that the end product is more stable since the catalyst is not active at room temperature.

The presence of the uretonimine cycloadduct in MDI depresses its melting point. Therefore carbodiimide and uretonimine reactions are used in making modified isocyanates having the advantage of being liquid at room temperature. Since uretonimine is a trifunctional species it can be used as a branching agent in prepolymers and further increases in cross-link density can be obtained if higher polymeric oligomers of uretonimines are formed.

Other reactions of isocyanates

There are many other reactions of isocyanates that can influence the polyurethane process and a few special cases are illustrated in Figure 7-11.

With HCl a carbamoylchloride can be prepared, stable at room temperature that dissociates at 80 to 100°C. This equilibrium reaction plays a role in phosgenation of amine to isocyanate. Another reversible reaction is possible with oximes, which can be used as blocking agent.

Figure 7-11   Other isocyanate reactions

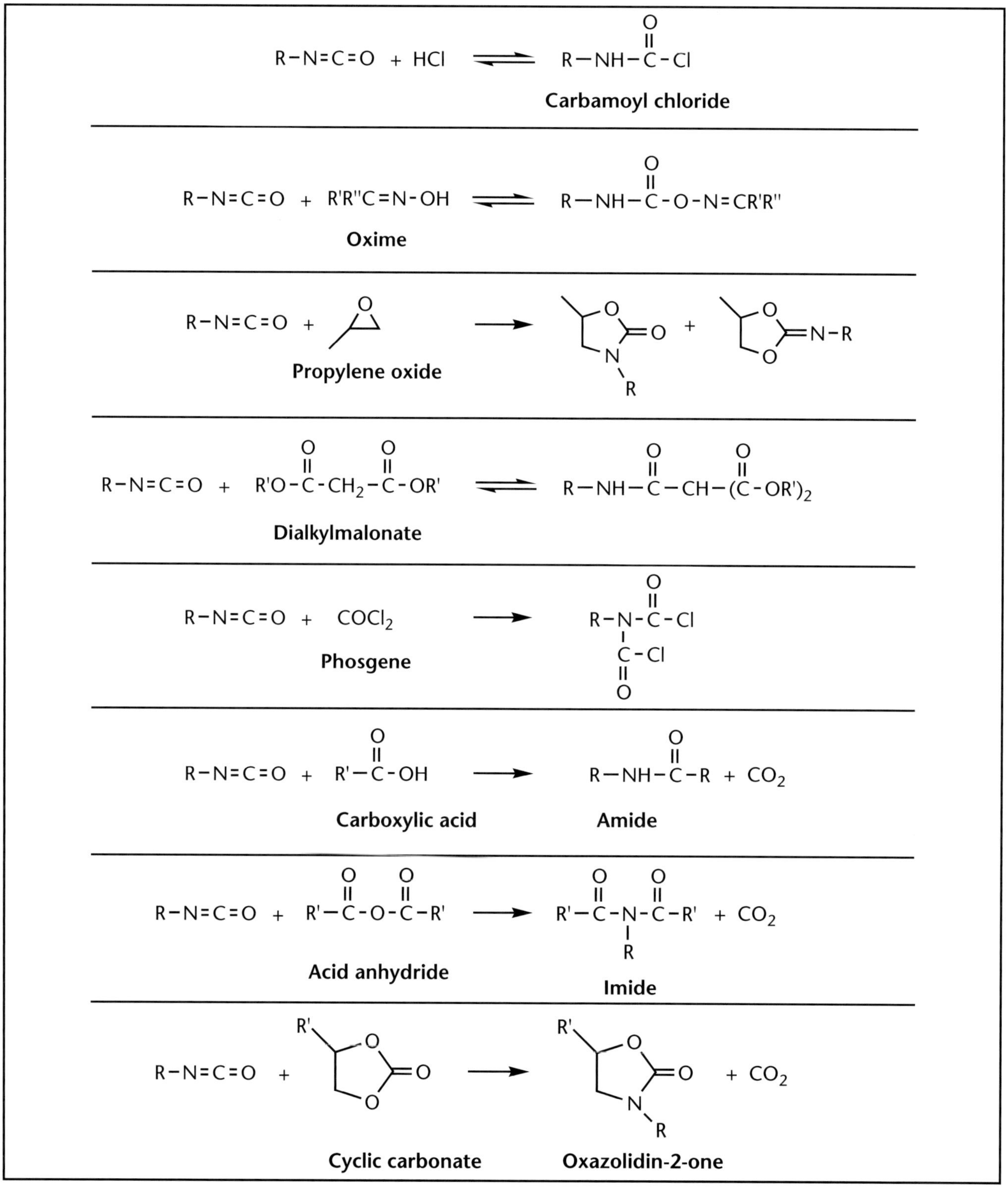

$$R-N=C=O + HCl \rightleftharpoons R-NH-\overset{\overset{O}{\|}}{C}-Cl$$
Carbamoyl chloride

$$R-N=C=O + R'R''C=N-OH \rightleftharpoons R-NH-\overset{\overset{O}{\|}}{C}-O-N=CR'R''$$
Oxime

$$R-N=C=O + \underset{\textbf{Propylene oxide}}{\triangle\!\!\!\!\!-\!\!\text{O}} \longrightarrow \underset{R}{\overset{}{\text{oxazolidinone}}}\!=\!O \;+\; \text{cyclic iminoether} = N-R$$
Propylene oxide

$$R-N=C=O + R'O-\overset{\overset{O}{\|}}{C}-CH_2-\overset{\overset{O}{\|}}{C}-OR' \rightleftharpoons R-NH-\overset{\overset{O}{\|}}{C}-CH-(\overset{\overset{O}{\|}}{C}-OR')_2$$
Dialkylmalonate

$$R-N=C=O + COCl_2 \longrightarrow \begin{array}{c} \overset{O}{\|} \\ R-N-\overset{}{C}-Cl \\ | \\ \overset{}{C}-Cl \\ \| \\ O \end{array}$$
Phosgene

$$R-N=C=O + R'-\overset{\overset{O}{\|}}{C}-OH \longrightarrow R-NH-\overset{\overset{O}{\|}}{C}-R + CO_2$$
Carboxylic acid   **Amide**

$$R-N=C=O + R'-\overset{\overset{O}{\|}}{C}-O-\overset{\overset{O}{\|}}{C}-R' \longrightarrow \begin{array}{c} R'-\overset{\overset{O}{\|}}{C}-N-\overset{\overset{O}{\|}}{C}-R' + CO_2 \\ | \\ R \end{array}$$
Acid anhydride   **Imide**

$$R-N=C=O + \text{cyclic carbonate} \longrightarrow \text{oxazolidin-2-one} + CO_2$$
Cyclic carbonate   **Oxazolidin-2-one**

Cyclo-addition reactions mostly take place over the reactive C=N double bond of the isocyanate and an important reaction is with residual propylene oxide that can chain stop an isocyanate.

Insertion reactions probably go through a stepwise mechanism and an example of insertion into a C-H bond is provided by dialkylmalonate. This reaction is reversible and takes place with all enolisable ketonates and diketonates. These compounds can be used as blocking agents deblocking in the range 130 to 160°C. Another important insertion is the C-Cl reaction that can take place during phosgenation of amines to isocyanates.

Nucleophilic addition and cyclo-addition reactions can result in unstable intermediates, leading to carbon dioxide release. These reactions can be used to create a blowing agent in situ, but can be dangerous due to pressure build-up in a closed container. The best-known reactions are with carboxylic acids, acid anhydrides and cyclic carbonates. Most of the end products in this series can react again with isocyanates.

Polyurethane degradation reactions

Polyether polyol

Polyether polyols are hydrolytically stable and decompose only at high temperature in the presence of strong acids. The reaction products include allyl and propenyl ethers, cyclic ethers, such as dioxanes and crown ethers. Breaking and recombination of chain fragments can occur and even acetals and cyclic acetals are suspected to be formed this way. Eventually, the polyol chain can completely degrade. These reactions are shown in Figure 7-12.

A useful feature of acid decomposition is the hydrolysis of propenyl ethers in the presence of acidic water leading to an alcohol and aldehyde, which minimises the amount of unsaturation in the polyol.

Aldehydes and ketones can react further to unsaturated species by aldol condensation reactions. One example is the condensation of two molecules of propionaldehyde to yield 2-methyl-2-pentenal.

Acetals can be generated from the reaction of aldehydes or ketones with alcohols, Figure 7-12. The first product is a hemi-acetal and a second alcohol can drive the reaction to acetals. These are equilibrium reactions and can only be pushed to the acetal end under dehydrating conditions. Cyclic acetals are possible and once an acetal is formed, exchange of alcohols between different acetals is possible as well. A wide range of unwanted by-products, of which some are low boiling and have strong objectionable odour, can be generated in this way.

Figure 7-12   Polyether degradation reactions

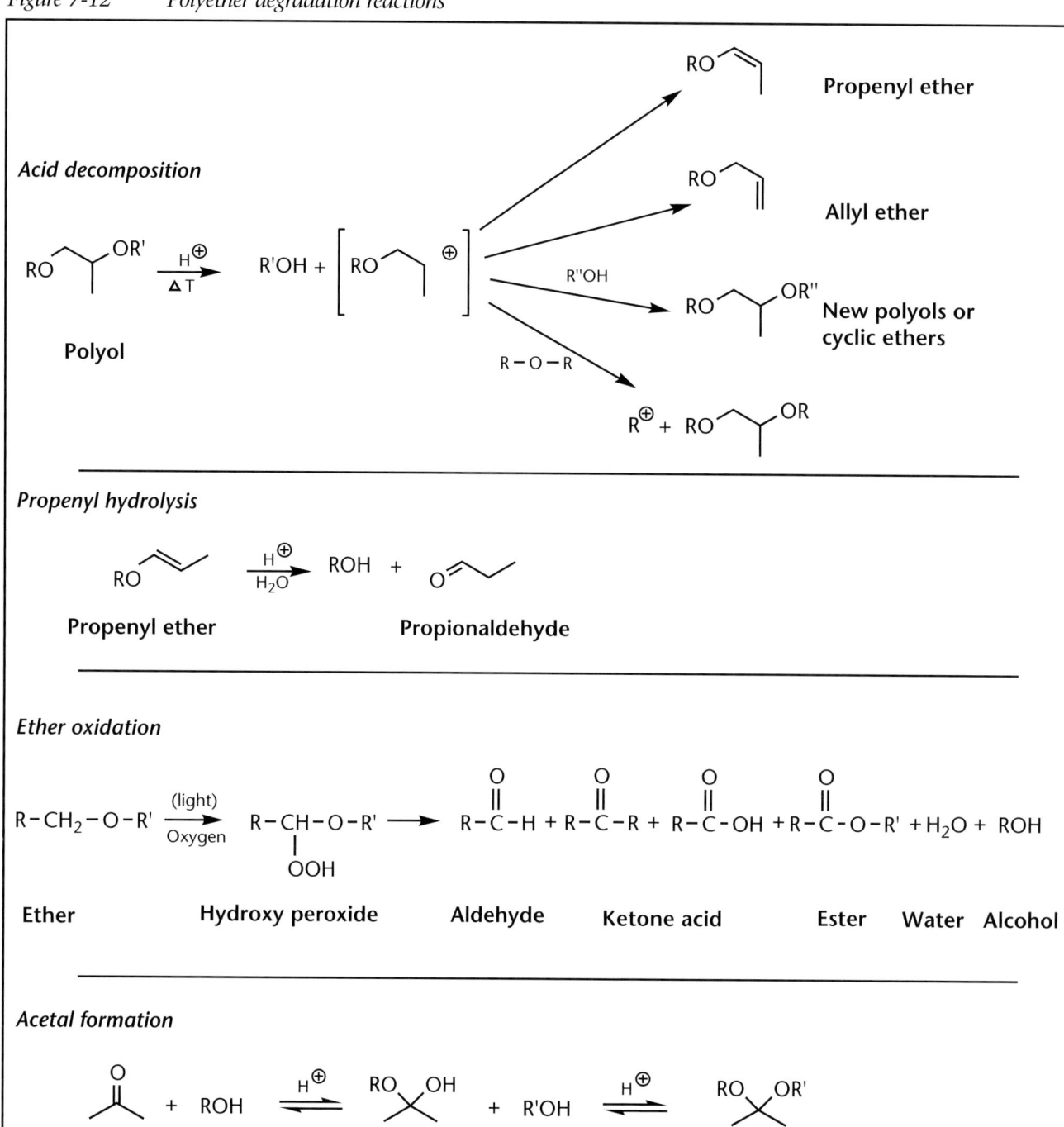

Ether oxidation leads to intra-chain hydroxy peroxides, which cause chain cleavage and decomposition to aldehydes, ketones, acids, esters and water. Light and/or strong acids and/or heavy metal ions can accelerate this process. The main oxidation products of propylene oxide are acetone, propionaldehyde, propanol and 2-propanol whilst ethylene oxide oxidises to acetaldehyde.

Oxidation is kept to a minimum by using nitrogen and/or the addition of an anti-oxidant such as butylated hydroxytoluene (BHT) either to the polyether polyol or to the polyol blend.

Polyester polyol

Esterification is an equilibrium reaction with water removal being the driving force so in contrast to polyether polyols they are sensitive to hydrolysis. This means that the inherently hygroscopic polyesters must be protected against moisture. Once hydrolysis starts the liberated acid will act as a catalyst to accelerate further hydrolysis. To protect against hydrolysis acid acceptors such as carbodiimides, can be added to polyester polyols and/or formulations to stabilise the polyol and final polyurethane product. Hydrolysis resistance increases with increased length of chain between the ester linkages and with increased branching and ether groups present in the product will make it more hydrophobic and therefore also more hydrolysis resistant.

Polyesters are also more prone to microbial attack than polyethers, but this can be overcome by adding anti-microbial compounds that normally contain arsenic. The actual breakdown route due to microbial attack is not entirely clear, but is thought to be related to hydrolysis and due to the microbes activating the ester group to attack by water.

Polyester polyols are also prone to transesterification, a reversible reaction, which occurs in polyol blends where the diol chain extenders react with the polyester polyol splitting the chain and forming shorter molecular weight species. There is no loss of hydroxyl groups involved, but if the transesterification goes to completion all the chain extender will be consumed. This reaction is activated by increased temperature and the presence of catalysts and can severely limit the storage life of polyester polyol blends. The reaction is also used in recycling when scrap polyester-based polyurethane, in the form of two to five millimetre chips, is heated to 220°C in the presence of a short chain diol leading to the breakdown of the polyurethane polymer with the result being a hydroxyl terminated low molecular weight product.

Polyesters are much less prone to oxidation attack than polyether polyols, but if necessary BHT can also be added to provide increased protection.

Ultra-violet radiation

Aromatic polyurethanes are susceptible to yellowing and ultimately to degradation due to UV exposure when used in exterior applications. This attack can be slowed down by the addition of a synergistic blend of light absorbing benzophenone and benzotriazole compounds with UV absorbers and free radical scavengers such as hindered amines. Pigments and fillers can also be added to mask the effect and in many cases this approach works well for short-term use.

For high-performance applications where long-term exposure demands non-yellowing behaviour and no degradation in performance, aliphatic polyurethanes need to be used. Examples of such applications are architectural glass for building windows, safety glass laminates for aircrafts and automobiles, exterior and interior components of automobiles, medical products, fashion and industrial textile coatings.

Physical chemistry

There are two fundamental types of chemistry involved in making polyurethane polymers. These are:

- Phase separated structures as found in flexible foams, thermoplastic polyurethanes, elastomers, adhesives and coatings.
- Highly cross-linked glassy amorphous material as seen in rigid foams and some composites.

In addition, because the chemistry of polyurethane is so versatile there are hybrids between the two, but most applications clearly fit into one or other of the above categories.

At room temperature the higher melting polar hard segments are incompatible with the less polar soft segments and phase separation or segregation occurs. This takes place more readily with polyether based polyurethane polymers than those made from polyester polyols. The hard segments form crystalline domains by hydrogen bonding and the resultant hard segment microphases are covalently linked with each other through the flexible soft segments. The flexible segments give the material its elasticity while the crystalline domains prevent permanent deformation of the soft segments as the polymer chains are stretched. Therefore, the elasticity and toughness of a polymer depends on the percentage of hard segments and the chain length of the soft segment.

Cross-linked glassy networks can be obtained using polyfunctional isocyanates, alcohols and amines with the precise level of cross-linking being dependent on the functionality and molecular weight of the starting material. Another option is to promote trimerisation of the isocyanate by addition of a catalyst. In all cases the cross-linking consists of non-reversible strong covalent bonds. It is possible, however, to achieve thermo-reversible cross-linking by promoting allophonate, biuret or uretonimine formation. This way, even the difunctional chain extenders such as diols, diamines and water can be effectively used as cross-linkers.

Full details of the influence of this physical chemistry are included in Chapters 11, 15 and 19.

8. Blowing agents

Sachchida N Singh

Until the late 1950s polyurethane foams were blown by carbon dioxide, liberated as a result of the isocyanate-water reaction. It was then the discovery and use of chlorofluorocarbons (CFCs), in particular trichlorofluoromethane, CFC-11, as a physical blowing agent that greatly accelerated the large-scale use of polyurethane foam. It led to the production of closed cell rigid foam with low densities, good mechanical properties and very low thermal conductivity and low-density flexible foams with tailored load-bearing properties used for many cushioning products.

By the early 1980s the flexible foam industry was already moving back to full water-blown technology for technical, environmental and cost reasons. However, this was not possible for the majority of rigid foams that are required to be dimensionally stable and most importantly need to have a low thermal conductivity. This, in conjunction with the requirement to maintain an essentially closed cell structure, has meant that physical blowing agents have been the only continuing option for rigid foams.

The first indication that there could be an environmental problem with CFCs came when Rowland and Molina published a paper in 1974 describing their research on reactions in the atmosphere, which could affect the ozone layer. Their paper postulated that CFCs were stable enough to reach the upper atmosphere, at heights of the order of 20 to 25 kilometres, where they could be broken up by ultra violet solar radiation. The chlorine atoms released could convert the ozone into oxygen and so deplete the ozone layer. The depleted ozone layer would allow increased levels of UV radiation to reach the surface of the Earth with consequent implications for human health and other biological systems.

Figure 8-1   Ozone 'hole' above Antarctica

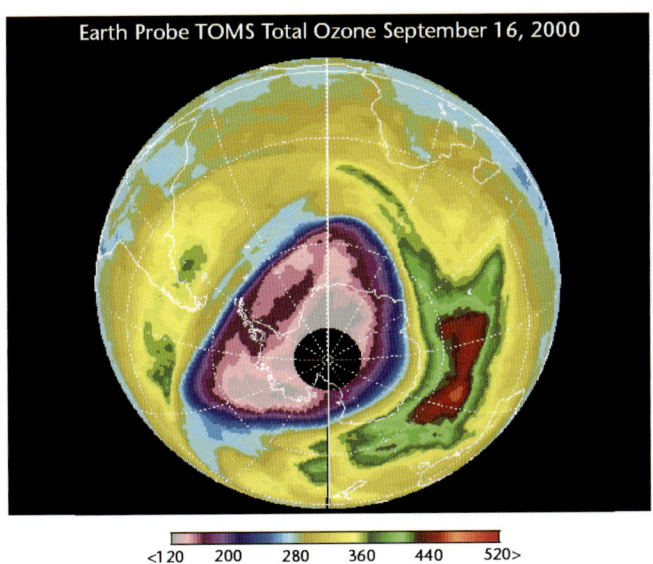

Certain nations took early precautions to reduce CFC emissions, for example the USA banned their use in aerosols from 1978, whilst some industries including the polyurethane foam industry began a search for alternatives. However, the pressure to move away from CFCs was low until the discovery, in 1985, that serious ozone depletion was a reality and that a 'hole' in the ozone layer was occurring over Antarctica. The 'hole' was measured using satellite imaging of ozone concentration, Figure 8-1. This led rapidly to the Vienna Convention later in 1985 and to the Montreal Protocol in 1987.

Montreal Protocol and other regulations

At a conference held in Montreal, Canada, under the auspices of the United Nations Environment Program (UNEP), 24 countries signed an accord in September 1987 to control the production and consumption of ozone depleting substances (ODS). The ozone depletion potential (ODP) of these substances is defined on a scale, where CFC-11 has a value of 1.0, to indicate the stratospheric ozone depleted by a unit mass of a given product. The Protocol was designed so that the control measures could be revised on the basis of periodic scientific and technological assessments. Following such assessments, the Protocol was amended at London in 1990, Copenhagen in 1992, Vienna in 1995, Montreal in 1997 and Beijing in 1999.

New control measures have since been added to the protocol, including accelerated phase-out schedules and the addition of other controlled substances. Whilst most nations, 181 as of November 2001, have ratified the 1987 Montreal Protocol, many have not ratified some of the amendments. The key element of the Montreal Protocol sets out the time schedule for the freeze and then reduction in the use of ozone deleting substances. It looks at countries, based on their annual per capita calculated consumption level of ozone depleting substances, as developed, Non-Article 5(1), or developing, Article 5(1). Table 8-1 shows the phase-out schedule of CFCs.

Table 8-1   Montreal Protocol phase out schedule for CFCs

Date	Control Measure (ODP-Weighted % Reduction)	
	Developed Non-Article 5(1) Parties	Developing Article 5(1) Parties
July 1, 1989	Freeze at 1986 level	–
January 1, 1994	75% of 1986 level	–
January 1, 1996	Phased out	–
July 1, 1999	–	Freeze at 1995 – 97 average level
January 1, 2005	–	50% of 1995 – 97 average level
January 1, 2007	–	85% of 1995 – 97 average level
January 1, 2010	–	Phased out

Non-Article 5(1) countries successfully eliminated the consumption of CFCs by the end of 1995 and Article 5(1) countries had to freeze their consumption by 1999 based on average 1995 to 1997 levels and complete their phase-out by the end of 2009. The Multi-Lateral Fund (MLF), set up by the United Nations with funding provided by developed countries, has been established to provide financial and technical co-operation, including the transfer of technologies, to the Article 5(1) countries to support the phase-out of CFCs.

Hydrochlorofluorocarbons (HCFCs) were developed as the first major replacements for CFCs as their ozone depletion potential ranges from 0.01 to 0.13. The HCFCs are viewed as transitional alternatives to be used while zero alternatives are developed and in 1992 the Copenhagen amendment defined the phase-out for HCFC consumption between 1996 and 2040, details in Table 8-2.

Table 8-2   Phase-out schedule for HCFCs

Date	Montreal Protocol Reduction[1] Developed Non-Article 5(1) Parties	Developing Article 5(1) Parties	US Reduction[4]	EU Reduction[8]
January 1, 1996	Freeze at cap[2]	–	–	–
January 1, 2003	–	–	141b – 100%[5]	100% appliances, & lamination[9]
January 1, 2004	35% of cap[2]	–	–	Phased out
January 1, 2010	65% of cap[2]	–	142b & 22: 100%[6]	–
January 1, 2015	90% of cap[2]	–	Phased out[7]	–
January 1, 2016	–	Freeze at 2015 level	–	–
January 1, 2020	99.5% of cap[3]	–	142b & 22: 100%	–
January 1, 2030	Phased out	–	Phased out	–
January 1, 2040	–	Phased out	–	–

1. Up to and including 1997 Montreal Amendment and applicable to production only
2. ODP-weighted HCFC consomption cap equals 100% 1989 HCFC + 2.8% (Montreal Protocol) or 2.6% (EU) of 1989 CFC consumption
3. Phased out except for service of existing refrigeration and air-conditioning equipment
4. As of January 1st 2002
5. Production only for domestic foam use
6. Except for use in equipment manufactured before January 1, 2010
7. Except for use in equipment manufactured before January 1, 2020
8. As of November 1, 2001
9. Except in refrigerated transport

Following the rapid introduction of the HCFCs other blowing agents with lower ozone depletion potential have been developed and the families, in decreasing order of ozone depletion potential, are: hydrofluorocarbons (HFCs); hydrocarbons (HCs); halons and liquid carbon dioxide. The choice of the optimum product is not simple as it depends on several factors such as: end application, polyurethane technology, machinery, legislation, cost and environmental factors.

In addition to the Montreal Protocol, other bodies such as the European Union and national governments/agencies have imposed more strict regulations and phase-out schedules, and in fact the European Climate Change Programme (ECCP) will result in proposals for a EU Regulation on fluorinated gases, including HFCs, by the end of 2002.

The USA Environmental Protection Agency, US-EPA, imposed a tax levy to limit CFC use in foam and HCFCs have been banned from use in all foams except those used for thermal insulation since January 1, 1996. They have accelerated the phase out of HCFC-141b, the HCFC with the highest ODP level, to January 1, 2004. HCFC-142b and HCFC-22 are currently allowed until January 1, 2010. However, this date is subject to change given the dynamics of the environmental and regulatory communities. In Canada the manufacture, import, sale or use of all HCFCs is banned after January 1, 2015.

The European Union has opted to phase out HCFCs for use as blowing agents by the end of 2003, significantly ahead of the Montreal Protocol timelines, and in Europe generally several states have even more rapid phase-out schedules, but with many of the phase-out dates being application dependent. In Japan, HCFC-141b will be phased out by the end of year 2003.

Considerable attention is being focussed on possible regulations relating to the use and emission of HFCs and in Europe some states have proposed bans while some countries, such as Denmark, have a use ban that started in May 2002. The Montreal Protocol also has a clause requiring CFCs to be replaced by the best options in the environmental sense. However, the criteria governing this choice have not been defined. In an attempt to help the polyurethane foam industry with all of the regulations, UNEP has set up the Flexible and Rigid foams Technical Options Committee (TOC). It has issued periodic reports detailing the available technical options that can be implemented by each foam type.

One of the major concerns in applying the Montreal Protocol is the identification of the best replacement technology suitable for small and medium-sized enterprises (SMEs). This applies to all foam sectors and to both Non-Article 5(1) and Article 5(1) countries. The choice is particularly difficult for rigid foam manufacturers who wish to convert from HCFC-141b to a zero ozone depletion potential option. Hydrocarbon technology is technically suitable in many cases, with the exception of spray foam, but the capital cost to convert the equipment to handle a flammable blowing agent can be prohibitive. Such enterprises may use HFCs despite the higher unit foam costs and the growing environmental pressures. Another area is the production of small flexible (box) foam where the manufacturers want to replace CFC-11, but are forced to use methylene chloride because the capital cost to convert to liquid carbon dioxide technology, the other alternative, is too high.

The 1987 Montreal Protocol and the subsequent revisions have had a profound effect on how polyurethane foam is made or used today. Figure 8-2 shows that the production of CFCs and HCFCs by AFEAS (Alternative Fluorocarbons Environmental Acceptability Study) reporting countries, weighted according to the ozone depletion potential of each compound, has been reduced by more than 93 per cent from the peak in 1988. The reduction has been even more rapid than mandated by the Montreal Protocol, which is remarkable considering that the polyurethane foam industry has been growing at two to three times global GDP (gross domestic product).

The measures in the Montreal Protocol to limit the use of all ozone depleting substances are predicted to counter the depletion of the ozone layer and return it to its early 20th Century state by 2060 to 2070 with the time scale governed by the atmospheric lifetimes of the substances.

This provides a brief snap shot of a complex subject that, whilst it has made excellent progress, continues to evolve and the polyurethane industry can be justifiably proud of the way in which it reacted so rapidly to such an important environmental issue.

Figure 8-2   ODP-weighted global production of CFCs and HCFCs

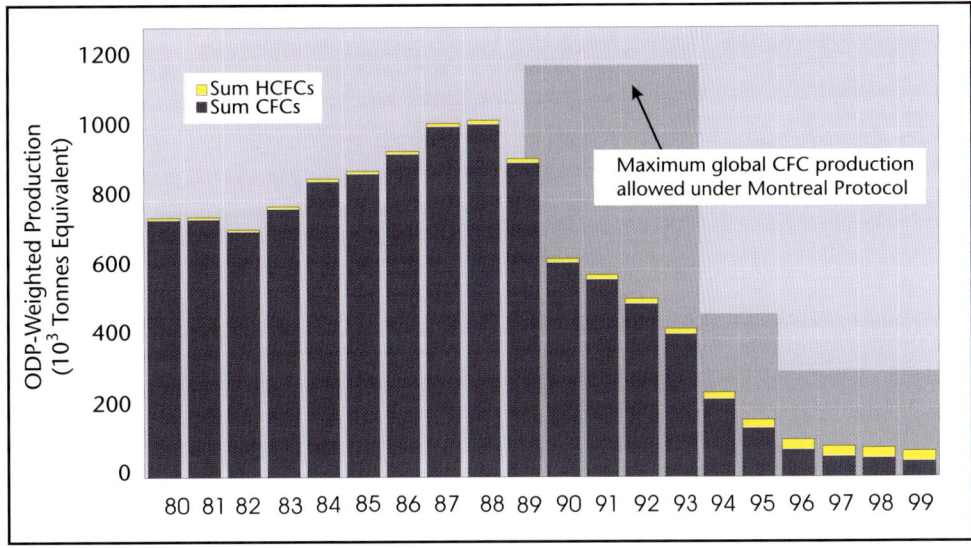

Blowing agents

CFC-11, having a low molecular weight, boiling point around room temperature, low toxicity, non-flammability and low thermal conductivity has the ideal characteristics for a blowing agent. This, along with excellent chemical and thermal stability and low cost, made CFC-11 the blowing agent of choice for all polyurethane foam, especially rigid thermal insulation foam. Another CFC, dichlorodifluorocarbon, CFC-12, saw some limited use in the manufacture of rigid polyurethane foams as a frothing and as a dispensing agent whilst methylene chloride was introduced in the middle of the 1970s for use in flexible and integral skin foam. The properties of these three blowing agents are compared to carbon dioxide in Table 8-3.

Table 8-3   Properties of CFC and other blowing agents

		CO_2	CFC-12	CFC-11	Methylene Chloride
Chemical formula		CO_2	CCl_2F_2	CCl_3F	CH_2Cl_2
Molecular weight		44	120.9	137.4	85
Boiling point, °C		-78.3	-29.8	23.8	40
Specific gravity at 25°C		n/a	1.31	1.48	1.33
Heat of vaporisation at BP, *kJ/mole*		6.8	20.0	24.8	28.0
Gas conductivity, *mW/(m.K)* at	10°C	15.3	9.2	7.4	n/a
	25°C	16.4	9.9	7.9	n/a
Vapour pressure, *kPa* at	10°C	4,500	420	60	30
	25°C	6,400	640	110	60
Flammable limit in air, *vol-%*		None	None	None	12 – 19
OEL, *ppm*		n/a	1,000	1,000	35 – 100
ODP (with CFC-11 = 1)		0	1	1	0.007
GWP (100-year., CO_2 = 1)		1	10,600	4,600	0.02
Atmospheric lifetime, *years*		120	100	45	0.5

OEL   Occupational Exposure Limit
ODP   Ozone Depletion Potential
GWP   Global Warming Potential

After extensive and thorough screening of other possible blowing agents a few different families of compounds have emerged as potential candidates and are described below.

Hydrochlorofluorocarbons (HCFCs)

Featuring at least one hydrogen atom in the molecule and a carbon hydrogen bond, HCFCs are chemically less stable than CFCs and tend to breakdown in the lower atmosphere into simple species, such as hydrogen halides and formyl fluoride. Consequently, HCFCs' ability to migrate to the stratosphere and to decompose into ozone-damaging chlorine is much lower than with CFCs. Also the breakdown products do not contribute significantly to the photochemical smog formation in urban areas. Thus, HCFCs have low ODP and are not a VOC.

The search for suitable HCFCs to replace CFC-11 in the late 1980s resulted in the selection of HCFC-141b as a liquid plus HCFC-22 and HCFC-142b as gaseous blowing agents. Table 8-4 lists the properties of these HCFCs and that of 2-chloropropane as its low ozone depletion and global warming potentials attracted much attention at one time, but it is not used widely.

Table 8-4   Physical and environmental properties of HCFCs

		HCFC-22	HCFC-142b	HCFC-141b	2-chloropropane
Chemical formula		$CHClF_2$	CH_3CClF_2	CH_3CCl_2F	$CH_3CH_2ClCH_3$
Molecular weight		86.5	100.5	116.9	78.5
Boiling point, °C		-40.8	-9.8	32.9	35.7
Liquid specific gravity at 25°C		1.19	1.12	1.23	0.87
Heat of vaporisation at BP, kJ/mole		20.2	22.4	25.8	26.5
Gas conductivity, mW/(m.K) at	10°C	9.9	8.4	8.8	n/a
	25°C	10.7	9.5	10	n/a
Vapour pressure, kPa at	10°C	680	210	50	40
	25°C	1,050	340	80	70
Flammable limit in air, vol-%		None	6.4 – 14.9	7.6 – 17.7	2.8 – 10.7
OEL, ppm		1,000	1,000	500	n/a
ODP (with CFC-11 = 1)		0.055	0.065	0.11	0.003
GWP (100-year., CO_2 = 1)		1,900	2,300	700	9.9
Atmospheric lifetime, years		11.8	18.5	9.2	0.75

Hydrofluorocarbons (HFCs)

These compounds do not contain chlorine and thus have zero ozone depletion potential. Though many HFCs have been studied, two liquid, HFC-245fa and HFC-365mfc, and two gaseous, HFC-134a and HFC-152a, products have been selected for large-scale commercialisation.

Table 8-5 lists the physical and environmental properties of the commercial HFCs. The breakdown products from these HFCs in the lower atmosphere have been studied and found to have insignificant effect on the photochemical smog

Table 8-5   Physical and environmental properties of HFCs

		HFC-134a	HFC-152a	HFC-245fa	HFC-365mfc
Chemical formula		CH_2FCF_3	CHF_2CH_3	$CH_3CH_2CHF_2$	$CH_3CH_2CF_2CH_3$
Molecular weight		102	66	134	148
Boiling point, °C		-26.5	-24.7	15.3	40.2
Liquid specific gravity at 25°C		1.20	0.90	1.32	1.25
Heat of vaporisation at BP, kJ/mole		22.1	21.7	28.0	26.2
Gas conductivity, mW/(m.K) at	10°C	12.4	n/a	12.5	10.6
	25°C	13.8	14.7	13.3	11.6
Vapour pressure, kPa at	10°C	415	273	83	26
	25°C	665	500	149	59
Flammable limit in air, vol-%		None	3.9 – 16.9	None	3.8 – 13.3
OEL, ppm		1,000	1,000	500	n/a
ODP (with CFC-11 = 1)		0	0	0	0
GWP (100-year., CO_2 = 1)		1,600	140	990	910
Atmospheric lifetime, years		14	1.7	7.4	10.8

formation in urban areas. The flammability limit of HFC-365mfc in air has led to the introduction of a non-flammable blend with HFC-227ea (CF_3CHFCF_3) in the weight ratio of 94:6.

Hydrocarbons (HCs)

The many advantages of hydrocarbons (HCs) as blowing agents, namely: low cost, ready availability, halogen-free, zero ozone depletion potential and nearly zero global warming potential, had been known for some time, but safety concerns, due to their high flammability, had blocked serious consideration by the polyurethane industry. This changed in the late 1980s when a great deal of development work was done on the technology and the machinery required to handle these flammable materials and three isomers of pentane, cyclo-, iso- and normal, that are all liquid at room temperature, and iso- and n-butane, which are gaseous, have been commercialised. Table 8-6 lists their physical and environmental properties. The pentanes have emerged as the dominant products; iso-butane and the other hydrocarbons are not used very widely.

The flammability limit of the pentanes in air is only about 1.4 to 8.0 vol-% and the energy of ignition is extremely low whilst the density of the vapour is about 2.5 times that of air. Therefore, the use of the pentanes as blowing agents needs a careful consideration of equipment and procedures in storage, handling, manufacturing and shipping. Though the extent of plant modifications required can vary significantly, dependent on local codes and regulations, in general, improved ventilation, explosion and fireproofing and alarm systems are required. Hydrocarbons are classified as volatile organic compounds and are subject to emissions control in many countries.

Table 8-6 Physical and environmental properties of HCs

		Iso-butane	Iso-pentane	n-pentane	Cyclo-pentane
Chemical formula		C_4H_{10}	C_5H_{12}	C_5H_{12}	$(CH_2)_5$
Molecular weight		58	72.0	72.0	70.1
Boiling point, °C		-12.0	27.8	36.2	49.3
Liquid specific gravity at 25°C		0.55	0.62	0.63	0.75
Heat of vaporisation at BP, kJ/mole		21.3	24.6	25.7	27.3
Gas conductivity, mW/(m.K) at	10°C	14.8	12.8	13.7	11.4
	25°C	16.2	14.3	15.0	12.8
Vapour pressure, kPa at	10°C	220	50	40	24
	25°C	350	90	69	43
Flammable limit in air, vol-%		1.8 – 8.4	1.4 – 7.8	1.3 – 8.0	1.4 – 8.0
OEL, ppm		n/a	600	600	600
ODP (with CFC-11 = 1)		0	0	0	0
GWP (100-year., CO_2 = 1)		11	11	11	11
Atmospheric lifetime, years		Few days	Few days	Few days	Few days

Liquid carbon dioxide and other physical blowing agents

With a critical temperature of 31°C and critical pressure of 7.38 MPa, liquid carbon dioxide has long been considered an attractive blowing agent for polyurethane foam, primarily for non-insulation applications. Readily available from the air and other natural sources, it is inexpensive and safe.

Methyl chloroform (CCl_3CH_3, ODP = 0.1, GWP = 140, boiling point = 74°C) and acetone (CH_3COCH_3, ODP = 0, boiling point = 56.1°C) are some of the other physical blowing agents considered for non-insulation foams.

Perfluorocarbons such as perfluoropentane (C_5F_{12}) and perfluorohexane (C_6F_{14}) have been evaluated as blowing agents and as co-blowing agents with HCFCs. They are not considered viable options as they are characterised by very long atmospheric lifetimes, of the order of hundreds to thousands of years and are very infra-red active and thus have high global warming potential.

Fluorinated ethers, such as HFE-245 ($CF_3CH_2OCF_2H$), HFE-356 ($CF_3CHFCF_2OCH_3$) and HFE-254mf ($CH_3CF_2OCCF_2H$) have been evaluated in the laboratory for rigid foam applications, but substantial development work will be required to determine their ultimate commercial viability. Similarly fluoriodocarbons, such as heptafluoro-2-iodopropane and hydrogen-containing fluoromorpholine, have been evaluated in the laboratory, but their high cost is a major barrier to commercial use.

Chemical blowing

Chemical blowing agents, such as azodicarbimides, are used extensively in many plastics where the processing temperature causes chemical breakdown of the blowing agent to form gases, for instance nitrogen in this case. Though this type of chemical blowing agent is not often used in polyurethane chemistry, another type of chemical blowing, carbon dioxide generated by the water-isocyanate reaction is widely utilised. At present all polyurethane foams, rigid or flexible, low or high density, are blown at least partially with carbon dioxide.

Isocyanate groups can react with another isocyanate group to form a carbodiimide and at the same time give off carbon dioxide, especially in the presence of specific catalysts. Though this has been evaluated, it is not practised widely. The hydrolysis of dimethylcarbonate and dialkyl dicarbonates, such as diisobutyl dicarbonate, to give carbon dioxide has been studied for the production of flexible foam, but has not been commercialised.

Selection criteria

The search for the blowing agent to reduce and/or replace CFCs in polyurethane foams has been a challenge, as many factors must be considered. Some factors are mandated by the environmental considerations, some are necessary by feasibility consideration whereas others have to be balanced from performance considerations. Many of the blowing agent selection considerations are common across all applications and geographical location whilst others are specific to an application or location.

Environmental considerations

Though stratospheric ozone depletion was the only reason to move away from the CFCs, many other environmental issues such as global warming, ground level air pollution, tropospheric degradation, long-term breakdown products, halogen content, acidification potential have to be considered when choosing a new blowing agent. Clearly, from an ozone depletion potential consideration, the lower the value of a product the more desirable the blowing agent, with zero being the goal.

Global warming and resultant climate change is an issue because it is known that some gases, present in the lower atmosphere, reflect infra-red radiation (heat) back to earth and thereby raise the earth's surface temperature. How much a given mass of a chemical contributes to global warming, over a given time period, usually 100 years, compared to the same mass of carbon dioxide is referred to as its global warming potential (GWP). The global warming potential of a blowing agent is a function of its lifetime in the atmosphere and its ability to absorb infra-red radiation with the atmospheric lifetime characterising the overall stability of the blowing agent in the atmosphere.

Some blowing agents may undergo photochemical reactions in the lower atmosphere and contribute to smog formation and thus are classified as volatile organic compounds. Their use is strictly regulated in an increasing number of countries in the world, and may require monitoring and control measures to limit emission during foam manufacturing. In cases where a foam product is exported, the environmental requirements of both the manufacturing country and the use country need to be considered.

Feasibility

Many factors are considered prior to selecting a blowing agent for commercialisation; these include checking for toxicity, flammability, environmental impact, compatibility with materials of construction, safe and economic manufacturing process. All blowing agents brought onto the market have already undergone extensive testing to ensure that there are no problems with toxicity or environmental issues.

Many of the alternative blowing agents to CFCs present varying degrees of flammability. To safely use flammable alternatives, it is necessary to evaluate manufacturing risks from ignition, storage and transportation of blowing agent and fire performance of foam products and finished product. Similar considerations apply to compatibility with materials of construction.

Performance

Boiling point, molecular weight, vapour pressure in the operating temperature range, heat of vaporisation, solubility in components and foam, compatibility with materials of construction are among the many attributes of a compound that must be considered when choosing a blowing agent. Though some performance attributes such as non-reactivity and compatibility with materials of construction are common to all, most depend on the final application of the foam product.

However, there are still some guidelines concerning preferred attributes that apply across all applications. For example, a lower molecular weight, or perhaps lower cost per mole of blowing agent that meets all the other performance criteria is desirable. This is because generally, the lower the molecular weight the higher the gas volume that can be generated per unit weight of the blowing agent. Blowing agents combining high blowing efficiency with a high heat of vaporisation are particularly advantageous since they reduce the maximum exotherm temperature during foaming, which could otherwise result in scorch or lead to a residual stress gradient in blown polyurethane composite products.

9. Catalysts

Robert L Zimmerman

Tertiary amines, organometallics (primarily tin compounds) and carboxylic acid salts are used to catalyse the reaction of isocyanates with water (blowing) and polyols (polymer gelation). The catalyst controls the relative reaction rates of the isocyanate with polyol and water.

Amine catalysts are generally considered to be predominantly blowing catalysts since they tend to catalyse the isocyanate-water reaction better than the isocyanate-polyol reaction. However, amines do catalyse both reactions, with the relative rates of each reaction being dependant on the specific amine catalyst used.

Organometallic catalysts are mainly seen as gelation catalysts although they do affect the isocyanate-water blowing reaction. Organotins are the most widely used, but organomercury and organolead catalysts are also used. The mercury catalysts are very good for elastomers because they give a long working time with a rapid cure and very good selectivity towards the gelation. The lead catalysts are often used in rigid spray foams. However, both mercury and lead catalysts have unfavourable hazard properties so alternatives are always being sought.

Potassium and sodium carboxylic acid salts and quaternary ammonium carboxylic acid salts are used to catalyse the trimerisation reaction and thus are used mainly in isocyanurate foams.

Preparation of amine catalysts

There is a large variety of amine catalysts available, with different structures, therefore several different manufacturing routes are required. These include condensation, reductive methylation, oxide addition plus the combination of two or more of these steps. Many of the starting materials, such as ethylene amines and morpholine are derived from other well-known processes.

There are two manufacturing processes for ethylene amines based on 1,2-dichloro-ethane and on ethanolamine, Figure 9-1. Both produce piperazine and diethylene-triamine, which are used in the manufacture of polyurethane catalysts. The piperazine can either be reductively methylated to give N,N'-dimethylpiperazine or reacted with ethylene oxide then cyclised to give triethylenediamine (1,4-diaza-bicyclo[2.2.2]octane). The diethylenetriamine can be reductively methylated with formaldehyde to give bis(dimethylaminoethyl)methylamine.

The morpholine process provides three important starting compounds – morpholine, bis(2-aminoethyl)ether, and 2-aminoethoxyethanol, Figure 9-2.

Figure 9-1   Reaction routes for the ethylene amines

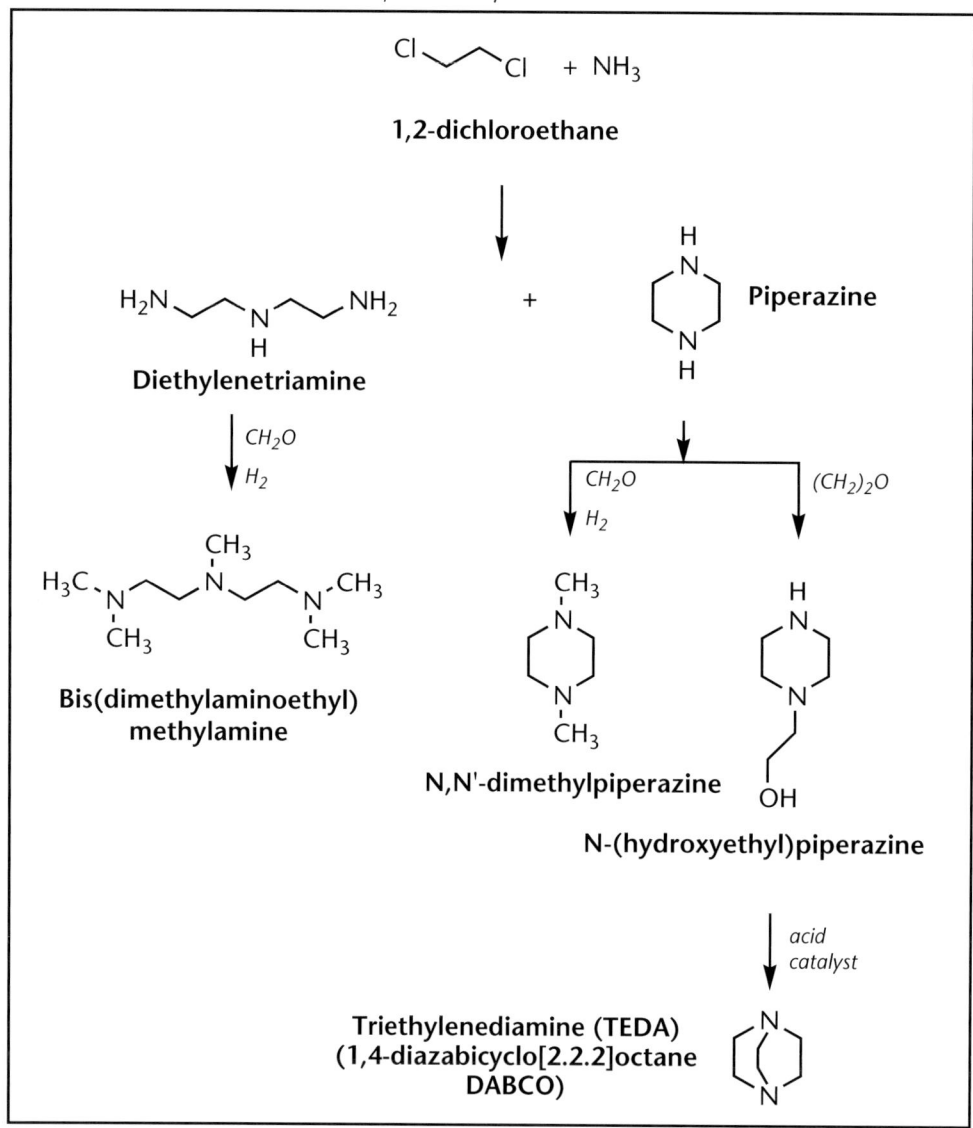

Morpholine can be akylated to give alkylmorpholines – such as N-ethylmorpholine; bis(aminoethyl)ether reductively methylated to give bis(N,N-dimethylamino-ethyl)ether, and 2-aminoethoxyethanol reductively methylated to give 2-(2-(dimethylamino)ethoxy)ethanol.

Catalysts such as N,N-dimethylethanolamine, that have a reactive hydroxyl group, are prepared by reacting N,N-dimethylamine with ethylene oxide whilst propylene oxide is used for the production of similar catalysts.

N,N-dimethylcyclohexylamine is a widely used catalyst and can be prepared by two different processes either by reductively methylating aniline or by reductively reacting cyclohexanone with dimethylamine.

Some common amine catalysts are listed in Table 9-1.

Figure 9-2   Reaction routes for the morpholine process

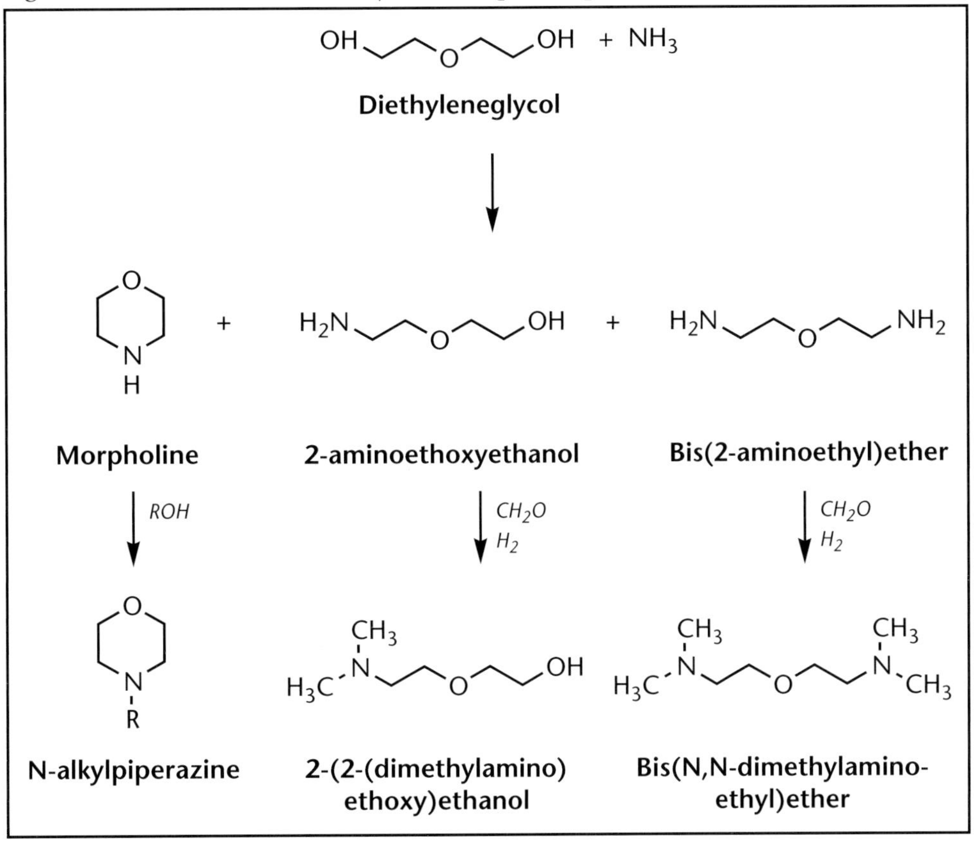

Preparation of organotin catalysts

Two general types of organotins are used as polyurethane catalysts, tin II (stannous) and tin IV (stannic). The major stannous compound used is stannous 2-ethylhexanoate, more commonly referred to as stannous octoate, and is prepared by reacting either stannous oxide or stannous chloride with 2-ethylhexanoic acid.

The main tin IV compounds used are dialkyltin dicarboxylates or dialkyltin mercaptides. They are prepared by basic hydrolysis of dialkyltin dichlorides to give dialkyltin oxides, which are then reacted with a carboxylic acid or mercaptane to give the desired dialkyltin dicarboxylate or mercaptide. The general formulae for the dialkyltin dicarboxylates and mercaptides is R_2SnY_2 and two typical catalysts are dibutyltin(IV) dilaurate where R = butyl and Y = $C_{11}H_{23}COO$ and dibutyltin(IV) dilaurylmercaptide where R = butyl and Y = $C_{12}H_{25}S$.

Some common organotin catalysts are listed in Table 9-2.

Table 9-1   Tertiary amine catalysts and their applications

Catalyst	Formulae	Characteristic and use
N,N-dimethylethanolamine (DMEA)	$(CH_3)_2NCH_2CH_2OH$	Inexpensive, used in flexible foams and in rigid foams. Acid scavenger for rigid-ester foams and fire retarded foams.
N,N-dimethylcyclohexylamine (DMCHA)	$C_6H_{11}N(CH_3)_2$	Inexpensive, has a high odour, is used mainly in rigid foams.
Bis(N,N-dimethylaminoethyl)ether (BDMAEE)	$(CH_3)_2NCH_2CH_2OCH_2CH_2N(CH_3)_2$	Excellent blowing catalyst used in flexible, high resilience and cold moulded foams.
N,N,N′,N′,N″-pentamethyldiethylenetriamine (PMDETA)	$(CH_3)_2NCH_2CH_2N(CH_3)CH_2CH_2N(CH_3)_2$	Good blowing catalyst used in isocyanurate board stock and moulded rigid foams.
1,4-diazabicyclo[2.2.2]octane (DABCO) (Also referred to as triethylenediamine (TEDA))	$N(CH_2CH_2)_3N$	Very good amine gelation catalyst. Used in all types of foams.
2-(2-dimethylaminoethoxy)-ethanol (DMAEE)	$(CH_3)_2NCH_2CH_2OCH_2CH_2OH$	Reactive catalyst used in low density packaging foams.
2-((2-dimethylaminoethoxy)-ethyl methyl-amino)ethanol	$(CH_3)_2NCH_2CH_2OCH_2CH_2N(CH_3)CH_2CH_2OH$	Excellent reactive low odour blowing catalyst used in high resilience and flexible foams. Low vinyl staining.
1-(bis(3-dimethylamino)-propyl)amino-2-propanol (Also referred to as N‴-hydroxypropyl-N,N,N′,N′-tetramethyliminobispropylamine)	$(CH_3)_2N(CH_2)_3N(CH_2CHOHCH_3)(CH_2)_3N(CH_3)_2$	Low odour reactive catalyst used in rigid and high resilience foams. Replaces DMCHA in spray and is low vinyl staining.
N,N′,N″ tris(3-dimethylamino-propyl)hexahydrotriazine	$(NRCH_2)_3$ where R = $(CH_2)_3N(CH_3)_2$	Isocyanurate catalyst that provides back end cure. Decreases demould time of appliance foams.
Dimorpholinodiethylether (DMDEE)	$(O((CH_2)_2)_2N)(CH_2)_2O(CH_2)_2(N((CH_2)_2)_2O)$	Low odour catalyst used in one-component foams and sealants.
N,N-dimethylbenzylamine	$C_6H_5CH_2N(CH_3)_2$	Characteristic smell used in polyester-based flexible foams, integral skin foams and for making prepolymers.
N,N,N′,N″,N″-pentamethyldipropylenetriamine	$(CH_3)_2N(CH_2)_3N(CH_3)(CH_2)_3N(CH_3)_2$	Strong ammoniacal odour used for polyether-based slabstock foams and in semi-rigid foam moulding.
N,N′-diethylpiperazine	$CH_3CH_2N(CH_2CH_2)_2NCH_3CH_2$	Low odour balanced blow cure catalyst for flexible and semi-flexible systems.

Table 9-2   Organometallic catalysts and their applications

Catalyst	Characteristic and use
Stannous octoate	Slabstock polyether-based flexible foams, moulded flexible foams.
Dibutyltin dilaurate (DBTDL)	Microcellular foams, elastomers, moulding systems, RIM.
Dibutyltin mercaptide	Hydrolysis resistant catalyst for storage stable two-component systems.
Phenylmercuric propionate	Delayed action catalyst for elastomers.
Lead octoate	Rigid spray foams.
Potassium acetate/octoate (KA/KO)	Isocyanurate foams.
Quaternary ammonium formates (QAF)	Isocyanurate foams.
Ferric acetylacetonate	Cast elastomers system especially those based on TDI.

Reaction mechanisms

Amine catalysts

Tertiary amines are the most widely used polyurethane catalysts. Two mechanisms have been proposed for amine catalysis. In the one proposed by Baker, Figure 9-3, the activation starts by the amine using its lone pair of electrons to coordinate with the carbon of the isocyanate group. This intermediate then reacts with an active hydrogen from an alcohol to produce a urethane group.

Figure 9-3   Baker mechanism amine catalyst

$$R-N=C=O + R''_3N \longrightarrow R-\overset{\delta^{\ominus}}{N}\!\!=\!\!\underset{\underset{R''\overset{|}{\underset{R''}{N^{\oplus}}}R''}{}}{C}\!\!=\!\!\overset{\delta^{\ominus}}{O} \xrightarrow{R'OH} R-HN-C\!\!\underset{O-R'}{\overset{O}{\diagup}} + R''_3N$$

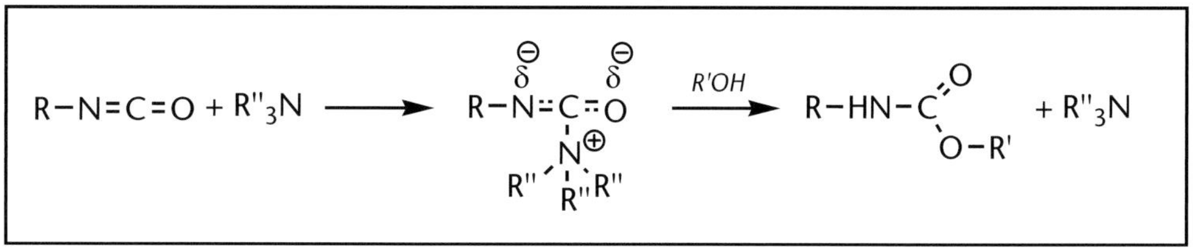

In the second mechanism, proposed by Farka, Figure 9-4, which is supported in the more recent literature, the activation starts by the amine interacting with the proton source (polyol, water, amine) to form a complex, which then reacts with the isocyanate.

Figure 9-4   Farka mechanism amine catalysts

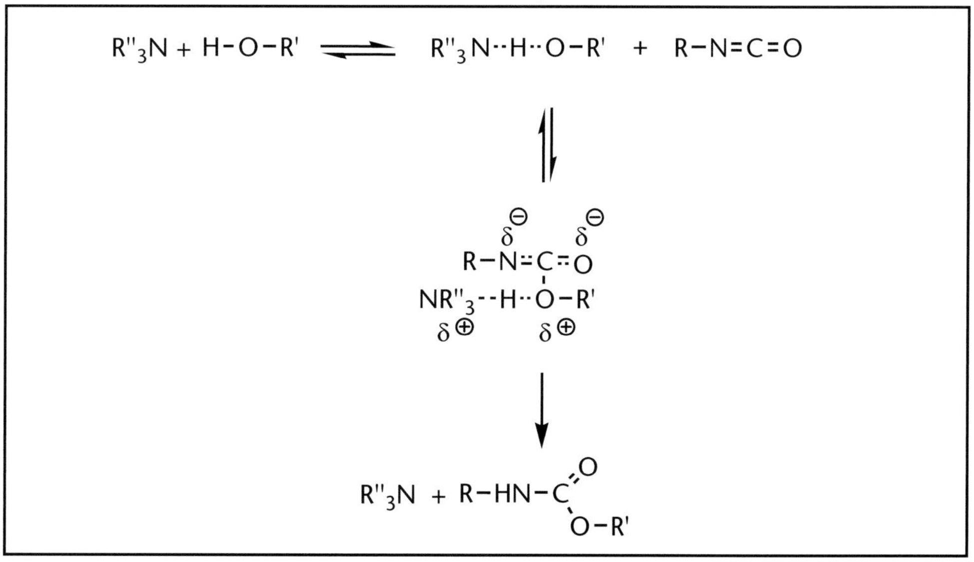

$$R''_3N + H-O-R' \rightleftharpoons R''_3N\cdots H\cdots O-R' + R-N=C=O$$

$$\Updownarrow$$

$$\underset{\underset{\delta^{\oplus}}{NR''_3}\cdots\underset{\delta^{\oplus}}{H\cdots O-R'}}{R-\overset{\delta^{\ominus}}{N}\!\!=\!\!C\!\!=\!\!\overset{\delta^{\ominus}}{O}}$$

$$\downarrow$$

$$R''_3N + R-HN-C\!\!\underset{O-R'}{\overset{O}{\diagup}}$$

Factors that affect the catalytic activity of an amine are nitrogen atom basicity, steric hindrance, spacing of heteroatoms, molecular weight, volatility and end groups. The level of basicity is determined from the pKa value, defined as the pH at which the concentration of unprotonated and protonated forms, of an ionisable group, are equal.

If two structures are similar the one with the higher pKa will be more reactive. This is shown by comparing the relative reactivity of N-methylmorpholine and N-methylpiperidine, using a model system of phenyl isocyanate and butanol in toluene (full details under 'Reaction kinetics'). The N-methylpiperidine, which has a pKa of 10.1 is six times as fast as N-methylmorpholine which has a pKa of 7.4.

Two examples of the effect of steric hindrance are seen from the comparison of N,N-diethylcyclohexylamine and N,N-dimethylcyclohexylamine compared to the classic example of triethylenediamine. The N,N-diethylcyclohexylamine and N,N-dimethylcyclohexylamine have about the same pKa, 10.0, but the diethyl substitution provides more steric hindrance on the amine than the dimethyl substitution. As a result, the dimethyl compound is about 10 times more active than the diethyl. Triethylenediamine has a lower pKa, 8.8, but is much more active (34 times that of the diethyl) than either of the cyclohexylamines because the bi-cyclic structure of triethylenediamine fully exposes the lone pair of electrons making it very active as a catalyst.

The spacing of the heteroatoms has a definite effect on the gel-blow ratio of an amine catalyst. Amines with an ethylene linkage between heteroatoms are much better blowing catalysts than amines with a propylene linkage between heteroatoms. A good illustration is a model study where bis(N,N-dimethylaminoethyl) ether and N,N-dimethylaminoethyl N,N-dimethylaminopropyl ether are used to catalyse the reactions of phenyl isocyanate with water and butanol, Table 9-3.

Table 9-3   Comparison of gel and blow reactivity for N,N-dimethylaminoethyl N,N-dimethylaminopropyl ether (DMAEDMAPE) and bis(N,N-dimethylaminoethyl)ether (BDMAEE)

Time (s)	% Gel DMAEDMAPE	BDMAEE	% Blow DMAEDMAPE	BDMAEE
0.5	7	10	3	11
1	11	16	4	21
2	18	26	8	36
4	32	41	15	54
8	52	57	27	69
15	69	72	41	82
30	84	86	58	92
60	96	93	75	98
100	100	97	86	100

The data shows that the gel reactions for both are about the same while the blow reaction for BDMAEE is much faster than DMAEDMAPE. A theory that explains the difference is that the heteroatom (nitrogen or oxygen) hydrogen bonds with the water by using an unshared pair of electrons. With an ethylene linkage the water is held in close proximity so that when the amine and isocyanate interact the water is favoured over the polyol. With a propylene linkage the atomic distance and degree of freedom of the extra tetrahedral linkage is too great for any significant interaction. Thus, the water and polyol will compete more equally in the reaction with an isocyanate.

The volatility of a catalyst, which is controlled by the molecular weight and functional groups within the molecule, also has an effect on the performance of catalysts. High molecular weight molecules will be less volatile than low molecular weight molecules. Thus, low molecular weight catalysts such as N-ethylmorpholine or N,N'-dimethylpiperazine will migrate to the surface of a flexible foam resulting in improved surface cure. If a compound contains a hydroxyl group this not only increases the boiling point, but the hydroxyl group will react with the isocyanate, thus limiting its migration.

Organotin catalysts

For the tin II salts the following mechanism has been proposed, Figure 9-5. The isocyanate, polyol and tin catalyst form a ternary complex, which then gives the urethane product. Two routes, not shown, to the complex have been proposed. In the first one the tin first adds to the polyol then the isocyanate. In the second one the tin adds to the oxygen of the isocyanate then reacts with the polyol.

Figure 9-5   Mechanism for tin II salts

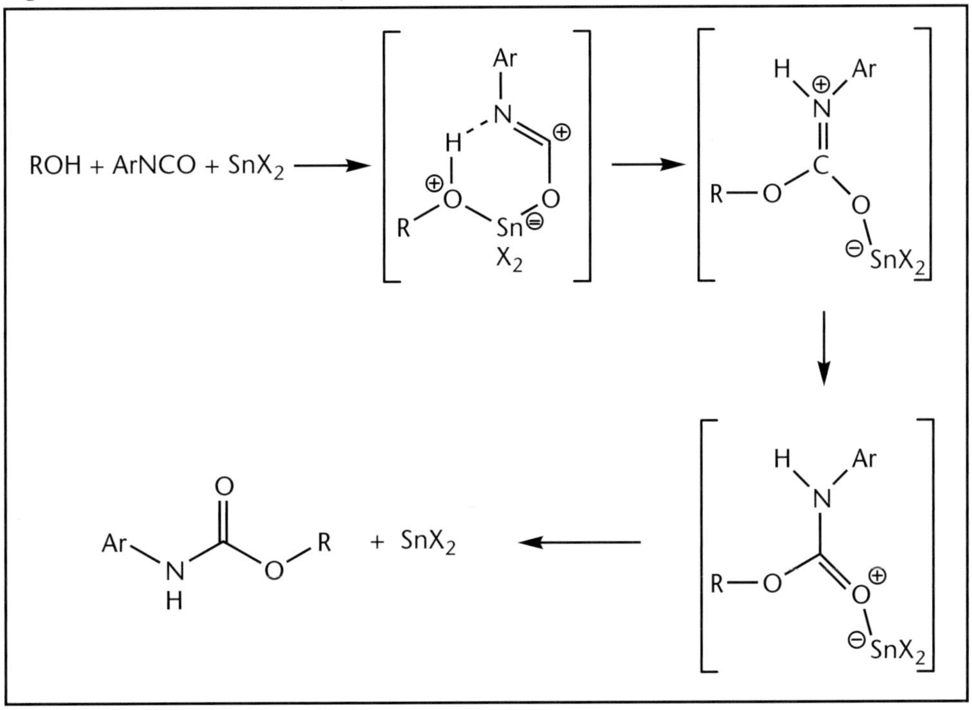

The proposed mechanisms for tin IV catalysts, dialkyltin dicarbonates and dialkyltin dialkylthiolates, is the reaction of the tin with a polyol forming a tin alkoxide, which can then react with the isocyanate to form a complex, Figure 9-6. Transfer of the alkoxide anion onto the co-ordinated isocyanate affords an N-stannylurethane, which then undergoes alcoholysis to produce the urethane group and the original tin alkoxide.

Figure 9-6   Mechanism for tin IV compounds

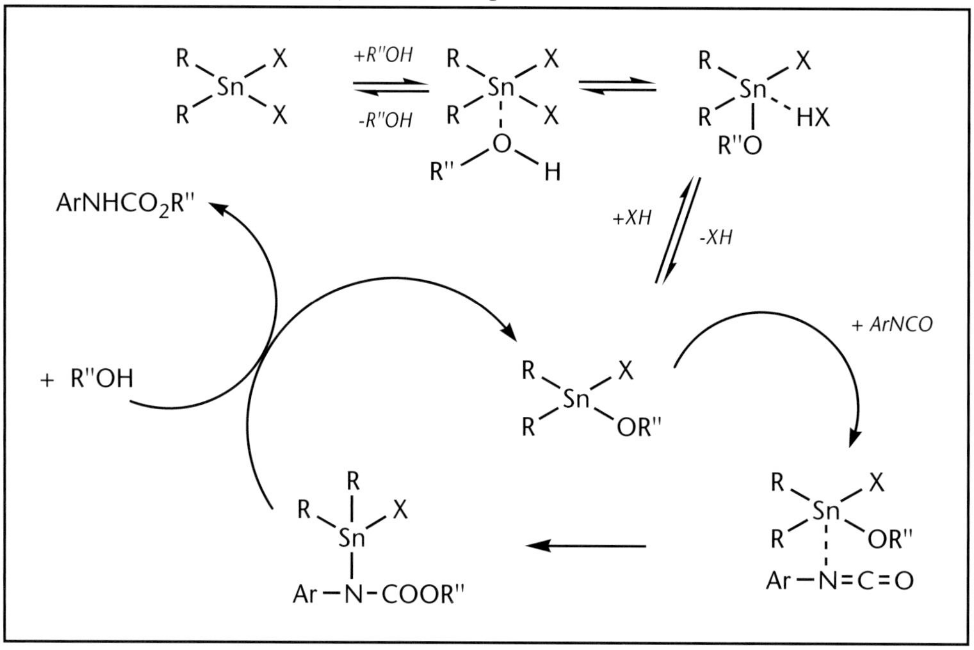

For convenience the tin alkoxide has been shown as being monomeric, but in solution dialkyltin alkoxides are usually mixtures of oligomers with the alkoxide group acting as a bridging ligand. Different oligomers most likely exhibit different levels of catalytic activity. This results in extremely complex kinetics, which is why after 30 years with many kinetic studies being conducted there is still not a generally accepted mechanism.

A synergy between tin and amine catalysts is also observed. One of the proposed mechanisms is given here, Figure 9-7.

Figure 9-7   Mechanism of tin-amine synergism

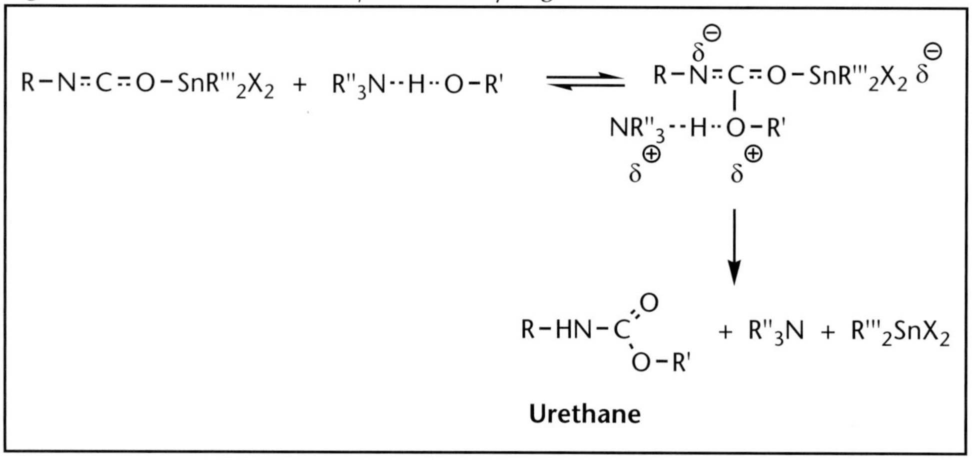

Tin II compounds are used in flexible slabstock foam with the most common being stannous octoate. These are used because they become inactive during the foaming process, thus do not give any dry heat degradation. When tin IV compounds were used in this application, severe thermal oxidation occurred resulting in foams having much poorer physical properties.

Tin IV compounds are used in moulded HR foams and in rigid foams. The most common is dibutyltin dilaurate. However, this and other dialkyltin dicarboxylates suffer from hydrolytic instability. Thus, polyol blends that contain water and the tin IV catalyst cannot be stored for extended periods of time without their reactivity being greatly affected. Tin mercaptides are more hydrolytically stable and are used when this is required.

Isocyanurate catalysts

Isocyanurate catalysts can be divided into two broad categories, carboxylic acid salts and tertiary amines. The most often used carboxylic acid salts are potassium octoate, potassium acetate and quaternary ammonium salts of formic acid. Only a few amines demonstrate isocyanurate catalysis with tris(N,N-dimethylaminomethyl)phenol and tris(N,N-dimethylaminopropyl)hexahydrotriazine being two of the more widely used. Some authors have stated that amines with three nitrogens present in the molecule will show isocyanurate activity. A proposed mechanism is shown in Figure 9-8.

The catalytic activity of quaternary ammonium carboxylates increases with the nucleophilicity of the active centre –COO⁻. The presence of a hydroxyl group in the quaternary ammonium cation significantly increases the catalytic activity of the catalyst. This effect is probably associated with the formation of a carbamate group bonded directly to the cationic part of the catalysts.

The urethane group has been found to be a strong co-catalyst for the isocyanurate reaction probably caused by the induced polarisation of the isocyanate group. The reactivity is also affected by the hydroxyl group present in the quaternary ammonium carboxylate: primary greater than secondary greater than tertiary. Quaternary ammonium carboxylates with two hydroxyl groups are faster than those with one or three.

Reaction kinetics

Having discussed the mechanisms and what affects the catalytic activity of a compound, we need to determine the effect the catalyst has on foam systems and there are currently four methods available for doing this:

- Traditional studies.
- Model systems.
- Fourier transform infrared spectroscopy.
- Ion viscosity.

Figure 9-8   Mechanism for isocyanurate formation

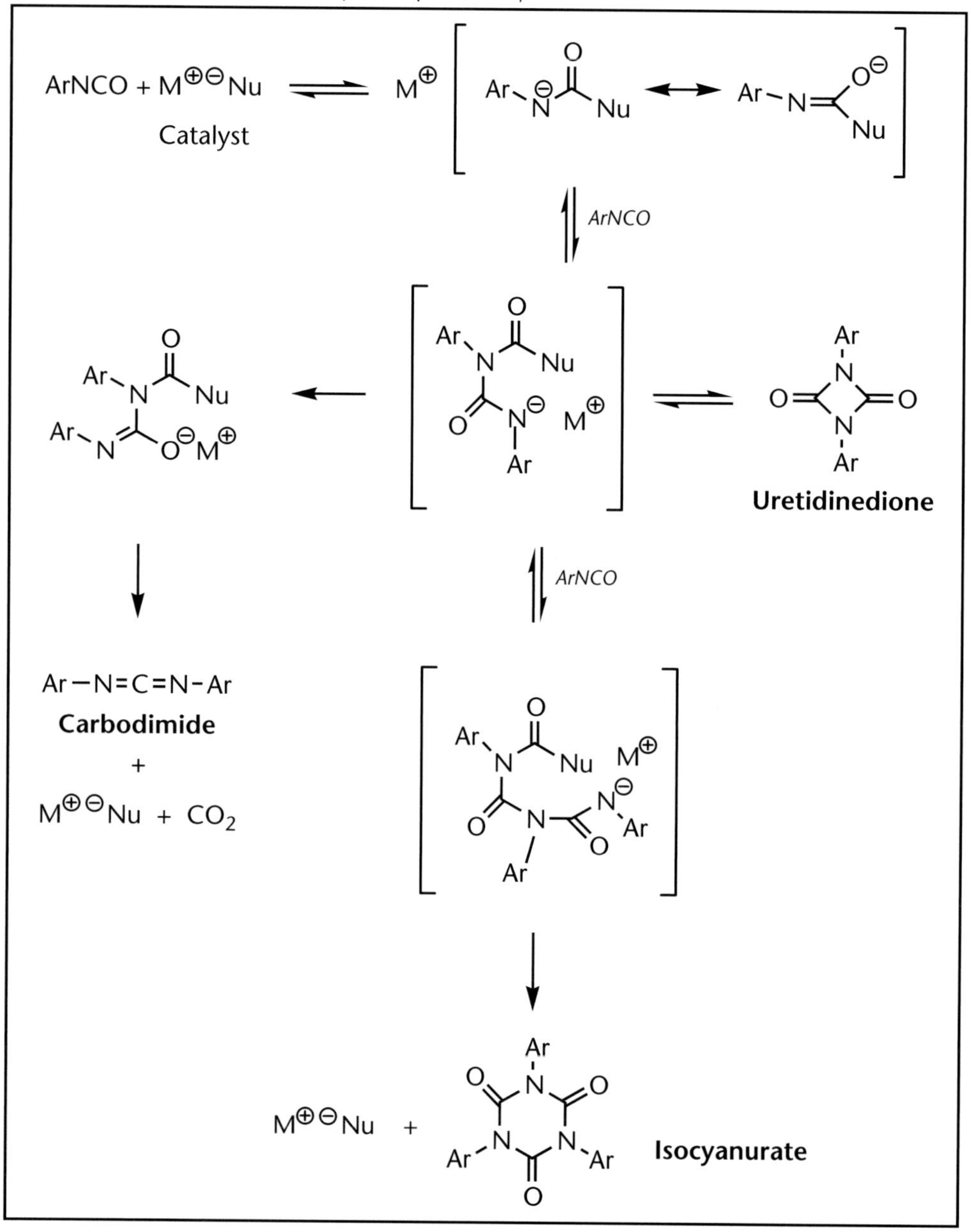

Traditional studies have looked at altering the catalyst content in conjunction with varying amounts of other components, water for instance and observing the difference in processing characteristics. More complex work looks at model systems where phenyl isocyanate is reacted with water and butanol in a solvent. The newest tool available, that enables the researcher to look more closely and deeply at real reacting foams is FTIR (Fourier Transform Infrared Spectroscopy) with programmable heated probes. The programmable heated probe helps to eliminate the problem of the probe acting as a heat sink.

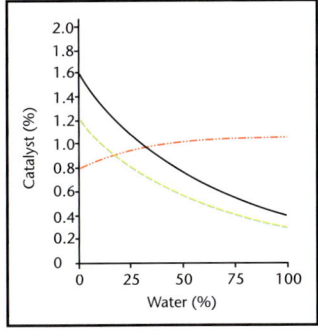

*Figure 9-9
Comparison of catalyst concentration versus percentage water blowing to achieve a 90-second string gel time*

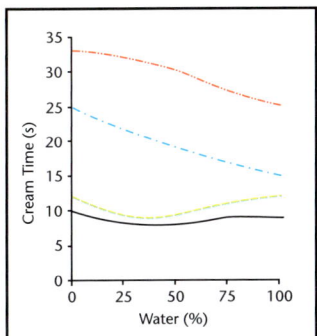

*Figure 9-10
Comparison of cream time at 90-second gel time versus percentage water*

Traditional experiments

An illustrative experiment looks at the blowing reaction in rigid foam where the blowing system was varied from 0 to 100 per cent water blown. This was monitored by determining the amount of catalyst required to achieve a 90-second string gel time. Plotting the amount of catalyst required to achieve this, versus the amount of blowing due to water, it was observed that catalysts like bis(N,N-dimethylaminoethyl)ether (BDMAEE) and hydroxyethyltrimethyl-bisaminoethylether (HETMBAEE), which have an open ethylene linkage between hetroatoms, were greatly affected by how much water is in the system. Other catalysts like triethyl-enediamine were fairly insensitive to the amount of water present. Another interesting point is the cream times of these systems as both TEDA and DMCHA have a much longer cream time than BDMAEE or HETMBAEE. Thus, the time between cream and gel is much shorter with the former two, which leads to some practical advantage when designing a catalyst package for a foam system. The results of these experiments are shown in Figure 9-9 and Figure 9-10.

Model systems

More detailed studies have been conducted using model systems. In these studies phenyl isocyanate is reacted with water and butanol in a solvent such as toluene or acetonitrile. Aliquots of the reaction are taken at specified times, quenched and analysed by high-pressure liquid chromatography.

With this technique the amount of urethane reaction, water reaction, biuret formation, allophonate formation and isocyanurate formation can be determined as a function of how much isocyanate has reacted. By dividing the amount of product observed by the maximum amount possible in the reaction the percentage of possible urethane (Gel) or water reaction (Blow) can be determined and plotted against time or isocyanate concentration.

This procedure was used to measure the efficiency of a range of catalysts determining per cent gel, per cent blow and per cent phenyl isocyanate against time in each case, Figure 9-11.

The comparison of triethylenediamine (TEDA) and bis(N,N-dimethylaminoethyl) ether (BDMAEE) shows that at the same amount of un-reacted isocyanate, TEDA gives more gel reaction then BDMAEE.

A similar comparison for N"-hydroxypropyl-N,N,N',N'-tetramethylimino-bispropylamine (HPTMIBPA) and N,N-dimethylcyclohexylamine (DMCHA) shows they are very similar with HPTMIBPA having slightly more gelation.

The effect of replacing a methyl group, with a hydroxyethyl group is shown by comparing BDMAEE with N-hydroxyethyl-N,N',N'-trimethyl bis(aminoethyl) ether (HETMBAEE). It can be seen that whilst the latter catalyst is slower, its use gives greater gelation at any given isocyanate conversion.

Figure 9-11   Per cent gel, blow and phenyl isocyanate level versus time for a range of catalysts in a model system
A: Per cent gel formation   *B: Per cent blow*   *C: Per cent phenyl isocyanate*

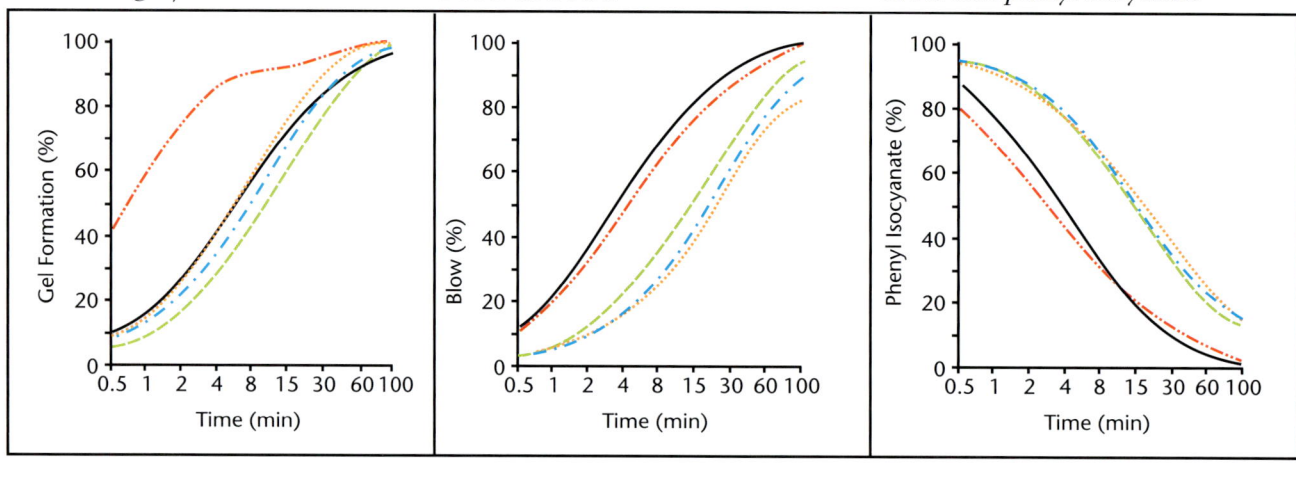

The preceding data is all based on a model system. However, results from the model can vary with the solvent used, the concentration of reactants and catalyst and temperature.

FTIR experiments

The challenge has been to determine what is actually occurring in the foam during the foaming process. Recently, FTIR equipment with programmable heated probes has become available which allows the monitoring of the reactions occurring in the foam during the foam process. The heated probes can be programmed to follow the reaction exotherm, thus ensuring that they do not act as heat sinks, which locally perturb the reaction kinetics. By identifying IR peaks and following their increase or decrease the reactions and rate of reaction can be determined.

Figure 9-12   Typical FTIR plot

A typical FTIR plot is shown in Figure 9-12 and this technique can be used, monitoring the 1400 cm^{-1} to 1800 cm^{-1} band that includes the very interesting carbonyl region, to follow the varying effect of several catalysts in a water-blown rigid foam.

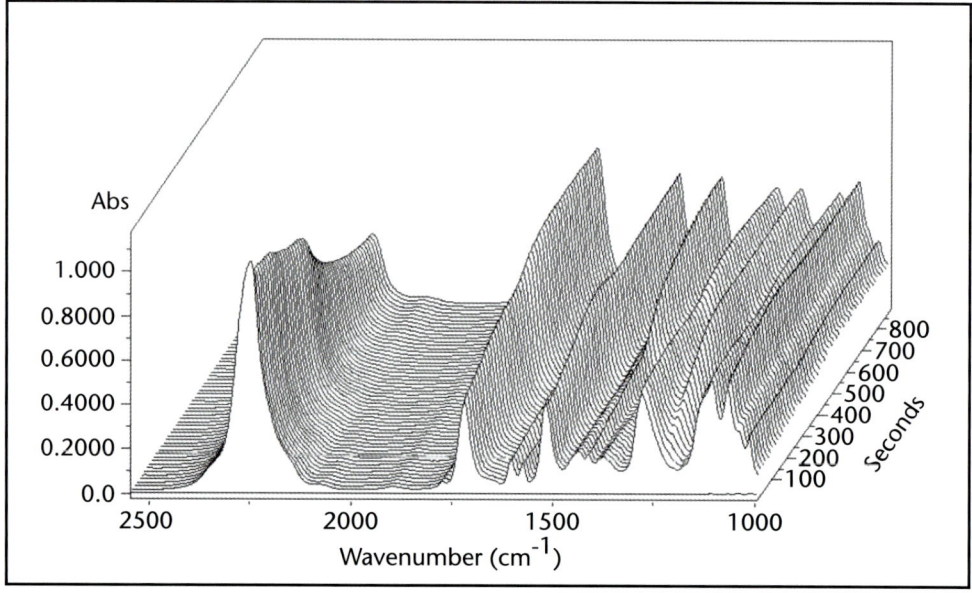

Figure 9-13   Urethane development with time in MDI high resilience foams comparing TEDA and BDMAEE catalysts from FTIR plots at 1733 cm⁻¹

Figure 9-14   Isocyanate conversion with time for high resilience foams using TEDA and BDMAEE catalysts

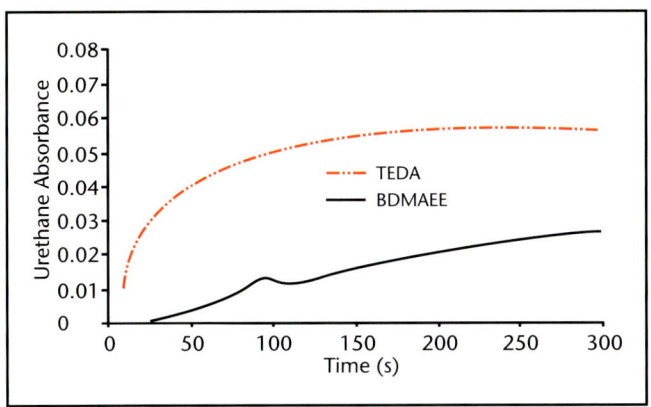
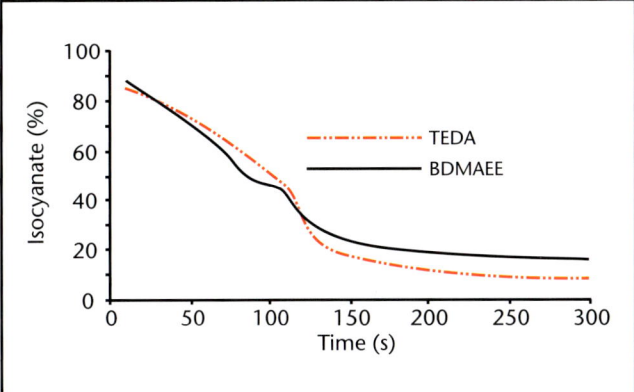

In a second example the urethane absorbance peak (1733 cm⁻¹) was monitored and plotted versus time, Figure 9-13, for an MDI-based high-resilience flexible foam system catalysed by TEDA or BDMAEE The isocyanate conversion for these two systems was also plotted versus time, Figure 9-14. Both foams had the same rise time.

This comparison shows, like the model study, that TEDA is a much better gelation catalyst than BDMAEE.

FTIR has also been used to study polyisocyanurate foams by measuring the isocyanurate peak at 1409 cm⁻¹ for a range of catalysts, Table 9-4.

The data show that catalysts like potassium acetate, potassium octoate, quaternary ammonium formate are very good isocyanurate catalysts while PMDETA which is often used in polyisocyanurate foams and BDMAEE are less active isocyanurate catalysts.

Table 9-4   Comparison of catalysts for the formation of isocyanurate links (trimer peak at 1409 cm⁻¹)

Time (s)	Temp (°C)	Trimer peak height (mm) at 1409 cm⁻¹				
		BDMAEE	K Acetate	K Octoate	PMDETA	QAF
0	26	0.0	0.0	0.0	0.0	0.0
100	43	1.0	0.0	0.0	0.4	0.0
200	60	1.3	0.3	0.6	0.9	0.4
300	77	1.3	0.6	1.0	1.0	1.2
400	94	1.2	1.2	1.2	0.8	1.8
500	111	1.1	4.0	2.9	0.8	3.5
600	128	1.0	5.0	3.3	0.8	4.0
700	145	1.0	5.5	3.5	0.8	4.1
800	162	1.0	5.8	3.6	1.0	4.1
900	180	1.0	5.8	3.6	1.1	4.0

Ion viscosity

All this data still does not provide us with a missing piece of information: how rapidly does a foam cure or to what extent has the polymerisation been completed. This can be determined by measuring the ion viscosity of the polymer since the more a polymer is cured the higher the ion viscosity. All materials have an ionic dipole although some are very small. The ion viscosity is a measurement of the ability of the ions to align themselves with an electric field. As a polymer cures the ions are more resistant to alignment and thus the ion viscosity increases. A comparison of two catalyst systems is shown in Figure 9-15. The TEDA-HETMBAEE system, as shown by the higher ion viscosity, provides a faster surface cure than the TEDA-BDMAEE system.

This brief overview of polyurethane catalysis shows that an understanding of the reactivity of catalysts and their effects on the blowing and gelation reactions enables better choices of catalysts to be made in order to achieve desired effects for a particular foam systems.

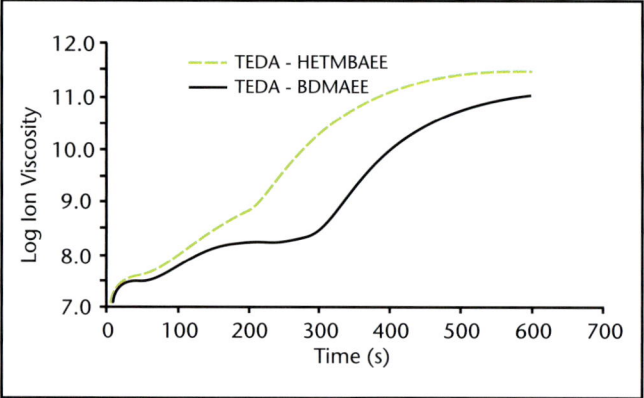

Figure 9-15   *Ion viscosity data for two catalyst systems*

10. Additives

Alan Hamilton

In addition to the basic materials needed to make polyurethanes, isocyanates and polyols, a wide range of other chemicals can be added to modify and control both the polyurethane chemical reaction as well as the properties of the final polymer. The main additives are blowing agents, catalysts, surfactants, fire-retardants, cross-linking agents and fillers.

Blowing agents

A full description of the structure and chemistry of the blowing agents used to make foamed polyurethanes is provided in Chapter 8.

Catalysts

Details on the production, structure and chemistry of the amine and metal catalysts used to promote polyurethane reactions is provided in Chapter 9, but a careful selection is required in order to achieve specific performance characteristics for the different applications.

Flexible foams

When producing flexible foams, the catalyst choice is important since as well as promoting and controlling the different rates of the foaming reactions, it is necessary to ensure that polymer growth and blowing remain in balance. This can only be achieved by selecting both the correct combination and concentrations of catalysts. Other factors that have to be taken into consideration when choosing a catalyst are: odour, vapour pressure, their influence on final product properties and cost.

Many flexible foam production processes require the use of tin (II) catalysts to ensure that the rate of polyol and isocyanate reaction is fast enough. Stannous octoate is mainly used since the residue left in the final foam has little or no effect on the foam properties after ageing.

A list of typical catalysts used in flexible foams is given in Table 10-1.

Table 10-1   Flexible foam catalysts

Chemical description	Application and comments	Trade Names
Tin octoate	One-shot slabstock foam, organometallic	Jeffcat™ T9, Dabco™ T9, Fomrez™ C-2, Kosmos™ 29
Triethylenediamine	General purpose gel catalyst	Jeffcat TD-33A, Dabco 33LV, Niax™ A-33
Bis(N,N-dimethylaminoethyl)ether	Blowing catalyst for most flexible applications	Jeffcat ZF-22, Niax A1, Dabco BL11
N,N-dimethylethanolamine	Low odour general purpose catalyst	Jeffcat DMEA, Dabco DMEA, Niax DMEA
N-ethylmorpholine	Cure catalyst for polyester foams – volatile	Jeffcat NEM, Dabco NEM
Dimethyl-2-hydroxy(propyl) 1,3-propylene diamine	Low odour balanced catalyst, good flow, reactive	Jeffcat DPA
N,N,N-trimethyl-N-hydroxyethyl bis(aminoethyl)ether	Blowing catalyst, low odour in foam, used in foams moulded against PVC skins	Jeffcat ZF-10
Pentamethyldipropylenetriamine	Balanced catalyst for moulded HR foam, low odour, smooth rise profile	Jeffcat ZR-40, Polycat™ 77
N,N-dimethylpiperizene	Latent cure catalyst for foam	Jeffcat DMP,
Acid blocked bis(dimethyl aminoethyl)ether	Delayed action blow catalyst, acid blocked	Jeffcat ZF-54, Niax A107, Dabco BL-17
N,N-dimethylhexadecylamine	Surface cure catalyst, can lead to fogging	Dabco B-16, Armeen™ DM-16

TDI-based slabstock foam formulations typically contain two amine catalysts that control the gas generation reaction and support the cross-linking reactions with additional tin catalysts present to ensure that the polyaddition and cross-linking reactions are in balance. Too high a level of tin catalyst can result in 'foam shrinkage' whilst too little leads to foam recession or even internal foam splits.

The choice of catalyst combination for moulded foams depends very much on the process used. Hot-cure moulding tends to use similar catalyst types and combinations to slabstock formulations, as the relatively unreactive polyols used again require the use of tin catalysts. The cold-cure process uses more reactive polyols, which means that only tertiary amine catalysts are required. It is not unusual to use three different amine catalysts to ensure the correct blow, flow and cure of the foam systems. Delayed action catalysts are often used to obtain optimum demould times with good foam processability.

Rigid foams

The density range for rigid polyurethane foams varies tremendously from 8 to 1,000 kg/m^3, which leads to a wide range of processing and reaction conditions. For example, spray foams have to react almost instantaneously on a cold substrate at temperatures as low as 0°C, whilst other foam formulations may have reactions that take several minutes to start.

This wide range of products and applications has led to the identification and development of a range of catalysts for rigid foam that can be assigned into four groups:

- Aliphatic and aromatic tertiary amines.
- Organometallic compounds; these are normally tin (II) compounds, although lead and bismuth compounds are also used at times.
- Alkali metal salts of carboxylic acids and phenols.
- Miscellaneous compounds such as triazine derivatives.

The choice of catalysts from the above groups is very much dependent on the types of rigid foam application and the processing method being utilised and the commercial catalysts commonly used are listed in Table 10-2.

Table 10-2   Rigid foam catalysts

Chemical description	Application and comments	Trade Names
Dimethylcyclohexylamine	General purpose catalyst for most rigid applications	Jeffcat™ DMCHA, Dabco™ DMCHA, Polycat™ 8, Niax C-8
N,N-dimethylethanolamine	Reactive mid-range catalyst – good for block foam	Jeffcat DMEA, Dabco DMEA
Pentamethyldiethylenetriamine	Useful catalyst which gives cure and flow – used in appliance formulations	Jeffcat PMDETA, Polycat 5, Toyocat™ DT
Triethylenediamine	Cure catalyst – good for water-blown foam	Jeffcat TD-33A, Dabco 33LV, Niax 33A
Bis(N,N-dimethyl-3-amino propyl)amine	Reactive catalyst, good for water-blown foam, used in packaging foam	Jeffcat ZR-50B, Polycat 15
1,3,5-tris(3-(dimethylamino)propyl) hexahydotriazine	Strong urethane catalyst, good PIR co-catalyst, used in lamination foams	Jeffcat TR-90, Polycat 41
Bis(dimethylaminoethyl)ether	Blowing catalyst, good for fine cells	Jeffcat ZF-22, Niax A1, Dabco BL 11
Potassium octoate in glycol	Good PIR catalyst, used as co-catalyst, used in PIR foam block and lamination	Dabco K 15
Dibutyltin dilaurate	General purpose tin catalyst, cure catalyst for lamination and spray foam	Dabco T 12
Dibutyltin diisooctylmercapto acetate	High reactivity tin catalyst, used in spray foam	Dabco 120

As in flexible polyurethane foams, it is important to maintain the correct balance of the water-isocyanate and polyol-isocyanate reactions to ensure the optimum processing and product properties. Catalyst balance is also important in the production of polyisocyanurate foams where the reactions are mainly between isocyanate and polyol and trimerisation of the isocyanates themselves. Most commercial rigid foam systems require two or three different catalysts to obtain the optimum balance for processing.

Coatings, adhesives, sealants and encapsulants

This diverse area of polyurethane technology demonstrates the full versatility of polyurethanes and makes many demands on catalyst technology. Besides those catalysts used in flexible and rigid foam production, special grades of catalysts have been developed for this area and some of these are listed in Table 10-3.

Table 10-3 Other catalysts for polyurethanes

Chemical description	Application	Trade and General Names
Triethylenediamine/Glycol blends	Catalyst of choice for most applications	TEDA, Jeffcat™ TD-33A, Dabco™ EG
Tin salts (various)	Used in many elastomer and coating applications, different salts give varying reactivity and stability	DBTDL, Tin salts - UL1, Dabco 120
Bismuth iso-octoate	Less toxic alternative for coatings and elastomers, hydrolyse fairly easily	–
Lead naphthenate	Catalyst for many spray applications, being phased out for environmental reasons	–
Ferric acetylacetonate	Catalyst for TDI based cast elastomers	–
Phenyl mercury salts	Catalysts for potting compounds and adhesives, very specific to isocyanate-hydroxyl reaction.	–
N,N-dimorpholinodiethylether	Catalyst for moisture-cured systems, gives better shelf life	Jeffcat DMDEE
Tris(dimethylaminomethyl)phenol	Weak PIR catalyst, co-catalyst in PIR foam	Jeffcat DMP 30
Tris(dimethylaminopropyl)amine	Less active elastomer catalyst, gives better flow than TEDA	Jeffcat Z 80

Polyurethane coatings exist in a wide range of options from one- or two-component thin films, which are in effect thin layers of elastomer, to cellular coatings for carpets, to special lacquers and high-performance hard protective coatings. There are three general classifications for coatings, all of which require different catalyst technology to some extent:

- Two-component coatings (includes blocked systems and prepolymer).
- One-component moisture-cured coatings.
- Oil-modified or alkyd coatings.

Two-component systems use a range of catalysts both to accelerate the isocyanate-polyol reaction and to achieve the correct drying time and final coating properties. This is of particular importance for coatings based on the less reactive aliphatic isocyanates, where combinations of organic tin and zinc compounds are often used in combination with tertiary amines.

Blocked prepolymers generally have much lower catalyst levels, as the temperatures required to unblock the isocyanate also tend to ensure adequate cure or drying times.

One-component systems cure either by water vapour reacting with the terminal isocyanate group or by heat that promotes further cross-links and final cure. Although it is possible to add catalysts to these coatings to promote cure this is only normally done for the less reactive isocyanates as catalyst addition reduces storage stability and pot life. Typical catalysts used are metal salts such as cobalt and tin. It is possible to add the catalyst just prior to use and thus increase cure rates. These catalysts would be similar to those used in two-component coating systems.

Oil-modified coatings or 'Urethane Oils' are resins with hydroxyl end-groups rather than isocyanate groups. The cure of these coatings is via air oxidation of the unsaturated linkages in the polyol and catalysts that aid this 'drying process' are metal salts, typically of cobalt or lead.

Polyurethanes form a large and versatile family of adhesive polymers that have the ability to provide excellent adhesion at low temperature with rapid curing and high cohesive strengths. Applications vary from low-density expandable gap filling to lamination and binders for organic and inorganic materials. Polyurethane-based adhesives can be one- or two-component, both of which require different catalyst technology to enable the correct processing and property needs to be met.

One-component reactive adhesives consist of prepolymers that react with moisture from the air or a substrate and amine catalysts are added to achieve the correct cure level. The choice of catalyst depends on the type of isocyanate and prepolymer, the substrate temperature, volatile residue issues and the level of cure required. Organometallic tin (II) catalysts are also used, usually in combination with amine catalysts. The selection of the catalyst combination depends not only on obtaining the correct reaction rate of the adhesive, but also in ensuring that the storage life of the one-component system is as long as possible. Some organometallic catalysts are incorporated into blocked isocyanate formulations where they carry out the dual role of catalysing the unblocking of the isocyanate and the subsequent moisture curing reaction.

Elastomeric materials are an even more diverse category ranging from the high-density end of flexible foams, via footwear and RIM microcellular elastomers, through solid elastomers and potting compounds to thermoplastic polyurethane (TPU). This wide divergence of products leads to a large variation in the type of catalyst required, dependent on the raw materials, the process technology and final product. These can be split into three key groups:

- Tertiary amine catalysts.
- Tin catalysts.
- Other metallic catalysts.

Tertiary amine catalysts

These are generally used in two-component polyurethane formulations especially for footwear and RIM applications with the standard catalyst being triethylenediamine (TEDA), which is normally used as a solution in the relevant chain extender. Polyester footwear formulations are often formulated using TEDA as the sole catalyst. There has been a general trend to use the blocked amine catalysts for elastomer applications to improve demould time and maintain processing performance as these catalysts give enhanced 'back-end' cure without shortening the cream time.

Tin catalysts

Tin catalysts are often used in polyether-based shoe soling and RIM formulations to maximise cure, normally in combination with tertiary amine catalysts. Typical examples include dibutyltin dilaurate and dibutyltin dimercaptide; the latter has an improved hydrolytic performance. Conversely, residual levels of certain tin catalysts can contribute to the poor hydrolytic stability of the final polymer.

Other metallic catalysts

Other metallic catalysts include lead, bismuth and mercury compounds and the specific choice of these depends very much on the application. These catalysts are usually octoate and naphthenate salt compounds. Mercury catalysts such as phenyl mercury propionate are widely used in elastomer technology because of their ability to give good cure and physical properties at relatively low temperatures. These catalysts are also very selective toward the isocyanate-polyol reaction. Lead catalysts, historically, have been used in spray applications, but are less popular today because of toxicity concerns.

Bismuth catalysts are becoming more widely used where reduced catalyst toxicity is desired. The wide application of elastomers has led to the development and use of a number of special effect catalysts. Amongst these are ferric, zinc and nickel acetyl acetonates, these being selected to give the desired controlled reactivity and selective polymerisation. Titanium-based catalysts are reported to be used for applications where rapid cure and improved thermal stability are required.

For the production of TPUs, the residual catalyst in the polyol, in conjunction with their primary hydroxyl content, ensures an adequate reactivity. Also, any additional residual catalyst can lead to problems such as property degradation during both the melt stage of production and in final use.

Surfactants

Surface-active additives are used in the manufacture of most polyurethane foams and carry out two roles. First, they help in mixing incompatible components and, second, they have a critical role in stabilising the early stages of the reacting foam structure until sufficient polymerisation has occurred to form a self-supporting polymer network. The most important function of the surfactant in foams is bubble stabilisation. To achieve this it is necessary to combine a controlled rate of surface tension reduction, whilst achieving the overall reduction required, but also to have a degree of elasticity in the rising liquid foam.

During the early part of foam rise, carbon dioxide diffuses into the air bubbles, which act as nucleation sites. As the foam expands the amount of interface between liquid and gas increases lowering the concentration of surfactant at the interface leading to an increase in surface tension. This is compensated for partly by the diffusion of surfactant molecules from the bulk of the liquid, and partly by the migration of surfactant along the surface from areas of high surface concentration to those of low concentration. This migration or surface transport draws underlying layers of bulk material with it reducing the tendency for cell windows to thin and thus delays cell rupture. These phenomena were explained by Plateau, Gibbs and Marangoni in the 1870s and more recently by Ewers and Sutherland in 1952.

To obtain the most efficient bubble stability and prevent coalescence the correct balance of foam reactivity and surfactant activity must be obtained. In most polyurethane systems the surfactant must act within a few minutes since if this balance is not optimised the surfactant will not be able to do its job and surface film rupture and defoaming can occur.

Early polyurethane foams used one or more organic, usually non-ionic surfactants such as substituted nonyl phenols, fatty acid ethylene oxide condensates or alkylene oxide block co-polymers. These surfactants are still used for some specific semi-rigid foams and low-density polyester-based foams. Most polyurethane foams are now made using silicone-based surfactants either simple silicone oils, polydimethylsiloxane (PDMS) with no pendant side chains, or more complex polyoxyalkylene polysiloxane copolymers with varying levels of pendant side chains.

Structural parameters of silicone surfactants

To perform as a surface-active agent in a given polyurethane system the polyoxyalkylene polysiloxane copolymers must be partially soluble in the system. This is achieved by obtaining the correct balance of the molecular structure and molecular weight of both the polydimethylsiloxane backbone and the polyoxyalkylene pendant groups. Generally, the following parameters can be varied to obtain the desired surfactant effect:

- Ratio of siloxane backbone to pendant oxyalkylene chains.
- Final molecular weight of the copolymer.
- Polydimethylsiloxane functionality.
- Ratio of ethylene oxide/propylene oxide in the polyoxyalkylene chain.
- Terminal group of the polyoxyalkylene chain, reactive or non-reactive.
- Type of bond between siloxane and polyoxyalkylene chains.

Generally, silicone surfactant copolymers are classified as hydrolysable or non-hydrolysable depending on the chemical bond between the siloxane (backbone) chain and the oxyalkylene (pendant) chain. In hydrolysable copolymers the chains are linked by silicon-oxygen-carbon bonds whilst in non-hydrolysable copolymer they are linked by silicon-carbon bonds. The generic structures of the polyoxyalkylene polysiloxane copolymers are shown in Figure 10-1.

Figure 10-1   Structures of polyoxyalkylene polysiloxane copolymers

Silicon-oxygen-carbon bonds

P—O—Si(Me)(Me)—[O—Si(Me)(Me)]$_n$—[O—Si(Me)(O-Si(Me)(Me)-O-Si(Me)(Me)-O-P))]$_m$—[O—Si(Me)(Me)]$_n$—O—Si(Me)(Me)—O—P

Silicon-carbon bonds

Me$_3$SiO—[Si(Me)(Me)—O]$_n$—[Si(Me)([CH$_2$]$_3$-O-P)—O]$_m$—SiMe$_3$

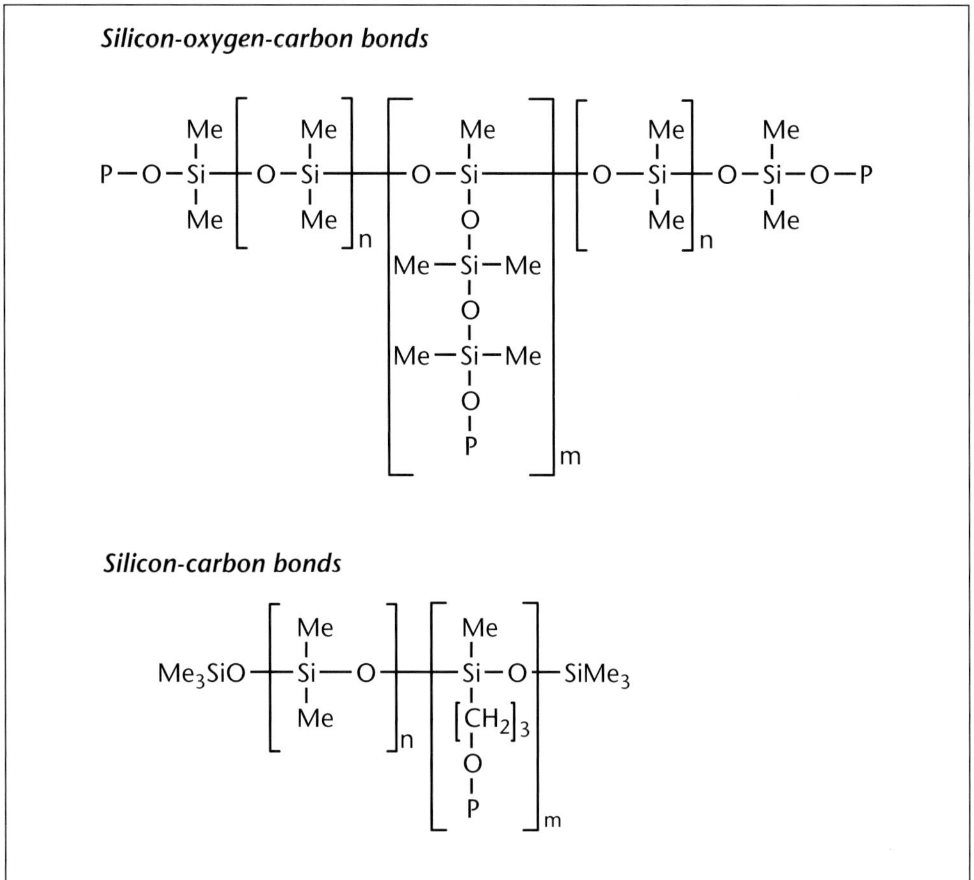

P = (C$_2$H$_4$O)$_x$ (C$_3$H$_6$O)$_y$-R   R = alkyl or H

Flexible foams

In flexible foams, surfactants not only stabilise the foam, but help control the degree of cell opening and increase the processing latitude by widening the operating margin between cell collapse and foam shrinkage caused by too high a closed cell content. Collapse can occur when cell opening takes place prior to sufficient foam strength being obtained. Shrinkage and high closed cell content occur when the cell opening process has been delayed too long. In flexible polyether foams the cell opening mechanism is also greatly dependent on the stannous octoate catalyst level. It is, therefore, necessary that for each foam formulation the correct balance of surfactant and catalyst level is established to ensure the broadest processing latitude.

Currently, polyoxyalkylene polysiloxane copolymers are the only class of surfactant found to be effective in producing flexible foams. It is possible using such surfactants to lower the surface tension of a polyol from approximately 0.034 to 0.022 N/m, while organic surfactants cannot achieve this.

The surfactants for conventional flexible foams are normally non-reactive polyoxyalkylene polysiloxane copolymers with typical polysiloxane contents of the order of 20 per cent and a molecular weight of 5,000 to 15,000. Reactive surfactants link to the polyol tending to lead to a closed cell structure. The greater reactivity of a high resilience flexible foam formulation means that foam stabilisation requirements are less, but cell structure control and cell opening needs are greater so the surfactants for these systems have higher polysiloxane level, typically of the order of 35 per cent and a molecular weight of 10,000 to 30,000.

With flexible foams based on polyester polyols stabilisation is less important than effective foam emulsification so organic and silicone-based surfactants are used. Silicone surfactants for polyester foams are polyoxyalkylene polysiloxane copolymers with relatively low molecular weights of around 2,000 and polysiloxane levels of 45 per cent. The current practice in polyester flexible foam production, of blending the water, catalyst and surfactant into a single component has led to the development of surfactants that exhibit good solubility characteristics in water and amine catalyst blends.

Rigid foams

Rigid foam formulations are substantially different to flexible foams in the following ways:

- The surface tension of the polyols is higher, typically 0.040 to 0.055 N/m, so polyoxyalkylene polysiloxane copolymers with different activity are required.
- The higher viscosity of the foaming mixture, in part due to the high viscosity of the polyols, inherently leads to a more stable bubble structure.
- A closed cell structure, not open, is required.

Because of these differences, the specification for the surfactants has to be different and whilst the structure is similar to the flexible foam surfactants there are important differences:

- The ethylene oxide content of the polyoxyalkylene chain is higher, by 50 to 100 per cent.
- The polysiloxane content is higher, by 40 to 50 per cent.
- Hydroxy functional compounds are used in the majority of cases as they generally have better solubility in the polyols.

An additional trend seen recently is that the introduction of different blowing agents has led to a greater range of polyoxyalkylene polysiloxane copolymer surfactants for rigid foam, each one being developed to suit the blowing agent technology being used. Surfactants are now available which enhance pentane solubility, reduce voids when using low-boiling blowing agents or ensure stable emulsions of insoluble blowing agents. Some silicone surfactants have also been developed which give improved fire performance characteristics when compared with other surfactants in small-scale fire tests.

Elastomers and other applications

The wide family of elastomeric compounds ranges from high-density solid materials at one end to higher density versions of flexible foams at the other. Organic and silicone surfactants are often added to the latter to promote emulsification and mixing, but stabilisation during foaming is not normally required, as the rate of polymer build up is fast enough to ensure stability. Besides these foams only a few of the elastomeric applications require the use of surfactants with the other major area being the production of shoe soles.

Polyester shoe sole formulations are usually based on typical rigid foam silicone surfactants, with high ethylene oxide levels, to help emulsify the chain extender in the polyol and thus prevent physical separation of the polyol blend in the machine tanks. On the other hand, shoe sole systems based on polyether polyols, in which the chain extenders are soluble, only need weakly stabilising surfactants, of the type used in high resilience foam systems, to promote mixing and to ensure stability of the foam whilst it flows and fills the mould.

Organic and silicone surfactants are often added to sprayed elastomers and reactive two-part coating systems for two reasons: first, they promote mixing and the emulsification of the two components, secondly they reduce the surface tension sufficiently to allow air and carbon dioxide to escape from the reacting matrix without foaming. Without this release of gas a foam structure would form resulting in loss of performance and physical properties.

Fire retardants

The addition of fire retardants to polyurethanes reduces the level of fire, flame and smoke combustion products when checked by specified combustibility test methods. There is an extensive range of products available from solids such as melamine, exfoliated graphite or aluminium trihydrate to low-viscosity liquid compounds. These latter products may be reactive or non-reactive, usually contain bromine, chlorine or phosphorous with the latter present as phosphate, phosphite or phosphonate groups.

The general view is that halogen fire retardants function in the gas or vapour phase by interfering with the free radical process, which is associated with combustion, whilst phosphorous retardants are thought to act in the solid phase and promote the formation of protective char. Because these retardants work by different mechanisms a combination of both types can have a synergistic effect, Figure 10-2.

The choice of fire retardant additive depends on a number of key factors: the specific nature of the fire test, long-term ageing performance plus environmental, health and safety issues. The latter has been the driver to move away from bromine-containing compounds to those containing phosphorous or chlorine and also accounts for the increasing use of reactive fire retardants. The range of fire retardants used in polyurethane formulations is listed in Table 10-4.

Figure 10-2   *Synergy of bromine and phosphorus fire retardants on fire test performance (DIN 4102 B2)*

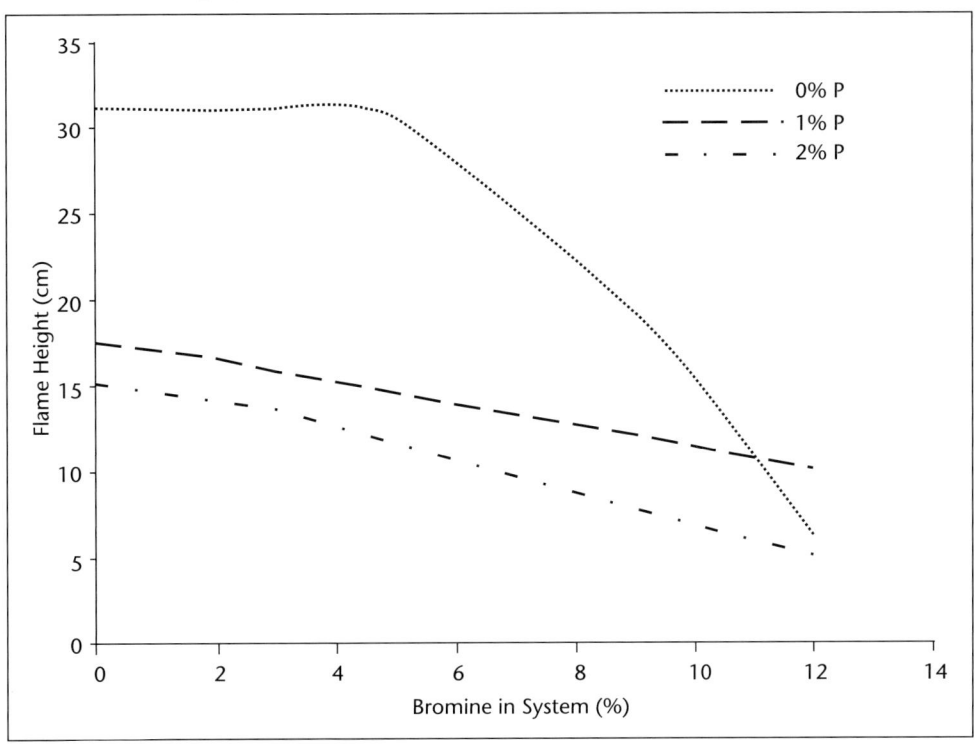

Table 10-4   *Fire retardants used in polyurethane systems*

Product	Description	Application	Active Component (%)			
			N	P	Cl	Br
TCPP	Non-reactive phosphate ester	Most polyurethane applications		9.5	32.5	
TEP	Non-reactive phosphate ester	Most polyurethane applications		10.8	37.3	
DEEP	Non-reactive phosphate ester	Mostly rigid applications		18.7		
Diethyl bis (2-hydroxyethyl) amino methyl phosphonate	Reactive phosphonate, OHv 450	Rigid and flexible foams		12.1		
Brominated phthalic anhydride based ester	Reactive diol OHv 215	Rigid foams				46
Dibromoneopentyl glycol	Reactive diol, OHv 420	Rigid and flexible foams				61
Brominated polyether polyol	Complex mixture, OHv 330	Rigid foams			6	31
Melamine	Solid	Flexible foams	67			
Ammonium polyphosphate	Solid	Flexible foams and elastomers		30.5		
Aluminium trihydrate (ATH)	Solid	Flexible foams and elastomers, used as a smoke suppressant				

Flexible foams

Historically, the fire resistance of flexible foams was obtained by a combination of non-reactive phosphorous and halogen-containing fire retardants at a level of 5 to 10 per cent on the overall formulation. Reactive fire retardants were not used, as these were often difficult to process, gave undesirable foam properties and cost more. However, such systems do not perform in some of the current more demanding fire tests for furniture applications.

Therefore, solid fillers such as exfoliated graphite, surface treated melamine or antimony trihydrate are utilised to give improved performance. These fire retardants increase the oxygen index of the foam and enable the formation of a stable surface char on the foam surface during the early part of the test protecting the remaining bulk of the foam. Typical levels of these solid additives are 15 to 30 per cent on the final formulation.

Rigid foams

Both reactive and non-reactive fire retardants can be used to improve the fire test performance of rigid foams. The selection depends very much on the specifics of the fire test, cost and processing requirements. The most widely used fire retardant is probably trichloropropyl phosphate (TCPP), this has a high phosphorous and halogen level and as a stable low viscosity liquid is easy to process. Triethyl phosphate (TEP) and diethyl ethyl phosphate (DEEP) are also used where halogen-free systems are required. Typical use levels are 3 to 10 per cent on overall formulation. Two reactive compounds commonly used, which both contain bromine at relatively high levels, are: brominated polyether polyol and brominated phthallic anhydride based ester.

Solid fire retardants are sometimes used in rigid foam formulations, but the added processing inconvenience has restricted their use. The most common solid fire retardants are ammonium polyphosphate and encapsulated red phosphorous. The addition of fire retardants into rigid foam inevitably results in the generation of more smoke during combustion that can cause problems with passing some specific fire tests. In these cases the correct balance between fire retardant level and isocyanate index is necessary.

Other applications

Most other polyurethane applications, foam and solid, which require some level of fire retardancy use the additives already mentioned plus solid fillers such as calcium carbonate, barium sulphate and aluminium trihydrate that work by absorbing heat leading to a delay in the onset of ignition.

Cross-linking agents, chain-extending agents and their reactions

These are low molecular weight polyfunctional compounds, reactive with isocyanates and are also known as curing agents. Chain-extenders are difunctional glycols, diamines or hydroxy amines and are used in flexible foams, elastomers and RIM systems. The chain-extender reacts with an isocyanate to form a polyurethane or polyurea segment in the polyurethane polymer. Through reactions with excess isocyanate, allophonates and biuret can be formed, transforming the chain-extender effectively into a thermo-reversible cross-linker.

Simple diamines are, in general, too reactive for a high level of addition and special amines have been developed such as aromatic amines with bulky substituents ortho to the amino group. A widely used chain-extender in RIM applications is DETDA (diethyl toluene diamine). The steric factors are responsible for lowering the reactivity of the amino groups as compared to TDA (toluene diamine). The reaction of one amino group with the isocyanate introduces a urea substituent on the aromatic ring, which lowers the reactivity of the second amino group. A more recent development is themoreversible chain extension by hydrogen bonding through polyols containing special end groups. Typical chain-extending agents are shown in Figure 10-3.

Cross-linkers, Figure 10-4, have a functionality of three or more and are used to increase the branching or cross-linking of polyurethane networks through the formation of urethane bonds. They are mostly used in rigid polyurethanes and coatings and adhesives formulations.

Other additives

Adhesion promoters

Adhesive promoting additives are used to facilitate adhesion of the polyurethane to a substrate. These promoters can either be coated on the surface or are often incorporated into the polyol or isocyanate component. Special polyesters and siloxanes are often used to assist the bonding to substrates such as PVC.

Anti-static

Where polyurethane is used to produce items with reduced electrical resistance, for example in safety shoes, special anti-static additives are added. The most commonly used additives are tetraalkylammonium sulphates, these are often used in combination with special grades of carbon black dispersions. The combination of these two approaches can reduce the surface resistivity from 10^{14} down to as low as 10^6 Ohms.

Figure 10-3   *Chain-extending agents*

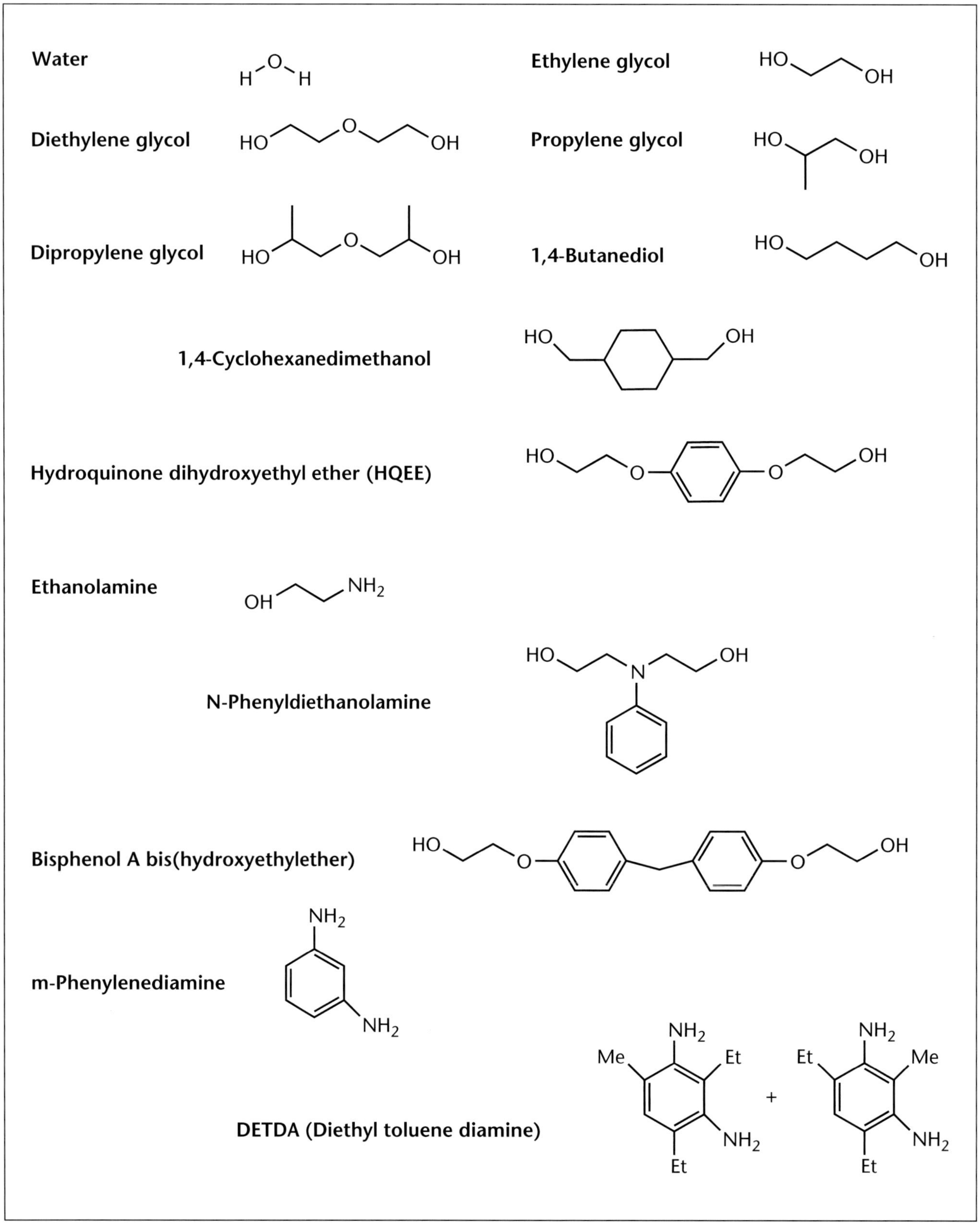

Figure 10-4   Cross-linking agents

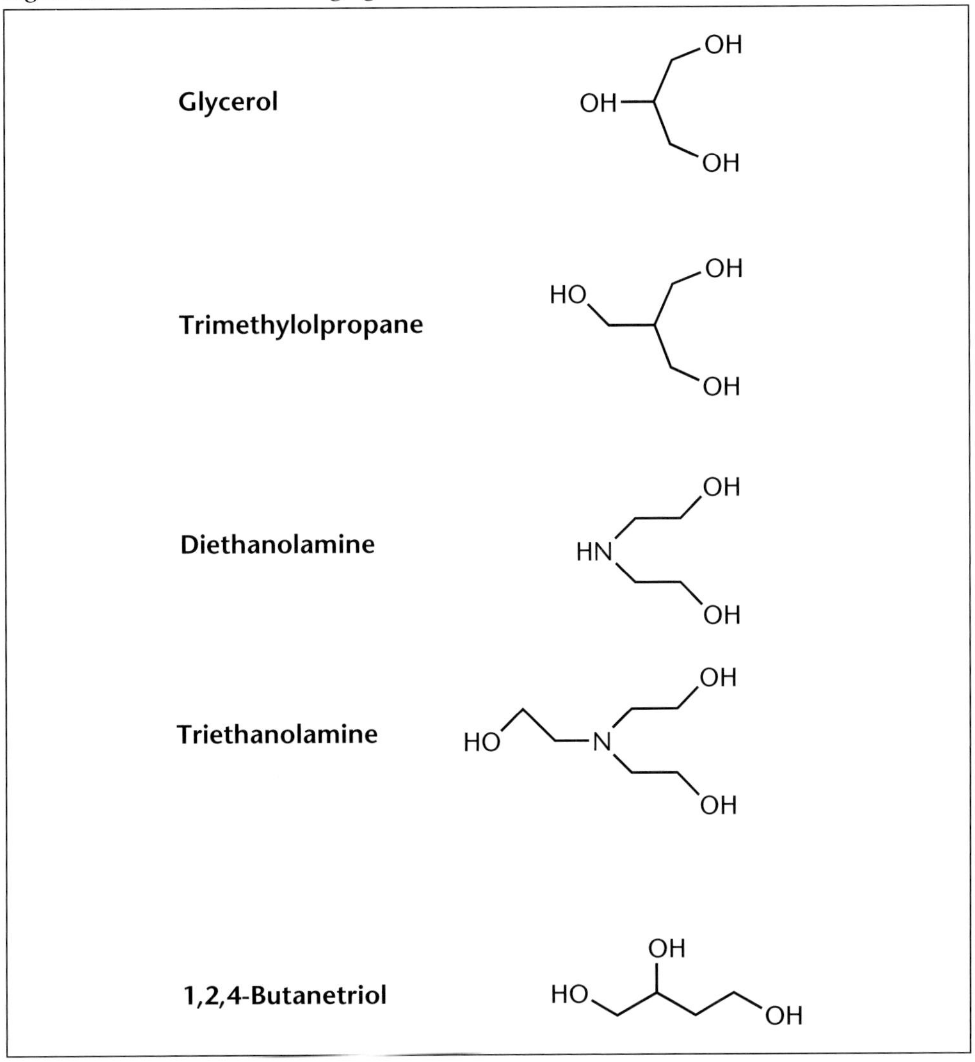

Anti-oxidants

Polyether polyols are prone to degrade by thermal oxidative attack and require the addition of anti-oxidants, which also protect slabstock foam from discoloration or scorch. There are many products that can be used, most of which are mixtures of sterically hindered phenols, diphenylamines or benzofuranone derivatives. The most widely used stabiliser has been butylated hydroxytoluene (BHT). Polyester polyurethanes are much less susceptible to oxidation.

Fillers

Fillers are used to reduce cost, increase the stiffness of the part or increase the temperature stability. Special fillers can also be used to improve the resistance to water ingress in the foam or to act as cell openers. Particulate and fibrous fillers are used depending on the application and the end-use of the product. Examples are shown in Table 10-5.

Table 10-5   Fillers and their application in polyurethanes

Filler Type	Typical Application	Level (%)	Comments
Calcium carbonate	Flexible & semi-flexible foams	5 – 30	Cost saving
Barium sulphate	Sound absorbing foams	5 – 50	Technical performance
Glass fibres	Elastomeric RIM and rigid foams	10 – 50	E-Glass most common
Carbon fibres	RIM and speciality foams	5 – 30	Limited use
Micro-spheres	Rigid foam and elastomers	2 – 30	Increased strength
Silicas	Elastomers and sealants	1 – 10	Dryers and matting agent
Melamine	Flexible foams	5 – 40	Fire retardant
Carbon black	Elastomeric compounds	1 – 5	UV absorber or anti-static

Particulate and mineral fillers have particle sizes of 5 to 100 microns and are added as dispersions to the polyol blend. These products need to be dried to a consistent water content to ensure satisfactory processing. Solid fire retardants and reinforcing polymer polyols are also solid fillers.

Glass fibre is the most commonly used fibre. Each typical glass fibre strand contains over 200 filaments, each of about 10 microns in diameter and a roving or tow will contain up to 6,000 filaments. Pre-formed mats are also used, these are more expensive than chopped fibres, but are easier to use and require less sophisticated equipment.

Carbon fibres have not been used in the past due to high cost, but falling prices are leading to more developments as they provide major improvements in both stiffness and tensile strength.

A growing area for the use of fillers is in recycling. Ground or finely chopped foam is mixed into the polyol stream and re-used. Levels of re-work can be as high as 25 per cent, although 5 to 10 per cent is more common.

Hydrolysis

Carbodiimides that act as acid scavengers preventing further autocatalysis are commonly used as anti-hydrolysis agents and typical levels are one to two per cent of the overall formulation.

Lubricants

Lubricants in the form of waxes, soaps and other processing aids are often added to thermoplastic polyurethane and elastomeric compounds. These additives improve the flow in the mould and assist in demoulding finished parts. These materials can also act as internal mould release agents reducing the application of external release agents and improving the economics of the moulding process.

Anti-microbials

Polyester-based polyurethanes are prone to microbial attack by enzymatic hydrolysis of the ester linkage. A range of stabilisers, usually metal organic derivatives based on antimony, copper or arsenic, can be used and it is important to select the right type and level of stabiliser to protect against the attack by specific microbes. Microbial attack is not usually a problem for polyurethanes made from polyether polyols unless the application requires special attention. Examples of this could be in carpet underlay, coated fabrics or some bedding applications.

Pigments

Pigments and reactive dyes are commonly used to colour polyurethane parts and can be inorganic or organic in nature usually dispersed in polyol or plasticisers. Important inorganic pigments include titanium dioxide, chromium oxide, carbon black and iron oxide. Organic pigments are often phthalocyanines or dioxazines. Unreactive water-soluble dyes are not used, as they tend to 'bleed'. Typical levels of addition are one to four per cent on the overall formulation.

Viscosity reducers

A number of additives are used in polyurethane systems to reduce viscosity and aid processing. The need for these products has especially increased in the low-density rigid foam area, where the move to systems with lower levels of physical blowing agent and higher water requires viscosity minimisation and additives that by plasticisation reduce foam friability. Improved flow is usually obtained when plasticisers are used. The most common chemicals used are low molecular weight glycol ethers, dimethyl AGS esters or propylene carbonate. Although the latter product is very effective in reducing viscosity, care should be taken when using this product as instability and pressure build-up, due to decomposition, can occur in the presence of metallic iron and certain catalysts.

UV resistance

Aromatic isocyanate-based polyurethanes show surface yellowing when exposed to light. Although this effect is aesthetic and does not affect physical characteristics, certain applications, such as shoe soling, require a consistent colour to be maintained. Where white or light coloured polyurethane parts are needed, the onset of discoloration can be delayed by the addition of a synergistic mixture of three stabilisers: a blend of light absorbing benzophenone and benzotriazoles compounds with UV absorbers and free radical scavengers such as hindered amines. Typical addition levels are one to two per cent of the overall formulation. When long-term light stability is required using aromatic isocyanates, the only acceptable solution is to paint the component, which can be done after the part has been manufactured or by using an in-mould coating. Aliphatic isocyanates are inherently light stable.

11. Introduction to flexible foams

Gaby Verhelst
Alain Parfondry

Flexible foams are broadly characterised as low-density cellular materials, with a resistance to compression that is both limited and reversible. The most widely used flexible foam types are water-blown slabstock and moulded flexible foams. Both are used for automotive and furniture/bedding cushioning applications. In addition, technical products such as semi-rigid, packaging and viscoelastic foams are used in special applications.

Markets

Today, flexible polyurethane foams meet most of the requirements for cushioning in furniture (office and domestic), bedding (mattresses and pillows) and automotive (seating and interior trim). The steady growth over the past 50 years has been so consistent that other cushioning materials, based on fibres or foams, now have only a marginal position. The versatility of polyurethane chemistry, allowing a wide range of material properties and processing conditions, has been the major driver for growth and a variety of isocyanates and polyols can be used to make polyurethane flexible foam with a wide spectrum of densities and hardnesses (load-bearing ability).

Flexible foams can be divided into two main categories, dependent on the method of production, as slabstock and moulded foams and the overall market size in 2000 and the key subsets are shown in Figure 11-1.

Raw material history

The history of flexible foam dates back to the early 1950s, Table 11-1, when the first slabstock foams were made from prepolymers based on TDI and polyester polyols. The latter were soon replaced by polyether polyols, first based only on propylene oxide, then, to improve the control of polyol reactivity, on propylene oxide/ethylene oxide mixtures. These, in conjunction with strong tertiary amine catalysts and to a lesser extent the development of silicone-based surfactants, allowed one-shot foams to be produced with lower densities and better foam stability.

The hot-cure moulding process was an extension of this conventional slabstock technology, but required polyether polyols with a higher reactivity. This was achieved by putting a short ethylene oxide tip on the polyol, either propylene oxide or propylene oxide/ethylene oxide based, to achieve a primary hydroxyl level of 40 to 50 per cent.

Figure 11-1   Global flexible foam market (2000)

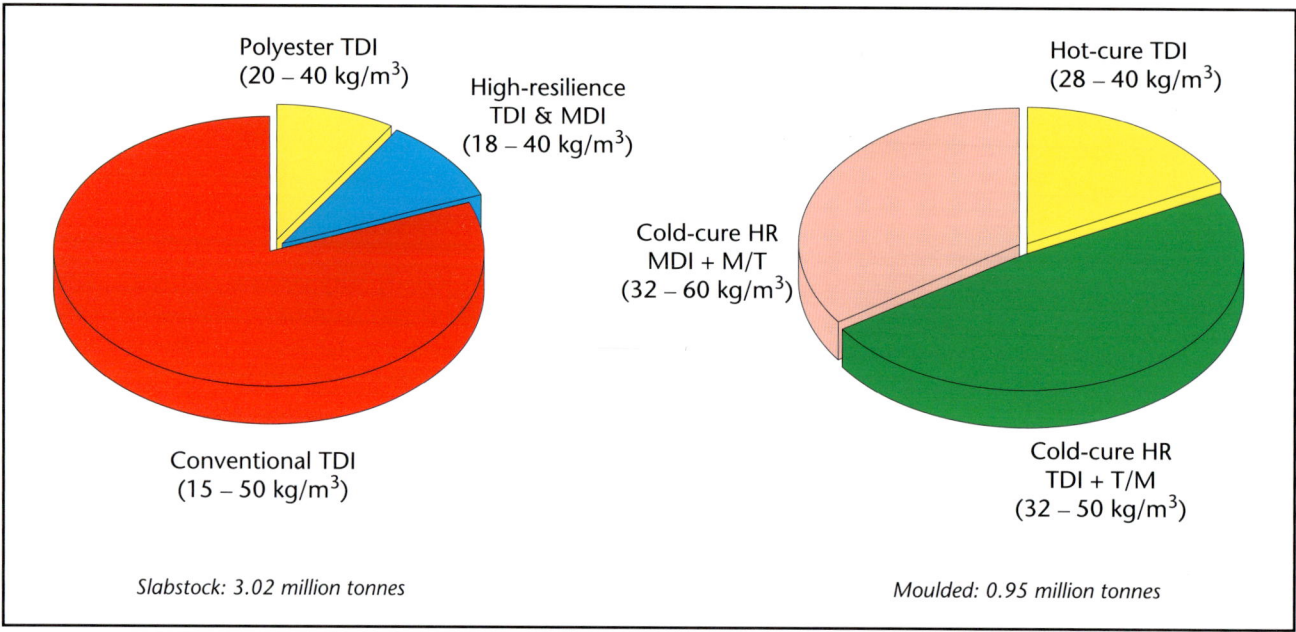

Table 11-1   Historical development of flexible foams

Introduced	Polyol	Isocyanate	Chemical process	Foam technology Slabstock	Moulding
1952	Polyester	TDI	Prepolymer	Polyester	—
1952	PO-based polyether	TDI	Prepolymer	Conventional	Hot-cure
1953	PO/EO random polyether	TDI	One-shot	Conventional	Hot-cure
1955	Polyester	TDI	One-shot	Polyester	—
1958	PO + EO tip polyether	TDI	One-shot	High resilience, soft	Cold-cure (furniture)
1968	PO + EO tip polyether	TDI/MDI	One-shot	—	Cold-cure (furniture)
1974	PO + EO tip polyether & SAN polymer polyols	TDI	One-shot	High resilience	Cold-cure (automotive)
1978	PO + EO tip polyether & PHD polymer polyols	TDI	One-shot	High resilience	Cold-cure (automotive)
1980	PO + EO tip polyether	mod TDI	One-shot	High resilience	Cold-cure (automotive)
1980	PO + EO tip polyether + PIPA/SAN optional	MDI	Quasi-prepolymer	—	High and low resilience
1983	PO + EO tip polyether & PIPA polymer polyols	TDI	One-shot	High resilience	Cold-cure (automotive)
1993	PO + EO tip polyether	MDI	Prepolymer	High resilience	High-resilience
1996	PO + EO tip polyether	M/T blends	One-shot	—	High-resilience
1999	PO + EO tip polyether	MDI	One-shot	High and low resilience	High-resilience

	Primary (OH%)	MWt
Slabstock	0 to 5	3,000 to 3,500
Hot-cure	40 to 50	3,000 to 3,500
HR	80 to 90	4,500 to 6,000

The next major step was the development of more reactive polyols with higher levels of ethylene oxide tipping, 80 to 90 per cent, which allowed cold moulds to be used and also led to the production of foams having higher elasticity (high-resilience or HR foams). Blends of TDI and polymeric MDI (T/M) were also introduced to provide robust processing.

In the early 1970s polymer polyols, initially based on acrylonitrile and then on styrene/acrylonitrile blends (SAN), were used to further improve the load-bearing ability of high-resilience foams, whilst maintaining elasticity and resilience. This work led to the development of moulded foams for automotive seating. Further diversification continued, first with the extension of SAN polymer polyol technology to conventional and hot-cure foams, then with the introduction of alternative technologies to high resilience polymer polyols, such as PHD and PIPA (see Chapter 6 for details).

Most of the developments until this point were focused on variations to polyols and additives with TDI as a constant. The first MDI-based flexible foams for moulding applications were introduced in 1980. These were based on quasi-prepolymers made from low functionality MDI's and flexible polyols. It was now possible to produce cost effective automotive seating foams with a broad processing latitude, fast-cure, and good foam hardness control, without resorting to polymer polyols.

In 1993, a prepolymer process based on MDI and an ethylene oxide-tipped polyol was introduced in Europe for high-resilience slabstock foams. The use of blends of MDI with small amounts of TDI (M/T) was introduced in 1996 for automotive seating in an attempt to combine the benefits of each isocyanate type in low-density foams.

Most of these technologies are still in use today, and, even in the presence of well-established preferences for specific applications, there are no clear signs of an overall dominant technology in the market.

Basic flexible foam chemistry

Flexible polyurethane foams consist essentially of two chemically inter-linked polymers:

- A urethane polymer formed by the reaction of a high molecular weight polyol that behaves as a soft segment and an isocyanate that may be di- or poly-functional. The resulting elastomeric polyurethane network gives the foam its stability, elasticity and mechanical strength.
- A urea polymer from the reaction of the isocyanate and water. The carbon dioxide generated acts as the blowing agent that forms the cellular network. The urea groups, physically linked through strong hydrogen bonds, phase separate forming hard segments that contribute to load-bearing properties.

To obtain a low-density foam a high level of water, in molecular terms, is required. Thus, the urea reaction dominates the urethane reaction and is the main contributor to the overall foam exotherm, Table 11-2.

Flexible foams can be made either by a one-shot process, where the urethane and urea reactions occur simultaneously, or in a two-step process. In the latter the polyol is first reacted with an excess of isocyanate and the resulting isocyanate prepolymer reacted in a second step with water and the other additives. This effectively segregates the urethane and urea steps of the reaction, thus avoiding direct competition between the polyol hydroxyl groups and water. Because of its inherently higher cost the two-step process is reserved for more demanding applications.

MDI has a higher molecular weight and a higher isocyanate equivalent weight than TDI, so a higher hardblock content is reached for a given amount of water. This results in foams with a higher density or higher hardness. High-resilience foams, both as slabstock and moulded, are therefore now produced from MDI as this provides significantly more versatile formulation technology than TDI. This has been due to the introduction of specially designed MDI grades and polyol blends that include the use of:

- MDI variants modified with low amounts of polyols.
- Higher molecular weight polyols.
- Special ethylene oxide rich polyol additives as softeners.
- A wider isocyanate index range.
- Lower filler levels (polymer or inorganic solids).

The balance between the urea blowing and urethane gelling reactions controls the foam stability, Figure 11-2, and the various stages of foam development are shown in Table 11-3.

The rate of gelation depends on the polyol and isocyanate structures and the level and types of catalyst used whilst the level of blow is controlled by the amount of water used. The volume of carbon dioxide also depends on the temperature reached in the foam with thicker foams reaching higher temperatures and therefore lower densities. A typical plot of rise-height, pressure and temperature for rising foam is shown in Figure 11-3.

The critical step for flexible foams is cell opening which depends on a large number of formulation related parameters, such as polymer viscosity, surfactants, solid particles (from polymer polyols), ethylene oxide additives (in high-resilience foams), and, more generally, the miscibility of the various raw materials. In polymer terms, these parameters control phase separation, the first step in the development of the final polymer morphology. This morphology is characterised by a relatively amorphous soft phase with hard segments aggregated into small domains (less than 100Å) and/or larger urea balls (0.1 to 1µ). These larger aggregates may promote rupture when positioned in the thinner portions of the cell windows.

Table 11-2
Competing reactions in flexible foams

	Urethane	Urea
Weight, %	20	80
Volume, %	20	80
Enthalpy of formation, kJ/mole	90	125
Energy of formation, %	15	85

For a typical MDI-based foam of 35 kg/m^3 density

Figure 11-2   Stages of foam development

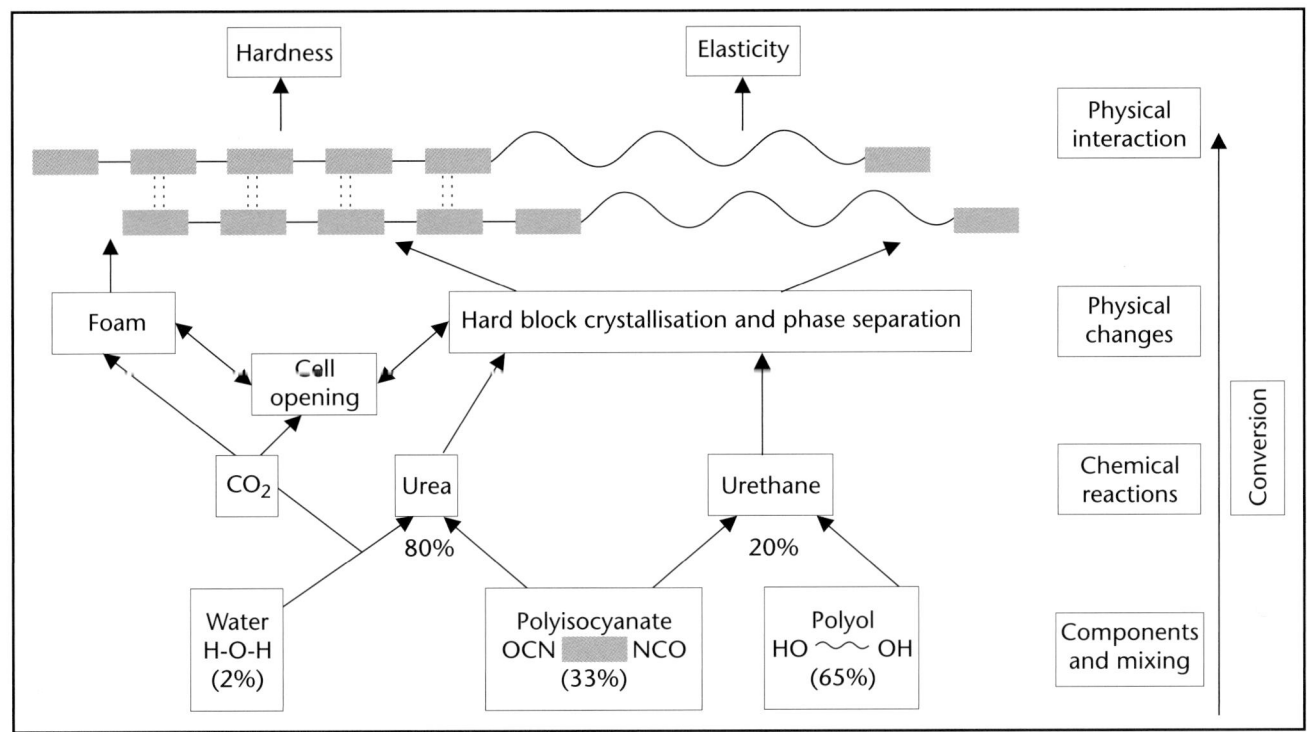

Table 11-3   Timeline of foam development

Time	Foam development	Physical event	Chemical event(kinetics)	Morphology
0 seconds	—	Raw material mixing, air nucleation	—	—
5 seconds	Cream, initial rise	Froth formation	Urea reaction, CO_2 saturation	—
20 seconds	Rising foam	Viscosity increase, bubble expansion and coalescence, window thinning, gas diffusion	Urethane reaction, molecular weight increase	Beginning of hard domain formation
60 – 90 seconds	End-of-rise,	Cell opening blow-off	Hydrogen bond formation	Phase separation (urea aggregate formation)
3 hours > 1 day	Cure	Hardness development	—	—

Cell opening is also related to process parameters such as component temperature, gas nucleation, internal pressure, liquid flow and mould temperature. To produce stable foams, cell opening should take place when most of the blowing gas has been produced. If it takes place too early, an unstable recessing, splitting or even collapsing foam is obtained. If it is delayed until after the end of foam rise, a pneumatic foam (too little cell opening) or even a shrinking foam (no cell opening) can result. Both situations are obviously unacceptable.

Figure 11-3   Rise-height, pressure and temperature versus time for a flexible foam

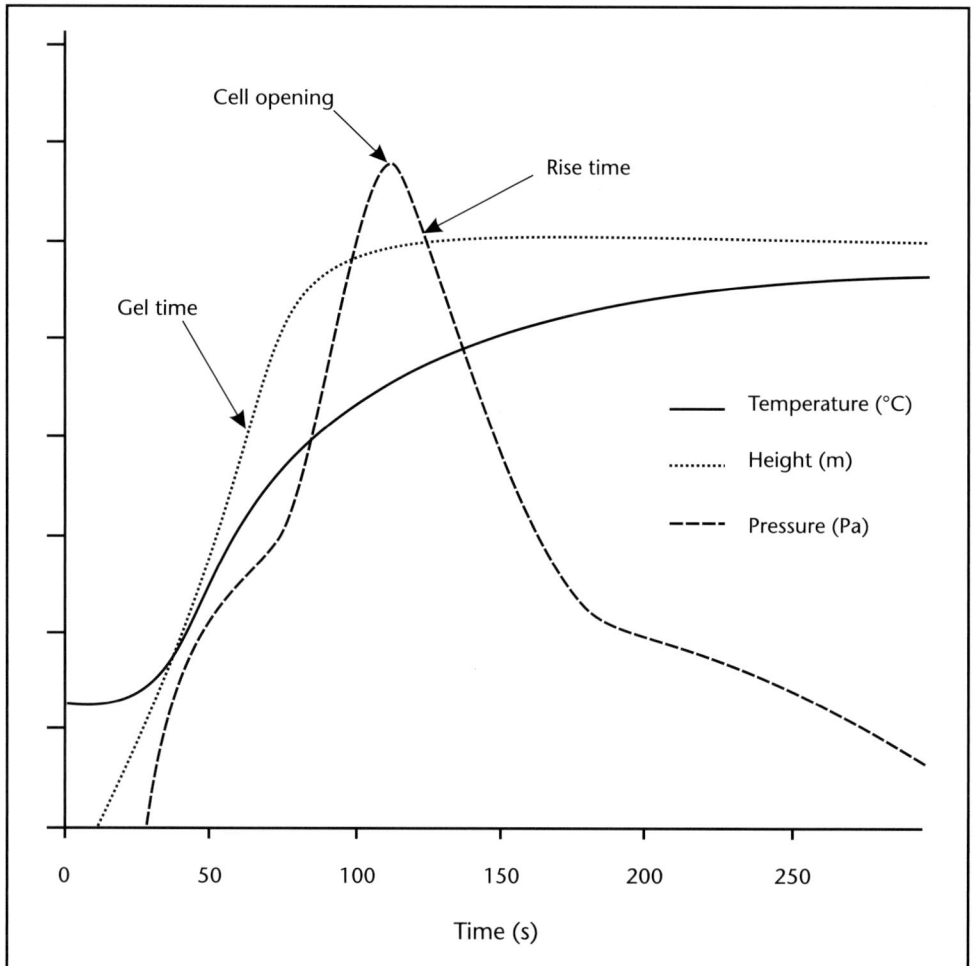

The stability of conventional slabstock foams, as they are based on less reactive secondary hydroxyl polyols depends heavily on the activity of the gelation catalyst (such as stannous 2-ethyl hexanoate), and strong silicone surfactants. High-resilience foams are less sensitive, because of the inherently stronger polymer gel obtained from the more reactive high-resilience polyols. MDI, due to its higher functionality, imparts further foam stability, and therefore robustness in processing.

Foam and polymer morphology

The properties of flexible foams are determined by their cellular structure and the morphology of the polymer in the struts and cell windows. These are both dependent on the chemical composition of the raw materials, their physical miscibility and chemical reactions that, overall, result in a complex balance which needs accurate control during processing.

Cellular structure

Cellular structure is traditionally characterised directly by visual observation with cell size and distribution assessed by the 'cell count' along a given axis (BS 4443, part 1, method 4). This is common practice with conventional foams having a relatively narrow cell size distribution, but difficult to apply to high-resilience foams which have a very broad range of cell sizes, covering several orders of magnitude. In this case, the degree of cell anisotropy may be quantified by counting cells both in the direction of the foam flow and perpendicular to foam flow.

In addition, image analysis can be used, but this requires specialised equipment including stereoscopic lenses, cameras with digital output, and image analysis software, with special algorithms designed to capture the most relevant geometric parameters. The more recent developments in this area include either special sample preparation or advanced techniques such as computer-generated X-ray tomography to generate true three-dimensional images. Airflow measurements can also provide additional, indirect, information on foam cell morphology.

Polymer morphology

A range of methods, such as thermal analysis, electron microscopy and X-ray spectroscopy, exist to determine the morphology of flexible foam polymers.

Thermal analysis
Dynamic Mechanical Analysis (DMA) has been traditionally used to study the viscoelastic behaviour of elastomers and it has now developed to the high level of sensitivity necessary to record the foam polymer response to very small low frequency and low amplitude deformation cycles over a wide range of temperatures. The sharpness of the glass transition (representative of the urea hard segments), the shape, level and temperature limits of the rubbery plateau (defined by the polyol structure) are used to characterise and compare phase separation of different foams. Differential Scanning Calorimetry (DSC) gives similar and complementary information to DMS, but is simpler to apply.

Microscopy
The overall structure of flexible foam can be examined at the micron level using Scanning Electron Microscopy (SEM) that also provides a detailed insight into the mechanisms of cell opening through study of the cell strut and window microstructures.

Transmission Electron Microscopy (TEM), with a magnification of up to 80,000, allows the visualisation of electron rich areas in the foam struts such as the larger scale agglomerates or inclusions found in some foams. Atomic Force Microscopy (AFM) is complementary to TEM, and it can give two-dimensional scans of the local hardness profiles on the surface of the foam struts, and therefore also of the hard segment distribution.

Spectroscopy

Small Angle X-ray Spectroscopy (SAXS) gives information on the electron density, and therefore also on the morphology of the hard segments in the foam polymer. Domain size in the nanometer range, hard/soft interlayer thickness, and total hard segment content may also be calculated from SAXS data. Wide Angle X-ray Spectroscopy (WAXS) looks at the diffraction patterns of X-rays to define the degree of ordering (crystallinity) of the electrons in the hard domains. Domain sizes in the micrometer size range may be calculated from the patterns observed.

Figure 11-4   Model of flexible foam morphology

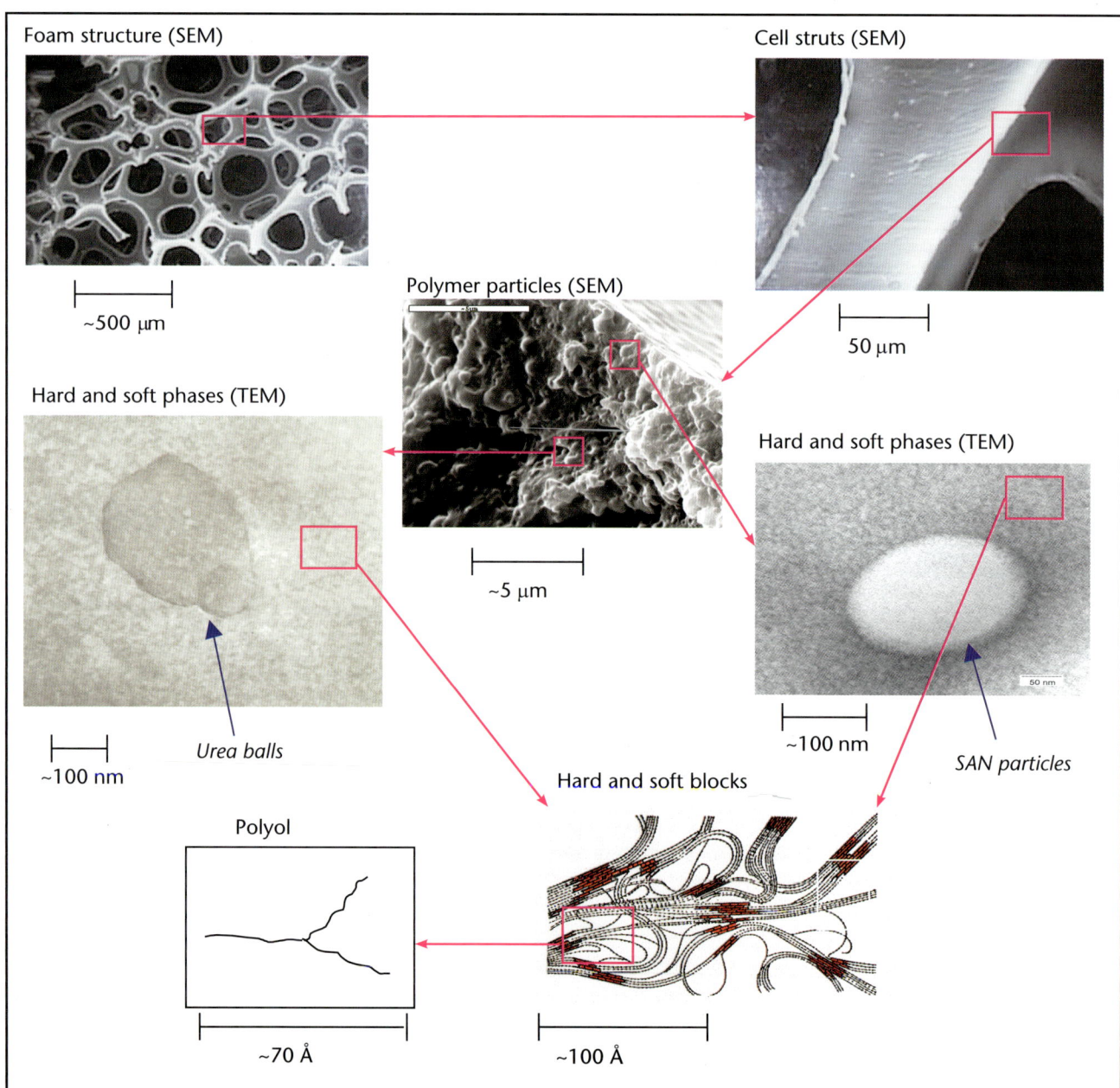

The results from these characterisation techniques can be combined to produce an overall model of flexible foam morphology from the millimetre to the angstrom level as shown in Figure 11-4.

Functionality and performance tests for flexible foams

In most applications, the main function of flexible foam is based on some form of energy management. This energy may be applied in a static mode, as in furniture and bedding, or dynamically, with relatively large amplitude and low frequency as in automotive seating. It may also consist of a complex combination of very low amplitudes and high frequency range as in sound absorption foams. Packaging and energy absorption foams are a special foam class designed to manage impact energies, under non-cyclical conditions of very high rate of deformation. In seating or bedding applications, the first determining parameter for optimal comfort is foam hardness and this should be adapted to the weight and size of the user. The relationship between deformation and frequency for the different flexible foam applications is shown in Figure 11-5.

Figure 11-5   Dynamic testing of flexible foam

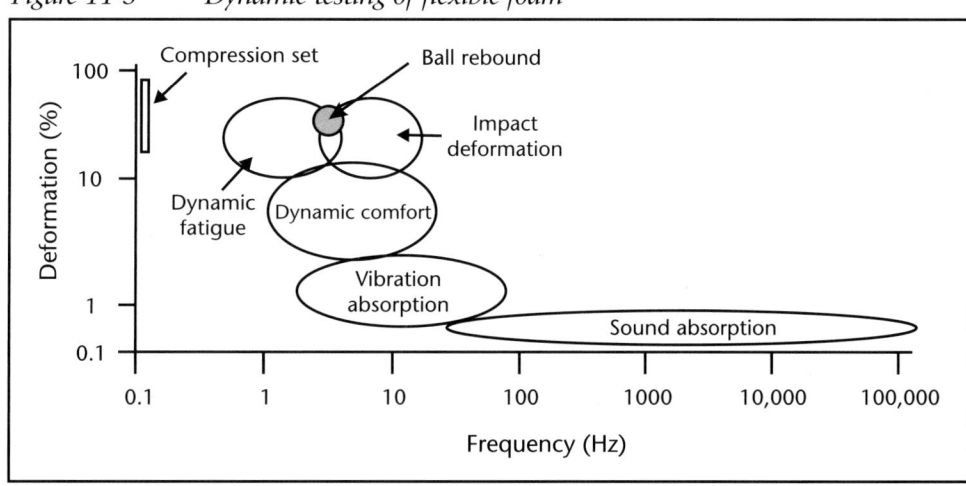

Static comfort, as defined for seating or bedding applications, is primarily related to the distribution of pressure on the human body. If the foam is too hard, local pressure peaks exceeding about 3.3 kPa may restrict flow in the capillary blood vessels under the skin, and induce discomfort. Sensitive pads containing an array of pressure sensors are commercially available to test mattresses and seats. When these are inserted between the body and the cushion or mattress, a two-dimensional pressure distribution map is generated, which can be used to quantify the differences in perceived comfort between different foams.

For automotive seating foams, in addition to static comfort, dynamic considerations become very important. Depending on the application, flexible foams may be tested under a variety of dynamic regimes characterised by a combination of amplitude and frequency. Obviously, high frequencies are associated with low amplitudes (sound), and low frequencies with higher amplitudes, as in automotive seats. The least comfortable vibrations are in the 5 to 12 Hertz area, because they tend to cause unpleasant resonance effects with parts of the human body.

When compared to other foams (and/or foam combinations in layers) with the same nominal hardness, high-resilience foams are usually preferred because their shallower stress/strain curve at high strains translates into a more even pressure distribution in use. In addition, they exhibit a smoother and thinner hysteresis profile, which contributes to an overall feeling of better support. Direct hand feel of the foam, although more subjective, usually confirms this impression.

These effects are enhanced after extended periods of use. In comparison with conventional foams, high-resilience foams tend to resist fatigue better and exhibit less 'bottoming out' due to hardness loss.

Specific test methods have been developed to characterise the properties of flexible foams. They can be grouped in six key performance areas:

- Basic properties determining functionality and cost effectiveness: density and hardness.
- Durability and comfort under a variety of conditions: heat, humidity, static and dynamic stresses.
- Property changes with ageing.
- Mechanical properties mainly related to the handling of foams in production.
- Fire properties.
- Environmental properties.

The most common are described in the recommended ISO and ASTM methods, Table 11-4. They are preferred both for product characterisation and for easier communication to a wide audience.

Tests for finished articles such as complete seats, mattresses or other composites are specific to each industry, and will only be briefly mentioned here.

Because most flexible foams are cured in production with only a minimal level of external heat (and most of the time none at all), it is important to ensure that samples have reached a stable level of cure before testing. This is usually specified at a minimum of three days under ambient conditions (see below). Exceptionally, accelerated cure at up to 120°C can be considered if necessary.

Flexible foams have a high specific surface and therefore can easily absorb/desorb humidity from their environment. Therefore, in addition to the curing conditions mentioned above, foam samples must be preconditioned, typically for a minimum of one day, typically at 23°C ± 1°C, 50 ± 2% relative humidity (RH), and 86 to 106 kPa atmospheric pressure (as specified in ISO 291). Other, less common, conditions include 20°C and 65% RH, or, for tropical conditions, 27°C and 65% RH, depending on the end-use application.

Table 11-4   List of major ISO, ASTM and JIS standards

Standard	Global/Europe ISO/EN	America ASTM	Japan JIS
Sample dimensions	1923	D1622	—
Sample conditioning	291	—	K 6400-4
Density	845	D3574-test A	K 6400-5
Compression hardness	3386	D3574-test C	K 6400-6
Indentation hardness	2439	D3574-test B	K 6400-6
Compression set	1856	D3574-test D	K 6400-7
Wet compression set	DIS 13362	—	—
Resilience	8307	D3574-test H	K 6400-9B
Tensile strength and elongation at break	1798	D3574-test E	K 6400-10
Tear strength	8067	D3574-test F	K 6400-11
Accelerated aging tests	2440	D3574-tests J, K	—
Flex fatigue	5999	D3453/D3770	K 6401
Dynamic flex fatigue	3385	D3574-tests I1-to I3	K 6400-8B
Foam laminates (general specifications)	6915	—	—
Flex fatigue (mattresses)	EN 1957	—	—
H Point method	6549	—	—
Air flow	7231	D3574-test G	K 6400-13B
Air flow	4638	—	—
Air flow (acoustic materials)	9053	—	—
Water vapour transmission	11092	—	—
Impedance/absorption of materials (impedance tube method)	10534	C 384	—
Impedance/absorption of materials (frequency analyser method)	—	E 1050	—
Colour fastness	4582	E 313	—

In addition, for specific tests, flexible foam samples may need to be mechanically conditioned before testing (e.g. gently rolled or squeezed) to remove any remaining internal stresses, which could lead to erratic results. This is especially recommended for compression set tests.

Basic properties

Apparent density
This test (ISO 845) is obviously important for load-bearing and cost reasons. A carefully cut parallelepiped of foam is weighed and its dimensions are measured. The weight to volume ratio is expressed in kg/m^3.

Tensile strength/elongation at break and tear strength
Flexible foams are usually not subjected to tensile stresses, so these properties are not essential for regular use. However, they provide useful information on the ability to handle foams during post-processing operations.

The tensile strength, usually expressed in Pa or kPa, is obtained by pulling a standard dumb-bell sample to its breaking point with the elongation at break derived from the same test. The tensile modulus (or modulus of elasticity, or Young's modulus) is found from the initial slope of the stress strain curve. The tear strength is the force needed to extend a slit cut in a small foam sample by slowly (typically 100 millimetres per minute) pulling both sides apart from each other. It is reported in Newtons per minute, because the force is proportional to the actual slit width.

Durability cushioning/comfort tests

These tests are essential tools to compare foam property changes during use, and, in some cases, to predict service life. They may be static (compression set) or dynamic (flex fatigue) creep tests.

Hardness/hysteresis

The load-bearing properties of flexible foams are usually essential to their function. They may be measured by simple compression (when the compression plate is larger than the foam sample – ISO 3386) or by indentation (if the compression plate is smaller than the foam sample – ISO 2439).

In compression mode, a small sample, typically measuring 10 x 10 x 5 centimetres, is placed between the plates of a compression test machine. These plates are perforated to facilitate airflow. The sample is then slowly pre-compressed three times to 70 per cent and allowed to recover. During the fourth compression cycle at a constant rate (typically 50 millimetres per minute), the thickness corresponding to 50 per cent compression is held for 30 seconds before the load is recorded. Pre-compression is important to minimise the effects due to viscoelasticity and/or closed cells that may prevail during the first cycles. The result is divided by the compression area and expressed as compression force deflection (CFD) at 50 per cent in kPa. Values at 25, 40 or 65 per cent compression may also be recorded.

The complete stress strain curve is also recorded during the last cycle. The ratio of the areas under the loading and unloading curves is then used to measure the energy absorbed during compression. This ratio is called hysteresis and provides a simple measure of foam elasticity/resilience, Figure 11-6.

In indentation mode, a large sample, measuring 40 x 40 x 10 centimetres, is compressed with a standard indentor smaller than the foam surface (typical diameter 20 centimetres). The tensile forces around the edges of the indentor have a significant effect on the overall test results so that they cannot be directly compared to compression data. The overall procedure is otherwise similar to the compression test, also with three pre-compression cycles. Indentation tests are commonly used in production environments because they more accurately predict performance during use.

Figure 11-6   Stress strain curves for high-resilience and conventional foams

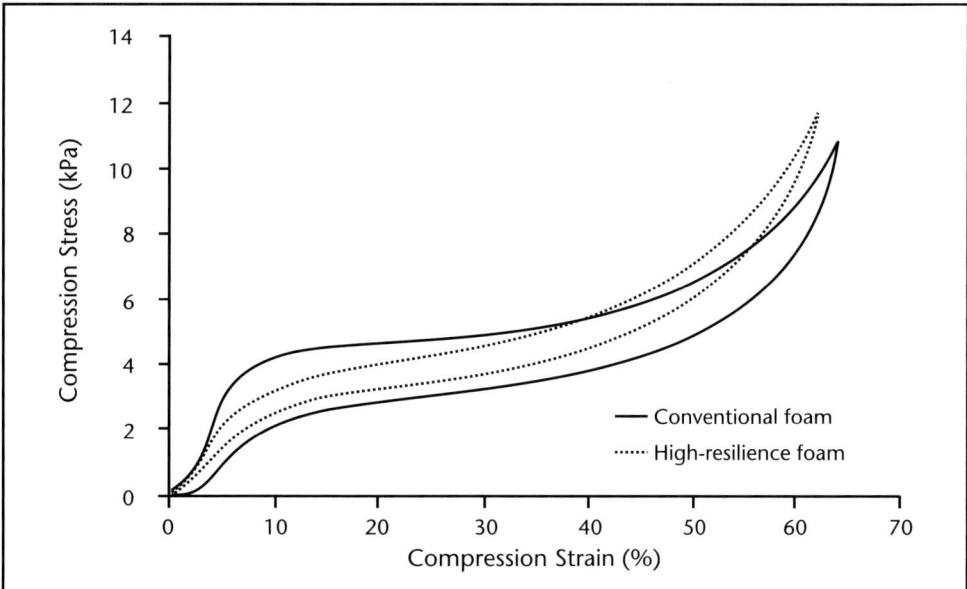

Other parameters of interest, which may be derived from the indentation test, include the SAG factor (sometimes also called comfort or support factor), which is the ratio of indentation forces at 65 and 25 per cent. A high value, typically around 2.7 to 3.0, suggests the high comfort expected from a high resilience foam.

Another, less common, measure of comfort via surface feel is obtained from the ratio of indentation hardnesses at 25 and 5 per cent after one minute, this is the initial hardness factor. Finally the guide factor is the ratio of the indentation value at 25 per cent after one minute normalised to the density.

For automotive seating, foam hardness must be controlled within very tight limits, because it determines the exact position of the driver, and therefore his view of the road through the windscreen and mirrors, which are all essential for safety. This can be measured on a finished seat assembly by the Hip-point or H-point method, using a standardised 'dummy' representing the 'average' car user. The Society of Automotive Engineers has defined a standard human body shaped dummy (SAE J 826 or ISO 6549, Figure 11-7) as a basis for various tests. This is now well accepted both in the Americas and in Europe, but a different standard is used in Japan (JIS D 0024). This rig is also used by the OEMs for some dynamic (vibration, fatigue, impact,) tests.

Compression set

Compression test ISO 1856 is an accelerated creep test that measures the residual deformation of a small foam sample after it has been compressed for 22 hours at 70°C to 50, 75 or 90 per cent of its original thickness. Sample size is usually 5 x 5 x 2.5 centimetres, and the smaller axis (determining the direction of compression) is taken in the main foam blowing direction. The test must be followed by a

Figure 11-7 H-point dummy details

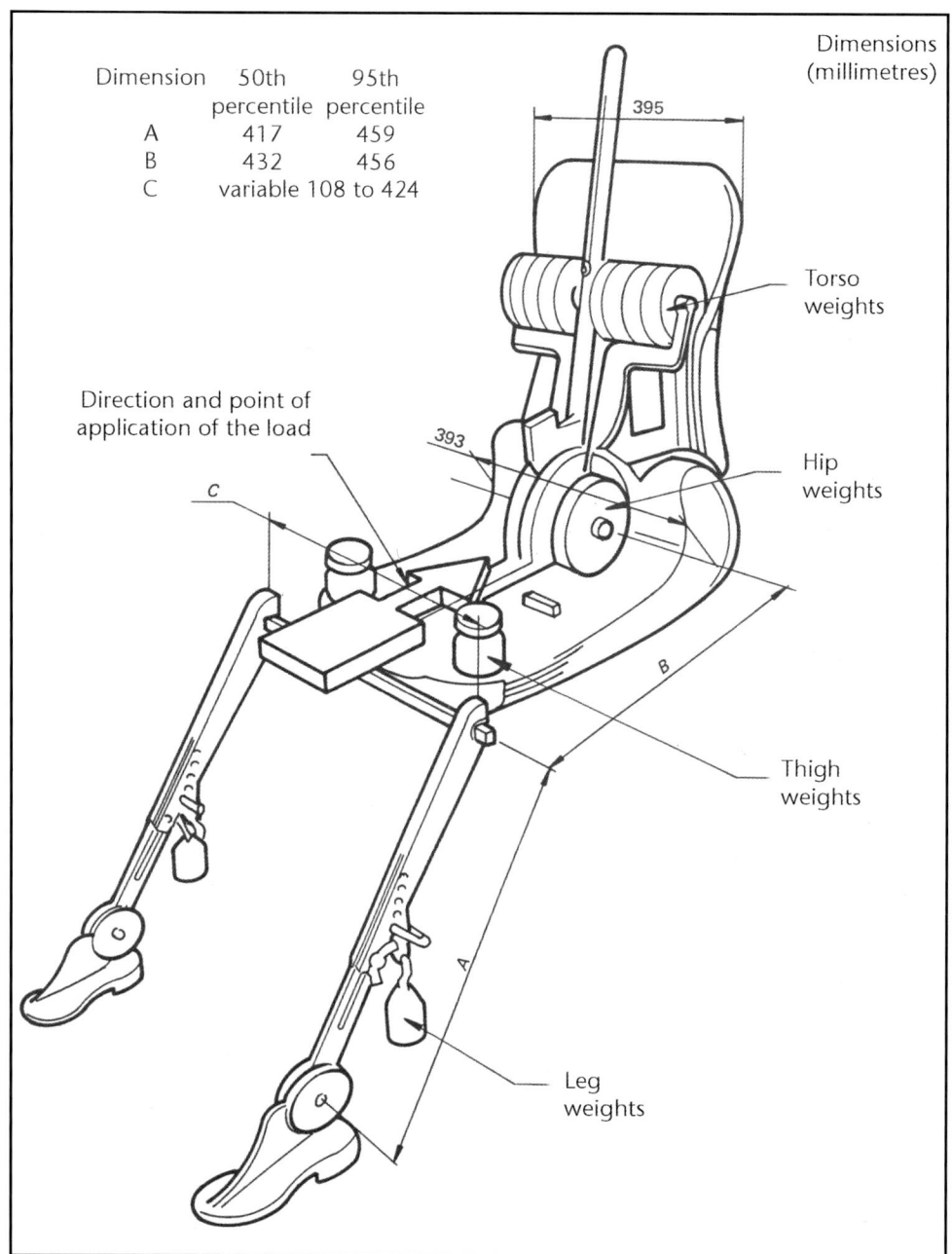

standard period of recovery (usually 30 minutes at 23°C) after decompression. This recovery period must be controlled accurately because during this period, thickness can change rapidly as a result of the release of internal stresses accumulated during the compression stage of the test.

The result is the percentage thickness loss, expressed by reference to the original thickness (C_t, usually for slabstock foams), or by reference to the compressed thickness (C_d, more common for moulded foams), Figure 11-8.

Figure 11-8   Calculation of compression set

$$C_t \text{ (Original thickness)} = \frac{(T_o - T_f)}{T_o} \times 100$$

$$C_d \text{ (Compressed thickness)} = \frac{(T_o - T_f)}{(T_o - T_s)} \times 100$$

T_o = original thickness
T_f = thickness after 30 minutes recovery
T_s = thickness of compressed foam sample during the test

Flex fatigue tests

Static fatigue tests, similar to the compression tests described above, but performed at room temperature, may provide a quick measure of foam durability (ASTM D 3574 – part I), but dynamic tests are much more commonly used because of their better correlation with useful service life.

The most widely used test is the ISO 3385 pounding test, where a 40 x 40 x 10 centimetres foam slab is submitted to 80,000 load cycles at one Hertz with a flat cylindrical indentor (diameter 20 centimetres, load 750 Newtons). In the original test design, during each cycle, a pneumatic actuator lifts the indentor before being dropped free, so that the load cycles are not sinusoidal. Newer test machines use hydraulic power to actuate the indentor in a similar way.

After the test, the losses in indentation hardness and foam height are measured. This test has been shown to correlate well with use, and it has been employed, among other criteria, as the basis for foam classification for light, average, severe, very severe, and extreme use conditions (ISO 5999, or ASTM D 3453).

There are many variations around this test, including the roller shear test (ASTM D3574 test, which includes an additional shearing effect) used mostly in the USA, and many proprietary tests developed by individual automotive manufacturers. One of these tests, jointly designed by Renault and the PSA group, specifies that small (10 x 10 x 5 centimetres) foam samples should be compressed 100,000 times by between 25 and 75 per cent of their original thickness. As the test proceeds, foam thickness decreases, and therefore the actual stress during the test also decreases, thus reducing the overall severity of the test. The main advantage of this test is to use a simpler machine design and to provide more data with a single compression plate, since several samples may be tested in parallel.

Specialised tests (EN 1957 for mattresses, ISO 6915 for foam sheets for laminates) have been defined to determine the main requirements of semi-finished or finished articles, where flexible foam plays a major role. These are designed to support the use of authoritative quality standards, and therefore the marketability of flexible foams.

Dynamic creep

The non-linear response of flexible foams to compression load is characterised by a median zone with a relatively low local modulus. In this zone, which corresponds to the strain experienced during normal use, hysteresis loss is relatively high, so that a larger degree of damping can be observed.

Small sinusoidal load cycles applied on a pre-compressed foam in this region can simulate actual ride conditions, Figure 11-9. Under this regime, the changes in dynamic modulus, loss factor, and foam height under load can be studied as a function of time. It has been shown that dynamic modulus increases, while foam height under load decreases during such tests – as during actual rides, resulting in the so-called 'bottoming-out' effect, experienced as a loss of comfort. These effects are accelerated under hot and humid conditions, especially when cyclic conditions (hot/cold and/or dry/humid cycles) are applied. The optimum requirement is a plateau ride effect where the foam has the ability to react rapidly to loading without the load/unload hysteresis ellipse degrading with time.

These effects are somewhat similar in nature to those observed during the flex fatigue tests described above, but, because they are measured for relatively shorter periods, they remain largely reversible.

Airflow properties

Airflow, or porosity, is also related to comfort because air movement through the foam helps in transporting heat and humidity from the body surface. It is

Figure 11-9   Principle of dynamic creep test

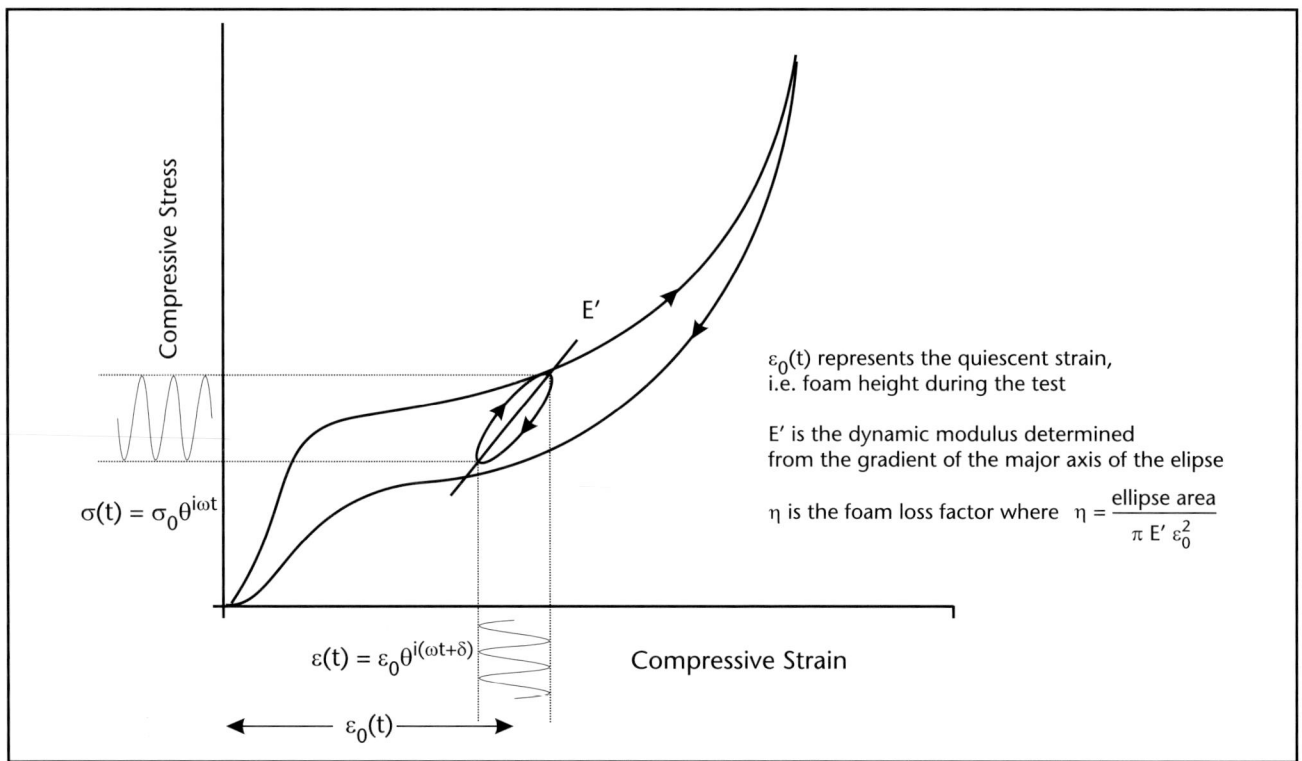

$\varepsilon_0(t)$ represents the quiescent strain, i.e. foam height during the test

E' is the dynamic modulus determined from the gradient of the major axis of the elipse

η is the foam loss factor where $\eta = \dfrac{\text{ellipse area}}{\pi E' \varepsilon_0^2}$

therefore one of the most basic properties of open-cell foams. Several methods are available to quantify it.

The simplest method, described in ISO 7231, and widely used in the slabstock industry for quality control purposes, measures the airflow, which generates a standard pressure drop equivalent to a water column of one centimetre, through a standard sample size.

In test ISO 4638, both airflow velocity and pressure drop through the sample are varied; plotting pressure drop (normalised to a standard thickness) versus air flow velocity generates a linear plot whose slope is proportional to air flow resistance, inversely proportional to permeability.

Another method based on the same principle has been designed for materials used for acoustic absorption (ISO 9053). In this test, a uni-directional or alternating airflow (at a frequency of two Hertz) can be used.

Water vapour transmissibility (as defined in ISO 11092) is less commonly used, but can help to quantify another significant element for comfort: the contribution of an open-cell foam to diffuse humidity generated by the user's skin surface.

A high porosity, as exhibited by high-resilience foams, also contributes to improved moisture and heat dissipation away from the body surface. Additionally, HR foams usually have slightly higher moisture equilibrium than conventional foams, which also contributes to a better moisture management.

Dynamic testing (energy absorption and damping tests)
In addition to their initial comfort ('showroom feel') based on a static assessment by the user, automotive seats may be characterised by their dynamic comfort, which more accurately represents their vibration performance under ride conditions.

The ball rebound test is the simplest measure of foam comfort, and it is widely used in all applications. It is primarily used to differentiate conventional foams from high-resilience foams, the latter usually having a ball rebound resilience above 50 per cent. A standard steel ball with a diameter of 16 millimetres (mass 16.3 grammes) is dropped from a height of 250 millimetres onto a standard foam sample. Its rebound height is measured and expressed as a percentage of the drop height.

Similar tests (called falling weight tests) based on the same principle, but practiced with specially designed test machines, are used to characterise the energy absorption of packaging foams.

Vibration energy management
The non-linear modulus behaviour of the foam affects the transfer of vibrations to the human body, which is experienced by the passenger as a change in

comfort after a few hours' drive. This may be demonstrated in a test where a free weight is loaded onto a flexible foam sample positioned on a vibrating plate. The input vibration amplitude is kept constant but the frequency is varied continuously, to cover the whole spectrum representing a car ride. Both the acceleration amplitude of the free weight oscillations and the resonance amplitude and frequency are recorded. These parameters are related to ride comfort as experienced by the passenger. Flexible foams act as a damping element at frequencies above about 10 Hertz, but amplify vibrations at low frequencies. It is now generally accepted that the best foams combine the following features:

- Low resonance frequency, less than 4 Hz.
- Low amplitude at resonance.
- High damping at a frequency of 6 to 20 Hz.
- Good retention of these properties during use.

The typical response of a flexible foam to dynamic testing is shown in Figure 11-10.

Figure 11-10   Typical response of flexible foam in dynamic comfort test

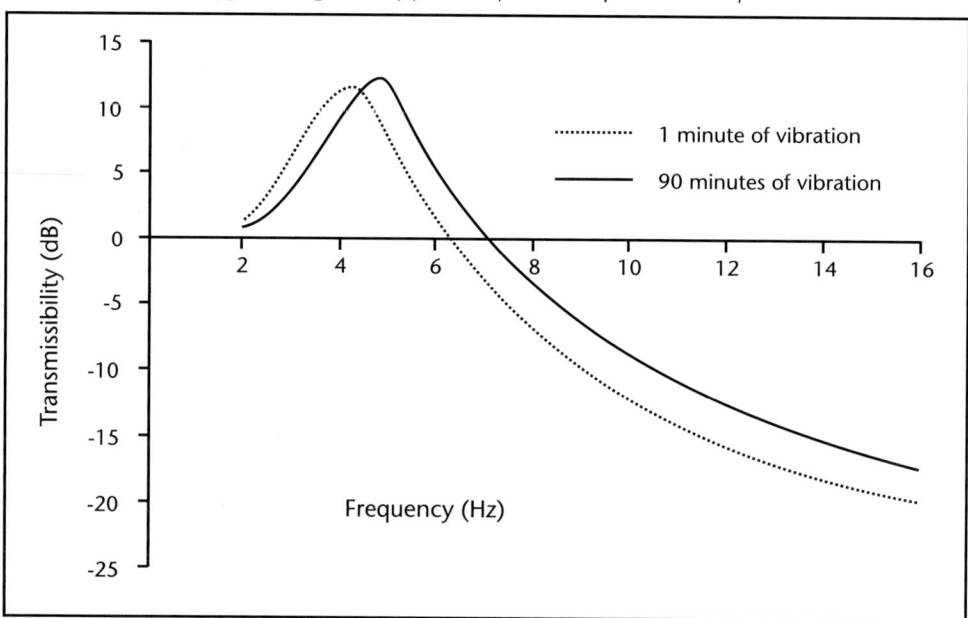

Acoustic performance

In the acoustic area, a commonly used test is the impedance tube test (ASTM C 384) which uses a standing sound wave produced in a tube equipped with a microphone and a loudspeaker to measure the impedance ratio and the sound absorption coefficient of foams used for acoustic purposes. An alternative standard (ASTM E 1050) specifies the use of two microphones to measure the decomposition of the standing wave in two locations. These are connected to a frequency analyser to determine the transfer function of the microphone signals and calculate the sound absorption coefficient and impedance ratios of the acoustic material.

Ageing tests

Humidity/heat

Flexible foams (both automotive seats and furniture or bedding articles) may be exposed to a wide variety of hot and humid conditions during use, Table 11-5.

Table 11-5   Common ageing tests used for flexible foams

Test	ISO	ASTM	Description	Comments
Wet compression set	DIS 13362		22 hours at 40°C and 70% compression, 95 to 100% RH	Polyether foams
Humid ageing	2440		20 hours at 85°C 100% RH	Polyester (slabstock) foams
Autoclave ageing	2440	D3574 J*	5 hours at 120°C (or 3 hours at 105°C)* with saturated humidity level followed by compression set test	Polyether foams
Heat ageing	2440		22 (or 16, 72, or 168) hours at 140°C	Automotive foams

Tropical conditions are defined as 40 to 50°C and 80 to 100 per cent RH. These conditions are used for so-called wet compression sets, as well as dynamic modulus and flex fatigue tests.

Autoclave conditions are 105 to 120°C, saturated steam and simulate sterilisation conditions used, for example, in hospitals for foam mattresses. They are also used by some automotive manufacturers for seating foam specifications, even though these test conditions far exceed those experienced during actual use.

Flexible polyurethane foams based on polyether polyols maintain their cushioning performance at temperatures down to about minus 40°C, which is useful in extreme winter conditions. In the high temperature area, automotive undercarpet foams due to the proximity of the engine or a catalytic exhaust converter can experience conditions up to about 120°C. In such cases, tensile properties should be tested after ageing under these conditions.

In most other applications, flexible foams are used in intermediate temperature conditions (0 to 100°C) for extended periods, without suffering property deterioration.

Other ageing tests

Depending on the expected usage conditions, various other tests can be considered for testing flexible foams in simulated environments. The most widely used standard tests are summarised below. Caution should be used when interpreting the results of these tests, because the ageing conditions can alter the relative rates of the chemical ageing processes, and also introduce phenomena that would not occur in practice. It is best to compare the test samples (at least three specimens to avoid erratic effects) with reference specimens stored at room temperature.

- Most flexible foams based on aromatic isocyanates yellow on exposure to light and oxygen and the degree of yellowing is measured using ISO 4582 or ASTM E313.
- The solvent resistance of flexible polyurethane foams is quite good. Polyether foams are more stable towards aqueous acids and bases than polyester foams. Solvent swelling, however, is lower for the latter foams, a useful feature for foams used in clothing. This can be tested using BS 4443 part 4, method 10.
- The sensitivity of polyurethane foams to microbial attack can be tested by ASTM E 21 with the test being most relevant to polyester foams, which may need to be protected against microbial attack.

Fire test methods

There are many fire tests available to assess the performance of materials, composites or finished articles under a variety of scales, from smouldering cigarettes to kerosene burners, or even full-scale room tests. Each of these is designed to address a specific aspect of fire development. Some, based on low-energy sources such as a cigarette or a small flame, focus on ignitability. Others use a larger source, such as a gas burner, a wood crib, a large calibrated heat source, and therefore are designed to assess the rate of burning or fire spread, or the emission of toxic gases. They may be specified by international or national institutions, by industry associations, or individual companies. A distinction should be made between the tests designed to compare material performance under simulated conditions, and those defined for materials or finished articles by legislating, regulating or purchasing bodies to define minimum requirements. The latter are usually more sophisticated and can only be performed in specialised and approved laboratories.

For fuller details see Chapter 4.

Environmental performance

Environmental performance in a broad sense is taken more and more into account when selecting materials for comfort applications, both for automotive and domestic use. The main and most objective criteria are material and energy requirements, as derived from life cycle analysis, including recycling routes. In the case of cushioning materials, volatile emissions from polyurethane or other foams are also a key factor, mainly because the high surface to mass ratio of the foamed polymer greatly amplifies the issue.

See Chapter 3 for a fuller discussion.

12. Moulded foams for automotive seating, sound insulation and furniture

Nick Duggan
Alain Parfondry

The major advantage of moulded foam, compared to slabstock foam, is an improvement in productivity, a reduction in the level of waste and the ability to produce a wide variety of shapes and foam types in a combination of hardness, density and other specified properties required for each end-use application.

The majority of moulded foams are made using the 'cold-cure' process, which operates at a mould temperature of 30 to 60°C. This technology is normally based on ethylene oxide tipped, high molecular weight polyols (typically 5,000 to 6,000), and TDI or MDI isocyanates, alone or in blends. This process, besides offering improved productivity, also provides formulation versatility and excellent foam quality and consistency. The majority of cold-cure formulations are high resilience foams.

The 'hot-cure' process is based on TDI and activated polyols (compared to those used for conventional slabstock foam), with moulds heated above 100°C to complete the foam cure. This process is now less popular due to its higher energy requirements, higher waste production plus the process and design limitations. Its use is now limited to soft, low density pads with simple shapes.

Applications

Automotive interiors on average contain 18 to 20 kilogrammes of polyurethane products, of which 13 kilogrammes consists of flexible foam in various forms such as seating, head rests, arm rests and sound insulation The major application areas are shown in Figure 12-1.

Figure 12-1   Blow-up of polyurethane components in a car

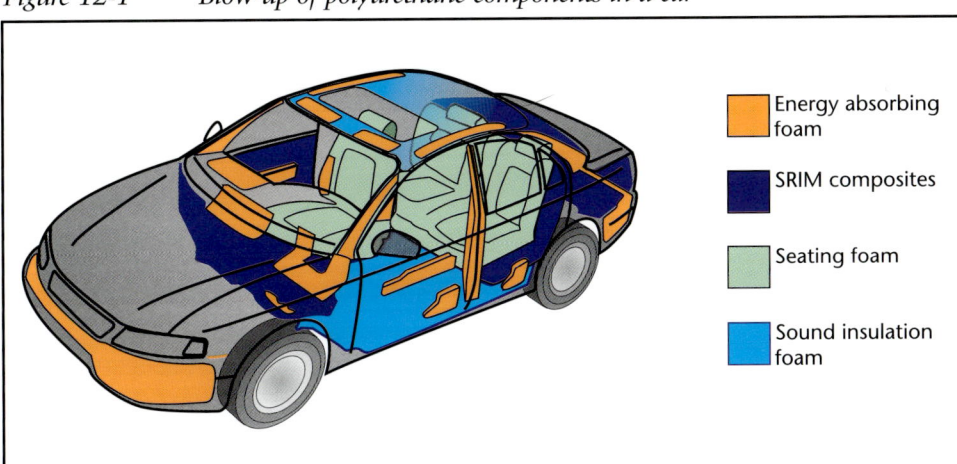

The main market trends in recent years have been towards an increased focus on comfort and comfort retention, weight savings, improved recyclability and reduced volatile emissions. A typical example is the 'corporate average fuel efficiency' or 'CAFE' regulation in the USA, which calls for reduced petrol consumption in cars. In the trim area, lower part weights, which are now achievable through the use of lower density foams, can significantly contribute to this. Advances in technology and new generations of flexible polyurethane foams have successfully supported all these ideas.

Increased competition in the automotive industry between original equipment manufacturers (OEMs) and their seat and foam suppliers, has led to major structural changes in the industry. This has resulted in a reduced number of OEMs and a smaller number of worldwide tier-one suppliers with large production volumes, but relatively small individual seat production units designed for just-in-time delivery to nearby car assembly plants.

The largest application for flexible foam in the automotive industry is in seating where it has assumed a major position since its introduction in the early 1960s. Competitive materials for this application are based on fibres, traditionally coconut fibres or animal hair bonded with latex and more recently polyester fibres. In comparison with these materials, polyurethane foam offers improved design freedom, easy processability, low waste and energy requirements, cost effectiveness, and very good durability and comfort, so it is likely to remain a major cushioning material in the foreseeable future.

Flexible moulded foam also plays an important role for acoustic attenuation in automotive interiors. In the early 1980s it was first introduced in Western Europe as a felt replacement in floor mats and dashboard insulation. In the mid-1980s this technology was extended to the American market and led to the development of foam-backed automotive carpets that have now become a standard global technology.

In the last few years, flexible foams have also been introduced nearer the engine, under the bonnet, to dampen noise and vibrations.

Finally, furniture cushions (domestic armchairs and sofas), bedding (orthopaedic mattresses and pillows), or office furniture, also represent interesting markets for flexible moulded foams.

Automotive seating

The two main functions of moulded seating cushions are body support (load bearing) and comfort (vibration damping) and both properties should be preserved after long periods of use, to provide durability and high resilience respectively. Foam hardness, usually expressed as indentation force deflection, is therefore the first parameter to be defined by seat designers; see Chapter 11 for details of test procedures.

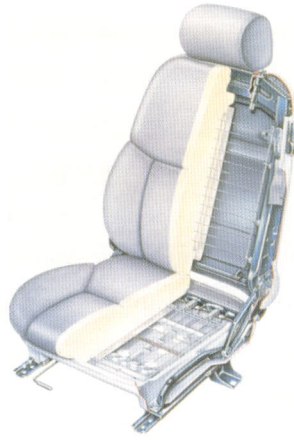

*Figure 12-2
Cut-away view of seat construction*

Seat design

The shape of the seat cushion should be adapted both to the average anatomical features of the end-user and also to the different car design principles and styles prevailing in different regions of the world. In recent years, however, global OEM companies have started to design car models around standardised 'world' specifications. The key elements of an automotive seat are shown in Figure 12-2.

The main trends have been for smaller cars, sports or off-road vehicles with elaborate suspension systems designed to cope with rougher terrain and an increased use of diesel engines. This calls for the design of cushions with more elaborate shapes allowing good lateral support. In addition to stringent durability criteria, foam property specifications often include vibration absorption in the low frequency area for improved dynamic comfort.

Two distinct technologies are used: either a full foam seat, often positioned on a metal shell (dead pan) which requires harder and higher performance foams; or cushions on a spring base which calls for foam cushions, with simpler shapes, and less emphasis on vibration absorption.

In addition to these basic principles, other refinements such as the use of 'dual hardness' cushions and foam-in-fabric (in the USA this is referred to as 'pour-in-place') technology were introduced in the mid-1980s. Examples of seat designs are shown in Figure 12-3 and Figure 12-4.

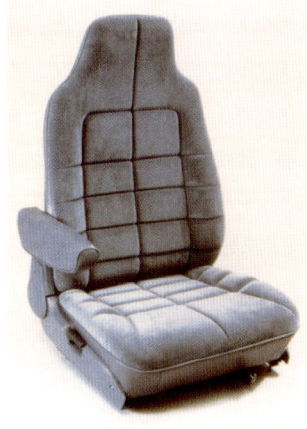

Figure 12-3

Comfort

The first requirement is static or 'showroom' comfort, which is based on a combination of:

- Load-bearing ability (expressed as ILD hardness).
- Even pressure distribution on the human body.
- Heat and moisture diffusion properties of the cushion.

Metal wire inserts are commonly used to further improve the overall seat shape retention, as anchor points for the textile cover and to attach the seat onto the base frame. The pressure distribution is a function of foam hardness, hysteresis and the shape of the seat. Many developments in seat design have been introduced to optimise this feature, like textured or structured (channelled) foam surfaces, and textile/foam composite covers.

Figure 12-4

Sound insulation foams

The current standard sound insulation system for automotive interiors, which sits directly on the metallic floor pan, consists of a two-layer construction. This composite is made up of a layer of flexible polyurethane foam in contact with the floor pan, and an upper layer of carpet textile. The thin flexible foam layer acts as a vibration insulator and decoupler. Alternatively, to improve acoustic performance a triple layered construction can be used. In this construction a mass of elastomeric polyurethane sits as a layer in between the foam and the textile layers.

The major benefits for the OEM, over traditional materials such as felt, are that the system provides optimal design freedom such as: variable thicknesses and shapes, tuneable viscoelasticity to specific vibration spectra, weight reduction and finally high productivity and low waste.

The end-user benefits from the high sound insulation performance obtained from the neat fit to the floor pan shape that leaves no air space for sound waves to escape and in contrast to felt it produces minimal levels of volatile compounds in the car interior, Figure 12-5.

Sound insulation foams designed for use as bulkhead insulation, or directly in the engine compartment, take full advantage of the above benefits and these are growing application areas, Figure 12-6.

The noise spectrum ('signature') generated by specific automotive engines, coupled with the vibration transmitted from the road surface, covers a wide frequency range (typically 50 to 2,000 Hz). Soft grades of cold-cure foam with very open cells have traditionally been used for the foam interlayer which, in conjunction with the elastomeric mass and textile cover, have been optimised in terms of the key physical parameters – thickness, modulus, density, and air flow.

In the last decade there has been an increasing interest in the development of viscoelastic foam formulations for sound insulation, especially for middle to upper priced European cars.

Figure 12-5   Floor mat

Figure 12-6   Bonnet liner

Moulded furniture foams

Flexible foam cushions for domestic furniture – settees, armchairs, sofas, dining chairs and other items – are usually thicker and softer than for automotive seating and, in general, the physical property requirements of the foam are less demanding. However, they must have a pleasant surface feel and compression set requirements may be very stringent, depending on the furniture manufacturer. In the case of office furniture, foam grades need to be adapted to a wide variety of styles and designs, but thinner and harder foams tend to be used.

Raw materials and formulations

The choice of cold-cure flexible foam formulations for seating and sound insulation continues to be driven by the trend for lower density, which has increased the need for foams with higher hardness. For TDI systems this cannot be met simply by increased water levels, as the higher urea levels produced would lead to semi rigid foams. The new developments have therefore focused on systems based on:

- MDI and MDI/TDI blends.
- Polymer polyols.

MDI contributes more to foam hardness than TDI and improved MDI grades have been developed to tune foam properties to the exact requirements of each OEM for seating and sound insulation applications. Also, blends of TDI with small amounts of MDI (T/M blends) and more recently, blends of special MDIs with small amounts of TDI (M/T blends), have been introduced.

The need for faster cycle times and for lower scrap rates in production has also favoured MDI systems as they are more reactive and their slightly higher viscosity leads to less leakage on moulding. Also, additional foam performance requirements, especially in the areas of comfort and durability, can be met more easily with MDI systems, which are inherently more tolerant to formulation additives.

Environmental pressures have favoured the use of isocyanates with lower vapour pressure, such as MDI and the need for foams with lower volatile emissions is easier to satisfy when using MDI because of its intrinsic higher reactivity leading to lower catalyst usage. New MDI types with higher isocyanate content, lower viscosity, and easier metering on high-pressure machines have been developed to satisfy these new requirements.

The demand for higher hardness also triggered the development of high solids polymer polyols that are increasingly used in these cold-cure formulations.

Against this background, the use of hot-cure foams has steadily declined.

Cold-cure foams

Cold-cure foams were first produced in the mid-1960s when polyols with higher reactivity, made commercially for the first time, were reacted with blends of TDI and polymeric MDI (up to 40 per cent). The initial foams were very soft and could only be used for furniture cushions.

In the mid-1970s the first polymer polyols, based on polyacrylonitrile and blends of polystyrene and polyacrylonitrile (SAN), were introduced and led to the production of foams with higher hardness and good processing that could be used for automotive seating. This was followed by a range of developments such as the use of other types of polymer polyols based on polyureas (PHD) or polyurethanes (PIPA), TDI 65/35 and blends of TDI with polymeric MDI. Also, modified TDIs, which could be used without MDI to make foams with higher resilience, hardness and improved processing, were developed.

These were followed by modified MDIs, with lower functionality, a higher proportion of 2,4' isomer and partial pre-reaction (prepolymerisation) with special polyols. These improved system compatibility as well as foam processing and properties so that MDI could compete with TDI in the automotive seating market.

Since 1980, additional developments have further widened the scope of cold-cure foams and there is now a range of other options available:

- Special MDI foams can be made from prepolymers that have excellent processability (fast cure, wide isocyanate index range) and very good properties (hardness, durability).
- Flexible foams made from blends of MDI and TDI where a low functionality MDI is the major component and TDI is used at less than 25 per cent.
- A range of polyols with molecular weights of 5,000 to 6,000 and ethylene oxide tipping of 12 to 20 per cent.
- Specially designed additives further fine-tune both foam processability and properties. These include ethylene oxide rich polyols that act as cell openers to minimise the level of closed cells on demould.
- Alkanolamines such as diethanolamine and triethanolamine improve foam stability and compression set properties, mainly in TDI foams.
- Various low molecular weight chain extenders such as glycerol, trimethylolpropane and their ethylene oxide or propylene oxide adducts improve specific properties.

Typical cold-cure formulations are given in Table 12-1.

Table 12-1   Cold-cure formulations and properties

	TDI	T/M	M/T	MDI		TDI	T/M	M/T	MDI
Formulation, wt-%					**Processing conditions**				
Glycerol PO/EO (OHv 35)	38.7	–	–	–	NCO Index	100	100	100	100
Glycerol PO/EO (OHv 32)	–	44.2	–	–	Mould temperature				
Glycerol PO/EO (OHv 28)	–	–	41.0	47.8	at pour, °C	55	55	40	40
Polymer polyol (43% SAN)	31.7	24.8	–	–	Demould time, min	6	5	4	3
PIPA polyol (48%)	–	–	18.2	9.8					
Catalyst package	0.5	0.5	0.5	0.7	**Properties**				
Surfactant	0.7	0.7	0.5	0.2	Density, kg/m^3	45	48	48	47
Diethanolamine or					ILD hardness (25%), N	210	211	220	218
Triethanolamine	0.8	0.8	0.5	0.1	CLD (40%), kPa	5.3	5.6	5.7	5.8
Water	2.1	2.1	2.3	2.2	Tensile strength, kPa	160	160	154	142
TDI 80/20	25.4	–	–	–	Elongation, %	125	108	103	102
TDI/MDI	–	26.9	–	–	Tear strength				
MDI/TDI	–	–	37.0	–	(Die C), N/m	325	375	350	285
MDI	–	–	–	39.1	Compression				
					set (50%), %	6.3	3.3	3	4.4
					Wet compression set*, %	11	10.6	8	11

* Toyota method TSM 7100G: 50°C, 95% RH, 50% compression

The urethane reaction in these cold-cure systems is faster than the polyurea reaction and, due to this, the viscosity increases rapidly to encapsulate the carbon dioxide gas evolved from the urea reaction.

The reactions can, therefore, be driven simply by a combination of two or more amine catalysts to promote blowing and gelling, and with the exception of TDI foams organometallic catalysts are not required. High foam stability is inherent to these foams, so that the addition of weak surfactants based on low molecular weight silicone oils at low concentration or siloxane-polyether copolymers are sufficient merely to control cell size.

Automotive seating

TDI systems are usually formulated on the basis of 5,000 molecular weight polyols and relatively high levels of polymer solids (up to 20 per cent of the total polyol) and they usually need low levels of cross-linker, such as diethanolamine for acceptable processing. Low levels of metal catalysts, such as dibutyltin dilaurate, are also used to optimise cure. Controlled processing conditions are required to ensure low reject rates and key items are: tightly sealed moulds, accurate temperature control plus consistent metering and mixing. The resulting foams have high mechanical properties, but their resilience is limited, as is their resistance to changing compression set, especially under humid conditions. Such systems are commonly used in the USA, mostly for relatively soft, deep seat cushions positioned on springs.

TDI/MDI systems (where polymeric MDI is used at low levels, usually less than 30 per cent) are based on the same principles as TDI systems, but they tend to minimise the problems, without compromising on essential mechanical properties. These systems are preferred in Europe and Asia mainly because of their higher hardness contribution, broader processing latitude, faster cure and higher resilience.

MDI systems are usually formulated with 6,000 molecular weight polyols that produce softer foams, but because MDI contributes much more to hardness than TDI, the overall requirement for polymer polyol is lower, sometimes none at all. There are a wide variety of systems classified according to the functionality of the MDI used. High functionality variants are more suitable for headrests, while MDI variants with functionalities of 2.15 to 2.3 give foams with a high hardness, robust processing and cure properties that are particularly applicable to dual hardness and foam-in-fabric technologies. Lower density foams, still with equivalent properties, can be formulated using MDI variants with higher 2,4' isomer content and with a functionality below 2.15.

In addition to their good processing and cure characteristics and partly as a consequence of it, MDI-based foams can be formulated over wide density and hardness ranges. They also display very low static and dynamic creep properties, especially under humid conditions and they can be tuned for specific dynamic comfort requirements.

To extend the low-density opportunities, systems based on MDI/TDI (MDI-based, but containing small amounts of TDI) and/or the use of added carbon dioxide as a physical blowing agent have been developed.

Sound insulation foams
Foam formulations for sound insulation are characterised by their very good flow properties combined with fast cure, low hardness and high porosity. MDI variants are preferred because they can be processed over a wide isocyanate index range to reach low hardness, while maintaining good cure without requiring high levels of catalysts. As a consequence, volatile components in the foams can be more easily controlled. TDI systems are seldom used.

Viscoelastic foams have become more popular, especially in Europe for mid-size and luxury cars. A special type of viscoelastic foam with slightly adhesive skin properties has also been introduced, based on a blend of special polyols. The main advantage of these 'tactile' foams is their perfect fit to the car floor pan, which prevents noise 'leaks' through air gaps.

Moulded foams for furniture
Moulded foams for furniture are usually made in small runs from simpler (cheaper) moulds made of epoxy or polyester resins that impose specific processing requirements. The moulds have a much lower thermal conductivity, leading to poor temperature control, so that they give a different surface finish (more open skin) than metallic moulds. They also tend to have a poorer fit than metallic moulds requiring fast gelling or cross-linked formulations for good processing to avoid excess leakage.

Formulations based on TDI/MDI blends have been traditionally used, but there is a clear trend towards the use of MDI variants alone. This is mainly because of the improved processing characteristics of a wide index and hardness latitude, tolerance to mould temperature variations, good mould filling ability without leakage, and fast cure. An example is the use of MDI foam systems for spraying onto substrates to improve their surface feel such as special mattresses in conjunction with a spring base.

Another recent trend has been the design of formulations with lower odour, which again favours MDI-based foams since they tend to need lower levels of amine catalysts, one of the main sources of foam odour.

Hot-cure foams

Hot-cure foam systems have traditionally been based on TDI and triol polyether polyols with molecular weights of 3,000 to 4,000, which are less reactive than the polyols used in the cold-cure process. They are similar to those used for conventional slabstock foams but with a short ethylene oxide tip. Therefore, a high heat input from the process and the use of metal catalysts, such as stannous 2-ethyl hexanoate, are required to force the isocyanate to react with both the water and polyol at the same time. Strong silicone surfactants, similar to those used for conventional slabstock foams, are necessary to ensure early foam stability. These formulations are usually processed on multi-component machines, as the metal catalysts are very sensitive to hydrolysis.

Foam density and hardness depends basically on the level of water and isocyanate used and, to a smaller extent, to the degree of mould over-packing. A typical formulation is shown in Table 12-2.

Table 12-2   Hot-cure formulation and properties

Formulation, wt-%		Processing conditions	
Glycerol PO/EO (OHv 55)	70.0	NCO Index	105
Amine catalyst	0.5	Mould temperature at pour, °C	40
Tin catalyst	0.1	Oven temperature, °C	140
Silicone surfactant	0.7	Mould temperature (max.), °C	125
Water	2.1	Demould time, *min*	10
TDI 80/20	26.6		
		Properties	
		Density, kg/m^3	33
		ILD hardness (40%), *N*	180
		Tensile strength, *kPa*	110
		Elongation, %	185
		Tear strength *N/m*	240
		Compression set, % 75%	7
		50%	5

To improve foam hardness TDI (65/35) is sometimes used, but polymer polyols (usually SAN) are often preferred and diols can also be added to increase mechanical properties. High-resilience polyols may also be used to improve processing.

Compared to cold-cure foams, hot-cure foams have higher mechanical properties, but poorer comfort and resistance to creep and flex fatigue. They also require a flame retardant additive to meet the standard automotive (MVSS 302) fire test.

Hot-cure technology does not meet all automotive seating performance requirements and in addition, the process requires high amounts of energy, generates more waste and is generally more difficult to control for the production of foams with consistent properties.

Processing of flexible foams

During the early years of their development on an industrial scale, flexible moulded foams were mainly processed using low-pressure machines, with two or more metered streams and a high-speed mixer. Solvent flushing at regular intervals during production was required and process and product consistency were limited, mainly because of machinery wear-and-tear plus limited metering accuracy. Such machines are still used for the production of small series or prototype parts, but the needs of mass production have led to the replacement of low-pressure machines by the more reliable high-pressure machines.

On high-pressure machines, two or more components are mixed by direct impingement at pressures of 12 to 20 MPa and the mixing chamber is cleaned after each shot. These machines are versatile, robust, need little maintenance and can handle a wide variety of throughputs without modification. Their main limitation is the metering of high viscosity streams, for example when high levels of fillers are used, but this can be overcome by the introduction of metering via lance pistons, where the pumps operate on hydraulic oil, which is then transferred to an equal volume of polyol or isocyanate. The key elements of a high-pressure mixing head are illustrated in Figure 12-7.

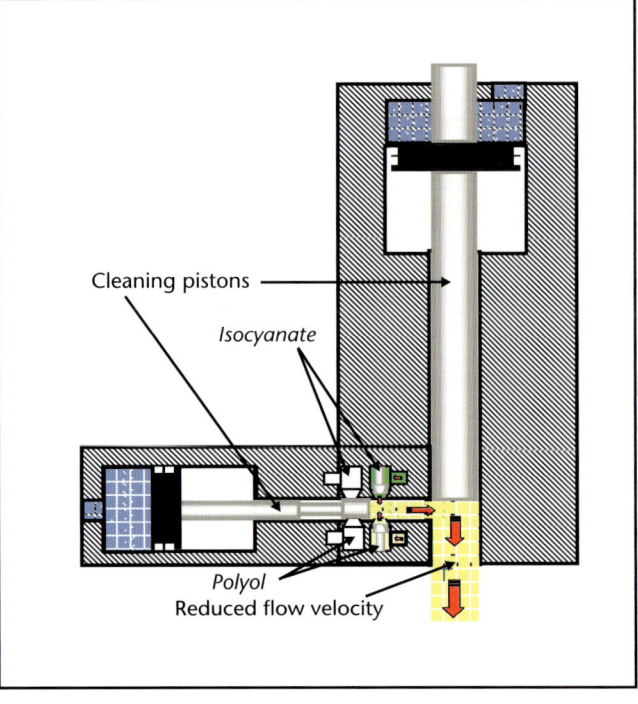

Figure 12-7   High-pressure mixing head

High-pressure machines can only handle a limited number of streams in contrast to low-pressure machines where the number of streams is limited only by mixing head size, which in practice means up to 10 streams. This is not a real limitation, however, as most moulded foam formulations can be produced as two to four streams, for instance by premixing additives into the polyol stream. To ensure uniform cell structure the liquid streams need to be 'nucleated' by the introduction of small air bubbles.

Most machines are now computer controlled in order to record all chemical, machine and mould process parameters during production to ensure accurate shot weights; early fault detection leading to correction.

Productivity can be maximised using several mould handling methods, the choice of which depends on part and production numbers. Fixed moulds fed in turn by a moving mixing head are excellent for large parts that require specific pour patterns for each mould and individual moulds can be changed without interrupting production. The traditional mould set-up for the volume production of hot-cure moulded parts consists of large 'racetrack' or 'paternoster' carriers, but these have the disadvantage that each mould change or maintenance check stops production.

A more efficient approach, which combines productivity and versatility, is to use small carousels linked to a fixed machine. The number of moulds required is reduced, an economic advantage with the foam-in-fabric process, which requires the use of complex cover positioning mechanisms, vacuum drawing and pneumatic actuators. Carousels are more and more favoured for the production of a wide variety of automotive cushions for 'just-in-time' delivery to car assembly plants.

Cold-cure moulding

As the cold-cure process relies on the use of polyols with highly reactive end groups the moulds only need to be heated to 30 to 60°C, by the use of hot water. Moulds are usually made of cast aluminium with the minimum number of vent holes drilled, to let air escape, but minimise foam leakage.

Pour patterns may be as simple as injecting into a closed mould or a fixed pour pattern for an open mould. More complex moulds may require more elaborate pour patterns; the most difficult having to be handled by robotic controlled mixing heads. These hydraulic powered robots also ensure the most consistent part quality and can reduce reject rates to less than 0.5 per cent.

Significantly more material is added to the mould than the minimum required to just fill the mould after expansion, over-packing may be as high as 50 per cent, and the mould pressure can reach 200 kPa. The degree of over-packing is an important factor in extending the range of foam hardnesses obtainable under a given set of conditions. However, to ensure the production of defect-free parts, with no decompression voids, air bubbles or other defects, it is essential to have tight-fitting, well-vented moulds. Various mechanisms are available to keep moulds tight, such as the use of silicone rubber-based seals in the parting line, coupled with the use of air filled bags under the clamped mould to ensure even and tight contact all around the parting line.

To ensure easy demould and defect-free parts a release agent needs to be evenly applied to clean moulds. Traditional release agents were based on low concentrations of waxes, soaps or oil solutions in hydrocarbon solvents, sometimes with the addition of silicones. However, for environmental reasons these solvent systems are being replaced by high solids water-based suspensions or very high solids liquid silicone systems. The selection of an appropriate release agent and the type of application system depends on the part, the mould material and mould temperature.

After demould, parts still contain closed cells and need to be mechanically crushed between rollers or under vacuum to avoid shrinkage caused by the cooling of carbon dioxide in the foam cells. This gentle crushing ruptures some of the cell windows, but has no significant effect on foam properties. Shrinkage of 0.5 to 1 per cent takes place after demould as a consequence of the crushing operation and needs to be taken into account in mould design.

The demould time for cold-cure systems varies from one to eight minutes, but is commonly around two to five minutes, dependent on the foam formulation used and the part complexity, leading to an overall production cycle, including mould cleaning and preparation, of three to ten minutes. This is significantly shorter than for hot-cure foams. Demoulded parts reach their final hardness in two to six hours depending on the system used.

Dual hardness
The use of dual-hardness cushions has become popular for a number of automotive designs and includes several variations of hard and soft foams. These dual hardness cushions were originally made in several steps, or by using rebonded foam inserts. They can now be produced in one operation when a harder foam system is injected, simultaneously with the main softer portion, into the lateral parts of the mould. This may be done, using two mixing heads, or a single head and a programmed shot sequence. A wide range of hardnesses can be produced from a single MDI-based formulation by simply changing the isocyanate index whilst with TDI-based systems, two different polyol formulations are usually needed, which requires a more complex dispensing system. Lateral support is improved and comfort preserved due to the softer central foam pad. This technology has also been extended to seat backs.

Foam-in-fabric
In the traditional 'cut-and-sew' approach for an automotive seat the foam cushion is moulded alone before being inserted into a textile cover. In the foam-in-fabric approach, the foam is moulded directly into a specially designed cover. These covers are usually triple laminates comprising:

- A top textile layer.
- An interlayer of flexible foam.
- A thermoplastic polyurethane film.

The flexible foam interlayer, normally polyester based as it is usually flame laminated or glued onto the textile, ensures heat and moisture control during use that preserves comfort whilst the air tight thermoplastic film prevents penetration of the foaming liquid into the foam backing and textile layers.

Simplified foam-in-fabric processes are commonly applied to headrests with systems being foamed directly onto textile covers, without film backing. This is because fast reacting foam formulations are used, which polymerise before penetrating into the foam-backing layer.

Moulding complete seat cushions directly in their textile covers is a cost effective technology because it allows a sharp reduction in the labour costs usually needed for seat assembly. In its most automated form, used for large series of small or mid-size cars, the cast aluminium moulds are fitted with rows of small holes connected to a vacuum accumulator. Very little or no release agent is needed. The cover is positioned over the mould and vacuum is applied to draw it into the mould. Special side clamps are used to adjust the cover tension evenly around the edges. Metal inserts are then positioned in the mould before the flexible foam is poured in. The seat cushions can be demoulded in the usual cold-cure conditions (mould temperature and demould times), and trimmed by robots. This process is particularly well adapted for efficient just-in-time delivery to a car assembly plant.

Carbon dioxide as physical blowing agent for flexible foam

One of the most successful new technologies for low density parts has been the use of carbon dioxide as a physical blowing agent since it provides additional blowing without increasing the polymer modulus and therefore retains elasticity and creep resistance properties. In contrast, the use of water as a chemical blowing agent to generate carbon dioxide forms urea linkages, increasing foam hardness and changing other properties.

With modern high-pressure machines, the metering of liquid carbon dioxide in addition to or instead of nucleating air, can take place in several ways; either in the machine tanks which gives very consistent results, but is less versatile because the level is limited by the tank pressure, and fixed for all the foam shots, or on-line into the polyol or isocyanate stream or separately into the mixing head. The latter two methods require machines to be modified and need very accurate metering and, therefore, controls, to ensure foam consistency shot-to-shot and within a shot.

The major change is that foaming liquids are replaced by physical froths with less 'wetting' of the mould surface. This is partly compensated by the slower gelling profiles of these systems and by the fact that this froth is in fact pre-expanded foam that fills a larger proportion of the mould volume. Overall, these systems can therefore be processed with only minor adjustment for reactivity, pour pattern, and mould positioning.

Hot-cure moulding

The hot-cure process requires the moulds to be heated to about 120°C in order to complete the foam cure. Consequently, to facilitate heat transfer, they need to be made from thin iron or aluminium sheet, less than two millimetres thick, or cast aluminium less than five millimetres thick. This results in moulds that have a light construction and need many vent holes, to avoid a build-up of pressure that could buckle the moulds. So, the use of over-packing to adjust foam hardness through density variations is precluded. As very low pressures are generated, then simple and light mould closing mechanisms can be used. Some material is lost when foam escapes through the mould vents and shrinkage can be as high as two per cent, which needs to be taken into account in the design of the moulds.

Typical steps in a production cycle include:

- Spraying a wax or silicone-based release agent into the mould.
- Adding metal inserts and other accessories into the mould.
- Pouring the liquid foaming mixture into the mould, cooled to 30 to 40°C, taking care to avoid excessive splashing, and following an optimised pour pattern.
- Closing the mould.
- Heating the mould for 8 to 15 minutes at 120 to 140°C usually in an oven at 140 to 160°C, or direct mould heating using a heat exchange fluid circuit.
- Demoulding the part after 10 to 15 minutes – no crushing is necessary to open cells.
- Cleaning the mould vents as excess foam may amount to two to five per cent of the total.
- Cooling the mould before the next cycle.

Total cycle time can be between 20 to 30 minutes and demoulded parts reach their final hardness after a further one to two hours.

The temperatures reached in this process are similar to those reached by the foaming polymer under adiabatic conditions and therefore the entire part reaches a similar level of cure, with little or no skin formation and an even, open cell structure. This older hot-cure process is normally now restricted to the manufacture of softer, low-density foams.

13. Slabstock foams

Alain Parfondry

Slabstock foams represent about two-thirds of the total flexible foam market and although this is now a mature technology it continues to meet new challenges, especially in the demands for increased production efficiency and reduced environmental impact. The slabstock industry is adapting to the changing needs of the markets through developments in both chemical and process technology. Historically, the majority of slabstock foams have been made using a free-rise continuous process, but a small proportion is also produced by a discontinuous 'batch-block' process. The end result of both processes is blocks of flexible foam that are semi-finished products that require further cutting, shaping and fabricating into finished articles.

The applications for polyurethane flexible foams are well established after more than 40 years of development and now have a dominant position in many market areas. The major areas of demand for slabstock foams are in cushioning applications with bedding, mattresses and pillows, plus furniture, domestic and office, having the largest market share. In the mattress market polyurethane foams are appreciated for their cost effectiveness and their low weight, especially when compared to mattresses produced from either metal springs or latex. Flexible foams are also used in a wide range of other applications such as industrial filters, textile linings, sponges, shoe components, toys, packaging and many others with the limitation on end-use being only the designer's imagination.

The family of flexible foam includes hyper soft, high load-bearing (HLB), semi-rigid, viscoelastic and high-resilience (HR) with densities from below 12 kg/m^3 to above 100 kg/m^3 (the majority being between 18 and 30 kg/m^3) and with cell sizes that can vary, independent of density, from very fine to coarse.

Chemically, there are two families of slabstock flexible foam. They are based respectively on polyether polyols, by far the largest section at 90 per cent of the market, or polyester polyols. TDI is the predominant isocyanate for both although a small, significant and growing portion of flexible foam is now produced using MDI.

Foam technologies

The main demand for polyester polyol-based flexible foams is for applications where specific properties are required, such as the partial melting of the surface for flame lamination or where a very small and uniform cell size is necessary as in filter applications.

The most widely used polyether foams are the conventional types, with resilience up to 50 per cent that can be produced over a wide range of density and hardness. High-resilience foams, on the other hand, are manufactured with much narrower density and hardness ranges. The high-resilience foams are targeted at high-quality furniture and bedding applications, which need improved durability and comfort characteristics.

Slabstock foams can be modified with additives in a variety of ways to impart special properties, such as improved flame resistance, resistance to combustion, increased dielectric properties, resistance to the growth of micro-organisms. All these types of foams, as well as the conventional and high-resilience grades, can be cut and shaped to make finished articles.

Raw materials and formulations

Most flexible slabstock foams are chemically blown with carbon dioxide generated by the water-isocyanate reaction. Additionally, for the softest and lowest density grades, small amounts of physical blowing agents are used. The CFCs which were used until the early 1990s have been replaced in the last 10 years by HCFCs, with HCFC-141b being the most common; even solvents such as dichloromethane, acetone and methyl formate have been used. These solvent-based blowing agents are increasingly being replaced by alternatives, such as liquid carbon dioxide, added under pressure via the raw material streams. Another route to lowering density is the use of a partial vacuum to assist foaming where the foam is poured and rises in a chamber with lower than atmospheric pressure which ensures a greater expansion of the foam for a given amount of blowing agent. It is notable that foam plants located at high altitudes tend to make lower density foams, for a given formulation, because of the lower atmospheric pressure and this can give them a commercial advantage.

The overall properties of flexible foams are dependent on the type of polyol and isocyanate used. The key factors for the polyols are molecular weight and functionality, with higher molecular weight giving softer and more elastic foams whilst a higher functionality leads to better foam stability during processing and improved tear strength and compression set.

The hardness of flexible foams is dependent on two key elements; first, on the amount of isocyanate added to the formulation, which is related to the water level and, second, on the addition of polymer polyols, such as polyacrylonitrile/polystyrene (SAN) dispersions, polyharnstoff dispersion (PHD) polyols, polyisocyanate polyaddition (PIPA) polyols, to the polyol stream. Inorganic fillers, such as talc, calcium carbonate, or barium sulphate, can also be used to lower formulation costs and increase the hardness and comfort of flexible foams. However, mechanical and creep properties are rapidly impaired at higher filler levels, so this approach is not as common as it used to be.

Atmospheric conditions also affect foam hardness since warm humid conditions require a higher isocyanate index to provide more isocyanate groups to account for the extra water absorbed from the air. If formulations are not adjusted to take account of the varying levels of atmospheric moisture at different seasons then an imbalance in isocyanate index will occur leading to variations in the level of early foam hardening.

The cell size of all flexible slabstock foams needs to be carefully controlled as it influences foam properties such as porosity (important for comfort), hardness-hysteresis balance (larger coarser cells giving higher resilience and lower hysteresis) and surface feel. The overall distribution of cell size is due to a combination of factors linked to the process employed and the properties of the chemical streams used.

The most common technique for controlling cell size in flexible slabstock is to use nucleating air. This can be introduced at low levels directly in the mixing head, or by consistent stirring of a main raw material such as the base polyol in its storage tank. Special nucleating pumps can be installed on a raw material re-circulating loop. Coarse air bubbles can be avoided by degassing the polyol under vacuum prior to foaming. This is important when foams are made for flame lamination, where pinholes must be avoided, or for filtration applications, where large cells can cause unreliable flow.

Polyester foams generally have a smaller cell size than polyether foams. The higher viscosity and faster reactivity of the polyester systems results in slower window drainage and less cell coalescence during foam formation. Cell size can be further reduced by high formulation reactivity and the correct surfactant choice.

The very wide cell size distribution of high-resilience foams contributes, to some extent to foam resilience, because cells of different sizes buckle under different stress levels and this ultimately leads to a smoother and thinner hysteresis curve and also to improved flexural fatigue resistance and the higher airflow obtained is a positive contributor to comfort. MDI high-resilience foams can be processed to produce fine cells, using the prepolymer process, or coarse cells, by other routes, over a wider range than TDI foams.

Viscoelastic foams, on the other hand, tend to perform better when formulated and processed for smaller cells.

Foam stability during processing is controlled by a combination of catalysts to promote both gelling (mainly organometallic salts) and blowing (tertiary amines) reactions to the right level with the choices dependent on the reactivity of the polyol. Excess blowing results in foam splitting and eventually collapse, while excess gelling results in shrinking foams. This shrinkage occurs because the internal pressure within the closed cells of the foam is substantially less than atmospheric after cooling. In addition, the carbon dioxide within the cells diffuses out at a greater rate than air can diffuse in, thus causing an even greater pressure difference. Within the acceptable processing range, open cell content, and therefore airflow through the foam, can still vary widely. Most balanced formulations reach maximum height at the same time as cells begin to rupture to release the blowing gas ('blow-off' time).

The surfactants used to stabilise the rising foam structure are of three general types: polydimethylsiloxanes for high-resilience foams, siloxane polyether copolymers for conventional and high-resilience foams whilst metal soaps are used for polyester foams. A wide variety of surfactants are available from specialised suppliers.

Flexible foams can be coloured in several ways with the two commonest being based on pigment dispersions and soluble dyes. In both cases the colour is normally added via a separate injection point direct into the mixing head stream so that the colour can be rapidly changed with formulations as separate colours are commonly used for easy grade identification. A wide range of pigment colours is available from the basic white titanium dioxide and carbon black to the brightly coloured mineral salts. All of these pigments are dispersed in liquid carriers that need to be compatible with the flexible foam formulation. Soluble dyes, from modified polyols bearing chromophoric groups, are used whenever a more uniform colour and improved colourfastness is required.

Polyester foams

Polyester foams are produced from TDI and polyester polyols, with typical formulations shown in Table 13-1. The majority of polyester-based foams are produced using TDI 80/20, but higher hardness foams can be made with the TDI 65/35 grade whilst the polyester polyols consist of low molecular weight, 1,500 to 3,000, glycol adipates which have a low level of functionality, typically 2.05 to 2.2. This low level of functionality is introduced to generate cross-links in the foam polymer and to improve foam stability and properties.

Table 13-1   Formulations for TDI polyester foam

Formulation, wt-%	A	B	Processing conditions	A	B
Polyester polyol	60.7	67.9	NCO Index	105	90
Non-ionic emulsifier	0.6	0.7	Polyol temperature, °C	35	35
Catalyst	1.5	0.9	Isocyanate temperature, °C	25	25
Surfactant	0.6	0.6			
TDCP	4.9	–	*Properties*		
Water	2.2	2.4	Density, kg/m^3	29	30
TDI 80:20	29.5	27.5			

A = Technical
B = Lamination

Unlike the polyether polyol systems, the polyester-based foams need only low levels of simple amine catalysts to control foam reaction and stabilisation. This is for a number of reasons:

- As water is very miscible with these polyols the blowing reaction proceeds smoothly.
- Polyester polyols have a higher viscosity than polyether polyols, which contributes to early cell stabilisation during foam rise.
- Polyester polyols have primary hydroxyl end groups promoting early gelation during foaming.

For many of their applications, especially when the final foams are destined for automotive use, polyester-based flexible foams need to have particularly low volatile emissions. For this reason special low volatile grades of polyester polyol are now produced and used in conjunction with de-volatilised silicone surfactants and low-odour or isocyanate-reactive amine catalysts.

Flame lamination of polyester-based foams, for textile lamination for instance, is another major application and obviously the fire resistance properties are of prime importance. These can be improved through the use of halogen or phosphorus additives whilst for higher levels of fire retardancy additives such as ammonium polyphosphate can be used.

Conventional polyether foams

Conventional foams represent the bulk of current flexible foam production, although their importance is slowly declining in favour of high-resilience and other special foams. They are made with a very wide density and hardness range of respectively 12 to 120 kg/m^3 and from 30 to 500 N (expressed as ILD at 40 per cent indentation). They are based on TDI, mainly TDI 80/20 and random propylene oxide/ethylene oxide polyols with up to 15 per cent ethylene oxide units and molecular weights between 3,000 and 4,000. A summary of the types of conventional foams available is given in Figure 13-1.

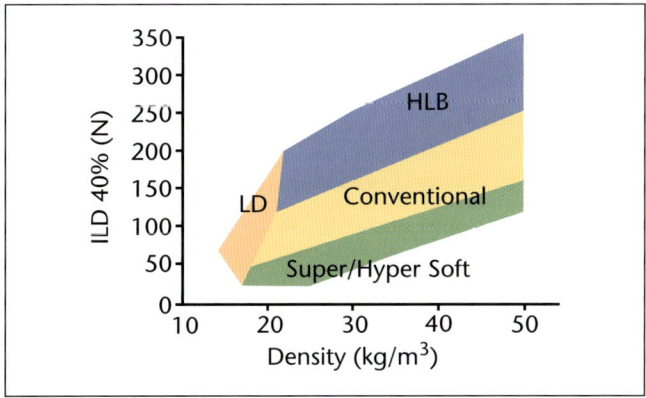

Figure 13-1 Hardness-density map of conventional and HLB TDI slabstock foams

Because these random polyether polyols have predominantly secondary hydroxyl groups, strong gelling catalysts such as stannous 2-ethyl hexanoate have to be used in conjunction with amine catalysts and powerful silicone-based surfactants to stabilise the rising foams. Typical formulations are shown in Table 13-2.

Low-density (LD), hyper-soft and super-soft foams as well as high load-bearing foams belong to this class, although they require specific formulation modifications. Their main characteristics, at least in the core part of their hardness density range, in comparison with the other conventional foam types, include high mechanical and compression set properties and high productivity. Super-soft grades are produced by the addition of specially designed softening

Table 13-2   Formulations and properties for TDI conventional and HLB foams

Formulation, wt-%	Conv.	HLB	Processing conditions	Conv.	HLB
Polyether polyol	63.0	9.7	NCO Index	108	105
Polymer polyol (SAN)	–	54.8	Polyol temperature, °C	35	35
Amine catalyst	0.2	0.1	Isocyanate temperature, °C	25	25
Tin catalyst	0.2	0.1			
Silicone surfactant	0.6	0.6	**Properties**		
Water	2.7	1.5	Density, kg/m^3	23	40
TDI 80:20	33.4	33.3	Tensile strength, kPa	130	90
			Elongation, %	280	210
			Tear strength N/m	500	420
			CLD (40%), kPa	3.5	–
Conv. = Conventional			Compression set (90%), %	5	6
HLB = High load bearing			Resilience, %	45	38

additives based on low molecular weight ethylene oxide rich additives whilst hyper-soft grades, as used for pillows, are produced directly from ethylene oxide-rich polyols, with a small amount of a propylene oxide rich polyol as stabiliser. These foams are more hydrophilic and therefore they can contribute to humidity management during use, an additional element of comfort.

Other special grades of conventional foams include weldable foams – where conventional foams are modified (additives used include low molecular weight glycols and plasticisers) to improve their melting behaviour under dielectric heat or high frequency fields; sponge-like foams – where very large cells are achieved by the use of special silicones (which promote local foam collapse); hydrophilic foams – where conventional foams (normally quite water repellent) are modified with soaps and other ionic additives to improve water absorption properties.

Conventional foams have limited fire resistance. This can be improved by the addition of halogen or phosphorus-containing fire retardants. Since foam mechanical properties are good, relatively high levels of additives can be used. The main disadvantages are their possible influence on foam processing, their softening effect, the risk of leaching into neighbouring materials and increased volatile emissions. These factors might lead to a loss of fire performance on aging or washing and it may be necessary to re-measure fire resistance after aging or washing if it is relevant to the application.

High-resilience foams

High-resilience foams have been more and more in demand over the past 25 years because their improved properties, mainly with respect to comfort and durability, make them attractive to consumers. They currently represent 10 to 15 per cent of the total flexible slabstock foam market. Their use is growing rapidly and this is the area where most innovation has taken place in recent years with development programmes still being actively persued. A summary of the types of high-resilience foams available is given in Figure 13-2.

Figure 13-2
Hardness-density map of high-resilience slabstock foams

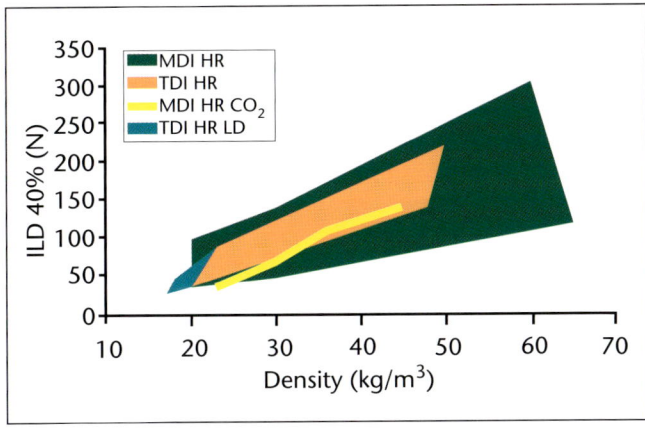

The main difference between high-resilience and conventional foam formulations is the use of ethylene oxide tipped polyether polyols with very high levels of primary hydroxyl end groups. The greater reactivity of these polyols means that much lower levels of the less active amine catalysts can be used and that very little organometallic catalyst is required for gelation. Typical formulations are given in Table 13-3.

The main isocyanate used is still TDI 80/20, but its predominance, at least in Europe, is being eroded by the introduction of special MDI grades that offer significant advantages to the processor whilst for the consumer the chief gain is the improved comfort.

Table 13-3   Formulations and properties for TDI and MDI high-resilience foams

	TDI-HR	TDI-CMHR	MDI-HR	MDI-CMHR	MDI-HR-CO_2	MDI Prepolymer
Formulation, wt-%						
Polyether triol (OHv 32)	35.8	39.2	–	–	–	–
Polyether triol (OHv 28)	–	–	54.0	38.4	56.9	4.5
SAN polymer polyol	35.8	16.8	–	–	–	–
Amine catalyst	0.4	0.3	0.5	0.4	0.5	0.5
Tin catalyst	0.1	0.1	–	–	–	–
Silicone surfactant	–	–	0.1	0.2	0.2	0.1
Cross-linker	0.7	0.8	0.9	0.7	–	0.2
Cell opener	–	–	1.2	0.8	–	2.8
Water	2.1	1.8	2.3	1.9	1.8	2.4
Melamine	–	16.8	–	17.9	–	–
Liquid fire retardant	2.1	2.8	–	2.5	–	–
CO_2	–	–	–	0.6	1.9	–
TDI 80:20	22.9	21.3	–	–	–	–
MDI prepolymer 1	–	–	41.1	36.8	38.7	–
MDI prepolymer 2	–	–	–	–	–	68.4
MDI prepolymer 3	–	–	–	–	–	21.0
Processing conditions						
NCO Index	108	110	95	103	115	–
Properties						
Density, kg/m³	33	31	30	28	19	33
Tensile strength, kPa	140	80	78	42	40	101
Elongation, %	135	140	145	70	128	102
Tear strength, N/m	340	400	200	95	143	187
CLD (40%), kPa	2.8	2.4	2.4	1.5	1.3	3.2
Compression set (75%), %	4	22	6.1	13.6	7.6	3.6
Resilience, %	45	51	58	53	55	55

CMHR = Combustion-modified high-resilience

MDI has lower volatility, compared to TDI, leading to less potential of exposure to isocyanate vapours. The greater reactivity of MDI leads to simpler formulations, with lower catalyst levels that provide foams with lower levels of volatile emissions. It also provides a wide processing tolerance whilst the improved hardness contribution from the MDI molecule leads to an extended hardness range without having to resort to polymer polyols. The combination provides the ability to change hardness independently from density and to easily formulate for finer or coarser foams with higher or lower resilience levels.

The most widely used technology for high-resilience slabstock, based on MDI, involves the use of two MDI streams that consist primarily of a 'full' prepolymer, (MDI prepolymer from Table 13-3) containing all or most of the flexible polyol, plus a second stream of a higher functionality MDI grade, with only water, catalyst and surfactant as additive streams.

More recently, the use of MDI has been extended to non-prepolymer routes, thanks to the development of MDI grades with higher levels of 2,4' isomer. These also give softer foams, which may be hardened through the use of polymer polyols.

The original formulations for high-resilience TDI foams used modified TDIs, usually with allophanate linkages, to enhance processing and more importantly to improve foam stability, but this technique has gradually been replaced by the use of an alkanolamine such as diethanolamine that achieves the same effect. However, such an approach whilst it softens the foams also reduces mechanical properties.

With MDI high-resilience foams, because of their broader isocyanate index latitude, it is possible to decouple water and isocyanate levels to some extent, leading to a nearly independent control of hardness (isocyanate content) and density (water level). This approach is, however, limited because of the deterioration of foam properties at very low isocyanate index and the effect is even more pronounced for properties that are measured after aging under heat or humidity conditions. TDI-based formulations do not have this advantage since their isocyanate index latitude, to achieve good processability, is relatively narrow compared to MDI.

Both MDI- and TDI-based high-resilience foams can be modified to improve hardness using polymer polyols, such as SAN, PHD or PIPA. The main differences between the various types of polymer polyols are related to the carrier polyols used or the nature of the solids themselves with 5,000 molecular weight polyols giving slightly harder foams, but lower resilience and comfort than 6,000 molecular weight polyols.

PIPA, and especially PHD, give improved fire performance and do not contribute to foam volatiles, because of the absence of monomer residues. Also, their larger particle size influences cell formation and tends to give coarser and more open cells and therefore improved foam resilience. Processing properties are mainly related to viscosity, which is linked to solids content, that can go up to 50 per cent depending on the type of reinforcing polyol used.

High-resilience slabstock foams are relatively more resistant to ignition than conventional slabstock foams, since they tend to melt and drip when exposed to fire. If necessary, additives such as halogenated compounds can be used to pass (low energy) ignition tests. Post-ignition behaviour is more systematically controlled with additives, which promote char formation, such as melamine, or exfoliating graphite.

Processing technologies

A typical continuous slabstock production unit consists of a complete factory made up of many components that is vastly larger in scale than the simpler machines used for moulded foam production. The typical elements of a flexible foam plant are briefly outlined below:

- Bulk storage for polyols and isocyanate raw materials under moisture-free conditions. This is usually achieved by feeding a blanket of dry air (typical dew point around -40°C) over the liquid surface.
- A temperature control system, usually provided by on-line heat exchangers, for the main raw materials.
- Metering pumps for each raw material stream – individual or master-batch. For slabstock applications, low-pressure gear pumps are commonly used. High-pressure piston pumps tend to need more maintenance and are reserved for special uses, such as for pumping low viscosity streams like TDI.
- A mixing head capable of accommodating a large number of streams. Low-pressure, but high-shear mechanical mixers are preferred for their consistent mixing performance over a wide range of conditions, mixing ratio, viscosity and overall throughput.
- A lay-down system to ensure laminar flow onto the foaming surface. This is the area where equipment suppliers have carried out most development and the details are discussed below.
- A curing station.
- Foam cutting equipment.
- A foam fabrication and conversion facility.

Most commercial foam production units have a digital control and recording system for data such as stream flow-rate, temperature and pressure. More modern production units also include an on-line foam density-monitoring device, which is based on the decay of ionising radiations (beta rays) through the foam. This can be coupled with on-line formulation adjustment for better foam bun consistency.

Other ancillary equipment includes temperature controls for both chemicals and conveyor, as well as an efficient ventilation system, with a scrubber for removal of isocyanate vapours, in the production hall.

Foam machinery

Slabstock foams are usually processed continuously by metering the mixed foam chemicals through a mixing head into a moving container formed by three parallel, paper lined, conveyor belts positioned at right angles, one bottom and two sides, forming a continuous trough. The foaming mix is prepared continuously in a low-pressure high-shear mixing head fed with the various polyol, isocyanate, catalyst, surfactant and other additive streams in precisely adjusted proportions. The conveyors are also moving at a controlled speed, until the foam bun has risen and solidified into a block ready for cutting into predetermined lengths and taking away from the line. The cuts into standard block lengths are made as the continuous bun is being produced with each standard bun length being rapidly moved away from the main production line.

The size of the final foam bun depends on the end foam application and needs to be greater than two metres wide for mattresses, but can be less than one metre for speciality grades. The final height of the foam bun can be well over a metre and the foaming mix can reach this height in one to two minutes. The length of an individual run can be several hundred metres and can use up to 50 tonnes of chemical dependent on the capacity of the chemical and final foam storage areas.

Many refinements have been added to this basic design to improve productivity and to produce foams with higher quality more consistently. An example is seen in the methods available to make blocks with a more rectangular shape that produce less foam waste after cutting. There are various routes to achieving such 'flat top' buns and material losses, after trimming the bun edges, may be reduced to 10 to 15 per cent by carefully controlling the process as described below.

The simplest method involves traversing the mixing head across the width of the bottom conveyor as it moves forward to improve liquid distribution on the bottom conveyor.

Draka/Petzetakis process
Dome-shaped buns arise mainly due to the friction generated by the rising foam dragging on the side-wall papers. This process minimises the problem by raising the side paper in sequence with foam rise thus greatly reducing friction drag.

Maxfoam process

In this method the foaming mix is poured into the bottom of a trough with the conveyor positioned downstream so that it can collect the froth as it overflows. The first part of the bottom conveyor plates are disposed as 'fall plates' at angles such that foam descends on the conveyor as it expands upwards, Figure 13-3. This process is particularly suited for use with high-resilience systems and can be operated very efficiently when foam system reactivity is precisely adjusted.

Figure 13-3   Schematic drawing of the Maxfoam process

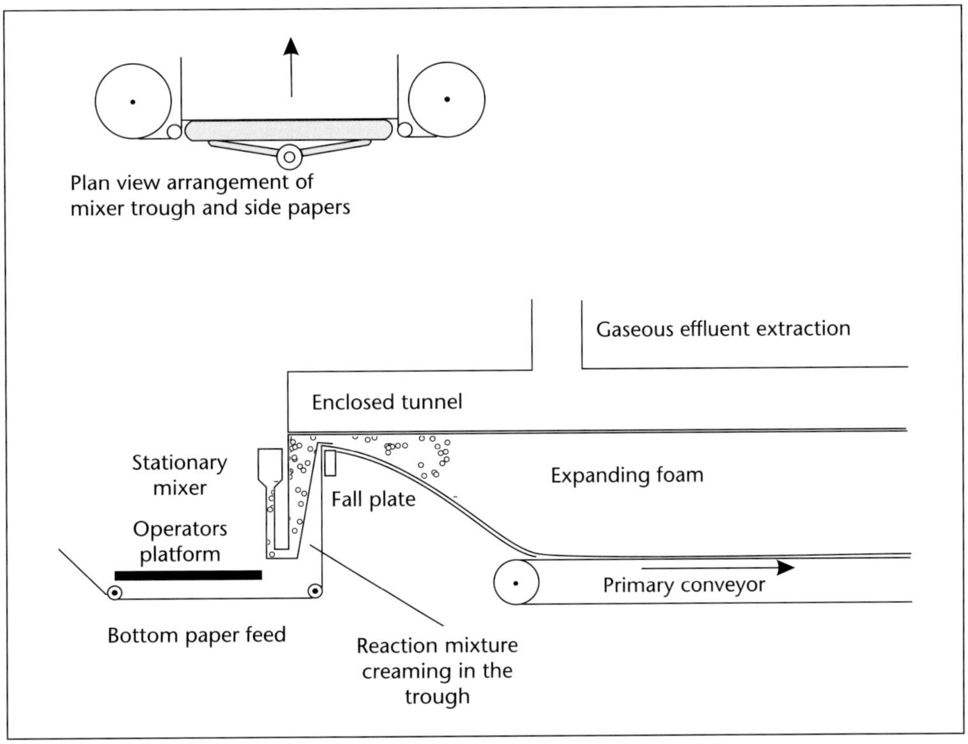

Planibloc

Another method of limiting the amount of curvature is to introduce a fourth release paper, by means of angled platens, that gently compresses the top surface as the foam is rising, Figure 13-4.

Figure 13-4   Schematic drawing of the Planibloc process

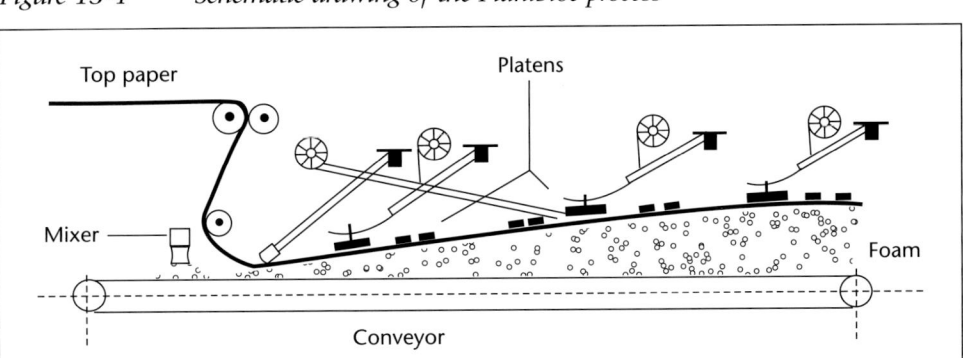

Vertifoam process

A totally different approach introduces the rising foam at the bottom of an enclosed vertical conveyor lined with release paper, which rises at a controlled rate to support the foam. This process is particularly efficient in reducing foam waste when changing grades without stopping production. It is also possible to produce large blocks at relatively low throughputs and with little waste by this process, plus in contrast to the horizontal foaming lines, these machines are very compact and less expensive, Figure 13-5.

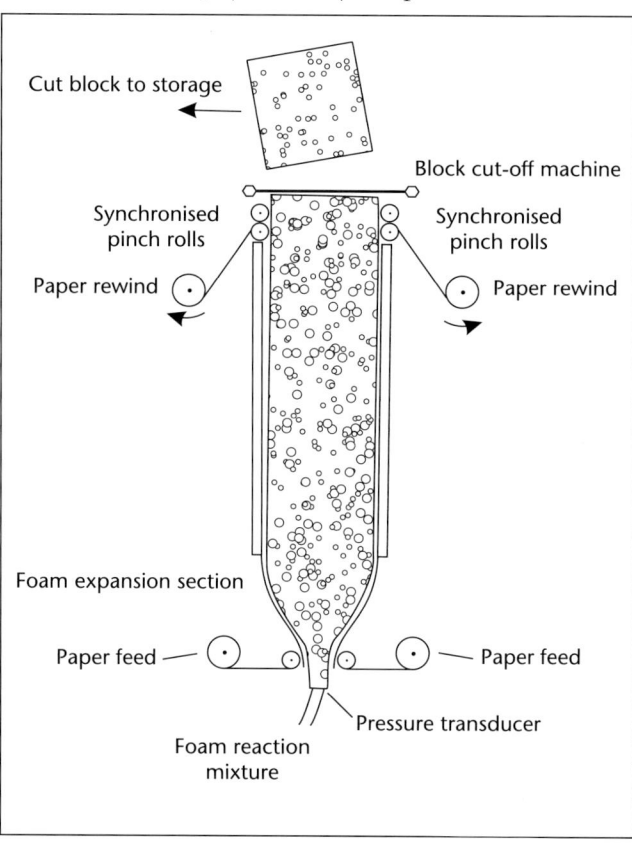

Figure 13-5
Schematic drawing of the Vertifoam process

Foam curing

Foam curing usually lasts about two days for most TDI systems, but can be reduced to a few hours in the case of MDI systems. During that period, it is essential to keep the foam buns in a well-ventilated area to ensure their rapid cooling. This curing station should be equipped with smoke and fire detectors and a water or carbon dioxide sprinkler system in anticipation of accidental scorching.

Carbon dioxide-assisted blowing

The addition of liquid carbon dioxide into one of the main streams, ahead of the mixing head, has become a well-established technology to make low density and softer foams, but both the carbon dioxide-carrying stream and the mixing head need to be kept under sufficient pressure to delay the expansion of the carbon dioxide until foaming. When this pressure is gradually dropped, a liquid froth is obtained, which needs to be laid down using specially designed fan-shaped applicators. The high initial viscosity of the foaming liquid obtained with an MDI system retains the added carbon dioxide very efficiently and provides a favourable combination that produces robust, fast-curing foam buns, down to about 20 kg/m³, with high and even hardness without the need to use polymer polyols. A typical plant for this process is shown in Figure 13-6.

Figure 13-6
Cannon CarDioT carbon dioxide-assisted blowing equipment

Variable pressure foaming

This technology was initially introduced in Europe and has now spread to the Americas and Asia. The key element is that the pressure is controlled in the section of the foaming line between the mixing head and the foam cure point. It is maintained at a reduced pressure, typically 0.7 to 0.8 bar, when the objective is to reduce the density or at a slight positive pressure for the production of special high-density grades. Foam buns are expelled through an airlock, which is the only discontinuity in the process. In the same manner as carbon dioxide-assisted blowing, it broadens the production range of foam grades making accessible difficult-to-mould products and it is particularly suited to the higher viscosity MDI-based systems.

Batch-block

When production volumes are low a discontinuous 'batch-block' operation is preferred. Isocyanate and polyol plus additive blends are metered separately to a cylindrical vessel where they are mixed using a high-shear agitator. On opening the bottom of this vessel, the foaming mix drops into a larger rectangular container equipped with a floating lid. Curing takes place in 10 to 20 minutes, so that fast recipe changes are possible.

Scorch

Scorch refers to the auto-oxidation during cure of freshly produced foams. This phenomenon is especially – but not exclusively – critical with polyether foams. A large amount of heat is generated from the isocyanate-water reaction during foaming, and this needs to be controlled to avoid secondary reactions at high temperature. Further temperature increase can follow from additional reactions during foam cure due to excess isocyanate groups and consequently the core of the foam block may then remain well above 150°C for several hours.

Auto-oxidation of the polyol ether linkages can be initiated at temperatures above 170°C. The reaction proceeds via a free radical mechanism, is highly exothermic and therefore auto-accelerates. Its early signs are a slight foam discoloration in the core. This effect should not be confused with foam yellowing due to some types of antioxidants, or from superficial foam yellowing from exposure to light. It can be followed by char formation in the foam core and even by a full foam ignition, with disastrous consequences.

Factors contributing to a high exotherm in the stored foam bun include both adverse chemical and unfavourable storage conditions:

- High water and isocyanate levels in the formulation that can be due to the intended formulation itself or to faulty metering by the machine.
- Unbalanced reactivity where low reactive systems leave un-reacted isocyanate groups until the foam reaches the storage area, where their late reaction with ambient moisture generates additional heat, which cannot easily be released.
- High bun stacks with little ventilation.
- Fresh foam buns compressed for transport.
- Local densification areas in the end parts of the buns.
- High closed-cell content can contribute to slow heat dissipation.
- Very open cells allows air (oxygen) to flow readily into the hot bun core and promote further oxidation and scorching.

The use of carbon dioxide or other auxiliary blowing agents can prevent foam scorching because their evaporation is endothermic. Other technologies for the rapid cooling of fresh foam buns, using cooled air, have also been introduced, especially in the USA. Other preventive measures include the optimisation of the level and type of antioxidant in the polyol and formulation. Specialist suppliers have developed synergistic blends of substituted phenolic compounds and secondary aromatic amines for this purpose. It is also important to ensure that both foam cure and storage take place in well-ventilated areas. The most modern and effective methods have fireproof storage chambers equipped with temperature sensitive probes and smoke alarms linked to automatic fire fighting systems.

Post-processing of slabstock foam

After curing, the foam buns are trimmed, with skin and edges removed, into regular rectangular blocks that can then be split, by horizontal sawing or splitting machines, into mattress size blocks, or into thinner sheets.

Long rolls, for lamination, can be produced by gluing together the ends of a long bun and peeling the resulting foam loop using a horizontal slitter. The resulting continuous sheets, normally wound into rolls, have thicknesses from 0.5 to 30 millimetres.

To produce articles with more complex shapes, mechanical or electronic templates can be used to guide the cutting blades and endless saw blades can also be replaced by hot wires, lasers, or high-pressure water jets, at least for shorter cutting lengths.

Compression cutting is applied in specific cases to obtain convoluted foam shapes, while profile cutting is used to produce corrugated foam sheets. In both cases, the foam is cut while being compressed between specially designed plates or rollers. On release of the pressure, complex three-dimensional components are obtained, Figure 13-7.

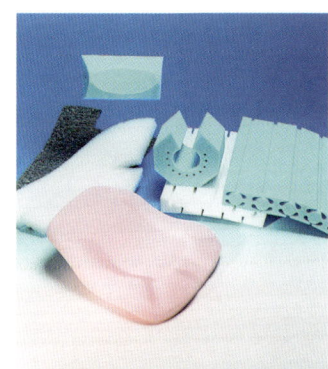

Figure 13-7   Complex shapes cut from slabstock

Computers are used to control the whole cutting operation in order to calculate the best cutting sequence and profiles and to minimise foam waste. The final process for complex shapes can involve buffing off edges or putting in specific grooves and details using abrasive wheels.

Other types of foam post-treatment, such as laminating, reticulating, impregnating, and rebonding, are described in Chapter 14.

14. Technical foams and applications

Alain Parfondry

The area of technical foams and applications covers the range of products that do not neatly fall within the conventional rigid and flexible foam or elastomer categories. This can be because of differences in polymer characteristics (hardness or density) or specific application requirements.

Semi-rigid foams

Applications

Semi-rigid foams have compression hardness values between those of flexible and rigid foams, but are normally closer to flexible foams. Like flexible foams, they have: an open cell structure, hydrophobicity, flexibility at low temperatures and high temperature stability. The main differences lie in their higher compression hardness and slower recovery rates after compression. This makes them particularly suitable for applications where energy absorption is a key requirement such as interior and exterior automotive applications. Typical examples are instrument panels, knee bolsters, side-impact protection pads, headliners and bumper infill.

Instrument panels
Instrument panels were originally designed simply as crashpads or dashboards. However, the introduction of safety belts and airbags has shifted the focus onto aesthetics and comfort. Semi-rigid foams are now used not only to bond and support the soft 'skin' material, usually PVC, PVC/ABS, polyolefin elastomer or thermoplastic polyurethane, but also to give an overall softer touch and ensure good adhesion to the rigid inserts. The design requirements have become very complex, to cope with the introduction of new electronic devices and air conditioning, which has lead to intricate mould geometries and consequently greater demand on foam liquid flow. A typical design is shown in Figure 14-1

Figure 14-1
Schematic cross-section of a dashboard

Bumpers
The formulation of semi-rigid foams can be modified to shift the balance of properties from flexible to rigid and thus modify the energy absorbing characteristics. This ability to tune the semi-rigid foam is used to produce automotive bumpers that are required to pass demanding low energy impact tests. Tough 'skins', made from polypropylene, polycarbonate or RIM polyurethane elastomers, are

filled with semi-rigid foam that not only has specific energy absorption properties, but also bonds to the rigid inserts used for mounting. The whole structure is lighter, simpler to produce and cheaper than metallic bumpers combined with hydraulic shock absorbers.

Manufacture

The external 'skins' can be produced by rotational casting, used for small parts, slush moulding, for larger parts such as instrument panels or bumpers, or by vacuum forming sheet material directly into the mould. The moulds are usually made of epoxy resin, using internal water heating coils, designed to withstand pressures up to about five bar. The actual pressure is dependent on foam density and degree of over-packing.

When the mould is ready, with the skin and inserts such as fastening studs in place, the liquid semi-rigid foam mix is poured into the mould using a two-component, high-pressure machine with the liquid distribution robotically controlled. With a mould temperature of 35 to 50°C, parts can be demoulded after two to five minutes and are normally fully cured within 24 hours. Well-sealed moulds with small, carefully located vents lead to the production of trim-free parts, further improving productivity. Small carousel machines or, less commonly, longer paternoster lines are used for production.

Formulations

The first semi-rigid foams were based on prepolymers made from castor oil and TDI that were cross-linked with glycerol and water. These formulations were difficult to process and gave poor foam quality. Improved processing and properties were then obtained by using prepolymers based on amine initiated polyols and TDI, cross-linking with urea and water. Despite the addition of emulsifiers, flow properties were still poor, because of the prepolymer viscosity and productivity was low due to cure times in excess of 10 minutes.

The latest generation of formulations is based on a one-shot technique using polymeric MDI and reactive polyether polyols with a high level of ethylene oxide tipping. These systems are now used in most car models, where they have demonstrated their versatility, high productivity, and cost effectiveness.

Besides raising productivity, another challenge that formulators faced, and resolved in the 1990s was the interaction between the PU foam and the PVC coverstock in instrument panels.

The exposure of the instrument panel to high temperatures, that can reach 100°C behind the windscreen, lead to discolouration, cracking and embrittlement of the PVC. This was, in part, due to the quality of the PVC, but was accelerated by the migration of residual amine from the polyurethane foam.

The mechanism of PVC degradation is complex and involves the emission of plasticiser, which causes fogging of the windscreen and embrittles the PVC. De-polymerisation, generating hydrogen chloride, causes discolouration and further catalyst breakdown. The degradation and fogging issues have been resolved through the use of non-migrating plasticisers in PVC, or its replacement by another polymer and the use of non-migrating amines, or metal-based catalysts in the polyurethane formulation. The need to further reduce emissions has led to the development of non-volatile raw materials: isocyanate reactive catalysts, polyols and surfactants modified to have lower volatile residues. A typical formulation is shown in Table 14-1.

Table 14-1   Semi-rigid foam formulation and properties

Formulation, wt-%		Processing conditions	
Polyether polyol	40.1	Mould temperature at pour, °C	45 – 60
Reinforced polyether polyol	21.6	Demould time, s	90 – 110
Polyester polyol	4.0		
Glycerol	0.6	**Properties**	
Tertiary amine catalyst	0.4	Foam density, kg/m^3	135
Black additive	0.3	Compression	
Water	1.4	hardness initial (40%), kPa	82
Isocyanate	31.6	Tensile strength (initial), kPa	295
		Elongation (initial), %	38
		Compression set (50%), % dry	14
		heat aged	31
		Compression set autoclave, %	22
		Adhesion (initial), N/2.5cm	3
		VOC, ppm	140
		Fogging, ppm	120

Foam properties

A key benefit of semi-rigid polyurethane foams is their excellent adhesion to most substrates: textiles, fibre mats, polymeric 'skins', metal and polymer inserts. In addition, they can be formulated to match a variety of hardness, density and elasticity requirements plus having the processing tolerance to fill complex moulds.

Their energy-absorbing characteristic, the main functional property, can be derived from the stress strain curve. As with flexible foams (see Chapter 11), a large area under the curve indicates a high degree of mechanical damping. This can be complemented by pneumatic damping, caused by restricted air circulation between cells, especially during very fast compression, as for example under impact conditions.

Viscoelastic foams

Viscoelastic foams have been known for many years, but their adoption by the foam industry is relatively recent. Moulded applications include soft pillows, orthopaedic cushions and even soft toys. Slabstock applications are mainly in the bedding area where new high comfort mattress designs often include a load bearing high resilience or conventional foam core and a viscoelastic upper layer designed to distribute body pressure over a wider area. Viscoelastic foams can also be used for dynamic comfort due to their good vibration absorption.

Viscoelastic foams combine the properties of resilient foams and viscous materials and they are generally characterised by a slow response to static or dynamic stress. In a typical stress relaxation test the stress needed to keep a constant strain decreases as a function of time. Conversely, in a typical creep experiment, a constant stress is applied and strain is shown to increase as a function of time. This type of behaviour can be described by mathematical models, which link stress and strain by equations where both modulus and time are parameters. The key models are: the Maxwell (spring/dashpot in series) or the Voigt (spring/dashpot in parallel), or their combination.

In the cushioning area, they are used to improve static comfort through a more even pressure distribution. Interestingly, the equations for modulus versus time and versus temperature have similar shapes, so that dynamic mechanical thermal analysis can be used to study and predict relaxation phenomena. Relaxation curves obtained at different temperatures may be superimposed on data obtained at a reference temperature by horizontal shifts along the time scale.

From a polymer viewpoint, viscoelastic foams are characterised by a glass transition temperature close to room temperature. This causes the foam to recover only slowly after compression and this relaxation is often used to characterise viscoelasticity.

There are many formulation routes to achieve viscoelastic foams, but most modify foam morphology to reduce the elastic component of the polymer in a specific temperature range around room temperature. Typical variables involve the use of the following either singly or in combination: low molecular weight cross-linkers, blends of high and low molecular weight polyols, mixtures of polyether and polyester polyols, mixtures of polyols with different ethylene oxide content, use of alkoxylated alcohols, variations of the isocyanate index, introduction of plasticisers, building interpenetrating polymer networks (IPNs) and blending polyurethanes with other polymers. A typical formulation is given in Table 14-2.

Table 14-2   Viscoelastic foam formulation and properties

Formulation, wt-%		Processing conditions	
Polyether polyol	49.4	Mould temperature at pour, °C	45 – 60
Reinforced polyether polyol	8.7	Demould time, s	300
Polyester polyol	2.9		
PEG 200	1.7	**Properties**	
Cross-linker package	0.7	Foam density, kg/m^3	72
Tertiary amine catalyst	0.8	Tear strength, N/m	187
Water	1.9	Tensile strength, kPa	138
Black additive	0.1	Elongation, %	107
Isocyanate	33.8	CLD (40%), kPa	6.4
		Hysterisis, %	40
		Compression set (50%) dry, %	28
		Compression set humid, %	20
		VOC, *ppm*	130
		Fogging, *ppm*	400

MDI has become the preferred isocyanate for viscoelastic foams because of its robustness to formulation change, wide isocyanate index tolerance, ability to accept the use of a variety of polyols, lower catalyst levels due to its higher reactivity and decreased surfactant because of its enhanced contribution to foam stability. All these factors provide a much lower risk of closed foams or collapse and overall decreased levels of additives. This is most important when formulating to satisfy new requirements such as low foam odour, pH or volatile content. These formulations can be adapted for both slabstock and moulding applications and, in general, their processing requirements are similar to those of other flexible foams.

Packaging

Polyurethane packaging foams are selected in preference to traditional materials, shredded wood, paper, films, or other foams (polystyrene or other polyolefins), for several reasons. The key factors are that polyurethane foams can be formulated to absorb specific energy, shock and vibration modes; they can be moulded directly in place and produced at low densities. They are definitely preferred for protecting delicate and valuable objects.

Another factor favouring polyurethane foams is the ability to design layers that are thin enough to prevent 'buckling' (uneven compression under load) whilst retaining impact effectiveness. Also, polyurethane foams are relatively unaffected by the effects of temperature and humidity during normal transport conditions.

Foam formulations need to be tuned to the type of product to be protected, its fragility, anticipated drop height and other design constraints. These factors usually lead to a specific foam compression, hardness and density for each key application.

Formulations are now generally based on polymeric MDI, due to the robustness in handling and processing that it provides plus its low viscosity and high reactivity. This is combined with high functionality, low molecular weight polyether polyols. Amine initiated polyols can be used to minimise the level of volatile amine catalysts, which can sometimes interfere with the packaged article and contribute to foam odour.

To optimise energy absorption under impact, high molecular weight polyols can be blended with the low molecular weight polyols and to reach the low foam densities required strong blow catalysts are used to promote the water-isocyanate reaction and generate as much carbon dioxide as possible. Silicone surfactants are also required, for the control of cell size and to ensure the uniformity of the foam structure.

Performance requirements may be specified using standard tests such as ASTM D4169 or an actual drop test.

Packaging foams are usually poured in-situ using a conventional two-component, low-pressure mixing machine fitted with a simple impingement dispense gun. The two-step procedure used is:

- Packaging box is lined with plastic film.
- First foam layer is poured into plastic film and allowed to rise and partially cure.
- The article (protected by a further film as required) is placed onto the uncured foam.
- The remainder of the chemicals are poured onto the article and the foam allowed to rise until the packaging box is filled.
- The packaging box is closed.

Items with a simple, regular profile can be packaged using slabstock foam cut to shape or specifically contoured so that parts can be fitted in slots in the foam and then backed top and bottom with foam sheets.

Polyurethane gels

Polyurethane gels are elastomers with a very low modulus and hard block content (as low as flexible foams) that have been known for a long time, but their commercial use started relatively recently. They are now being used for cushioning applications such as bicycle saddle pads.

Unlike flexible foams, they are not compressible, so spread laterally further increasing the contact area and therefore providing a smoother pressure distribution.

Traditionally, polyurethane gels have been produced using plasticisers or by running systems widely off ratio, but recent developments in MDI technology, combined with high molecular weight polyether polyols, have led to the production of fully reactive and reacted systems where formulations can be altered to provide the broad range of viscoelasticity and resilience effects required for different applications.

Hydrophilic foams

The major characteristic of hydrophilic foams is their ability to absorb and retain large volumes of water. They are typically made by reacting prepolymers, based on MDI and hydrophilic polyols, with a large excess of water. The major polyols used are high ethylene oxide polyether polyols, but polyester polyols or ionomers are also used. The use of an excess of water leads to foams with limited window drainage and therefore small openings between cells. In addition, microscopic blisters form on the cell walls. This combination leads to a remarkably enhanced ability to soak up and retain water by capillary action.

The prepolymer route ensures both robust and consistent processing whilst the water stream can be used as a carrier for all types of additives, such as:

- Mineral fillers to increase foam hardness.
- Super absorbent additives such as polyacrylate salts to improve water retention after absorption.
- Beads of crystalline high molecular weight polyethylene glycol (PEG) that melt and then re-crystallise to improve comfort in protective clothing for use under fast changing temperature conditions.
- Compost and soil additives for plant growth media.

A typical formulation, along with properties, is given in Table 14-3.

Table 14-3   Hydrophilic foam formulation and properties

Formulation, wt-%		Properties		
PEG	1.8	Foam density, kg/m^3		80
Water	32.7	Compression		
Filler	29.0	hardness (40%), kPa		3
Surfactant	0.2	Tensile strength, kPa	dry	55
Isocyanate	36.3		wet	40
		Elongation, %	dry	200
Processing conditions			wet	100
Prepolymer temperature, °C	25	Tear strength, N/m	dry	250
Prepolymer viscosity, mPa.s	6,000		wet	70
Water temperature, °C	15 – 45	Moisture at equilibrium, %		3
Curing time, s	120	Resilience, %		30

Hydrophilic foams have found many applications, including cosmetics (foam pads for make-up removal), paramedical or medical (wound dressings), industrial cleaning (sponges) or horticultural (plant growth media) markets. Some of these applications are required to meet proscribed standards or regulations. New applications in electrolytic water treatment and pollution industries are in development.

Textile laminated polyurethane foams

The comfort perception and surface feel of textiles can be greatly improved by laminating thin layers of flexible foam onto the back of the textile. The major applications are for automotive seating and a wide range of clothing. There are several processes for bonding the flexible foam to the textile, such as:

- Flame lamination.
- Adhesives.
- High frequency welding.
- Application of dielectric heat.

Flame lamination is the commonest technique and involves melting the surface of the foam, normally using a gas flame, before bonding the foam onto the textile. The concept of the process is shown in Figure 14-2.

Large continuous laminating machines that can run at 40 metres/minute are used in production and are set so that the laminates preserve the breathability and resilience of the original foam whilst forming a strong bond.

Polyester polyol flexible foams are normally used, but conventional polyether foams are sometimes preferred for their improved hydrolysis resistance. The

Figure 14-2   Schematic principle of flame lamination process

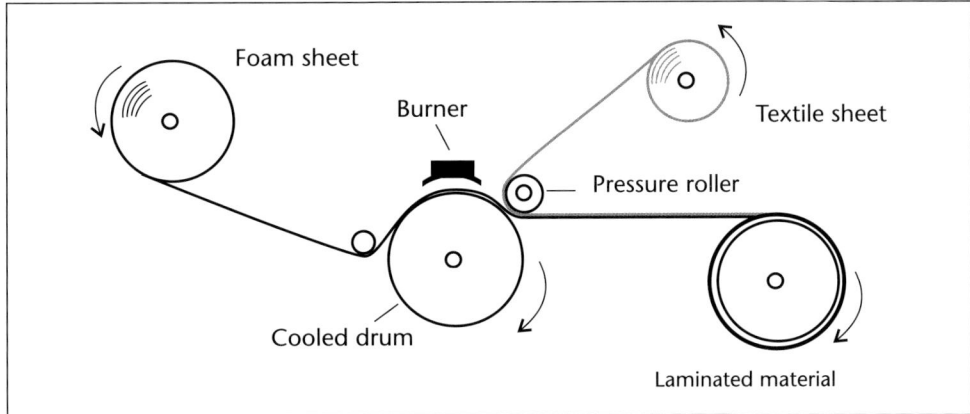

polyester foams can be laminated without the use of additives, but the raw materials require special vacuum treatment to reduce the generation of foam volatiles and flame retardant additives are required for certain applications. Conventional polyether foams always require the addition of flame-retardants and other additives, such as glycols, to improve laminability.

MDI-HR slabstock foams are flame laminable without modification and at a lower temperature, which results in less off gases during production and less volatile residues in the composite, a useful feature for automotive textile laminates.

Flame lamination is most common for automotive textiles such as seat covers based on polyester, acrylics, or even polyolefin polymers, while high frequency welding may also be used for non woven covers, including thermoplastic skins, which may then be used in a wide range of speciality applications.

Post-treated polyurethane foams

In addition to lamination, other post-treatments are used to achieve specific properties.

Reticulated foams

Reticulated foams, having no cell membranes, can be obtained by subjecting foam blocks to an explosive gas mixture in a pressurised container. The high heat generated for a short period melts the cell windows and only struts are retained. The resulting foams have the very high and consistent porosity required in specific application areas. Figure 14-3 shows a typical structure.

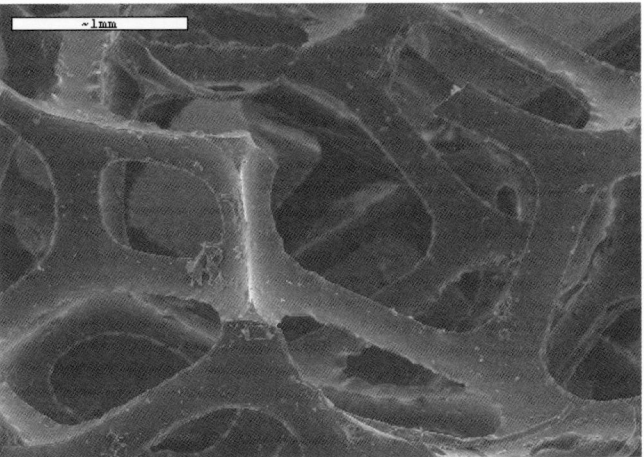

Figure 14-3
SEM micrograph of reticulated foam

The original foams must have very carefully controlled cell morphology, narrow size distribution, in order to yield reticulated foams with useful properties. This is achieved by carefully controlling the level of gas nucleation in the starting raw materials and in the mixing head to avoid pinhole formation.

Reticulated foams are mainly used for fluid management: water or air filtration and acoustic foams (used in loud-speakers). This is a fast expanding area of development for polyurethane foams.

Impregnated foams

Flexible foams – mainly slabstock foams, after cutting to shape, may be impregnated with various types of reactive slurries carrying additives designed to improve properties, such as fire resistance. A typical application is for aircraft seating. In this case, the foam part is dipped in a latex solution containing fire retardants such as melamine or alumina hydrate and the impregnated foam is then cured in an oven. Fire performance is greatly improved because the fire retardants are concentrated at the surface of the polymer, so that a char is more readily formed on exposure to fire.

Rebonded foams

Foam scrap from industrial foam production and, more recently, from selected post-consumer sources, can be shredded and rebonded into blocks or shaped articles. These foams are used as hard inserts in moulded seats, for packaging, as floor mats in sports facilities, for sound insulation pads in the automotive industry or for panels in the building industry. The stages involved in the process are shown in Figure 14-4.

Foam waste is first shredded in a granulator to yield particles ranging in size from 5 to 20 millimetres, any larger crumbs are eliminated because they would cause inhomogeneity in the finished product. The foam crumbs are then tumbled in a mixer and coated by spraying a prepolymer based on TDI or MDI and a flexible polyether polyol. To improve wetting-out, the prepolymer is sometimes diluted with a high boiling aliphatic or aromatic oil.

Water may be added, by moistening the crumbs with a water spray, to promote the moisture cure of the prepolymer and to generate a small amount of foaming. A catalyst, such as a low odour tertiary amine or an organo tin salt, may be added to speed up the cure. The resulting reactive blend is then pressed into large moulds and cured, using steam, before being cut to shape.

The hardness properties of the rebonded foam depend on those of the original foam and the degree of compression during curing. Foam density is between 48 and 130 kg/m³ and usually 100 to 300 per cent higher than for the original foam, and accordingly hardness is also higher.

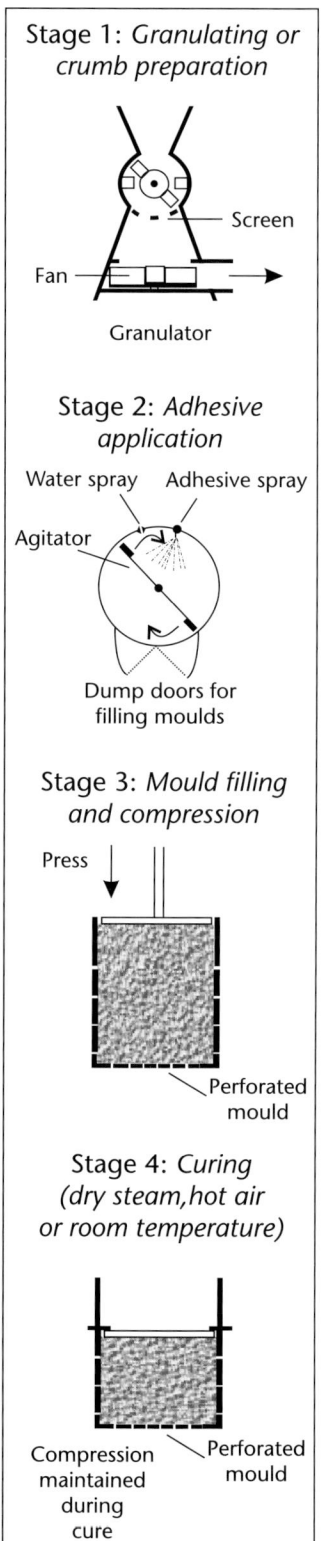

Figure 14-4
Stages in the manufacture of rebonded flexible foam

Sheets of rebonded foam, produced by a similar, but continuous process, are especially popular in the USA as indoor carpet underlay for residential use.

Carpet backing foams

Reacting polyurethane foam can be applied to the back of tufted carpets by spraying, direct coating or reverse coating. The process can be run on production lines similar to those designed for styrene butadiene rubber (SBR) latex. The direct coating process is the most widely practised, where the polyurethane foam is applied as froth onto the back of a carpet with the thickness controlled by the use of a coating knife, Figure 14-5.

Figure 14-5   Schematic of direct coating process for polyurethane carpet backing

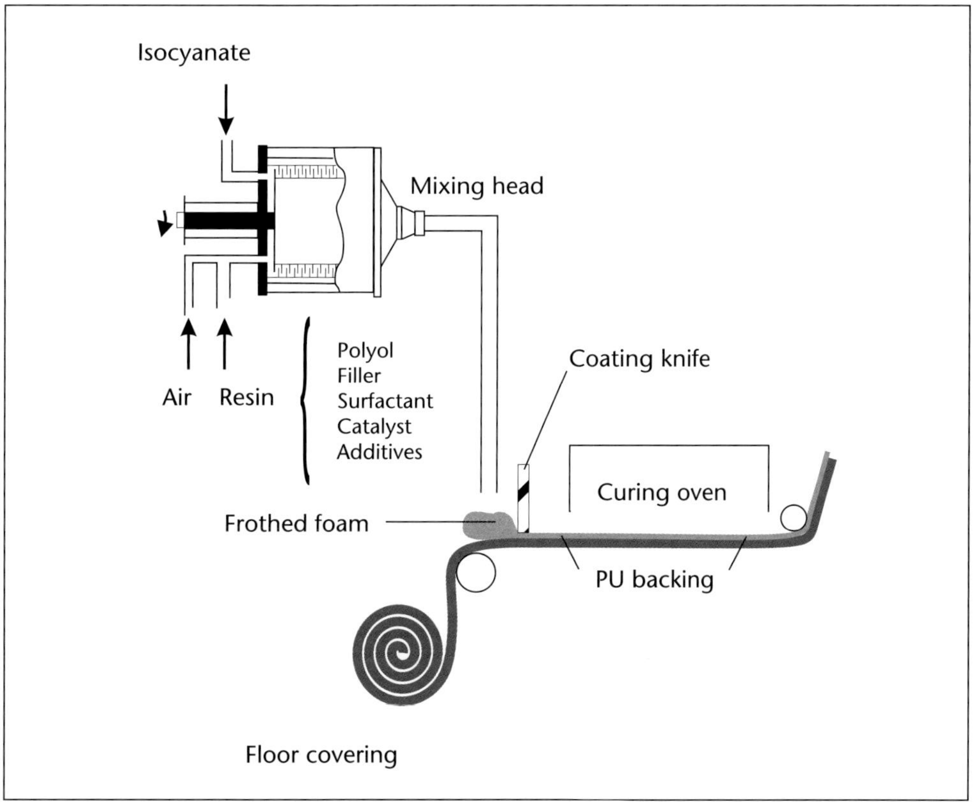

The formulations are based on elastomer technology involving a medium functionality MDI isocyanate, modified with a small amount of polyether polyol to improve miscibility and processing, in combination with a polyol blend consisting of a blend of a high molecular weight, high-reactivity polyether polyol and a cross-linking agent. The other additives used are: a heat activated catalyst, typically a metal diketonate salt, that operates during the oven curing stage, a strong emulsifier such as a high molecular weight polydimethylsiloxane to ensure retention of air and fillers such as calcium carbonate or alumina hydrate. A typical formulation is given in Table 14-4.

Table 14-4   Carpet backing formulation and properties

Formulation, wt-%		Processing conditions	
Glycerol-based polyether polyol	26.6	Density control	Froth
SAN polymer polyol	13.3	Oven temperature, °C	120
Diethylene glycol	1.3	Curing time, s	480
Catalyst	0.1		
Surfactant	0.3	**Properties**	
Filler	39.9	Foam density, kg/m^3	290
Isocyanate	18.4	Thickness, mm	7
		Tuftlock, N/lock	90
		Delamination (Tear strength), N/m	700
		Durability (Castor wheel test), cycles	25,000

During the manufacturing process large volumes of air are injected into the reacting polyol and isocyanate mix using a specially designed frothing mixing head, which incorporates high shear rotors with longitudinal teeth. The foam, strictly speaking frothed elastomer, is applied to the carpet back at the final density and is oven cured at about 120°C. The process results in foams with very fine cells and consistent thickness and quality.

15. Introduction to rigid foams

Kristof Dedecker

Of the 2.7 million tonnes of MDI produced globally in 2000 rigid polyurethane foams accounted for 45 per cent, mainly for thermal insulation, and with energy-saving regulations tightening the volume is expected to increase. The major use of rigid polyurethane foam is in the construction industry, where its superior long-term thermal insulation properties combined with good dimensional stability offer many benefits. The size of the global markets of the main applications are: construction, with 600,000 tonnes for boardstock and 450,000 tonnes for sandwich panels; refrigeration appliances, 700,000 tonnes; and technical insulation, 350,000 tonnes.

The demand for rigid polyurethane foam, whilst being mainly dependent on its low thermal conductivity, is also due to other factors such as good adhesion to facing materials and excellent mechanical strength at low density. Rigid foams can be made at densities from 10 to 1,100 kg/m^3, but most are used in the range 28 to 50 kg/m^3. Key developments in the past decade have focussed on the introduction of environmentally acceptable blowing agents and improvements in fire performance.

Basic science

Rigid polyurethane foams are prepared by mixing, under controlled conditions, MDI, polyols, blowing agents and a variety of additives such as catalysts, surfactants, water and, optionally, fire retardants. A wide range of polyols is used, normally in combination with polymeric MDI, and the additives are typically pre-blended into the polyol. The formation of a highly cross-linked homogeneous glassy network structure is crucial for the final properties of rigid polyurethane foams, as it leads to good heat stability, high compression strength at low density and good barrier properties.

Obtaining optimum processing and end properties at the same time cannot usually be achieved using a single polyol and the same holds for catalysts and blowing agents. Therefore, a large number of formulations are required and these are discussed in more detail in the three application Chapters, 16, 17 and 18.

One of the key properties of rigid polyurethane foam is low thermal conductivity, which is achieved by producing fine, closed-cell foam of the required density using water and a physical co-blowing agent. The physical blowing agent needs to have a low thermal conductivity as it is retained within the cells and contributes to the level and stability of thermal conductivity.

Chemical processes during foam formation

Full details of the reactions involved are provided in Chapter 7, but for rigid polyurethane foam the initial exothermic reaction is normally between isocyanate and water, leading to the formation of an amine and carbon dioxide; the amine then reacts with more isocyanate to form a polyurea. However, if an amine polyol is included in the formulation it will react first. The other key exothermic reactions are between isocyanate and polyol, producing polyurethanes, and isocyanate trimerisation. Appropriate catalysts are selected for specific functions, such as blowing, gelling or trimerisation, to control the overall rates and balance of these reactions.

Physical processes during foam formation

Once the isocyanate and polyol blend are mixed there is typically a 30-fold increase in volume upon reaction and the formation of individual cells is related to the nuclei present in the mixture. Cell formation and stabilisation is also related to choosing the correct surfactant, and an inappropriate choice can lead to a high degree of open cells. Further details on surfactants are provided in Chapter 10. The rigid foam polymer structure becomes self-supporting, once enough network formation has taken place, and the further expansion of the foam is then driven by the difference between internal cell gas pressure and external atmospheric pressure. The expansion stops when the foam has built sufficient strength to withstand the pressure difference or, in moulded foams, once the mould is filled.

The centre of the foam can reach temperatures as high as 190°C, due to the exothermic reactions, but the reactions are not complete at the end of foam rise and can go on for several hours. Similarly, it can take hours or even days for the centre of the foam to completely cool down to ambient temperature dependent on the dimensions and properties of the foam. Scorching at the centre of the foam can occur if a high temperature is maintained for a long time.

The stages in the foaming process are described by a number of characteristic times, with the start of mixing set at zero, referred to as cream, string or gel, tack-free and end-of-rise. The cream time is when the mix starts to foam, whilst the string time is identified as the point when strings of polymer can be withdrawn by dipping a pointer into the foam mix. The tack-free time is when the surface of the foam stops being tacky to the touch and the end-of-rise is the point of maximum foam height. The evolution of these four characteristics is shown graphically in Figure 15-1 and the physical changes in Figure 15-2.

Figure 15-1   Evolution of temperature and foam height versus time

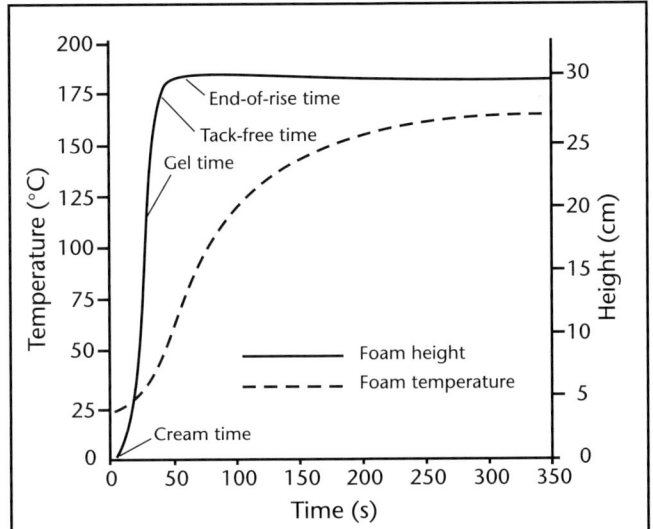

Figure 15-2   Laboratory preparation of rigid foam

Formulation design for optimum processing and end properties

The central issue in rigid foam development is balancing the relationship between formulation, processing characteristics and end properties. The key elements in processing are the compatibility between the different components, flow and reactivity whilst for final properties the main features are thermal conductivity, compression strength, density, dimensional stability, fire properties and adhesion. Two specific points need consideration:

- Choice of polymeric MDI and polyol.
- Isocyanate index.

Standard polymeric MDI has the right balance of viscosity, reactivity and physical properties for most applications, with high functionality polymeric MDI (HF) being used where improved physical properties are required. Low functionality MDI provides better flow during foam formation and results in improved impact resistance in structural foam.

The choice of polyols is much broader than polymeric MDIs with the key products being polyethers based on sorbitol and sucrose as they have a high functionality, four to seven, and are cost effective. Polyethers based on glycerol are added to modify the reactivity and to improve the processing whilst amine-based polyols are used for specific applications such as spray foam, where a high reactivity is required. Aromatic polyester polyols are widely used as they produce foam with better fire properties than polyethers, but their high viscosity limits their use to applications where flow is not critical. The choice of polyol also depends on the compatibility of the blowing agent.

The isocyanate index has a big impact on the processing and end properties and whilst polyurethane (PUR) foams are normally moulded over an index range of 90 to 130, with final properties relatively constant across the range, polyisocyanurate (PIR) foams are made with indexes in the range 200 to 350. Due to the large isocyanate excess isocyanurate ring formation takes place, but this will only happen at a late stage of the reaction when the temperature is high. This ring formation itself is exothermic causing the foam to rise again, a phenomenon known as 'second rise', shown in Figure 15-3, which can lead to processing difficulties. However, these can be limited by appropriate catalyst selection. PIR foams have an improved fire performance and reduced smoke generation compared to PUR foams, but at indexes above 350 they tend to be friable and have poor adhesion.

Figure 15-3   Second rise in PIR foams

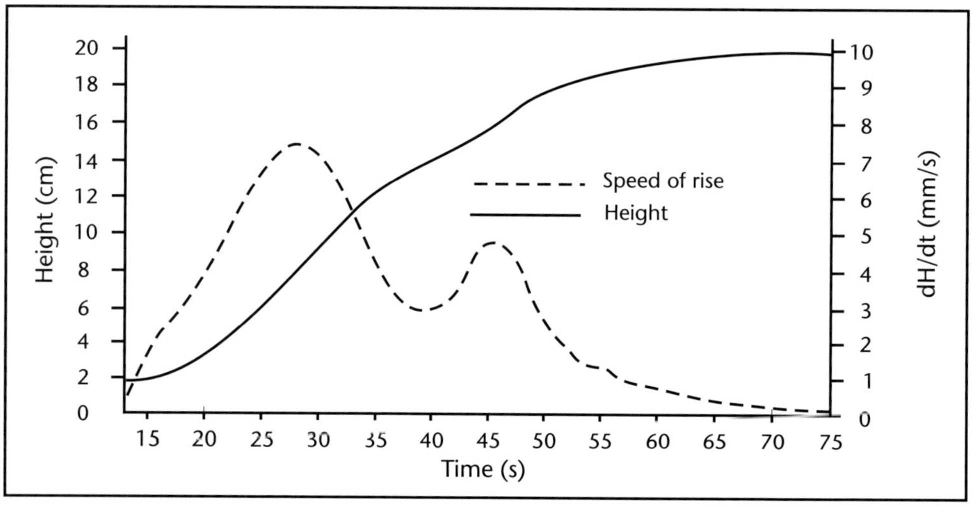

Blowing agents

The change in blowing agents has been one of the central themes within rigid foam development work for the past decade and full details of the changes are provided in Chapter 8.

Key criteria for blowing agent choice

Gas thermal conductivity
The gas thermal conductivity is an important criterion in the selection of a blowing agent and this is especially critical when the thickness of the insulation material is limited, as with refrigerators, for example, or for building insulation in markets with stringent energy-saving regulations.

Ease of handling
The specific properties of blowing agents lead to different machine requirements. Flammable blowing agents, such as pentanes, require suitable explosion-proof equipment, which has a higher cost than conventional equipment, together with an appropriate extraction system. Blowing agents with boiling points lower than room temperature require special cooling or pressure features. When a blowing agent is pre-mixed into the polyol blend care should be taken that phase separation does not occur during storage.

Behaviour within foam under service conditions
The condensation of the blowing agent in the final foam is undesirable since it can lead to a less favourable gas composition in the cells and potential shrinkage problems. The actual condensation temperature can be calculated from the boiling point, enthalpy of vaporisation, service temperature and amount of blowing agent present. For mixtures of blowing agents, the situation becomes more complex. The solubility of blowing agents in the foam matrix, which varies considerably, should be as low as possible since the combined effect of matrix plasticisation and lowered cell gas pressure can also cause dimensional stability problems. The diffusion rate out of the foam is low, but differs slightly for the different blowing agents.

Cost effectiveness
The potential use of a blowing agent is also dependent on its cost with one of the indirect influences being molecular weight, since the higher the molecular weight the more material is required to achieve the same gas volume.

Heat transfer

The low thermal conductivity of rigid polyurethane foam is one of its most important properties. It is determined from the rate of heat transfer through a material, which is proportional to the temperature difference divided by the thickness. The constant of proportionality, obtained from the equation shown in Figure 15-4, is called the thermal conductivity, also known as k-value or λ-value.

The thermal conductivity is measured using a constant temperature gradient and, being dependent on temperature, is usually expressed at the temperature relevant for the application. The thermal conductivity is normally expressed as W/(m.K) or mW/(m.K) although Btu.in/(ft².h.°F) is also used in the USA. A conversion table can be found in Appendix 2. Besides thermal conductivity, thermal resistivity or R-value is also used, especially in the USA. It is defined as the inverse of the thermal conductivity.

It is important to note that the thermal conductivity is related to the rate of heat transfer under steady-state conditions. At non-steady-state conditions, the density and heat capacity of a material will also influence its thermal behaviour. A low-density foam will equilibrate faster to the temperature of the environment than a high-density non cellular material despite the much lower thermal conductivity of the low-density material.

Figure 15-4   Equation for heat flow

$$Q/A = -\lambda \cdot dT/dx$$

Q (Watts) = Rate of heat transfer through a plane material
A (m²) = Surface area
T (K) = Temperature difference
x (m) = Thickness
λ = Thermal conductivity, known as the constant of proportionality and also called the k-value or λ-value

Fundamental aspects of thermal conductivity

The total thermal conductivity of a foam is the sum of four parts – radiative, solid, gas and convection. Due to the small cell diameter of rigid polyurethane foams the contribution due to convection can be ignored. This means that a cell can be considered to have a uniform temperature resulting in a still, non-convective gas. The split of the total thermal conductivity into the other three parts will be different from foam to foam and a few possible cases are given in Figure 15-5.

Figure 15-5   Split of thermal conductivity into three parts for freshly made fine celled foams

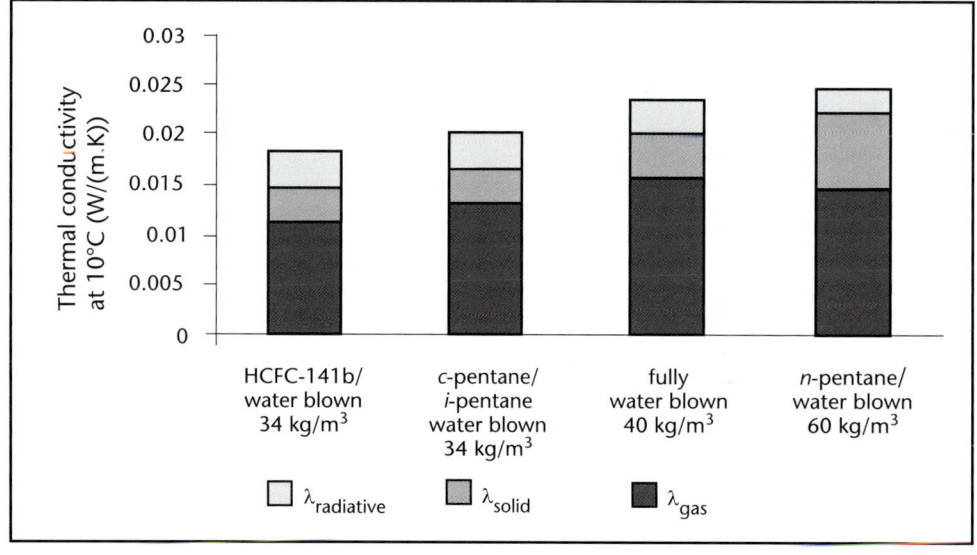

Radiative transfer ($\lambda_{radiative}$)

For a given temperature gradient between two walls there will be a significant transfer of radiation from the hot towards the cold wall. The presence of an insulating material will reduce the major part of this with the remaining radiation, which will still penetrate through the foam, defined as $\lambda_{radiative}$. It will decrease with increasing foam density, as more material is present to reduce the radiation. At a fixed density, $\lambda_{radiative}$ will decrease with decreasing cell diameter, Figure 15-6. In view of the large cell orientation often present, $\lambda_{radiative}$ can vary significantly depending on the direction in which the temperature gradient is applied and $\lambda_{radiative}$ is also dependent on the third power of temperature and increases significantly with increasing temperature.

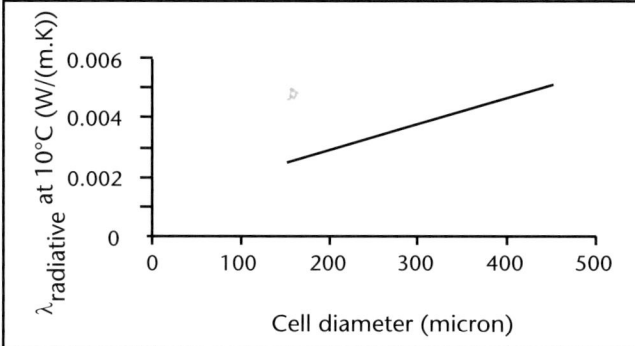

Figure 15-6   Relation between $\lambda_{radiative}$ and cell diameter for low density foams

Solid conduction (λ_{solid})

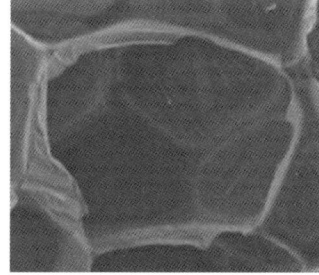

Figure 15-7   SEM photograph of rigid foam

A solid non cellular polyurethane has a λ_{solid} of about 0.22 W/(m.K). However, the solid polyurethane phase typically only accounts for about three to four per cent of the total volume of a low density rigid polyurethane foam and is distributed between struts and windows, both of which are only partially oriented in the direction of heat transfer. The resulting λ_{solid} will typically be around 0.003 to 0.004 W/(m.K) for a 35 kg/m³ foam and it will increase at higher density. A typical cell structure of rigid polyurethane foam is shown in Figure 15-7.

Gaseous conduction (λ_{gas})

The conductive part of the heat transfer through the gas phase is represented by λ_{gas}. All physical blowing agents as well as carbon dioxide have a much lower λ_{gas} than air. In general λ_{gas} decreases with increasing molecular weight of the gas, shown in Table 15-1 and Figure 15-8, but increases with increasing temperature.

In most cases the gas phase within foam cells is made up of several different gases and the λ_{gas} of the mixture will depend on the ratio of the mix, but cannot be linearly extrapolated from the composition and λ_{gas} of the individual gases. In general, a higher molecular weight gas will have a bigger impact on the λ_{gas} than a lower molecular weight gas. Predictive equations are available that can calculate the λ_{gas} of a gas mixture with reasonable accuracy. The total gas pressure has no significant impact on λ_{gas} for the practical pressure range within rigid foams.

Table 15-1   Relation between molecular weight and λ_{gas}

	MW (g/mol)	λ_{gas} at 10°C (W/(m.K))
CFC-11	137	0.0074
HCFC-22	86	0.0102
HCFC-141b	117	0.0091
HFC-134a	102	0.0123
HFC-245fa	134	0.0113
HFC-365mfc	148	0.0108
c-pentane	70	0.0115
i-pentane	72	0.0128
n-pentane	72	0.0137
i-butane	58	0.0148
CO_2	44	0.0153
N_2	28	0.0246
O_2	32	0.0249

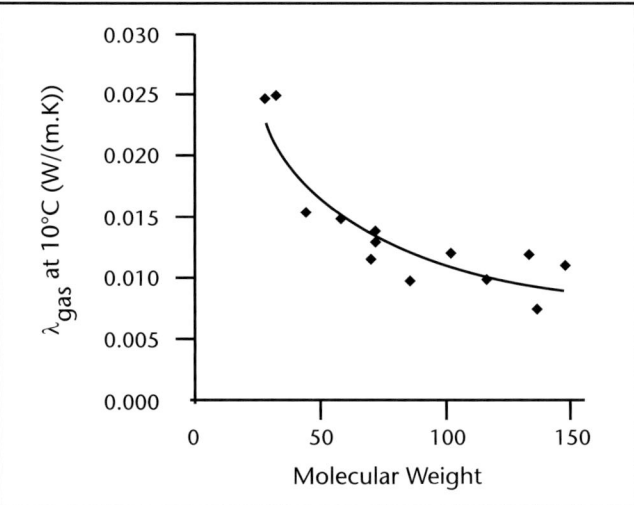

Figure 15-8   Relation between molecular weight and λ_{gas}

Several of the physical blowing agents have a boiling point well above the temperature at which the foam is used under service conditions therefore this part of the mixture would condense. This will affect the gas composition and hence also its λ_{gas}. Since the other gases present, carbon dioxide and air, have a higher λ_{gas} condensation than the physical blowing agent it will increase the λ_{gas} contribution to the total thermal conductivity.

Thermal conductivity ageing

All insulating foam materials that are blown with an insulating gas, such as polyurethane, polyisocyanurate, extruded polystyrene and phenolic foam, undergo changes in cell gas composition over time, accompanied by changes of the thermal conductivity.

Basic theory

For freshly made foam, the cell gas composition will consist of physical blowing agent and/or carbon dioxide with the amount of air in the cells limited to the small traces, which were dissolved in the chemicals. As the foam ages air will gradually enter the foam driven by the difference in partial air pressure in the foam cells and atmospheric pressure. Besides air ingress, the physical blowing agent and carbon dioxide will diffuse out of the foam, again driven by differences in partial pressure. The rate of diffusion will depend on many parameters, but as a general trend, the carbon dioxide loss will be a lot faster than the air ingress, which, in turn, is much faster than physical blowing agent loss.

An example of the typical evolution of the cell gas pressure is shown in Figure 15-9, with the thermal conductivity increase shown in Figure 15-10.

Figure 15-9   Possible evolution of cell gas partial pressures versus ageing time

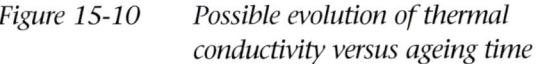

Figure 15-10   Possible evolution of thermal conductivity versus ageing time

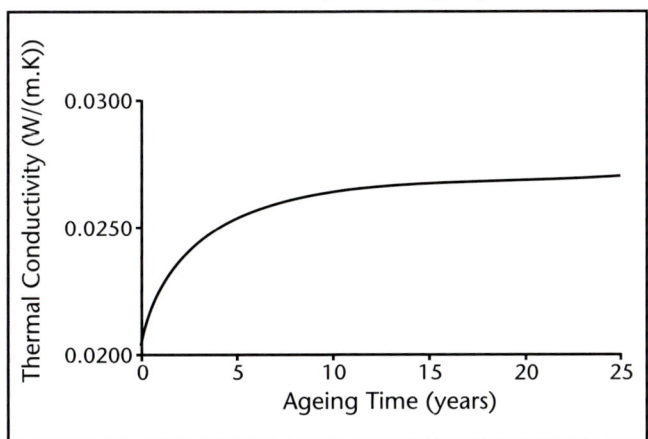

The changes in gas composition result in changes of thermal conductivity and for foams that are blown with a physical blowing agent and water the following stages can be identified.

- The outgassing of carbon dioxide should in theory lead to a slight decrease in the thermal conductivity since it has a higher λ_{gas} than all current physical blowing agents, but it is usually quite difficult to observe this. The difference in λ_{gas} between carbon dioxide and physical blowing agents is not large enough to significantly affect the overall thermal conductivity. Moreover, some minor ingress of air will already have taken place during this carbon dioxide loss, creating the opposite effect.
- The ingress of air is accompanied by a gradual increase in thermal conductivity since air has a much higher λ_{gas} value than the physical blowing agents. The air permeation will stop when the internal air partial pressure is equal to the external atmospheric pressure. Simultaneously, the thermal conductivity will also reach a plateau level. The difference between the initial thermal conductivity and the plateau level will typically be between 0.004 and 0.007 W/(m.K), depending on the physical blowing agents used. Oxygen permeates faster than nitrogen and will reach its saturation pressure earlier in the ageing process.
- The outgassing of physical blowing agents is usually a slow process, much slower than the air saturation. Due to this the ratio of air to physical blowing agent in the gas phase will increase and the thermal conductivity will also slightly increase. However, these thermal conductivity increases will be small compared to the thermal conductivity increase caused by the air increase. For many polyurethane products, the loss of physical blowing agent is a negligible process during the economic lifetime of the product.

Besides the outgassing of the physical blowing agent, it can also partially absorb into the polyurethane foam matrix. Depending on the operating temperature, a part of the physical blowing agent will be condensed, which will again affect the gas composition. This can have an important effect on the thermal conductivity, especially at low temperatures.

Thermal conductivity ageing under service conditions

The extent to which actual foam will age during its economic lifetime depends on many factors and the rate of the ageing process will depend on the foam dimensions, quality and the presence of any facings and the temperature.

The diffusion rate in a certain direction is inversely proportional to the square of the foam's thickness. This means that it takes four times as long to reach a certain ageing effect, such as air saturation, for an eight centimetre thick panel as it does for one of four centimetres.

The foam quality also affects the ageing rate, as any foam parameter that decreases the effective polymer path length through which gases have to diffuse will increase the ageing rate. Such parameters include cell size and orientation, open cell content, presence of blowholes or other defects, window thickness and density. Besides these foam morphology parameters, the chemical composition of the foam will also influence the ageing rate.

Facings can retard ageing significantly and even stop it completely. Semi-permeable facings such as paper or bitumen will slow down the ageing while impermeable facings such as steel, aluminium and several multilayer systems can stop it completely provided that the whole foam is covered and they are well bonded. Even the inherently fast diffusing carbon dioxide will not be able to completely diffuse out of such faced panels during their economic lifetime. Since facings are normally present, ageing of non-faced polyurethane foam in a laboratory can be seen as the worst-case scenario.

The ageing temperature can also play a significant role as the diffusion rate increases exponentially with temperature. The ingress of air has been determined to be 6 to 12 times faster at 70°C compared to room temperature. Similarly at low temperatures the diffusion processes will be slower than at room temperature.

Industry standards for thermal performance

For some applications, such as appliances, the thermal conductivity will not be specified directly on the end product, but will contribute indirectly to parameters such as energy consumption. For other applications, like boardstock or sandwich panels, the thermal conductivity of the foam will be stated and the declared lambda value will contribute significantly to the commercial attractiveness of an insulation product.

In general, the aged thermal conductivity value will be declared and not the initial one. Methods to obtain aged thermal conductivity values vary from country to country. The most frequently used methods include experimental ageing at elevated temperatures, experimental ageing of thin slices and addition of fixed increments to the initial thermal conductivity. A European norm has been developed, which will replace the national norms. In the US, a new norm based on thin slicing is being developed. The other key factor is that the mean temperature of measurement also varies, with Europe using 10°C and the USA 23.9°C (75°F).

Open-cell foam

For conventional rigid foam applications the level of closed cells should be as high as possible since open cells will contain air, which will increase the λ_{gas} and the overall thermal conductivity. Therefore, commercial rigid foam formulations are designed to give high closed cell content and are usually robust enough to maintain this property under slightly varying processing conditions.

However, when an appropriate surfactant is selected, or when extreme internal pressures are produced, then cells can open during foaming, which can lead to foams with a significant open-cell level or even fully open-cell foam. Although not desired for conventional rigid foam applications, there are some specific applications in which the properties of open-cell foam can be beneficial.

The difference between the internal cell gas pressure within the foam and the external atmospheric pressure can lead to dimensional stability problems. However, this pressure difference will be zero for open-cell foam and so dimensional stability problems will be avoided. Thus open-cell foams can be prepared at a very low density without shrinkage problems, which makes them suitable for applications such as packaging.

Another use for open-cell rigid foam is as the support medium in vacuum panels. This is a novel development in which open-cell rigid foam is encapsulated in a non-permeable membrane which is evacuated to a low pressure to produce a very efficient insulant.

As discussed earlier, the total thermal conductivity is the sum of $\lambda_{radiative}$, λ_{solid} and λ_{gas}. For conventional closed-cell foam, λ_{gas} is only marginally dependent on the total gas pressure for the typical pressure range present within the foam cells, 10 to 150 kPa. However, open-cell foam can be evacuated to much lower gas pressures, and λ_{gas} will gradually decrease and even become zero at pressures below 0.01 kPa and the overall thermal conductivity can then reach values as low as 0.007 W/(m.K). The thermal conductivity of open-cell foams, as shown in Figure 15-11, is dependent on achieving a low enough pressure, in this example 0.01 kPa, and this level is shifted to higher pressures for smaller cell sizes.

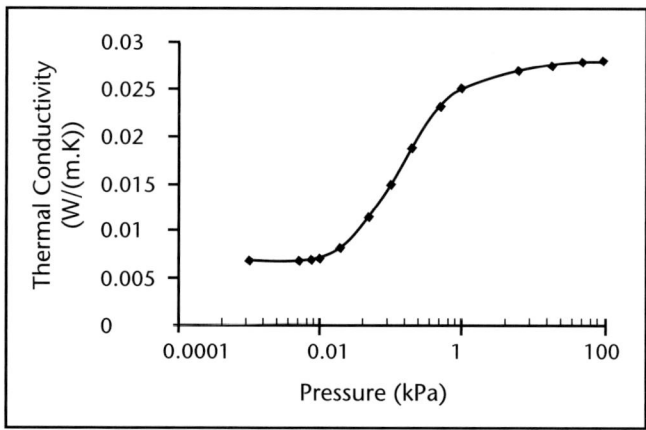

Figure 15-11
Thermal conductivity versus pressure for open-cell foam

To keep the thermal conductivity low the vacuum pressure within open-cell foams must be maintained and there are several processes for this. First, the lining material should be completely impermeable and aluminium foil is normally used. Secondly, closed cells will not be pulled to a vacuum within the evacuation period and diffusion of gases will take place from any closed cells present and over long periods of time, closed and open cells will equalise their pressure, affecting the vacuum. Therefore, the closed-cell content should be as low as possible. Foams with thick skins, which typically contain a higher number of closed cells, should not be used for rigid foam vacuum applications. Finally, the polyurethane foam matrix itself can release certain low molecular substances that affect the vacuum level and the addition of appropriate absorbers, known as getters, is required to eliminate this problem.

Key testing methods

The properties of rigid foam can vary significantly depending on changes in the formulation as well as on the processing conditions. These processing conditions include the type of mixing equipment, temperature of the chemicals and the flow pattern of the foam. For a fixed formulation and fixed processing conditions, changes in the ambient conditions, temperature, atmospheric pressure and relative humidity can also result in different physical properties.

Standardised test methods are available for all important properties of rigid foam and these are used to check the reproducibility of the foam in production or used in foam development work to evaluate the suitability for a certain application. The standard testing methods for rigid foam are given in Table 15-2.

Table 15-2   Testing methods for rigid polyurethane foams

Standard test method for physical properties	International ISO No.	European EN No.	USA ASTM No.
Determination of linear dimensions of test specimens	ISO 1923 (1981)	EN 822 (1994) EN 823 (1994) EN 12085 (1997)	D1622-98
Determination of the apparent density of cellular materials	ISO 845 (1988)	EN 1602 (1996)	D1622-98
Compression test of rigid cellular materials	ISO 844 (2001)	EN 826 (1996)	D1621-00
Bending test for rigid cellular materials	ISO 1209 (1990)	EN 12089 (1997)	D790-00
Determination of tensile properties of rigid cellular materials	ISO 1926 (1979)	EN 1607 (1996)	D1623-78 (1995)
Determination of shear strength of rigid cellular materials	ISO 1922 (2001)	EN 12090 (1997)	C273-00
Determination of apparent thermal conductivity by means of a heat flow meter	ISO 8301 (1991)	EN 12667 (2001)	C518-98
Determination of apparent thermal conductivity by guarded hot-plate	ISO 8302 (1981)	EN 12667 (2001)	C177-97
Test for dimensional stability of rigid cellular materials	ISO 2796 (1986)	EN 1603 (1996) EN 1604 (1996)	D2126-99
Determination of water absorption by immersion method	ISO 2896 (2001)	EN 12087 (1996)	–
Determination of water vapour transmission of rigid cellular materials	ISO 1663 (1999)	EN 12086 (1996)	E96-00
Determination of the closed-cell content of cellular materials	ISO 4590 (2002)	–	D2856-94 (1998)
Determination of friability of rigid cellular materials	ISO 6187 (2001)	–	C421-00

Thermal conductivity

Thermal conductivity is one of the most important and most measured property of rigid foam and there are two key methods – the heat flow meter and the guarded hot plate.

The heat flow meter method measures the heat flow through a known foam thickness under a specified temperature gradient. The foam specimen is placed between two isothermal plates and the temperature difference should be large enough to result in a heat flux that can be measured accurately. The heat flux meter is calibrated with samples of a known thermal conductivity, which will be air-filled to avoid any ageing effects. The heat flux through the foam needs to reach a steady-state condition, which requires some time, as the foam has to equilibrate to the applied temperature gradient. For foams, which have absorbed high amounts of water, it might take a longer time to reach the steady-state heat flux due to the high heat capacity of water and its eventual redistribution within the foam. Alternative set-ups, which require two identical pieces of foam, also exist.

In the guarded hot plate method the foam sample is placed between an electrically heated hot plate and another cold plate and at steady-state conditions, the electric current through the hot plate is a direct measure for the heat flux through the foam. Thermal conductivity is obtained from this heat flux, the dimensions of the foam and the applied temperature gradient.

Compression strength

The compression strength of rigid foam measures the degree of deformation, which will occur when a pressure is applied to the foam and the higher the compression strength, the lower the tendency of the foam to shrink or expand. For applications such as roofs, the foam compression strength should be high enough so that walking on it does no damage.

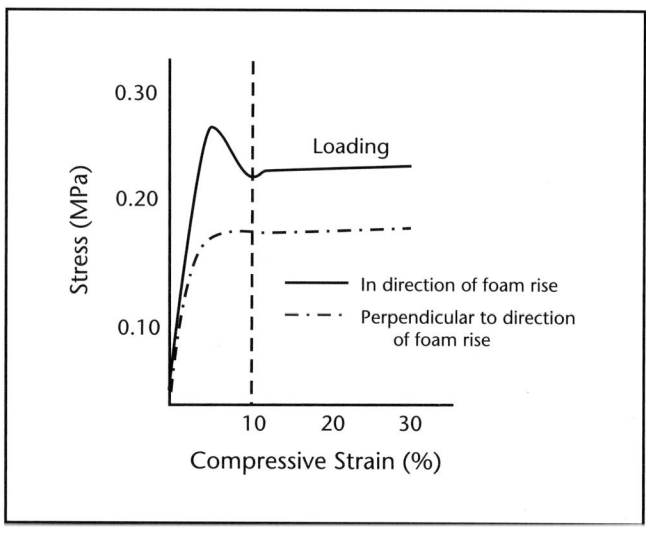

Figure 15-12　Stress-strain curve obtained during compression

Compression strength is measured on cubic foam samples placed between two parallel plates with a larger area than the specimen and the force required to compress the foam at a constant rate is measured. The test records the applied force and the displacement of the specimen to obtain a stress-strain curve and an example for low density rigid foam is shown in Figure 15-12.

The compression strength of rigid foam is the value of the maximum compressive force divided by the initial surface area of the test specimen, but only if the maximum compressive force is reached before the strain of the sample reaches 10 per cent. Otherwise the compressive force is recorded at 10 per cent strain. The slope of the initial straight-line part of the stress-strain curve represents the elastic part of the deformation and the ratio between stress and strain is called the elastic or Young's modulus.

The compression strength of rigid foam increases with increasing densities and when the foam cells are oriented in the direction of rise, the compression strength parallel to rise will be much higher than the one perpendicular to rise.

Dimensional stability

If the pressure difference between the internal cell gas pressure and atmospheric pressure becomes greater than the strength of the foam, it will start to deform and the dimensional stability test measures the tendency of the foam to shrink or expand under service conditions. The test consists of measuring the change with time of the three orthogonal dimensions of a rectangular foam slab held at a temperature and relative humidity relevant for its application.

At low temperatures, such as -20°C, the physical blowing agent can be almost fully condensed, which will increase the difference between the internal and the atmospheric pressure. However, the higher rigidity of the foam matrix at such low temperatures will also result in a higher resistance to pressure differences and hence partially compensate.

As diffusion of the different gases takes place, the internal cell gas pressure will change. Taking into account the relative rates of diffusion of the different gases, the total cell gas pressure will usually exhibit a minimum value when all the carbon dioxide has left the foam and only minor amounts of air have entered the foam. This critical point for shrinkage can occur at any time dependent on how fast the diffusion processes take place.

A test has been designed to cope with this problem where the carbon dioxide is first removed in a vacuum oven in order to create the lowest possible cell gas pressure. In a second step, the dimensional changes are measured at conditions relevant for the application. By creating the lowest possible cell gas pressure the test determines the greatest possible shrinkage and allows a safety margin for dimensional stability to be designed into the foam.

Closed-cell content

The closed-cell content of foam can easily be checked and is a useful quality control test. The foam specimen is placed in a closed chamber and the volume of the chamber is then increased by a known amount. The fall in pressure of the expanded air containing the test specimen is measured and compared with the pressure obtained by a similar volume expansion of the chamber without the foam specimen. The air displacement of the test specimen, which is proportional to its closed-cell content, is then calculated by the application of Boyle's law.

The largest error in the measurement will be produced by broken surface cells, which are counted as open cells. This could in principle be corrected for by measuring the apparent closed-cell content of several foam slabs with different surface to volume ratios. Extrapolation of the measured closed-cell content to a zero surface to volume ratio leads to the real closed-cell content. However, this rather time consuming method is usually not followed and comparative results are obtained by working with fixed foam geometry samples.

Water vapour transmission

Water vapour transmission properties are measured by a simple gravimetric test in which circular foam samples are sealed with wax into the neck of a circular glass beaker, containing a dry desiccant, which are then held at 38°C and 88 ± 1% relative humidity.

The beakers are weighed at fixed time intervals and from the measured change in mass per time, the vapour permeability of the sample can be calculated and related to the vapour permeability of air to give the resultant vapour diffusion resistance index, Figure 15-13. This dimensionless value expresses the rate at which water vapour can permeate through the sample relative to an equivalent thickness of air. The value serves as a useful comparative measure of the water vapour resistance of various materials, becoming larger as the barrier characteristics of the material increase.

Figure 15-13   Water vapour transmission

$$\mu = \delta_{air}/\delta_{foam}$$

μ = Vapour diffusion resistance index
δ_{foam} = Vapour permeability of foam
δ_{air} = Vapour permeability of air

This laboratory test is performed at a uniform temperature and the sample does not absorb a significant amount of water. In real-life applications rigid polyurethane foam invariably experiences a temperature gradient and water accumulation can take place at the cold side. Therefore, alternative experimental techniques, which measure water accumulation and its effect on the λ-value under a temperature gradient, have been developed.

However, the presence of skins and impermeable facings can retard or even stop water diffusion so, as with thermal conductivity ageing tests, measurements on plain foam samples represents a worst-case scenario and are in many cases irrelevant to end-use situations.

16. Appliances

Joris Deschaght

Generally more food is available at certain times than can be consumed, so to ensure that it is available throughout the year it is necessary to preserve the excess. All foods naturally contain micro-organisms or are subject to external attack of micro-organisms, fungi and moulds and will quickly deteriorate after harvesting unless steps are taken to stop or significantly slow down the degradation. In ancient times this was achieved by smoking, drying, salting or pickling the food, Figure 16-1, all of which worked on the principle of eliminating the water that the micro-organisms need to function.

Following the discovery by Louis Pasteur of the relationship between the growth of micro-organisms in food and its degradation, canning of food became popular. Food was heat treated in airtight cans, preserving it as long as the contents were protected from external influences, but whilst it killed all the micro-organisms it also destroyed many of the vitamins.

Figure 16-1   Smoked and salted food

Even in prehistoric times it was known that food could be preserved longer when kept cold, an example being the use of ice caves. Low temperature does not kill the micro-organisms, as the canning process does, but it stops and/or controls their rate of growth, so the conservation time will depend on the temperature of storage. The vitamins in food conserved by cooling or freezing remain unaffected.

It is difficult to say when the first domestic refrigerator was developed as it was a result of different inventions and improvements, but a key point was the development of the sealed compressor around 1900. Initially there were two types of mechanical refrigerator, based on gas or electric technology. Whilst both types still exist the electrical model now dominates the market with the gas type used almost exclusively in leisure applications, such as caravans or mobile homes where electricity is not available.

The first refrigerators were bulky models, Figure 16-2, as they were insulated with mineral wool and the compressors were very large. In 1963 rigid polyurethane foam was used for the first time to insulate refrigerators and freezers and has since become the dominant material, with close to 100 per cent market penetration. The latest development in the use of different temperatures to conserve food at optimal conditions are the multiple door refrigerator/freezers that have been introduced to the Japanese market.

Polyurethane adheres well to most metals and plastics so is used to bond together the inner plastic liner and outer metal components that make up the cabinet, neither of which have the structural strength to freely support themselves. The polyurethane foam provides the rigidity to the composite structure, which determines the final shape, stability and aesthetic look of the cabinet. Although the foam is not visible, it is very important that no voids are present at the foam/metal interface since these would create unacceptable sink marks.

Energy conservation is increasingly important and whilst great efforts are made to improve the efficiency of compressors, condensers and evaporators the run-time of the compressor has the biggest influence on the energy consumed. To reduce run-time the food compartment needs to be well insulated, to minimise heat ingress, and rigid polyurethane foam has some of the best insulating properties.

Figure 16-2   Early refrigerator

Appliance design

Refrigerators and refrigerator/freezers are constructed by joining an outer case, normally painted metal, and an inner plastic case, vacuum drawn from acrylic-butadiene-styrene (ABS) or high-impact polystyrene (HIPS) sheets. The cavity between the two cases is then filled with polyurethane foam to provide the insulation properties to the cabinet. To cool this insulated box, a cold circuit has to be added, as shown in Figure 16-3, with the key element being the compressor. The compressor, whose size depends on the volume and function of the refrigerator or refrigerator/freezer, is a sealed unit which contains the compressor pump, an electric motor, the refrigerant and lubricant oil, Figures 16-4 and 16-5.

Figure 16-3   Cold circuit for a refrigerator

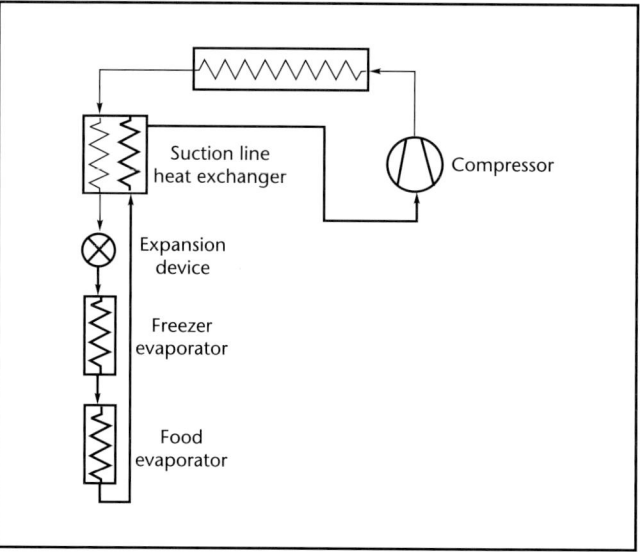

Figure 16-4
Refrigerator compressor: outer

Figure 16-5
Refrigerator compressor: inner

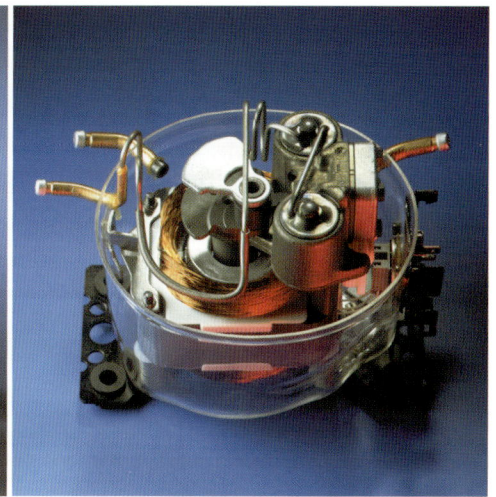

The other components of the cooling system consist of an evaporator and condenser that are connected with each other and to the compressor through copper or aluminium pipes. The common refrigerant liquids used in domestic refrigerators are R-600a (iso-butane) and HFC-134a.

To cool the inside of the refrigerator the compressed liquid refrigerant is evaporated through a pressure drop in the evaporator, normally positioned inside the refrigerator or in direct contact with the inner-liner in the foam-filled cavity, taking up heat during the process. The condenser is positioned on the back of the cabinet or inside the foamed cavity in direct contact with the outside metal wall and condenses the gaseous refrigerant back to a liquid, releasing heat.

A thermostat inside the cabinet, linked to the compressor, controls the temperature. In more sophisticated models the thermostat is supported by electronics that improve the efficient use of the compressor and so reduce the energy consumed by the refrigerator. The wiring between compressor and thermostat is hidden in the cavity between outer and inner case and held in place by the polyurethane foam.

The refrigerator door is built in the same way as the cabinet, from a painted metal outer sheet and an inner plastic sheet, with the cavity between the metal and plastic sheet filled with rigid polyurethane foam. To ensure that there is good seal between the door and the cabinet, the door is fitted with an integral gasket, which contains a magnetic strip to hold the door against the metal cabinet. As the air in the cabinet is cooled down the pressure decreases, which ensures a tight closure of the gasket.

The direct cooling system, mainly used in European-style refrigerators, uses a plate-type evaporator placed inside the cabinet or fixed directly against the inner liner and the heat is extracted from the cabinet through convection.

Indirect cooling, or frost-free technology, has been developed for refrigerators used in America and Asia. These areas tend to have higher humidity levels than Europe and this design prevents humidity present in the air from condensing and freezing on the sides of the food compartment. The key feature is that an extra cavity is created at the back of the refrigerator, which is fitted with a finned heat exchanger, as used in air conditioners, to act as the evaporator. A fan blows air over the finned evaporator and this cooled air is circulated through channels inside the food storage area of the cabinet. The channels are designed so that different temperatures can be created in the various compartments of the cabinet.

The key method of classifying appliances, both refrigerators and freezers, is by appearance and the following are the main style elements:

- Single door upright.
- Double door top- or bottom-mount.
- Side-by-side.
- Multiple door.
- Chest freezers.

Figure 16-6   *Single door refrigerator*

With a single door design the cabinet has one compartment in which the food is cooled or frozen and refrigerators can have a freezer cell integrated in the cabinet with the temperature obtained in the freezer indicated by a rating of one to four stars. These cabinets can vary in height from around half a metre, table-top or mini-bars for hotels, up to almost two metres. An example of a modern single door refrigerator is shown in Figure 16-6.

A double door top- or bottom-mount type has two compartments and doors, freezer and refrigerator, mounted on top of each other in a single cabinet. The size of the compartments can vary, but generally the refrigerator is bigger than the freezer. The total usable volume of both compartments together can be as great as 800 litres. Such big cabinets are mainly found in Asia, with most European models being smaller, a typical model is shown in Figure 16-7.

Two door, side-by-side cabinets are mainly found in America and Asia, but have started to appear in Europe. The freezer and refrigeration compartments are placed next to each other within one outer cabinet, with a vertical mullion separating them. These side-by-side cabinets always use no-frost technology and very often special features are integrated in the freezer and refrigerator doors, such as an ice dispenser, access to cold soft drinks without the need to open the door and a cold water tap; an example is shown in Figure 16-8.

Multiple door cabinets are mainly found in Japan and Korea and are upright models, which can have five or six different compartments. The doors can be of the drawer type and each compartment can be set to a different temperature, ideal for storing different types of food such as fish, vegetables, meat, soft drinks and milk products. Since only a small volume of the total cabinet is opened each time, this model, as shown in Figure 16-3, is also more energy efficient.

Chest freezers are constructed in a similar way to the upright models, but the inner liner is usually made from sheet metal or strong aluminium foil rather than a moulded plastic.

Figure 16-7 *Double door bottom-mount refrigerator* *Figure 16-8* *Side-by-side two door refrigerator*

Formulations and properties

The rigid polyurethane foam chemical systems used by appliance manufacturers – based on polymeric MDI, polyol blends and blowing agents – are usually delivered ready to use, with the exception of the USA where, in some cases, individual raw materials are supplied. The polyol blend consists of a mixture of polyols, normally based on sucrose and sorbitol, surfactants, catalysts and water with the latter acting as a chemical co-blowing agent. The selected blowing agent is normally mixed with the polyol blend at the appliance production plant.

The choice of blowing agent has been controlled by the Montreal protocol, which led to a phase-out of chlorofluorocarbons, such as CFC-11, and were replaced by the hydrochlorofluorocarbons over the last few years. The full details of the options now available are discussed in Chapter 8, but the following regional choices have been adopted by the appliance industry.

The European industry is using mainly alkane blowing agents and approximately 50 per cent of the formulations are based on a mixture of cyclo and iso-pentane, with the volume continuing to grow since it leads to lower foam densities combined with good processing characteristics. Another 30 per cent uses cyclo-pentane on its own with the last 20 per cent being based on a mixture of cyclo-pentane and iso-butane. The latter combination gives foam with similar densities to the cyclo and iso-pentane technology, but processing is more difficult since iso-butane is a gas and creates a frothing effect during mixing.

Although CFCs and HCFCs are still allowed in a number of Asian and South American countries most manufacturers have already switched to alkane-blown foam technology. The North American appliance industry is still using HCFC-141b as blowing agent, but this has to be phased out by the end of 2002; the leading replacement candidate is HFC-245fa although HFC-134a will also be used.

Typical formulations and properties for appliance foams based on a range of blowing agents are given in Table 16-1.

The cream and gel times are the critical measures for an appliance formulation since the cream time should be longer than the injection time and ideally the foam should expand and fill a cabinet in approximately the same interval as the gel time. After the gel time cells can only grow by stretching, this leads to anisotropy in the final foam, which is therefore weaker and less dimensionally stable. For appliance systems, the gel time normally occurs when the foam has reached about 85 per cent of its final free expansion. The difference between cream and gel time also determines the speed at which the foam will develop during its expansion process in the cavity. The reactivity profile chosen for a system reflects the injection technology being used and typical numbers are given in Table 16-2.

Table 16-1   Formulations and properties of rigid foams for appliances

	CFC-11	HCFC-141b	Alkanes	HFC-245fa	HFC-134a
Formulation, wt-%					
Sucrose polyols (OHv 440)	31.4	–	–	14.3	18.4
Sorbitol polyols (OHv 460)	–	10.6	18.2	–	–
Aromatic amine polyols (OHv 400)	–	20.8	14.0	12.5	14.0
Glycerol polyols (OHv 540)	1.7	–	3.0	1.1	–
Catalyst	0.7	0.8	0.8	1.1	0.9
Surfactant	0.7	0.8	0.8	0.7	0.9
Additives	0.2	0.2	0.8	0.4	0.4
Blowing agent	14.0	9.4	4.9	11.8	8.1
Water	0.7	0.9	0.8	0.7	1.1
Polymeric MDI	50.6	56.6	56.8	57.3	56.3
Processing conditions					
Cream time, s	9	6.5	4	Froth	Froth
Gel time, s	60	50	45	38	35
Free-rise density, kg/m^3	22	23	22.5	22	24
Polyol and isocyanate temperature, °C	22	22	22	22	22
Properties					
Density (core), kg/m^3	28	31	32	29	30
Thermal conductivity (10°C), $mW/(m.K)$	18	19.1	20.2	18.4	20.5
Energy consumption, %	100	102	109	103	109
Minimum compression strength, kPa	110	120	120 – 150*	110	110

* c-pentane, 150 kPa at 34 kg/m^3
c-pentane/i-pentane, 130 kPa at 31 kg/m^3
c-pentane/i-butane, 120 kPa at 30 kg/m^3

Table 16-2   Reactivity profiles for different injection methods

Injection point	Cream time (s)	Gel time (s)
Compressor step	5 – 7	40 – 50
Back of cabinet	6 – 8	50 – 60
Top flow	4 – 6	30 – 35

The free-rise density of the foam is also an important parameter since it will determine the expansion capability of the foam and hence the final overall density in the cabinet or door. The free-rise density of foam used to fill appliances is generally between 21 and 25 kg/m^3 dependent on the blowing agent used, with the final overall core foam density typically between 28 and 34 kg/m^3.

The choice of blowing agent also has an effect on the insulation properties of the foam with all the replacements to CFC-11 giving slightly worse insulating properties. The alkanes and HFC-134a are similar and just below HCFC-141b, which is similar to HFC-245fa. Although the insulation value of these foams has deteriorated slightly, polyurethane foam remains by far the best insulating material for appliances.

The insulating performance of a foam sample is normally determined by measuring the thermal conductivity, whilst the total insulation performance of an appliance can be expressed by the reverse heat leakage performance of the cabinet or by the energy consumption.

The thermal conductivity is measured on a piece of foam cut from the wall of a cabinet, a door or a laboratory foam sample. Foam samples used for these measurements have the foam skin removed and only provide the performance of the specific test sample. Since the foam quality in a cabinet will vary, the thermal conductivity will also vary and will be dependent on the position it is taken from within the cabinet; it can vary by up to 1.5 mW/(m.K) in one cabinet. Nevertheless, the thermal conductivity remains a good indication of the total insulating properties of the foam in a cabinet and is quick and easy to measure.

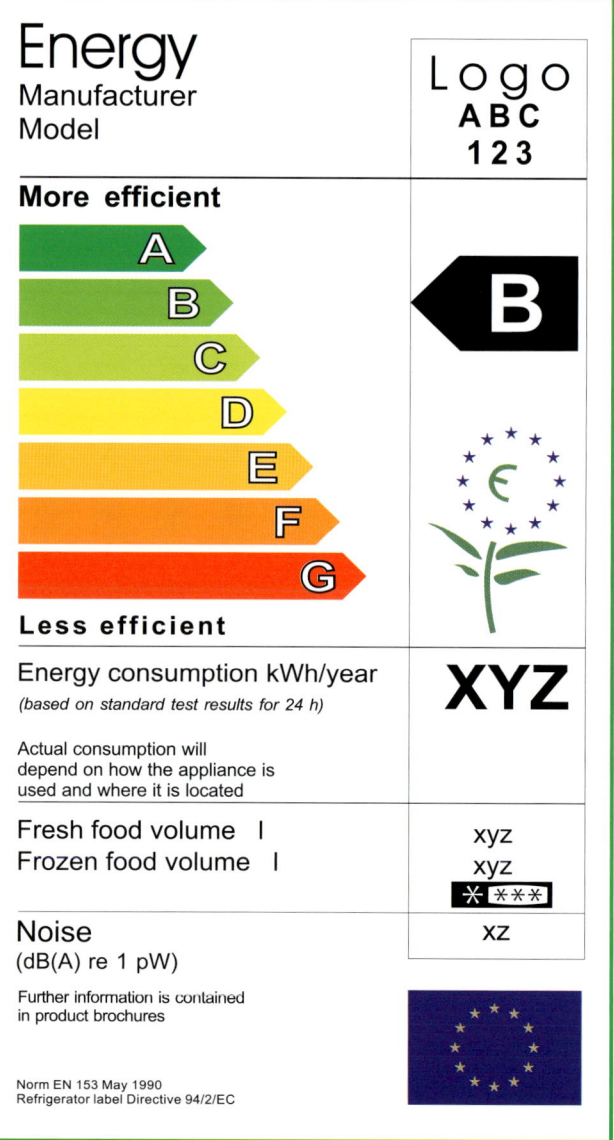

Figure 16-9
Example of the European energy label for appliances

Although the thermal conductivity of the foam has a big influence on the energy consumption, other parameters such as compressor efficiency, heat losses through the door gasket and edge effects are important and not dependent on the foam quality. The energy consumption is determined by measuring the electricity consumed by the compressor in a steady-state operating condition under specified DIN or ISO standards.

Within the European community each cabinet that is sold needs to have an energy label stating its energy class, energy consumption, producer, type of cabinet, energy consumed on a yearly basis and the noise level. An example of such a label is given in Figure 16-9. Most other countries in the world have similar labelling schemes as the European Community; see Chapter 3 for more details.

To measure the insulation performance of the total foam in a cabinet, excluding the compressor efficiency, the reverse heat leakage test was developed. As the name of the test implies the interior of the cabinet is heated while the cabinet is kept in a controlled low temperature environment. In one method the temperature inside the cabinet is maintained at 20°C, using heating elements, whilst the outside temperature is held at 0°C for refrigerators and at -20°C for freezers and the power consumed to maintain the temperature difference is measured. The reverse heat leakage test is not a standard test and conditions can vary from laboratory to laboratory so the data is not normally used as absolute numbers, but comparatively to compare the insulating performance of different foam systems in the same cabinet model.

Within the lifetime of an appliance, the foam must remain dimensionally stable and to determine the ability of a foam to achieve this the dimensions of precisely cut samples, taken from a cabinet, are measured before and after being held for specified times at different temperatures. Appliance foams are normally tested dry at -20 and 70°C and at 70°C in 95 per cent relative humidity and the change in linear dimensions should be below two per cent.

The compression strength of the foams is also used to judge long-term performance with recommendations for minimum compression strengths varying, dependent on the blowing agent used. The blowing agent can, over time, dissolve within the foam matrix weakening it so the minimum compression strength of freshly made foams must take account of this.

Process technology

The key steps in the production of an appliance involve preparing the rigid polyurethane system, setting up the mixing machine and ensuring that the inner and outer components plus the cooling system are securely positioned in the moulding jig. The reactive polyurethane mixture is then injected into the cavity and expands to over thirty times its original volume. During this expansion process it flows around the cables and pipes, which are present in the cavity, and fills up the smallest corners and undercuts of the complex cavity, this taking less than a minute. The polyurethane foam becomes stable enough within a few minutes for the cabinet or door to be removed from its jig without changing dimensions.

The polyol blend and polymeric MDI are normally supplied ready-to-use, so the main factor involved in getting the polyurethane system ready is the addition of blowing agent to the polyol blend. The simplest way is to use a batch process where the polyol blend and the blowing agent are charged, in the right ratio, to a stirred tank and mixed until homogenised. This method is not used anymore, because it can restrict production, is difficult to completely automate and is impractical for blowing agents that boil at temperatures below room temperature or that are flammable.

Since in most parts of the world the appliance industry now uses blowing agents that are flammable or have a boiling point below the operational temperature an in-line blending process is used. Polyol blend is transferred from the bulk storage tank, using a metering pump, and merged, just before an in-line static mixer, with the blowing agent, that is pre-pressurised to ensure it stays liquid as it is pumped and metered. The two components are thoroughly mixed in passing through the static mixer and are then transferred to an intermediate storage tank or directly to the day tank of a high-pressure machine. In-line blending has the advantage that the whole process can be carried out under pressure and in a closed loop with the day tank of the foaming machine. This allows much better control and containment of blowing agents that are flammable or which would froth if exposed to normal atmospheric pressure.

The use of low-pressure foam dispensers has almost completely disappeared in the appliance industry and high-pressure machines are now predominantly used. The key features of a high-pressure machine are described in Chapter 11. The original high-pressure machines were fitted with simple straight heads in which the polyurethane mixture leaves the mixing channel at a high speed with high turbulence. The high turbulence can create splashing in the mould cavity leading to imperfections in the foam surface. For this reason the L-shape mixing head was introduced which has a second channel, positioned perpendicular to the mixing channel. The 90° turn the mixture has to make to move from the mixing channel into the second channel decreases the speed and the mixture leaves the mixing head with laminar flow.

Besides its excellent insulation properties the other main advantage of polyurethane foam for appliances, over other insulating materials, is its ability to fill the complex shape of the cavity between inner liner and metal outer casing and to glue these inner and outer materials together to form a composite structure that becomes dimensionally stable within minutes. The key process characteristics that contribute to the success of polyurethane are:

- Flow.
- Demould time.
- Adhesion.

The flow of a system is exemplified by the minimum weight of foam needed to 'just fill' a cavity with foam. A production process normally uses the 'just fill' weight plus 15 per cent 'over pack' to ensure even density distribution throughout the cabinet or the door. If the foam were not over packed the density at the last area to fill would be too low and not dimensionally stable. The 'just fill' weight is influenced by the free-rise density, the type and viscosity of the components and the viscosity build-up during the reaction and rise of the foam. The better the flow of the foam system the lower the 'just fill' weight.

The reacting foam generates pressure due to the expansion of the blowing agent in a confined space and this increases with increasing 'over pack'. As the cabinet sides do not have sufficient strength to resist this pressure the cabinet or the door must be held in a mould during the foaming process. The demould time is therefore the time at which the cabinet or the door can be taken out of the mould without it expanding more than the set specification. If a cabinet or door is demoulded too soon the foam will continue to expand and the composite structure will swell. This swelling would lead to a poor fit between cabinet and door, also the shelves would not fit well in the cabinet and aesthetically the appliance would not be acceptable. The productivity of the line is thus determined by the demould performance of the foam and this is mainly influenced by the formulation and the design of the cabinet or door, as a rule of thumb the demould time required is one minute per centimetre of wall thickness.

The adhesive strength of foam to the liners is determined by the process conditions and the formulation, with the amount of water being especially critical. The temperature of the mould jig and liner materials is also important and it is recommended that a minimum of 35°C be used. Higher temperatures enhance the adhesion properties, but temperatures above 50°C can lead to surface defects in the foam. In production, a quality control expert manually peels the liner from the foam to judge the adhesive strength, but proper tensile peel tests are also carried out on samples cut from the composite structure.

As mentioned above, the inner and outer liners of the cabinets or doors need to be held in jigs during the foaming process. There are two ways that the jigs can be organised and moved during the production process, the static and carousel methods.

In the static process the jigs are in a fixed position and when the jig is open the empty cabinet is transported from the pre-assembly line via a belt in the jig. By lowering the male part of the jig into the food compartment of the cabinet and closing and fixing the sides, the jig is ready for the foam injection. In such a set-up the door opening is always up and the foaming mixture is injected at the compressor step at the back of the cabinet. Once the set demould time is reached the jig opens and the cabinet is moved out of the jig by a belt system to the final assembly line where the compressor, shelves and doors are fitted.

Each static jig can have its own mixing head or one mixing head can move from jig to jig, normally up to a maximum of six. A static jig set-up needs a large number of mixing heads, which whilst being an extra cost has the advantage that the mixing head settings can be optimised for the size of cabinet to be filled. The demould time of a polyurethane system is very important in this set-up since the faster a cabinet can be moved out of the jig, the higher the productivity of the production line becomes.

In a carousel line there is normally one mixing head for every carousel chain, which can have more than 20 jigs fitted. The bottom part of the jig is normally the male part jig to which the side plates are fitted. When the jig is open the empty cabinet can be lowered on the male part of the jig and the sides and back plate of the jig are closed and fixed with the door opening facing down. The injection of the foam mixture is then into the top of the jig at the mullion position of the cabinet (the division between multi cavities).

Another option, which is called top-flow, is to inject the foam from below the cabinet in which case the mixing head is positioned in the centre of the back and there is normally a distribution valve connected to the back of the cabinet into the cavity to be foamed. The role of this valve is to provide extra mixing and to distribute the injected liquid along the back of the cabinet. After the injection is finished the cabinet moves along the carousel chain through a heating oven, to keep the cabinet jigs at the required temperature, on to the extraction station.

At the extraction station the jig is opened and the foamed cabinet is taken out and moved to final assembly. From the extraction station the open jig is moved to the position where a new cabinet is loaded into the jig. In this set-up the demould time is determined by the chain speed, which again is determined by the number of jigs in the chain. In general, the demould time for a carousel line is longer than with static jigs because all the operations, loading, unloading and injection of the cabinet, require the whole chain to stop. To obtain the same output per hour for a carousel line, more jigs are needed overall compared to a static jig line, but only one mixing head is required, thus balancing the overall foam processing costs between a static jig and a carousel line.

A carousel arrangement is normally used for door foaming lines with the door jigs mounted on a chain that moves around through an oven. At the point where the jigs exit the oven the foamed doors are extracted, new door components placed in the mould and then injected with foam. The process can be injection into a closed mould or pouring in an open mould. Almost all recent lines use the pour process because a much better surface quality can be obtained.

The moulds for the doors are of the book type with the upper lid opening at a 90° angle to the bottom, horizontal part of the jig. In the latest designs the plastic inner liner, with the integrated door gasket already in place, is put in the horizontal part of the mould and the metal outer liner is positioned and held in place by magnets in the vertical lid. The mixing head moves along the open mould, dispensing foam so that the flow path of the foam in the length direction of the mould is much shorter, which results in better surface quality. As soon as the mixing head has moved out of the mould area the mould is closed and the jig moves further along the chain. This open mould pour technique is shown as Figure 16-10.

Figure 16-10   Open mould pour technique

There are two types of chain system, horizontal and a drum or tambour with the former taking up much more space than the latter. On the tambour line the moulds are mounted on a large vertical drum, which turns to bring the moulds in front of the injection position. Since it is not possible to have an oven mounted around such a drum the jigs are heated electrically or by water jackets, allowing a more accurate individual heating, which again contributes to the foam quality.

17. Construction

Martyn Barker

It is estimated that approximately 45 per cent of all carbon dioxide emission arises from climate control, heating or cooling, in buildings (see Chapter 3, Figure 3-4). Such control is expected to increase as a quality/comfort factor driven by increasing affluence, and social development in harsh climate zones. This is an area in which considerable energy savings can be achieved by the use of thermal insulation, and the energy reductions achievable through building design have been amply demonstrated. Even zero energy concepts are being considered.

The primary benefit of insulation materials lies in their ability to reduce energy consumption, which has grown rapidly over the last half century driven by increases in population and economic growth. Consequently, increasing emphasis is being placed on energy conservation issues to enable sustainable world development and insulation materials have a critical role to play in this area.

Insulation materials

A wide range of insulation materials is available each with its own specific property spectrum and although energy conservation is the primary driver in the use of insulation, the detailed requirements of any specific application determine material selection. Insulation materials can be divided not only into inorganic and organic materials, but also in terms of their physical structure and an outline of the range of commercially available insulation materials is given in Table 17-1.

Table 17-1   Range of insulation materials

Structure	Chemistry	Product
Porous	Inorganic	Mineral fibre
		Glass fibre
		Perlite
	Organic	Cellulose fibre
		Natural wool
Closed Cell	Inorganic	Foam glass
	Organic – Thermoset	Polyurethane (PU)
		Phenolic foam (PF)
	Organic – Thermoplastic	Polystyrene (PS)
		Polyethylene (PE)

The insulation properties of materials can be compared by examining the thickness required to achieve an equivalent performance, as shown in Figure 17-1. For instance 1,720 millimetres of brickwork or 200 millimetres of softwood is required to achieve the same insulation performance as 50 millimetres of rigid polyurethane foam. Such a thick layer of brickwork would be unacceptable owing to: cost, excessive weight, the resulting need for massive foundations, and potentially loss in the useable internal space of a building.

Rigid polyurethane and polyisocyanurate foam-cored, factory-made, composite boards are widely used for the insulation of the envelope of buildings – roof, walls and floor. When the composite has rigid facings, such as steel or aluminium for example, it can be made as prefabricated sections of walls and roofs that reduce erection time dramatically and result in a building that is well insulated even in the initial construction stage. Cut rigid polyurethane and polyisocyanurate foam slabstock can also be used as an insulant. Boards with thin impervious (such as 50 micron aluminium foil) and semi-permeable (such as treated paper) facings can be used to insulate roofs, as cavity lining boards and under-floor insulation. Boards faced with plasterboard can be used to line and insulate internal walls – particularly in multi-story apartment blocks and old buildings with non-cavity walls.

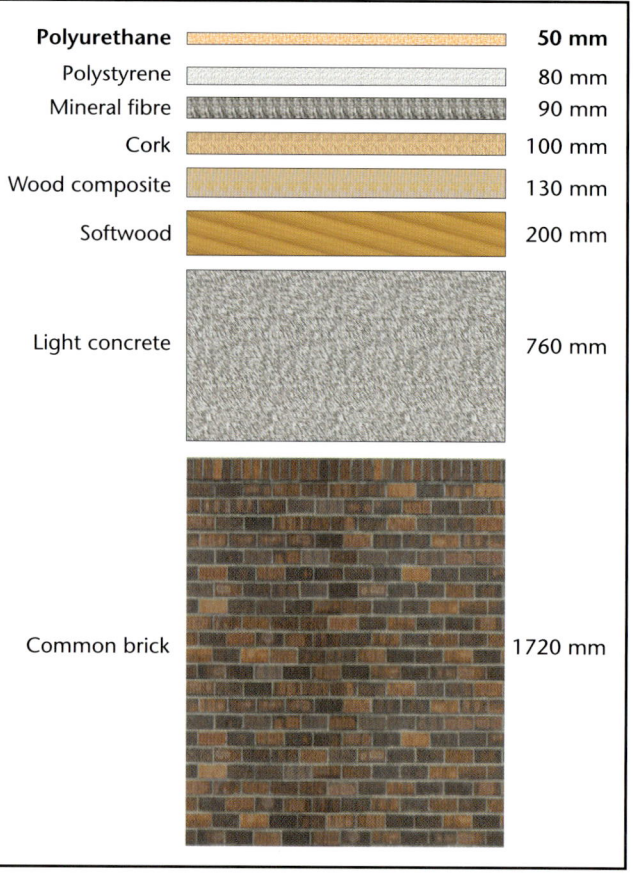

Figure 17-1

Thickness for equivalent insulation performance of common building materials

Formulation technologies

There are two fundamentally different formulation technologies being used in the construction area for the production of composite boards. The choice of which approach to follow is dictated by the local fire performance regulations, as described in Chapter 4, and the production economics, related to local raw material supply and market demand. Both are almost exclusively based on polymeric MDI.

In the first technology a polyurethane matrix is formed using an isocyanate index of 110 to 120, and low molecular weight polyols, often sucrose-based, with a functionality of 2.5 to 3.5. This is historically the most widely used technology, but flame retardants are sometimes needed to achieve specific fire performance standards. Flame retardants can add significantly to system cost and have a negative effect on physical properties such as compressive strength.

The second technology is based on using a large excess of isocyanate in the formulation, isocyanate index between 220 and 300. This excess of isocyanate is converted into polyisocyanurate using suitable trimerisation catalysts. This high index approach is usually used with low functionality aromatic polyester, often based on industrial waste streams such as dimethyl terephthalate (DMT) still bottoms.

Roofing panels in the USA have been almost exclusively based on high isocyanate index aromatic polyester polyisocyanurate foams for a number of years. The commercial acceptance of polyisocyanurate foams in some areas such as Europe has been more limited due to perceptions of processing disadvantages and availability of raw materials, but improvement in polyisocyanurate technology and greater experience in their manufacture continues to widen the range of such foams. These are now mature industrial products and this technology continues to spread into a wide variety of other applications. The main benefit of the isocyanurate link in polyisocyanurate foams lies in its higher heat stability compared to urethane or urea links. The basic polymer matrix of high isocyanate index polyisocyanurate foam therefore possesses an inherently better fire performance than the comparable polyurethane foam.

The overall balance between the factors means, however, that in practise polyisocyanurate foams only have a significant presence in applications where fire is a primary performance criterion in material selection.

The range of rigid polyurethane foam products and the applications that can be satisfied using these two basic sets of raw materials is very wide and large geographical variations exist driven by different climates, building code requirements and construction practises. The two common forms of these flat products are distinguished by the type of facing (facer), and this is used here to exemplify the use of rigid polyurethane and polyisocyanurate-cored composite materials in the building industry:

- Boardstock – has flexible facings.
- Sandwich panels – has rigid facings usually metal.

Boardstock

Flat sheet products made with flexible facings have names that differ regionally, so in this section the term 'boardstock' is used as a general descriptor even though it has a specific meaning for roofing panels in the USA. A wide variety of facing product is used to make boardstock reflecting the broad range of materials that can be bonded successfully with polyurethane. Products vary from paper-based facings, to polymer multi-layer laminates, thin metal foils, bonded fibreglass or bitumen/asphalt water vapour barriers. The latter are particularly common in roofing applications. The thickness of the boards can vary from 2 to 20 centimetres, dependent on the insulation performance required.

Although widely used in roofs or walls, polyurethane-cored boardstock is often not visible as it is hidden under waterproofing membranes or behind wall skins. Substantial volumes of these products are used with a market of over 190,000 tonnes for roofing applications alone in the USA in 2000.

Historically, a range of techniques has been employed in the manufacture of boardstock and panels, but the continuous lamination process is now dominant owing to its high production rates. The lamination process requires the following materials and components: basic polyurethane or polyisocyanurate system, blowing agents, mixing machine, facings, conveyors, cutters and stacker unit whilst running the process involves six different stages:

- Delivery of the facing, top and bottom, with any necessary pre-treatment.
- The lay-down stage in which the wet chemicals are spread on the facing.
- Movement of the foaming 'sandwich' into a heated conveyor press.
- Hardening of the foam.
- Cutting of the foamed boardstock or panel to the required length.
- Storage in a stack before shipment from the factory.

A primary factor in achieving high production rates is being able to restrict the time from lay-down to panel cutting to a few minutes. To achieve consistent production it is necessary to balance all the components and phases of the process with an obvious impact on the speed and control of the polyurethane reactions.

Formulations

The basic outlines of the polyurethane and polyisocyanurate technologies were referred to earlier and typical formulations and properties for the latter are given in Table 17-2.

Table 17-2 Typical formulations and properties of polyisocyanurate boardstock foams

Formulation, wt-%	A	B	Processing conditions		A	B
Aromatic polyester polyol (OHv 240)	27.1	28.4	Cream time, s		5	3
Silicone surfactant	0.5	0.9	String time, s		30	12
Catalyst package	0.9	1.4	Free rise density, kg/m^3		27	26
Fire retardant	4.0	3.4	NCO Index		300	300
Blowing agent: n-pentane	5.2	–				
cyclo/iso-pentane	–	7.1	*Properties*			
Water	0.3	0.1	Overall density, kg/m^3		34	34
Polymeric MDI	62.0	58.7	Core density, kg/m^3		32	32
			Compressive strength, *MPa*		0.16	0.16
			Water absorption, %		<5	<5
			Closed-cell content, %		>90	>90
			Initial thermal conductivity, $W/(m.K)$	10°C	0.022	
				75°F		0.023
			Dimensional stability, *vol-%*	70°C	<2.5	<2.5
A = *European flex faced laminate*				-20°C	<2.0	<2.0
B = *US Boardstock*				70°C/95%RH	<5.0	<5.0

The two approaches have inherently different processing parameters, but there are, however, specific key control factors which apply to production, one of which is accurate temperature control. Therefore, in modern, automated plants the chemicals are held in temperature controlled bulk storage vessels and charged directly at the required ratios to the laminator machine. The individual components can be blended directly into a holding tank or added and mixed in real time on-line. This is frequently undertaken with the minor, but influential, components such as catalysts and blowing agents. Good temperature control of the chemicals is essential as it has a direct influence on the reaction speed of the foam in the laminating process.

Blowing agents

Like all areas of rigid polyurethane foam, boardstock formulations were initially blown using CFC-11 on its own or in combination with water. However, these ozone depleting blowing agents have been phased out and replaced by more environmentally sound products. The full details of these changes are described in Chapter 9.

For boardstock applications, hydrocarbons, such as pentane, have developed a significant presence in the European market, whilst HCFC-141b is used extensively in the USA and CFC-11 is still allowed in a number of designated developing countries. In the longer term it is likely that the use of hydrocarbons will continue to increase although other blowing agents such as water, HFC-134a, HFC-245fa or complex mixtures will also be used specifically where there are highly demanding specifications.

Stringent precautions need to be taken, during the manufacturing process, in the handling, storage and blending of the hydrocarbon blowing agents due to the fact that they can form flammable or explosive mixtures. The handling issues with hydrocarbons apply only to the manufacturing process and the final panels can be used in an identical way to those produced using CFC, HCFC or HFC blowing agents. Generally, blends of pentane isomers are used, rather than single species, with the iso- and cyclo-pentane mixture featuring more prominently in boardstock than n-pentane due to its lower cost and higher volatility.

Pentanes are not normally completely soluble in the polyols, unlike the halogenated blowing agents, and therefore form an emulsion in the polyol blend. To improve the ease of handling of these emulsions the hydrocarbon is added to the polyol blend line as a separate stream, which then passes through an in-line emulsifier just before the machine dispense head. This minimises both the emulsion stability requirements and the physical amount of hydrocarbon present in the process outside of the closed-loop pentane handling system. Easy emulsification and adequate droplet stability can usually be ensured by careful selection of the surfactants in the foam formulation.

The extent of the problem concerning the release of volatile organic compounds to the atmosphere is dependent on factors such as the geographical location of the factory, the regulatory requirements for the country, the specific product being made and plant throughput. Normally about two per cent of the total hydrocarbons are lost to the atmosphere during the foaming process, mainly at the lay-down stage, but this can be controlled through the use of appropriate abatement systems.

Mixing

The isocyanate and polyol blends have viscosities ranging from a few hundred to a few thousand mPa.s and are mixed using either low or high-pressure machines. High-pressure mixing has become the more widely used technique and offers much easier cleaning without the need for solvent flushing. Most machine heads are of a two-component design processing isocyanate and polyol blend streams. Three- or four-component heads can be employed, however, for the addition of additives directly into the mixing head, but the complexity and cost of such heads restricts their use to specific applications.

In most cases, the polyol and isocyanate are not instantaneously miscible when initially mixed, but rapidly become fully miscible when some urethane reactions have occurred. The mixing technique, therefore, has to provide sufficient energy to the chemical stream to ensure good homogeneity and bulk distribution of the phases.

Facers

The facing materials need to be held at the correct tension, by pre-tensioners on the reels, to allow a consistent feed into the continuous process. In order to achieve good adhesion between the facings and the curing foam, infrared heaters are often employed to preheat the facings both to dry them and to ensure that they are at the required temperature. The lay-down table is also temperature controlled, again to bring the facing to the desired temperature, the facing being held tightly against the table by vacuum suction. The temperature profile of the lay-down table can be adjusted in order to achieve better control during the early stages of foaming and more even foam distribution across the panel width.

Lay-down

The application technique employed needs to ensure an even and controlled lay-down of the liquid chemicals into a thin film. This becomes more difficult for thinner panels as the amount of dispensed materials is decreased and the line speed rises. Of the various approaches employed, two common methods are:

- Liquid stream.
- Gas poker.

With the liquid stream process the mixed components are pumped through outlets at fixed points across the laminate so that the individual streams flow and spread into each other, Figure 17-2.

In the gas poker method the polyurethane mixture is passed through 10 to 100 individual holes in a sprinkler bar that moves across the laminate in a way similar to that of a moving watering can, Figure 17-3. The internal dimensions

Figure 17-2   *Liquid stream lay-down*

Figure 17-3   *Gas poker lay-down*

of the gas poker are carefully arranged to ensure an equivalent output for each orifice as the path length increases along the length of the poker away from the central point. The gas poker is driven across the width of the laminate by a traverse, which can be either motor or pressure driven, pneumatic or hydraulic. In most cases the width, speed and residence time of the traverse can be varied for fine control of the lay-down process.

The lay-down of the chemicals is a critical process in lamination and has to be carefully controlled to achieve the best final laminate properties. In the ideal case the rising foam has simply to move vertically with minimal lateral flow, although a cell orientation can be deliberately imposed on the rising foam if specifically required to improve insulation performance. Uneven distribution can cause problems in board stability, density gradients or compression strength variation. Generally, the onset of the initial foaming reaction, the cream time, is controlled so that it does not occur until the lay-down process has been completed.

Conveyor

The conveyor in a lamination process can be considered to perform three functions:

- Determine the size, shape and flatness of the final panel.
- Pull the laminate along the line.
- Provide sufficient temperature to accelerate the reactions and give acceptable facing adhesion for subsequent cutting and stacking.

The conveyor is therefore the motor of the whole lamination line. The rising foam is essentially liquid as it enters the conveyor, but sufficient reaction needs to take place within the conveyor to ensure that the foam is hard when it exits the conveyor. Therefore, it is essential that the chemical reactions are both closely matched to both the physical processing requirements and the achievement of the required performance properties of the final product. If insufficient cure takes place in the conveyor then the laminate will have insufficient strength to enable the conveyor to pull it along the line or to maintain the tension on the facings.

A wide range of individual conveyor designs is employed, but they can be broadly divided into two types:

- Fixed gap.
- Floating platen.

The fixed gap system, Figure 17-4, uses steel slats with edge pieces, which define the final panel thickness whilst the floating platen, Figure 17-5, has a flexible belt of smooth rubber or steel mesh to shape and pressure the rising foam. In both cases heating, either by heated platens or infrared heaters, is frequently employed in the conveyor section. Temperatures of 30 to 50°C are employed for polyurethane foam whilst temperatures over 60°C can be needed to ensure adequate facing adhesion in polyisocyanurate foams.

Figure 17-4   Schematic of fixed gap laminator

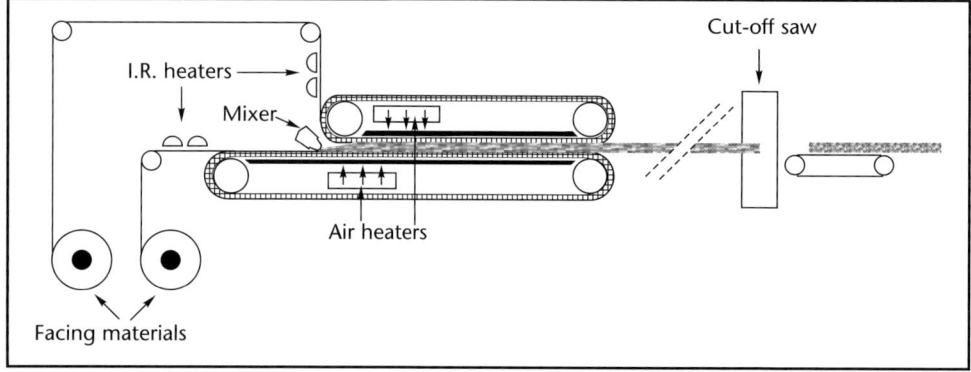

Figure 17-5   Schematic of floating platen laminator

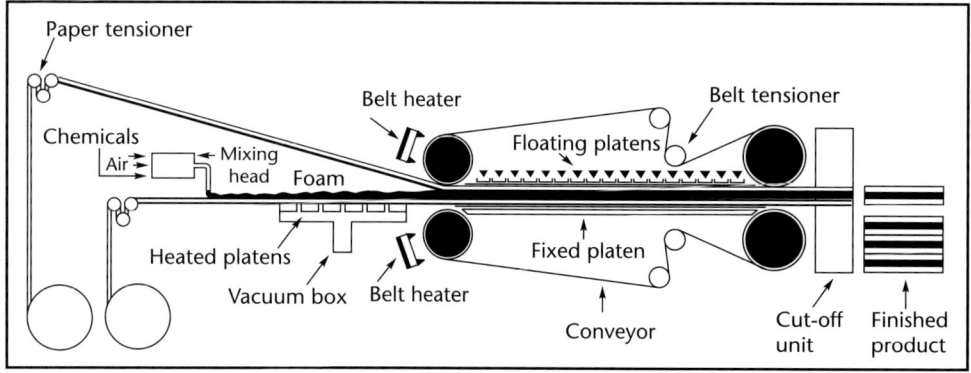

In a processing variant known as 'Early contact' the two facings are brought in contact with the reacting foam before entering the conveyor and a nip roller is employed to ensure even distribution.

Another conveyor design is the 'Inverse laminator', which although less frequently used than conventional lamination has specific benefits when handling mixed facings, one flexible and the other rigid, such as plasterboard or plywood. In this process the foam material is applied to the flexible facing and then the facing plus foam is inverted through 180° and brought into contact with the rigid facing as the fresh, rising foam reaches the string or gel point. This process is especially useful for very porous facings, such as a glass fibre mat, as it minimises penetration through the porous facing whilst still ensuring that good bonding is achieved.

Line speed – board thickness

Line speeds can vary from 10 to 15 metres per minute in Europe to above 40 metres per minute in the USA. There is an increasing tendency to higher line speeds to increase productivity and plant capacity and the chemistry of a given foam formulation, particularly the catalysis, has to be carefully adjusted to match the desired line speed. There is also a correlation between line speed and panel thickness, thicker panels tending to be run at lower overall speeds. Again, if the line speed is adjusted to suit the thickness of the panel, the chemistry of the foam has to be carefully matched to the process requirements.

Cutting and stacking

The final step in the lamination process involves cutting the continuous laminate into individual boards or panels and stacking them before dispatch. Cutting takes place shortly after the conveyor when the foam is only a few minutes old. The foam, therefore, has to have sufficient mechanical strength and adhesion to the facing material not to distort or delaminate at this stage. The cutting saw, band or circular, is driven at controlled speed diagonally across the moving laminate in such a way that a flat cut is achieved perpendicular to the panel edge.

Stacking is required for the curing reactions to reach completion and achieve the final panel strength. The highly exothermic reactions that occur provide sufficient heat for the temperature in the stack to be maintained above 100°C without external heating. Indeed, spacers to allow cooling ventilation to occur often separate stacked panels. Panels can normally be shipped following a day of storage for final curing. In a modern plant the production rates are such that the panel storage areas frequently occupy a significantly larger proportion of the total plant than the actual lamination line itself.

Sandwich panels

A polyurethane-cored composite with rigid facings is known as a sandwich panel and the major factors defining its use for construction are:

- High insulation performance.
- The ability to foam in place and bond facings together in one simple process step.
- Ease and speed of manufacture leading to high production rates.
- The quality/performance of the end product is essentially built-in under factory conditions and is not heavily reliant on the on-site workmanship or installation quality.
- The ability to create lightweight building envelopes together with reduced costs for the frame support structure of the building.
- Fast erection speeds.

Whilst rigid polyurethane foams are efficient insulation materials, insulation performance alone would not be sufficient to account for their high level of penetration in construction applications. One of the key factors is the simplicity of the factory-made polyurethane-cored panel versus the on-site construction required for a built-up panel, as illustrated in Figure 17-6. It can be seen that the built-up panel is a considerably more complex installation.

Figure 17-6   Schematic showing simplicity of polyurethane-cored panel versus a built-up insulation

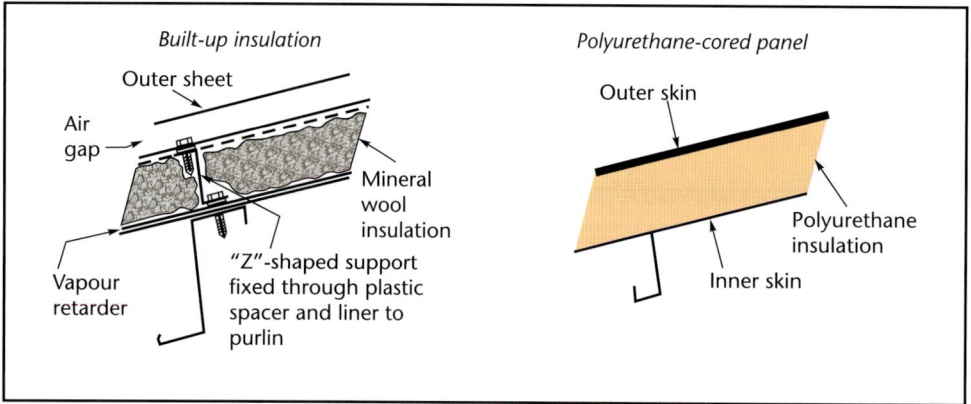

This extra complexity brings with it a number of issues related to the quality of the final construction. On-site installation is much more difficult to control than a factory environment and the skill and training of the workforce become critical issues. Building problems can occur due to either poor workmanship or design faults, with the former often being more dominant with conventional materials. Thus, the architect is often frustrated by the inability to achieve the expected performance of his original designs. On the other hand, factory-made rigid polyurethane foam panels are inherently simpler to assemble so that workers are able to handle and install modular panels of considerable size without the need for complex secondary assembly or lifting apparatus.

Important features of sandwich panels are:

- Freedom of design.
- Structural elements.
- Edge profile design.

The overall aesthetics of sandwich panels are important, as the final panel is visible in the building facade. This leads to the need to provide a wide range of colours and profiles to enhance the architects' design options. The final panels are structural elements within the building envelope, but are not usually considered as load-bearing elements.

The edge profile design is carefully controlled to achieve both quality and performance in the final application and can also be employed to incorporate additional functional features such as weather sealing, intumescent strips and other features.

This package of properties has led to particularly significant benefits in applications such as roofing or walls for buildings which possess large internal spaces with wide spans between the support elements. An example of the benefits of polyurethane-cored panels can be shown by the construction of a building for an artificial ski-slope, as shown in Figure 17-7. Both the walls and roof are constructed from polyurethane-cored panels, as can be seen in the external view, but it is the internal view which shows the primary benefit of being able to obtain a large open span without the need for supports.

Figure 17-7   Indoor ski slope building

Obviously, high-performance insulation is required in this application to minimise running costs, but the need for a clear, open internal space is also a critical requirement. Central support pillars in such a building would certainly be very undesirable. The steel spans required for roof support can be clearly seen and here the low weight of the polyurethane panels leads to considerable cost saving in the steel structures required. Finally, an attractive aesthetic appearance, both initial and long-term, is also highly important for such a building to create the right leisure time ambience; here the wide product range and durability of polyurethane composite panels play an important role.

The construction speed that can be achieved through the use of factory-made panels is clearly demonstrated by the example of a distribution centre built in the UK in 1996, with the stages of construction shown in Figure 17-9. The whole project, from the basic foundations to the construction of the final building envelope, was achieved within a period of only five weeks.

Figure 17-8   Construction stages for distribution centre

Foundations   *Steel framework*

Panel installation   *Completion of envelope*

Koschade has written an excellent review from the architects and building designer's point of view, which would be particularly valuable to readers who require in-depth information on the application of these materials.

Formulations

The formulation approach used in sandwich panels is principally dictated by the manufacturing method employed and tend to be dominated by the processing requirements. There are two manufacturing methods for sandwich panels, often considered as two different market segments:

- Continuous.
- Discontinuous.

Typical formulations and properties for these two approaches are given in Table 17-3.

Table 17-3   Typical formulations and properties of polyurethane sandwich panel foams

Formulation, wt-%	A	B	Processing conditions	A	B
Aromatic polyester polyol (OHv 250)	21.6	–	Cream time, s	9	12
Glycerol-based polyol (OHv 630)	5.4	–	String time, s	35	90
Sorbitol-based polyol (OHv 450)	–	26.5	Free rise density, kg/m^3	30	25
Amine-based polyol (OHv 440)	–	6.6	NCO Index	250	115
Silicone surfactant	0.5	0.3			
Catalyst package	1.0	0.3	**Properties**		
Fire retardant	5.4	5.0	Overall density, kg/m^3	37	40
Blowing agent: n-pentane	5.7	4.0	Core density, kg/m^3	34	35
Water	0.2	0.7	Compressive strength, *MPa*	0.18	0.20
Polymeric MDI	60.2	56.6	Water absorption, %	<5	<5
			Closed-cell content, %	>90	>90
			Axial shear strength (23°C), *MPa*	0.25	0.25
			Flexural strength, *MPa*	0.30	0.30
			Thermal conductivity, $W/(m.K)$	0.024	0.022
A = *PUR/PIR Continuous*			Dimensional stability, vol-% 70°C	2.0	<2.5
B = *PUR Discontinuous*			−20°C	<1.0	<1.0
			70°C/95%RH	<3.0	<5.0

Continuous lamination

Continuous lamination is generally used for the large-scale manufacture of long production runs with a standard design. There is a relatively high fixed cost investment, but low labour costs of production. A wide range of high-quality building elements can be created using metal-faced sandwich panels made by continuous lamination. They were first introduced on a large scale in the mid-1970s and their use has grown dramatically over the past decade as builders and architects become more familiar with their highly attractive total package of benefits. Typical applications in prestige buildings are illustrated in Figure 17-9.

Figure 17-9   Typical prestige buildings built using continuous lamination

Although conceptually the continuous lamination process for sandwich panels resembles that of boardstock, and all the factors listed under boardstock apply equally to sandwich panels, there are considerable differences in equipment design between the two processes, particularly with regard to the handling and preforming of the metal facings.

Comparing the process to boardstock lamination, the following factors can be noted:

- Higher investment is required, as the equipment needs to be considerably more robust to handle the stiffer facing materials.
- The preforming stage needed to profile the metal is a substantial proportion of the total investment.
- The heated conveyor has to be more strongly built than for boardstock .
- More extensive preheating of the facings is required due to the higher thermal conductivity and mass of the metal sheet.
- Edge profile design is important in the final application and sufficient foam flow is required to fully fill the complex profile.

Edge profile design varies between manufacturers, but is a critical performance factor in the final installation. The male/female interlocking provides weatherproofing and has an important effect on fire performance, but must also allow for easy installation on-site. Some examples of different edge profiles are shown in Figure 17-10.

Figure 17-10   Edge and profile designs

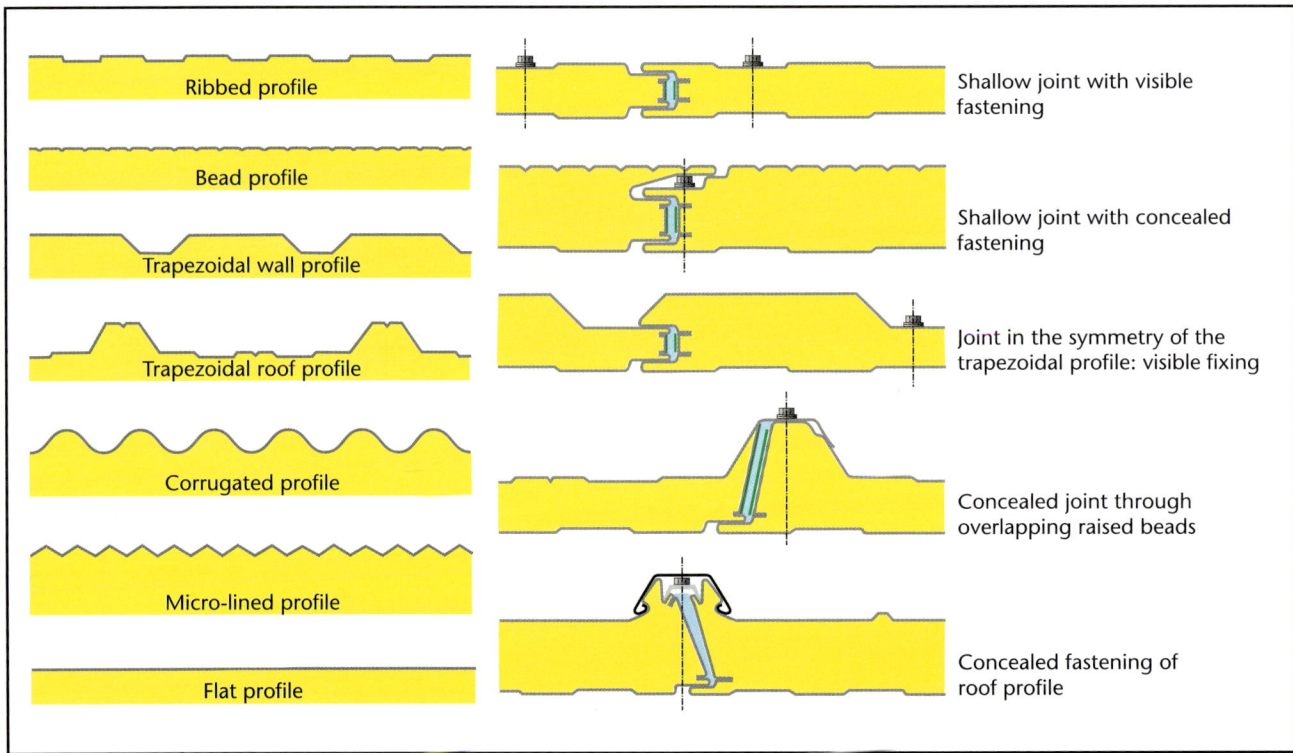

Discontinuous manufacture

The discontinuous manufacturing process is more flexible than continuous lamination and has lower fixed investment costs, but is more labour intensive. It is often used for shorter production runs or the incorporation of specific additional features. A wide range of sandwich panels can be produced, and although the final panels appear similar to those made by continuous lamination, the different manufacturing processes employed require substantial differences in the foam chemistry.

Foam attributes such as speed, flow and pressure are more akin to that required in appliance systems rather than the continuous lamination process. Discontinuous panels are made by single or multiple individual shots of reacting polyurethane chemicals into preformed moulds, and a number of individual processes are commercially employed. Because the rising foam has to evenly fill the panel along flow paths that can be several metres long, the rise profile is generally significantly slower than that which would be employed in a continuous lamination system. The formulation has to be carefully balanced to achieve fast enough cures for acceptable demould/press times whilst retaining good flow characteristics.

Hydraulic presses with single or multi-daylight openings need to be used to guarantee panel flatness although horizontal multi-daylight, single injection point presses have a dominant position. Temperature control is important, particularly to ensure facing adhesion and temperatures in the range 30 to 40°C are commonly employed. Press times increase with panel thickness, the press times being dictated by the minimum dwell period required to remove a physically stable panel.

For very long panels lance injection used to be widely employed, in which the reacting liquids were dispensed into the panel through a tube, which was gradually retracted during the injection phase. Little use of this technique is now made and it has essentially been replaced by dispensing into travelling open moulds.

The high versatility of the discontinuous process means that panel design is easily modified and features such as panel inserts or interlocks can be readily incorporated into the final design during the building frame assembly stage.

The single shot nature of the discontinuous process reduces the ability to incorporate additives on-line so full preblends are employed in the day tank and there is much wider use of pre-made full systems rather than bulk storage of individual components. Hydrocarbon blown foams generally required full solubility of the pentane isomers and little use is made of emulsion technology. Dispensing is almost exclusively undertaken using high-pressure impingement mixing.

18. Other construction applications

Tony AbiSaleh
Nick Hernandez
Alan Hamilton

In addition to the main applications for rigid foams, in construction and appliances, there are four other significant areas plus a whole range of niche uses. These are:

- Insulated pipe.
- Spray insulation.
- One-component foam (OCF).
- Domestic water heaters.
- Other niche applications.

Insulated pipe

Polyurethane pre-insulated pipes are widely used for the transport of oil and chemicals and for district heating/cooling systems. A high long-term thermal degradation resistance is required from the rigid polyurethane foam as district heating networks and oil transport systems can operate with service temperatures up to 140°C. A number of different manufacturing techniques can be used for the production of these pipes, which can be broadly classified as discontinuous and continuous production.

The discontinuous method, for which there are several filling techniques, is a batch process that uses short lengths of pipe. The continuous pipe production technique consists of two stages. In the first stage, the foam is applied to the inner pipe either by a moulding or spray operation. In a second stage, the casing pipe is extruded or wound around the pre-shaped foam.

The continuous techniques allow a fast and consistent production of a large volume of pipe of the same dimension at comparatively low variable costs. Although the initial investment may be higher, cost savings are achieved due to reduction of foam filling density and a reduced thickness of the high-density polyethylene casing pipe.

Each of the three technologies – discontinuous moulding, continuous moulding and continuous spray – requires a specific formulation in order to obtain the requisite foam reactivity, viscosity build-up and cure characteristics and typical formulations and properties for the three are given in Table 18-1, compared to the European Norm, EN253.

The foam systems used for the manufacture of pre-insulated pipes are normally blown using a mixture of water and a physical-blowing agent. Currently the main physical-blowing agent used is either HCFC-141b or cyclopentane. However, all water-blown systems or cyclopentane alone or in blends with iso- and/or n-pentane are also used, the latter especially by large-scale producers. It is

Table 18-1   Typical formulations and properties for insulated pipe foams

	Discontinuous moulded	Continuous moulded	Continuous spray	EN253 Requirement
Formulation, wt-%				
Polyols	32.1	42.9	32.9	
Catalyst	0.8	1.2	1.2	
Surfactant	0.3	0.4	0.3	
Water	0.6	0.5	0.2	
Cyclo-pentane	3.5	4.3	2.3	
Polymeric MDI	62.6	50.6	63.2	
Processing conditions				
Cream time, s	34	10	5	
String time, s	180	30	43	
Free rise density, kg/m^3	32	57	45	
Properties				
Density, kg/m^3	91	75	67	> 60
Core density, kg/m^3	78	65	61	
Compressive strength, *MPa*	0.45	0.21	0.41	> 0.3
Water absorption, %	8	4.5	4	< 10
Closed-cell content, %	> 90	> 90	> 90	> 88
Axial shear strength, *MPa* 23°C	0.42	n/a	0.21	> 0.12
140°C	0.25	n/a	0.13	> 0.08
Tangential shear strength (140°C), *MPa*	0.64	n/a	0.38	> 0.20
Flexural strength, *MPa*	n/a	0.85	n/a	
Thermal conductivity, $W/(m.K)$	0.027	0.027	0.027	< 0.033
Thermal lifetime, °C/30 years	144	144	145	> 120
Mandrel flexibility, *mm*	n/a	16	n/a	

likely that smaller producers who are currently using HCFC-141b will switch to HFC-245fa or HFC-365mfc/HFC-227ea when HCFCs are phased out due to the extra cost involved in running the pentane-based systems.

Discontinuous moulding

The main application for discontinuous pipes, which consist of an inner steel pipe and a slightly shorter outer high-density polyethylene (HDPE) pipe, is the construction of large diameter pipe networks.

There are various techniques used in discontinuous production to fill the gap between the HDPE outer pipe and the steel inner; these are:

- Top filling.
- Pour-and-rise.
- Mid-point.
- Lance withdrawal.
- Pull-through filling.

These differ mainly in the orientation of the pipe and the point of foam injection. However, in all these techniques the steel inner pipe is positioned centrally in a slightly shorter HDPE casing pipe. This is achieved by fixing distance spacers around the steel pipe at regular intervals and by fitting end-caps tightly around the steel and the HDPE pipes. These end-caps have holes for foam injection and air venting. In principle, pipes of any length up to 16 metres can be used, but the standard steel pipe lengths used in the industry are 6, 12 and 16 metres. The different methods of filling discontinuous pipes are shown in Figure 18-1.

The top-filling technique is the most widely used and is illustrative of the technology. The pre-assembled pipes are normally held at an angle of 1 to 15 degrees to the horizontal. The required amount of foaming mixture is injected

Figure 18-1   Schematics of discontinuous pipe production techniques

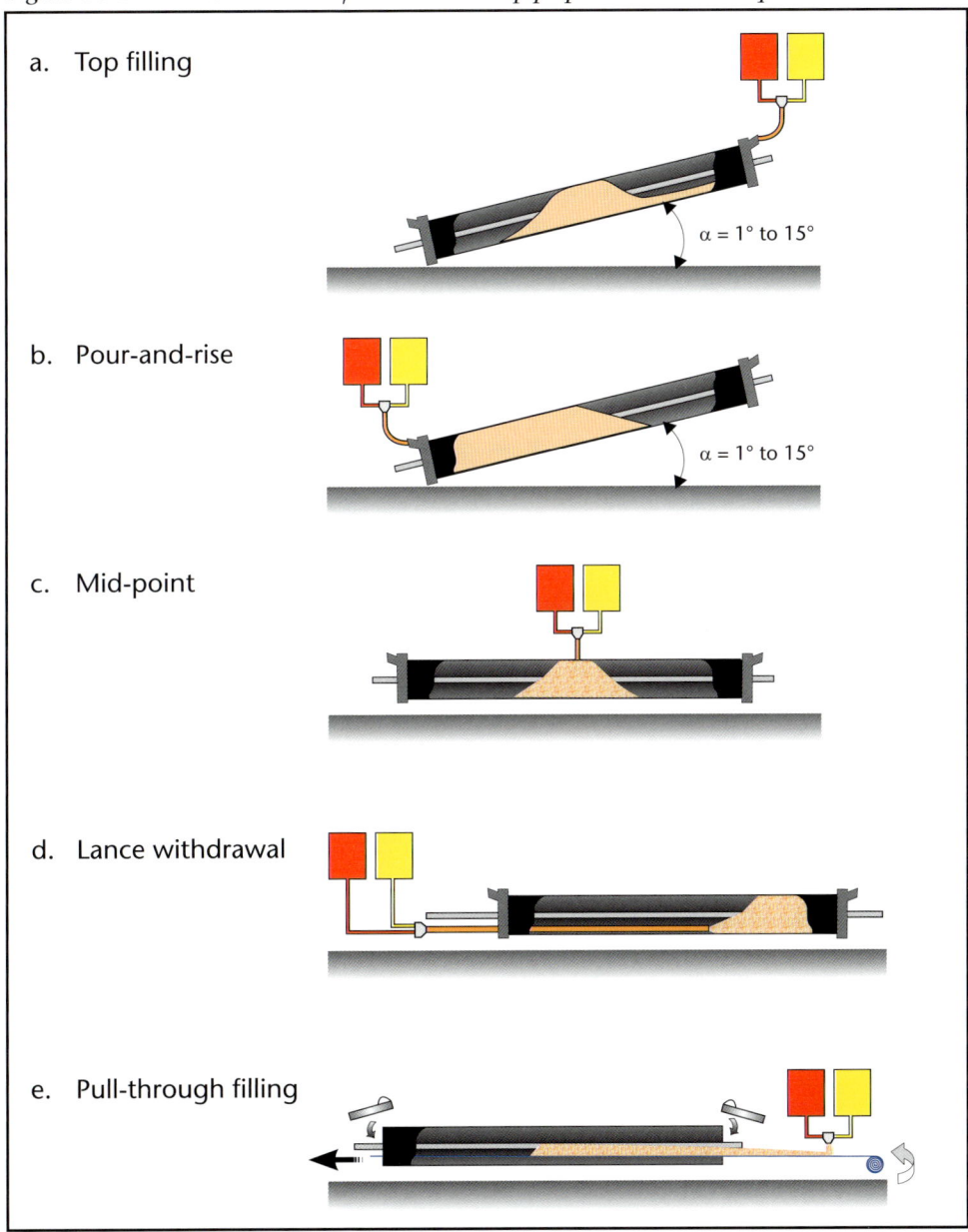

into the cavity between the steel and HDPE pipe via the hole in the end-cap at the top of the pipe assembly. Due to the force of gravity the mixture then flows down the pipe and the extent of this flow depends on the angle of the pipe assembly. This gives an initial distribution of the foam along the central area of the pipe and the foam then fills the cavity from the centre of the pipe to both ends. The optimal property distribution is obtained when the foam reaches the venting hole at the bottom end of the pipe some 20 seconds before it reaches the top end.

The initial material distribution in the pre-expansion phase reduces the path length the foam has to travel to fill the cavity completely, which leads to lower foam over pack or filling density and longer pipes can be filled more easily. A good foam distribution giving a narrow density distribution along the pipe can be achieved if the correct pipe angle is used. The key requirement from the formulation for this application is good flow, good adhesion to the inner and outer substrates, low foam thermal conductivity and thermal stability to withstand long exposure to service temperature of 130 to 150°C.

When an underground district-heating network is constructed, the 6 to 16 metre long pipes have to be joined together as they are being laid in the trench. This involves welding the steel pipes, welding a HDPE sleeve, to seal the outer HDPE casing and then filling the resulting cavity with polyurethane foam, Figure 18-2. The polyurethane components for joints are often mixed by hand, although small on-site foaming machines are also used.

Figure 18-2   Schematic of joint filling

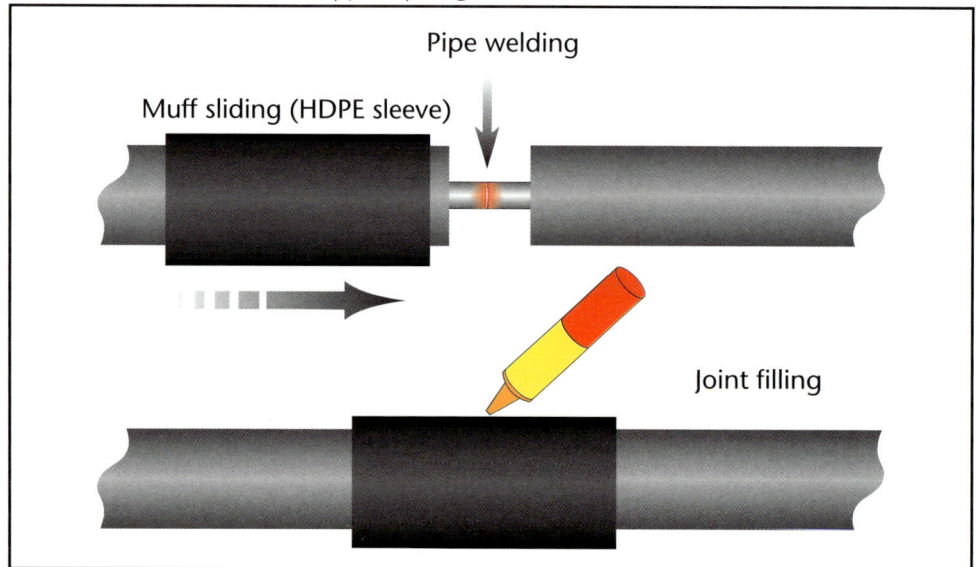

Sometimes pipe scales are used to fill a joint cavity or repair the insulation on an existing pipe network. They are produced in half-pipe shaped moulds or cut from foam blocks. In principle, the foam requirements of pipe scales are the same as for foam used in pre-insulated pipes, although there is obviously no adhesion between the pipe and foam.

Continuous moulding

This method is used exclusively for the production of small and medium diameter flexible pipes, which are normally made in lengths of several hundred metres and subsequently coiled up onto reels, as the technique imposes no restrictions on the length of pipe produced.

As flexible pipes can be laid continuously, fewer joints are required, which avoids the time consuming welding of the inner pipes and the subsequent joint filling. The trench profile for flexible pipes is usually narrower, since access for welding is not required and therefore excavation costs are lower. For the majority of flexible pipes, cross-linked polyethylene (PEX) inner pipes are used.

An important outlet for flexible pipes is the connection of district heating to individual houses. Other applications include drinking and waste water pipes in industry and agriculture, refrigeration plants, swimming pools and many other special fields.

In order to prevent the loss of integrity during coiling-up of the pipes after production and during the installation, a high degree of foam flexibility is necessary. However, the foam should also provide good mechanical properties and a high thermal resistance when used in district heating.

In the continuous moulding technique the reacting mixture is continuously laid down on a moving polyethylene film under the inner pipe. The film and rising foam is wrapped around the inner pipe and then pulled into a circular, temperature controlled, moulding section where a HDPE pipe is extruded around the cured foam and the finished pipe cut to the required length, Figure 18-3.

Figure 18-3   Schematic of continuous pipe moulding

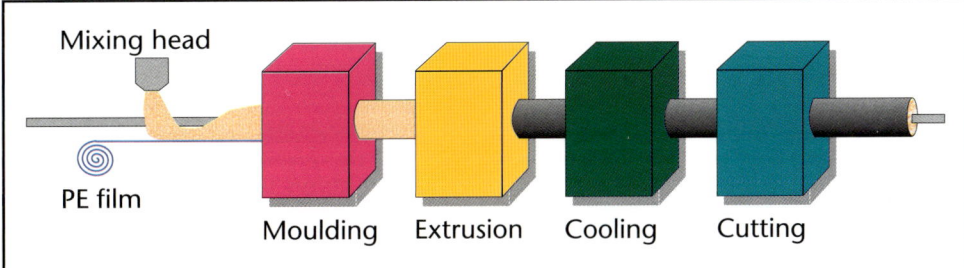

Continuous spray

The continuous spray technique is used for the production of medium to large diameter pipe and the benefits of this technology are the ability to produce pipes of any desired length at a low foam density leading to cost reduction. This route usually requires thinner outer pipe casings, which offer further cost benefits to the pipe contractor.

Medium to large diameter continuous spray pipes are often chosen for sub-sea and oil pipeline applications as the foam insulation and outer coating can be applied under difficult field conditions.

The reacting foam mixture is sprayed on the outside of the rotating pipe and very uniform foam is created over an extremely short path. By spraying one or several layers it is possible to build up any insulation thickness required and then the HDPE casing pipe can be extruded or wound around the insulation. The HDPE casing pipe can be thinner since it does not have to withstand the high foam pressure that occurs during conventional pipe filling thus saving cost. Alternatively, a polyurea coating can be applied as the outer casing again using a spray technique. The spray technique is particularly suited for large diameter pipes such as those used in under-sea oil pipelines.

Spray insulation

There is an increasing focus on reducing the heat loss from buildings and the codes of practice for the insulation of new buildings have been made more stringent in many countries. However, this leaves the problem of how to improve the insulation performance of existing buildings, both residential and commercial. The use of spray-applied rigid polyurethane foams is now seen as an important part of the solution as it provides a technique that can be applied in-situ.

The heat transfer through the building envelope is due to several factors, Figure 18-4, and the use of sprayed rigid polyurethane foams can reduce the heat loss in about 60 per cent of these areas.

As well as using spray foam in retrofitting/refurbishing roofs, walls, floors and windows of existing buildings it is also now finding increasing applications in new constructions such as commercial offices, industrial factories and warehouses, agricultural pig and chicken farms, and residential apartments and house buildings, again being used in the roofs, walls and floors. The technique is also utilised for the insulation of new and existing pipes, storage tanks and cold storage units.

Sprayed rigid polyurethane foams are closed-celled, air tight, resistant to mildew and fungal attack, provide no food value to rodents and have good vapour barrier properties. They are also durable with no settlement or thinning and can be used over a wide range of temperatures. This mix of properties provides a long service life.

Figure 18-4   *Distribution of heat loss in a typical medium sized building*

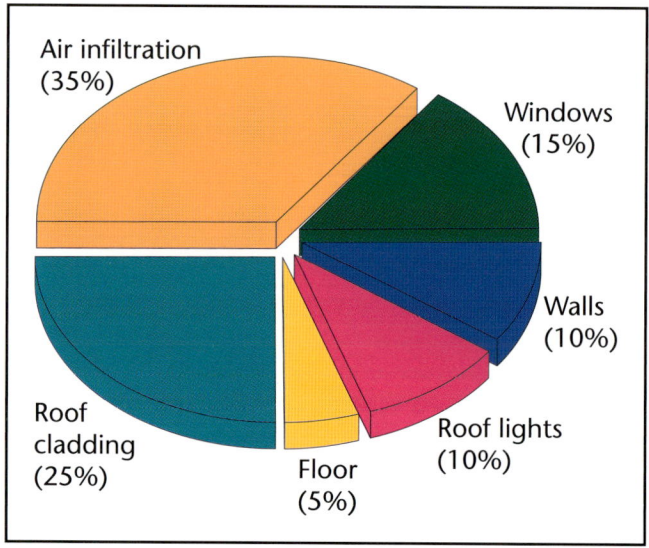

Spray technology combines polyurethane's insulation properties with ease of application and its unique adhesive ability to provide in-situ insulation on a variety of surfaces. Sprayed polyurethane foam also conforms easily to surface irregularities and projections such as pipes and conduits.

Sprayed foams provide a continuous covering on irregular surfaces, which leads to excellent crack and seam filling properties. This makes them an ideal solution for sealing gaps in new buildings to eliminate air infiltration, Figure 18-5, and also for gap filling problems in existing buildings.

The self-adhesive nature of sprayed foams means that there are no fasteners to act as heat bridges and the foam will adhere to many common building materials such as PVC, ABS, metal or wood to form strong composites. This adhesive characteristic adds to the structural strength of buildings when sprayed foams are used to coat walls and roofs, Figure 18-6.

The adaptability of the spray foam formulations, combined with the range of densities suitable for different applications, provides versatility for on-site application. It is possible to apply various thicknesses and densities depending on the insulation requirements. Manufacturing the insulation on-site also provides major savings in handling and labour costs since equipment is compact and mobile.

Those involved in the spraying process should be well trained with the appropriate certification in those countries where such schemes exist. Appropriate secure chemical storage on-site must be ensured and adequate personal protective equipment worn by operators. For external application weather conditions, especially wind direction, must be considered.

Figure 18-5
Spray between windows and trim

Figure 18-6
Coating a Stell dome roof

The need for in-situ application using hand-held pressurised spray guns, with polyol to isocyanate volume ratio fixed at 1:1, in a variety of climates imposes some unique requirements on the formulation and blowing agent. The systems need to have a high reactivity, to prevent dripping and the viscosity of the two streams should be below 2,000 mPa.s at the preferred line temperature of 38 to 45°C to achieve adequate atomisation. HCFC-141b is currently being used by all system suppliers in North America, Japan and by most European suppliers. CFC-11 is still being used in many (Article 5(1) countries) and HCFC-141b is expected to be the replacement blowing agent in these cases. A typical formulation and properties are given in Table 18-2.

Table 18-2   Typical formulation and properties for spray insulation foam

Formulation, wt-%		Processing conditions	
Aromatic amine polyol blend	23	Reactivity details (hand mix)	
Aromatic polyester polyol	10	Cream time, s	3 – 5
T.C.C.P. (fire retardant)	5	Rise time, s	20 – 30
Catalyst/Surfactant/Water blend	1.5	Tack-free time, s	8 – 12
Blowing agent: HCFC-141b	8.5	Free rise core density, kg/m^3	26 – 28
Polymeric MDI	52		
		Properties	
		Applied core density, kg/m^3	35
		Initial thermal conductivity, $W/(m.K)$	0.02
		Compressive strength, kPa	180
		Tensile strength, kPa	220
		Closed-cell content, %	92
		Water absorption, %	1
		Dimensional stability, vol-%	
		-20°C	0.5
		100°C	5
		70°C, 90% RH	3

Though many other blowing agent options have been evaluated, none appears to provide an acceptable solution for wide commercial use in major markets. Liquid HFCs, especially HFC-245fa, have been extensively evaluated, but high cost remains an issue. In an effort to reduce cost, HFC-245fa in combination with water (50 per cent) is being evaluated in North American roofing and Japanese wall applications. The system uses existing shipping containers and existing spray equipment. In Europe, blends of HFC-365mfc with HFC-227ea in weight ratio of 94:6 are being evaluated in addition to HFC-245fa. All water-blown systems have been evaluated especially in the Japanese market.

New developments are bringing the performance of such systems closer to that of the HCFC-141b, but there are still penalties regarding density and thermal conductivity. Though partially open-celled foam has been developed to offset the density penalty of all water-blown systems, its commercial viability remains in question due to the fact that many building codes require foams to be >90 per cent closed-celled.

Good adhesion between the substrate and the spray insulation is extremely important so all substrates should be clean, dry, and free of grease, oil, loose material or rust. Substrates that contain solvents should be avoided, but if a primer is used then adequate drying time should be allowed prior to application of the insulation. Each substrate needs to be prepared in accordance with the requirements of the spray insulation system.

The chemical components are mixed and sprayed using commercially available, fixed-ratio, and positive displacement pump machines, fitted with spray guns. A high (30 kilogrammes per minute) output system would be used for a roofing job whilst a small (10 kilogrammes per minute) output type is more suitable for stud wall insulation.

The equipment consists of proportioning pumps, either air or hydraulic driven, that transfer material from the individual drums of chemical components to a heat exchanger, raising the temperature to typically 40 to 70°C to lower the viscosity. This is necessary as the hoses from the pump to the spray head, that are also heated, can be up to 100 metres long. Most spray guns now use impingement mixing, so no solvent flushing is required. During spraying the spray gun should always be held perpendicular to the substrate being sprayed and an ideal thickness of foam, for both insulation and application reasons, is 25 to 40 millimetres.

The spray equipment is compact and many contractors have fitted out vehicles with the spray machine, hoses, tools, a compressor and the chemicals, normally in 200-litre drums, to make it easy to travel from one job site to another. Prior to starting a job it is advisable to spray a test pattern to ensure that proper equipment settings have been selected.

In interior installations, the exposed side of the sprayed polyurethane needs to be covered by a thermal barrier, such as one centimetre of mortar, according to applicable national and local codes whilst the long-term success of any external polyurethane foam is dependent upon the quality and integrity of its coating. Sprayed polyurethane foam must be protected to prevent surface degradation due to UV and atmospheric exposure, with SBS, butyl rubber or acrylic coatings being commonly used.

One-component foam (OCF)

OCF is a self-expanding, self-adhesive, moisture-curing gap filler with the main advantages being its portability and ease of application. OCFs are supplied to the building and do-it-yourself industries in pressurised cans fitted with dispensing nozzles. On dispensing, the mixture from the can expands to form a sticky froth, which will adhere to most surfaces without pre-treatment. The froth reacts with atmospheric moisture and expands to form semi-rigid polyurethane foam with good flexibility and with approximately 80 per cent closed cells. This expansion is useful in ensuring that the curing foam tightly fits the joint or gap being filled. When completely cured the foam can be cut, plastered and painted. Adhesion to damp surfaces, as well as to most building materials, is excellent.

The technology of OCF involves the manufacture of a prepolymer, made of a special grade of MDI, which is delivered in a pressurised can together with flame-retardants, plasticisers, catalyst, surfactants, propellants and other additives.

For cost reasons the European market has completely changed to hydrocarbon propellants, propane/butane in combination with dimethylether (DME), with the exception of the special fire retardant B2 foams where there is no cost advantage and HFC-134a or HFC-152a are still commonly used. In the Americas OCF formulations are still based on HCFC-22, but due to environmental legislations this will be phased out and replaced by the European options. Concerns about safety and the release of volatile organic compounds from the use of hydrocarbons have led to HFC-152a being considered and lately the use of gases such as carbon dioxide and nitrous oxide has also been proposed.

OCF is now widely used for on-site filling of joints, for example around window and doorframes and as a substitute for caulking. It is also used against air leaks and vapour diffusion, for insulation of sheet metal ductwork in ventilation systems and as a sealant for pipes through walls or other barriers. In industrial applications EHS requirements are similar to spray.

Water heaters

The major benefit of applying rigid polyurethane foam to water heaters (hot water cylinders) is in energy reduction, but it also provides a structural reinforcement as well. The water heater market is split into three main categories based on the heating method – electrical, gas or shell. These primary categories are then sub-divided by other manufacturing differences such as: installation style, wall or floor standing; orientation, vertical or horizontal; and shape, round, square or polygonal. The broad range of designs possible from the choice of options obviously has an impact on the application and performance of the rigid polyurethane foams that continue to dominate this market. The insulation of choice in the water heater industry for many years was fibreglass, but due to the energy conservation policies of the 1980s, it was replaced by CFC-11 blown rigid polyurethane foams.

The formulations were changed in the 1990s to eliminate CFC-11 blowing agent, but in order to meet the emerging energy standards, effective in the USA from 2004, water heaters foamed with all water, HFC-245fa, HFC-134a and pentanes have been compared for performance and cost. It appears that cyclopentane alone or in mixtures with lower boiling pentanes is the most cost-effective solution and will be the blowing agent of choice for larger manufacturers. HFC-245fa, HFC-134a and HFC-365mfc, each co-blown with large amounts of water are likely to be used by smaller manufacturers. Even though closed-cell all-water-blown foams have been found to meet the existing energy standards, their use is limited due to the relatively higher density obtained.

Insulation foams for water heaters must meet a number of physical and mechanical characteristics in order to satisfy the required performance standards. The most common requirements are low foaming pressure – 0.3 bar maximum due to the thin metal of the casing – good vertical flow, even density distribution, dimensional stability, compressive strength and thermal efficiency.

Other niche applications

Buoyancy

Rigid polyurethane foams are widely used in applications where flotation is required. The use of a low-density foam core provides the buoyancy, the water resistance and also provides strength in a composite structure. For flotation devices the inner and outer structures are usually preformed and assembled, then a relatively slow reacting MDI-based foam system with good flow properties is injected into the cavity to bond the inner and outer hulls. The choice of final density and properties of the foam system depends on the application and the stiffness required for the final composite structure. Most flotation foams currently utilise a blowing technology based on a mix of water and HCFC-22. However, the technology is changing to HFC-134a for pressurised dispensing systems and all water for conventional equipment.

Surfboards

Customised surfboards and windsurfing boards can be made by the process described above, but are usually made via a moulding/hand lay-up process generally using TDI-based foam rather than MDI as the white colour of the foam enables board designers greater design freedom and colour choice in the final board.

Aircraft propellers/windsails

Composite structures based on carbon-fibre reinforced foam cores provide strong, lightweight yet rigid structures and a three metre diameter aircraft propeller can show a weight saving of up to 60 kilogrammes when compared with the same product produced from conventional aluminium alloy.

Floral foam applications

Low-density rigid polyurethane foams are often used as a substrate for holding dry and silk flowers. The specially-formulated foam systems are usually made in large blocks or buns and the foam is cut and formed to the desired shape. Although most of these foams are blown with HCFC-141b, the phase-out of these products has seen a movement to a combination of water and hydrocarbon blowing. The use of water as the sole blowing agent has not been used as problems of foam scorch can occur at high block heights. The production of hydrophilic rigid polyurethane foam for floral applications has only been used for certain niche markets as the use of phenolic foams is more widely accepted on cost grounds. However, it is expected that the use of polyurethane may grow as the cost rises for the phenolics, due to their need to meet more stringent legislation on volatile organic compounds and landfill disposal.

Wall, soil and mine stabilisation

The excellent adhesive properties of in-situ applied low-density polyurethane foam has led to many applications such as the renovation of sewers, replacement of cavity wall ties and soil stabilisation. The use of foam pillows to protect pipelines has also been very successful.

Low-density semi-structural foams

Low-density rigid foams can be moulded to form complex shapes and structures, which find many uses in the furniture industry, where the freedom to design complex products economically makes them attractive to use.

High-density structural foams

High-density rigid foams are usually defined as having densities between 250 to 1,200 kg/m^3 and usually have a dense outer skin and lower density foam core, forming a stiff sandwich structure. Applications for these foams vary from automotive components to office equipment housings and the technology is used for making low volume, complex parts where the cost of thermoplastic tooling would be prohibitive.

19. Introduction to elastomers

Paul Mackey

The term 'polyurethane elastomers' covers a broad class of materials that can be distinguished from other polyurethanes by their elastic nature and density. Elastomers are similar to flexible foam in that they are rubbery materials that will resume their original shape after being deformed, but they are typically much more dense ranging from 200 kg/m^3 for foamed materials to densities greater than 1,000 kg/m^3 for solid materials.

The various polyurethane elastomers offer a broad range of processing characteristics, hardnesses and compositions depending on the specific processing and application requirements. The key areas can be defined as:

Thermoplastic polyurethanes
Available as granules or pellets suitable for conventional thermoplastic processing such as extrusion, calendaring or injection moulding.

Microcellular elastomers
Made by low-pressure mixing of liquid components into a mould to form a foamed article with applications ranging from footwear soling materials to automotive suspension components and steering wheels.

Cast elastomers
Made by low-pressure mixing and casting the liquid components into a mould to form a solid article with typical applications being high performance industrial wheels, seals, industrial rolls and skate board wheels.

RIM elastomers
Produced by high-pressure impingement mixing and injection into a closed mould with typical applications as automotive bumper fascias, body panels plus body components for a variety of other vehicles.

Spray elastomers
Made by high-pressure impingement mixing and spraying on to a substrate where the key applications are industrial protective coatings for bridges, pipes or linings for transport containers.

Elastomeric fibres
These are made by solution spinning directly from the polymerisation mixture.

Coatings
Various one- and two-component systems are available for making elastomeric coatings for textiles and other flexible substrates resulting in a wide variety of synthetic leather products.

Millable polyurethane gums
Available as slabs or crumb that are processed by the traditional methods used in the rubber industry.

The hardness range for polyurethane elastomers can vary from as low as 10 Shore A to greater than 75 Shore D. In general, the fully cured elastomers are tough, abrasion-resistant materials of high mechanical strength having good resistance to most solvents and chemicals. Only strong bases and acids, oxidising agents and a few strongly polar solvents affect them.

Polyurethane elastomers are typically produced by reacting an isocyanate, a high molecular weight polyol and a low molecular weight diol or amine chain extender. The major demand is met by the aromatic isocyanates such as MDI, TDI, or NDI, but during the last decade aliphatic isocyanates such as H_{12}MDI, IPDI, HDI, XDI and TMXDI have been increasingly used where transparency and light stability are important characteristics.

A variety of high molecular weight polyols are used such as adipate polyester polyols, propylene oxide and propylene oxide/ethylene oxide polyethers, polytetramethylene glycol polyethers, polycaprolactones, other specialised polyols and for spray and RIM applications amine-tipped polyols. These polyol materials all have the characteristic that they are 'soft' or flexible at ambient temperatures and as such are often referred to as the 'soft block'.

Glycols such as ethylene glycol, 1,4-butane diol and bis(hydroxyethyl) hydroquinone (HQEE) are the main chain extenders with amines such as DETDA used for the highly reactive spray and RIM systems and for certain cast elastomers.

Theory of polyurethane elastomers

The general class of materials that includes natural and synthetic rubbers, thermoplastic elastomers as well as polyurethane elastomers, all share a common mechanical behaviour in that they can be stretched many times their original dimensions and will then elastically recover to their initial shape when released. These elastomeric materials are able to perform in this manner because of certain common structural characteristics.

Their structure consists of long chain molecules that take on the conformation of a large random coil, much like a tangled mass of string. These long molecules are soft and flexible at room temperature and they will deform or uncoil when force is applied. The materials are interconnected or cross-linked in ways that allow them to return to their original shape once the deforming force is removed. The interconnectivity of the polymer chains is the key to this behaviour. Without such a feature the chains would simply flow when deformed. The polymer chains can be connected by chemical bonds, typical of thermoset materials or physical intermolecular interactions, typical of thermoplastic materials.

Polyurethane elastomers represent a special class of elastomer in that they can utilise both mechanisms of cross-linking, chemical and physical. Reactive systems typically use a combination of chemical and physical cross-links while thermoplastic materials rely almost exclusively on physical cross-links. The isocyanate and chain extender react to form a rigid or 'hard' sequence that is commonly referred to as the 'hard block'. These isocyanate-tipped hard blocks are linked through the polyol soft blocks to form the final polymer structures that can be described as a segmented block copolymer with the general structure $(AB)_n$, shown in Figure 19-1, having alternating hard and soft segments. These alternating hard and soft segments are the key feature that distinguishes polyurethane elastomers from other elastomer materials and is the primary reason that these materials have such good mechanical properties over a wide temperature range.

Figure 19-1   *Schematic of segmented block polymers and hard block chain packing*

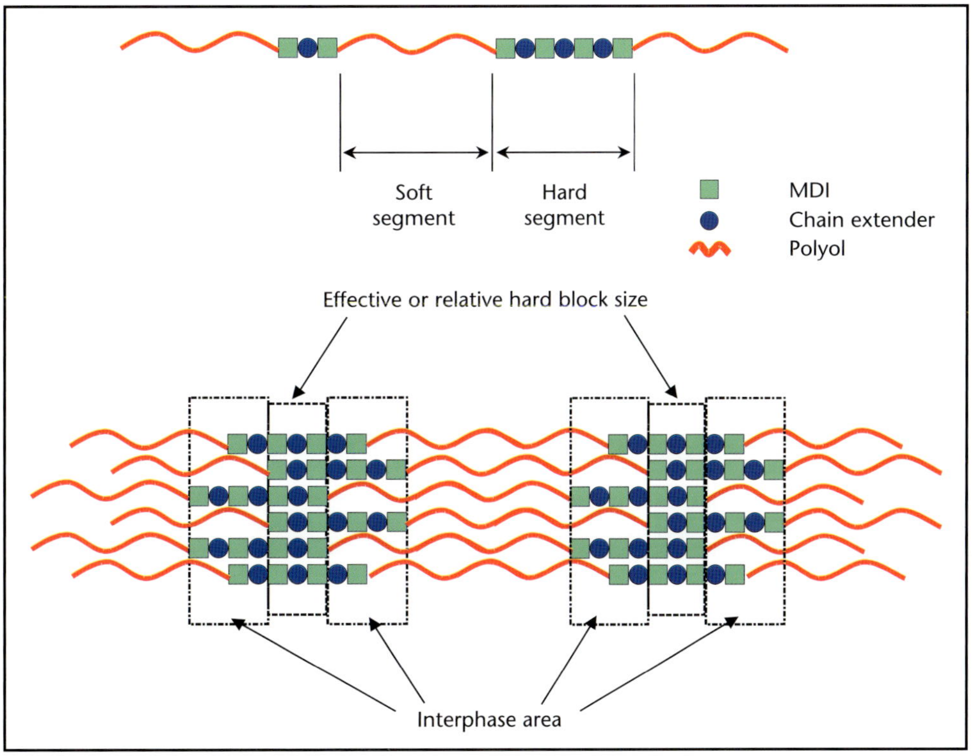

The chemical cross-links in the polyurethane elastomers are the result of using polyfunctional isocyanates or polyols whilst the physical cross-links are derived from the thermodynamic incompatibility of the relatively non-polar polyol or soft block and the polar urethane sequences or hard blocks. These incompatible sequences segregate into micro domains of soft and hard phases where the hard block domains function as multifunctional cross-links.

Within the hard block domain a secondary structure, or morphology, is possible depending on the degree of incompatibility of the hard blocks with the polyol and the affinity between the hard segments. Hard blocks organise by stacking and in some instances hard blocks develop sufficient order and packing density resulting in the formation of microcrystals, as shown in Figure 19-1, but in most instances the hard blocks do not form an ideal domain structure. Some portions of the hard block are found outside or at the interface of the micro domain. This material is commonly referred to as the interphase that is neither soft nor hard and as a result the hard block domain has an effective size that is smaller than the overall domain structure.

The hardness of an elastomer is thus dependent on the degree of phase separation, so for a given isocyanate level in a formulation a well-ordered hard block will produce a higher hardness than a structure with a high level of interphase material. Conversely, a high level of interphase material requires a higher percentage of isocyanate to reach an equivalent hardness level.

The hard block of an elastomer and its ability to phase separate is based both on its structure and the weight fraction of hard block in the elastomer. Since the isocyanate and the chain extender combine to form the hard block structure, the effect of the contribution of these components will be discussed seperately.

The affinity between hard segments depends to a great degree on the symmetry of the isocyanate and if an isocyanate is symmetrical and planar it has a greater capacity to pack tightly together, developing strong intermolecular interactions. The structures of NDI, MDI and TDI, with TDI being a mixture of the 2,4 and 2,6 isomers are shown in Figure 19-2.

Figure 19-2   Structure of NDI, MDI and TDI

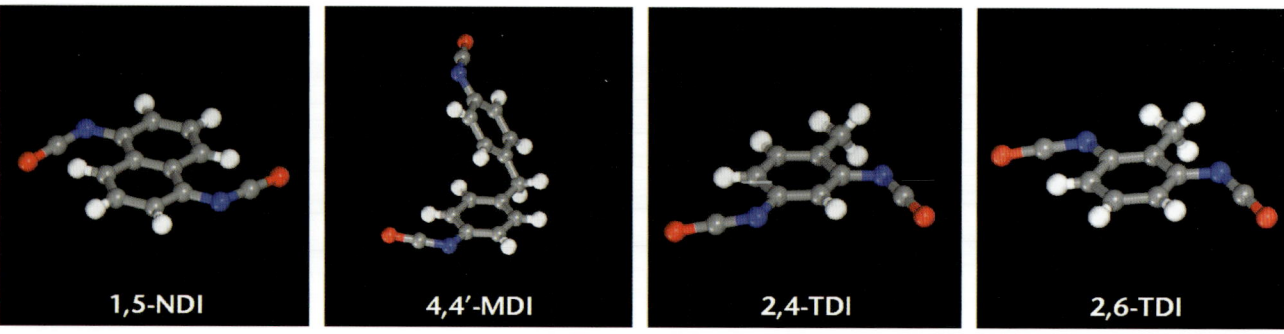

NDI is both planar and symmetric about a central axis whilst 4,4'-MDI is not planar, it is bent at the methylene bridge and slightly twisted, yet it is symmetric about an axis through the methylene bridge. TDI is planar, but only the 2,6 isomer is symmetric about a central plane, the 2,4 isomer is not, so because a mix of the isomers is always present with TDI a packing disruption is created. These observations would indicate that for a particular elastomer formulation the order in which the elastomers would form efficient phase separated structures with greater intermolecular interactions would be NDI > MDI > TDI. These observations are supported by the sinter/soften temperatures and melting points of hard blocks derived from these respective isocyanates and butane diol as shown in Table 19-1.

Table 19-1
Sinter/soften temperatures and melting points of hard blocks

	Sinter/soften temperature (°C)	Melting point (°C)
NDI	260	320
MDI	120	230
TDI	78	220

An example of how the isocyanate structure can influence the phase separation of an elastomer is shown in Figure 19-3, which compares the dynamic mechanical thermal analysis (DMTA) plots of solid NDI and MDI elastomers both with a Shore A hardness of 92 and based on the same polyester polyol. The NDI elastomer maintains a constant modulus over a broad temperature range indicating a stable, well defined, hard domain whilst the MDI elastomer modulus decreases rapidly with increasing temperature indicating the presence of a greater fraction of interphase material. Since the MDI elastomer has a greater degree of interphase material, its formulation requires a higher percentage of isocyanate, a larger more diffuse hard block, to match the hardness of the NDI elastomer.

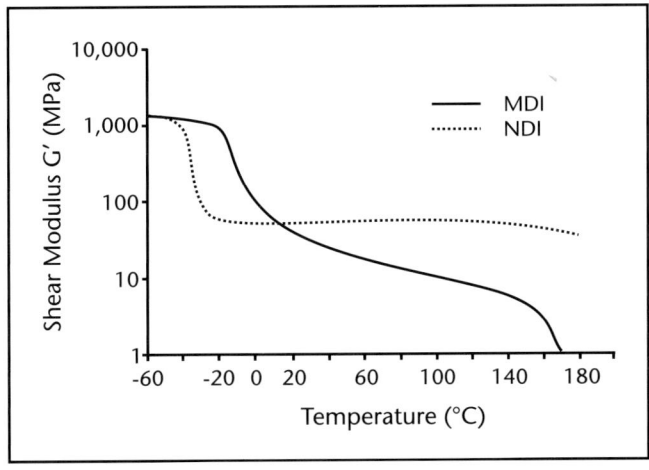

Figure 19-3   Dynamic mechanical thermal analysis of NDI and MDI elastomers with a Shore A hardness of 92 based on polyester polyol

The chain extender also has an impact on the structure of the hard segment. With regard to glycols HQEE yields the highest melting hard blocks followed by linear aliphatic glycols with an even number of carbon atoms. In these instances the hard segment domains consist of aligned chains, which are staggered to form straight hydrogen bonds between adjacent urethane groups in three dimensions. In order to achieve the same degree of hydrogen bonding with glycols having an odd number of carbon atoms the chains have to contract into a higher energy conformation. Hard segments derived from low molecular weight diamines or water form urea structures that are more polar than urethane increasing the driving force for segregation. Ureas can actually form bidentate hydrogen bonds that have melting points in excess of the decomposition temperature of the polymer: as a result they cannot be re-melted and further processing has to be through a solution technique.

As the weight fraction of hard block in an elastomer increases the hard block sequences become longer increasing phase separation and improving mechanical properties. The affect of varying the hard block content, from 20 to 40 per cent on the split tear and tensile strength for a series of microcellular elastomers based on MDI and a polyester polyol is shown in Figure 19-4.

Figure 19-4 *Split tear and tensile strength for a series of MDI polyester polyol microcellular elastomers versus hard block content*

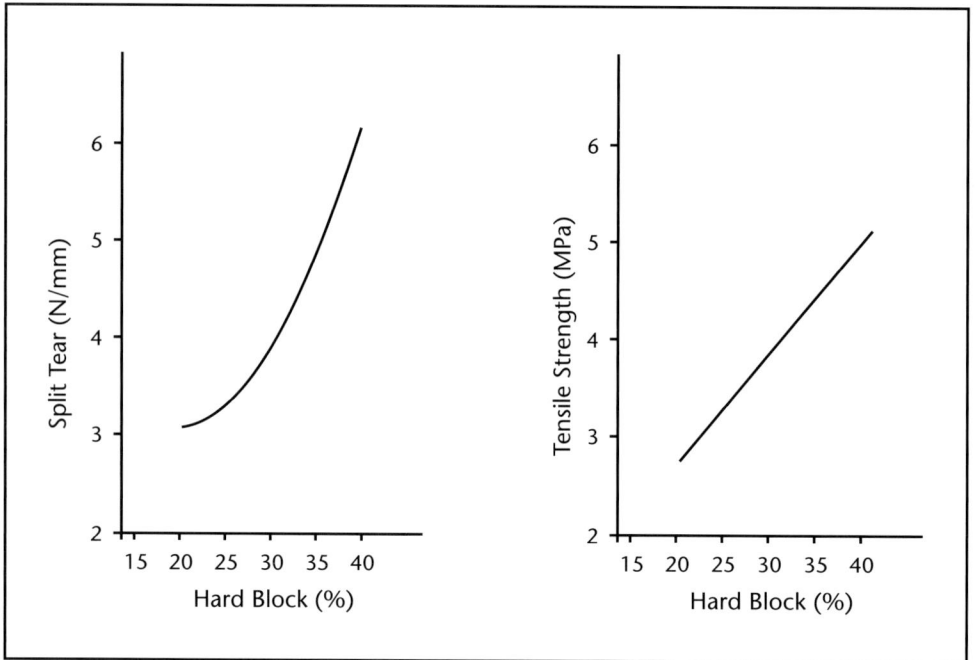

Although the relationship between mechanical properties and hard block content is essentially linear the dynamic properties are affected in a more complex way, as shown in Figure 19-5. The soft elastomers have the highest resilience, however, there appears to be an optimum level of hard block for the lowest hysteresis. Apparently, a specific amount of hard block is required for an elastomer to resist permanent deformation in a hysteresis test where the sample is compressed to 50 per cent of its original thickness and then allowed to recover.

Figure 19-5 *Resilience and hysteresis for a series of MDI polyester polyol microcellular elastomers versus hard block content*

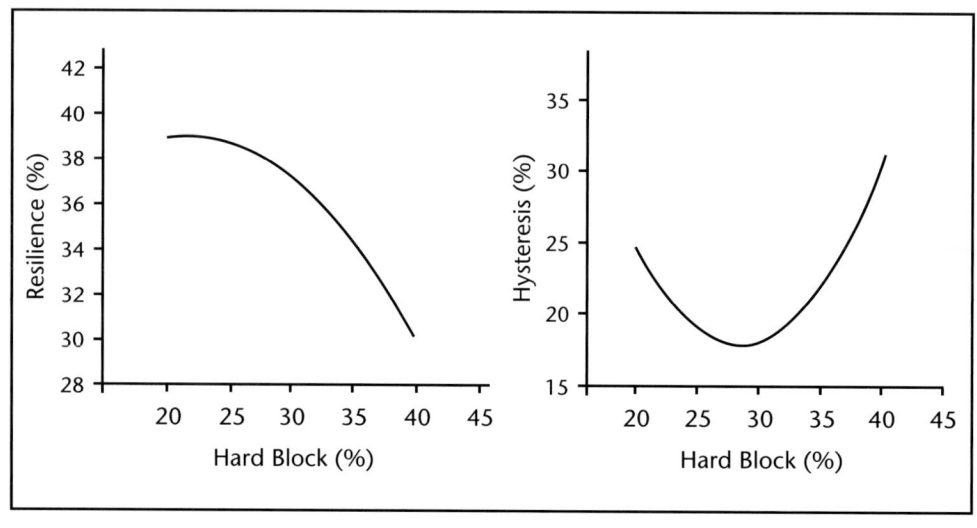

The effect of increasing the hard block content on both mechanical and dynamic properties can be investigated simultaneously using DMTA. Storage modulus and tan δ versus temperature for a series of microcellular elastomers, based on MDI and a polyester polyol, with increasing hard block content are shown in Figures 19-6 and 19-7 respectively. For temperatures above the glass transition temperature, Tg, about 25°C, the storage modulus increases for elastomers with a higher hard block content indicating increasing mechanical properties, however, as the hard block increases, the shape of the curve also changes. The storage modulus curve for the elastomer with 30 per cent hard block shows a distinct glass transition and a rubbery plateau while the curves for the elastomers with 40 and 50 per cent hard block show an increasingly broad glass transition and a steeper slope with increasing temperature. The broadening of the glass transition temperature is also evident in the tan δ where the peak shifts to higher temperatures and the curve becomes increasingly asymmetric. The broadening of the glass transition shown in these diagrams indicates an increase in the interphase between the hard block and soft block domains.

Figure 19-6
Storage modulus vs temperature for MDI polyester polyol microcellular elastomers with increasing hard block content

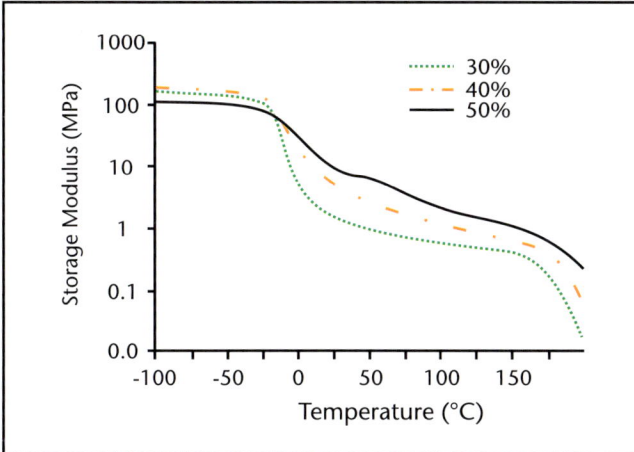

The effect hard block content has on the properties of an elastomer should be viewed from two perspectives. The effect on mechanical properties is essentially linear and is directly related to the volume fraction of hard and soft domains. The effect on dynamic properties is complex because these properties are dependent on the structure of the hard block domain. Optimal dynamic properties for a given elastomer composition are achieved where the hard block level is designed so the hard block domains are both efficiently organised and of sufficient mass to resist permanent deformation.

Figure 19-7
Tan δ vs temperature for MDI polyester polyol microcellular elastomers with increasing hard block content

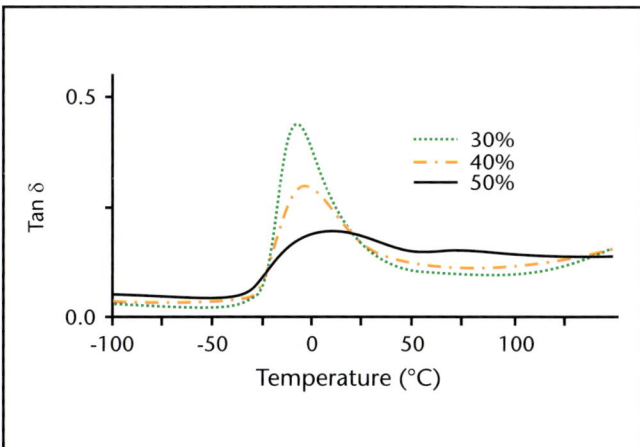

In order to obtain the ultimate dynamic properties of a polyurethane elastomer the material needs to be annealed and the most effective annealing temperature is generally just below the melting point of the hard blocks. Properties such as resilience, compression set and low temperature flexibility can all be improved with annealing. These improvements are obtained because there is a reduction in the amount of interphase between the hard and soft block domains. Since the interphase is a mixture of hard and soft blocks the material is able to rearrange and separate, becoming either part of an increasingly well structured hard phase or an amorphous soft phase as shown in Figure 19-8, with an increase, d_1 to d_2, in the size of the hard block domain.

Storage modulus (SM) is the modulus of the elastic portion of a material (stored energy).
Loss modulus (LM) is the modulus of the viscous portion of a material (energy lost)

$$\text{Tan } \delta = \frac{\text{LM}}{\text{SM}}$$

Figure 19-8   Schematic of a hard domain before and after annealing

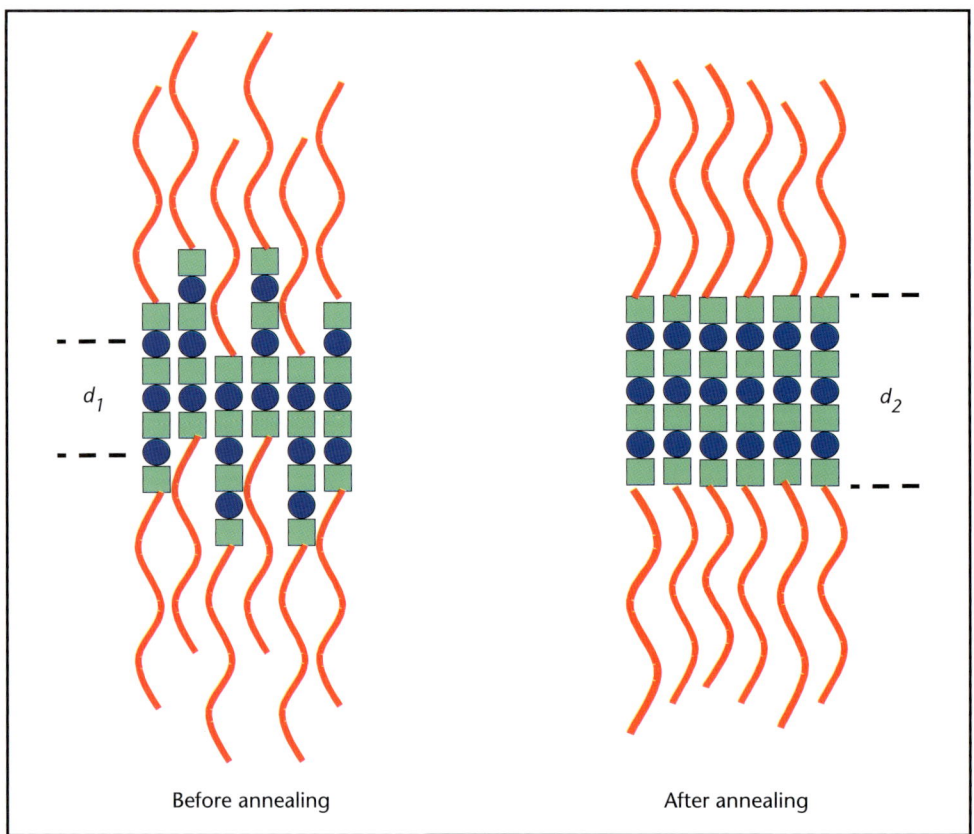

The structure of the polyol, the soft segment, affects polyurethane elastomers in several ways starting with its influence on the degree and type of hard segment aggregation. This has a direct effect on the dynamic and mechanical properties of the elastomers, which are also influenced by the inherent characteristics of the polyol structure. These effects can be summarised by a generalised comparative overview of the more important type of polyol materials and their respective effects on a variety of performance criteria as shown in Table 19-2.

Table 19-2   Properties of elastomers related to polyol structure

	Polyester	Polyether	Polyether	Polycarbonate
	Adipate or caprolactone	PO/EO	THF	Hexylene carbonate
Tensile strength	Excellent	Fair	Fair	Good
Tear strength	Excellent	Fair	Fair	Good
Rebound resilience	Excellent/Good	Good	Excellent	Good
Low temp. impact resistance	Good/Fair	Fair	Excellent	Good
Abrasion resistance	Excellent	Fair	Fair	Good
Weathering	Good	Fair	Fair	Good
Hydrolysis	Fair	Excellent	Excellent	Good
Microbial resistance	Good/Poor	Excellent	Excellent	Excellent
Solvent resistance grease, oil, water	Good	Fair	Fair	Good

Polyesters, in general, have superior mechanical properties whilst polyethers have better hydrolysis and microbial resistance plus good low temperature flexibility; polycarbonates, often referred to as a 'special ester', offer a compromise between the two.

The Tg of the soft flexible polyol chain limits the lower useful operating temperature for an elastomer with polyethers generally having the lowest values. The molecular weight of the polyol also has an effect on the Tg with the value decreasing as the molecular weight of the polyol increases, thus improving the dynamic performance of the elastomer particularly at low temperatures.

Polyethers, in general, are less compatible with MDI than polyesters and elastomers derived from them show a higher degree of phase separation. The polyurethane hard segment domains also tend to be larger and more complex than those found in polymers based on adipate polyesters. The latter usually contain many small, evenly dispersed hard block domains although over 50 per cent of the hard segments remain in the soft segment domains. Increasing the hydrocarbon length in both polyesters and polyethers will usually increase the tendency for phase separation because the polarity of the polyol decreases compared with the urethane hard segment.

Increasing the molecular weight of a polyol also favours phase separation as the thermodynamic compatibility of two polymers is inversely related to their molecular weight. As polymer molecules become larger they become less compatible. This is actually a compound effect since as the molecular weight of the polyol increases then for an elastomer formulation with a defined hard block content the sequence length of the hard block also increases. Therefore, incompatibility is not only driven by the increasing molecular weight of the polyol it is simultaneously driven by the commensurate increase in molecular weight of the hard block. The molecular weight distribution of the polyol will also have an effect and the wider the distribution the more interphase region is formed.

The effect of increasing molecular weight is shown in Figure 19-9 for a microcellular elastomer, based on MDI and a polyester polyol, at a constant hard block content. The compression hardness passes through a maximum, which shows that the mechanical properties improve with increasing molecular weight of the polyol. This effect is similar to that of increasing the overall hard block content, except that in this instance only the sequence length of the hard block is increasing.

This indicates that there is a specific hard block sequence length for this formulation that facilitates the efficient segregation of the hard segments into domains. Below some critical point the hard sequences are more dispersed in the polyol soft phase and thus do not contribute effectively to mechanical properties. Above this point the hard block sequences become too long to organise efficiently, most likely trapping soft block in the hard block domains.

Figure 19-9   *Effect of molecular weight on compression hardness, resilience, and hysteresis for an MDI polyester polyol microcellular elastomer*

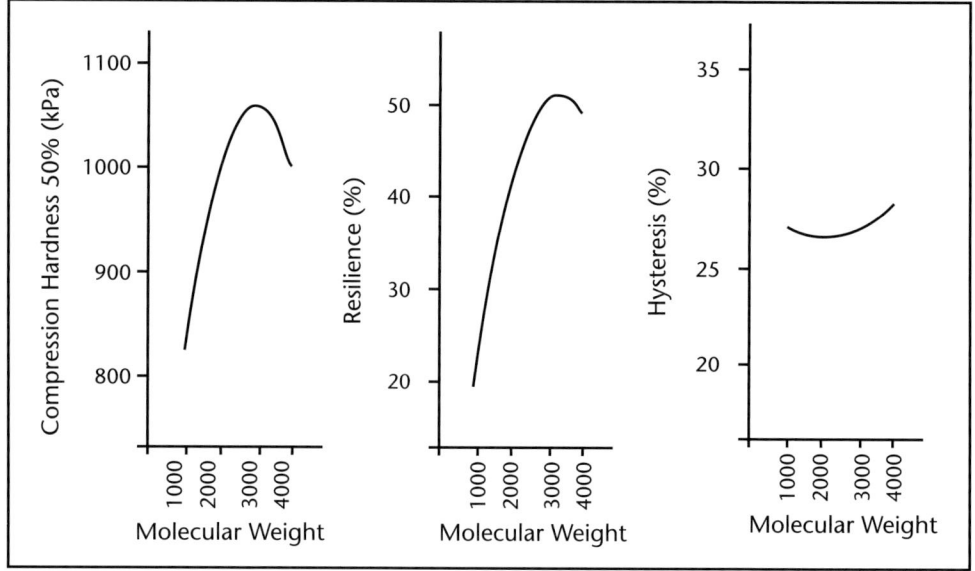

The resilience of the elastomer follows a similar trend. The resilience improves as the optimal sequence length for an efficiently organised hard block domain is approached and decreases as it is exceeded. The hysteresis of the elastomer passes through a minimum, as was noted earlier, for elastomers with increasing hard block content, however, the minimum is shallower. Evidently, hysteresis is affected to a greater degree by hard block content than the sequence length.

Test methods

The testing of elastomers like many other functional materials consists of both material property characterisation and a number of performance-based tests related to the specific use of individual materials. For polyurethane elastomers many of the standard mechanical property characterisation techniques are used and a list of some of the more important test methods is provided in Table 19-3.

Standard mechanical properties

Density
The density of a polyurethane elastomer is an important characteristic when evaluating the performance of the material and this is particularly true for microcellular elastomers. The dynamic performance and cost structure of elastomers is directly related to density. The density of these materials is measured by a water immersion method and expressed as kg/m^3.

Table 19-3   Test methods for polyurethane elastomers

Test	ASTM	DIN	ISO
Density			
Compact materials	D-792	53479	1183
Cellular materials		EN ISO 845	845
Hardness			
Durometer	D-2240	53505	868
Compression hardness	EN ISO 3386-2	3386-2	
Tear strength			
Rubber and thermoplastic elastomers, Graves	D-624	53515	34
Rubber and elastomer films	D-1938	53515	6383
Tensile properties			
Compact materials	D-412	53504	37
Cellular materials	D-412	EN ISO 1798	1798
Compression set			
Compact materials	D-395	ISO 815	815
Cellular materials	D-395	EN ISO 1856	1856
Rebound resilience			
Bayshore	D-2632		4662
Pendulum		53512	
Flexural fatigue resistance			
Flexing test		53522	
Ross flex test	D-1052		
Abrasion resistance			
Taber abrasion	D-3489		
Rotating drum abrader		53516	4649
Oil and solvent resistance	D471	53428	
Hydrolytic stability	D3137	EN ISO 2440	2440
Heat ageing	D573	53508	188

Hardness

The hardness of elastomers is measured using durometer indentation meters and is typically reported with reference to the specific device used to make the measurement, Shore A, Shore D, or Asker C hardness units. The durometer measures the resistance to the penetration, without puncturing, of a blunt indenter point pressed onto the elastomer surface against the action of a calibrated spring. A pointer moves across a scale to show the resistance to penetration. Scales on the durometer read from 0 to 100, with higher numbers indicating greater hardness. A comparison of the values for the commonly used durometers appears in Figure 19-10.

Compression hardness

The compression hardness of microcellular elastomers is determined using a standard size specimen with parallel flat faces that is compressed between two parallel plates larger than the specimen. The force required to compress the sample at a constant rate up to a specified compression ratio, typically 50 per cent, is determined and reported as kPa or MPa.

Tensile modulus and strength

To determine the tensile properties of any material a dumbbell 'dog bone' test piece is cut from a test sheet of the elastomer. The sample is then stretched at a constant rate of speed and the stress and strain measured. The tensile modulus, or the modulus of elasticity, is the slope of the initial, straight-line portion of the stress/strain curve. The terms 100 per cent modulus and 300 per cent modulus are also used for solid and microcellular polyurethane elastomers, but these values are not true moduli, simply the stress required to produce a strain of 100 or 300 per cent. The tensile strength is the maximum stress the material withstands before breaking and the elongation at break or ultimate elongation is the maximum extension at break. The moduli and tensile strength are all reported as kPa or MPa.

The tensile set of a material is a measure of the remaining extension after a sample has been stretched a specific amount and is determined by stretching a tensile sample to the designated elongation and then holding it for 10 minutes. The load is then released and the sample is allowed to recover for 10 minutes and the residual extension measured and reported as a percentage.

The tensile properties of a material can be used to classify the material and Figure 19-11 shows stress/strain plots that characterise four types of polymer materials. Elastomers generally fall into the category of tough materials with a range of hardness from hard to soft.

Tear strength

The tear strength of a material is determined by measuring the force required to propagate a cut and a number of different sample types, Figure 19-12, is used dependent on the elastomer material. Split tear and trouser tear are typically used for microcellular elastomers whilst Die C and a similar test, Graves, for solid elastomers. Tear strength is expressed as a unit of force per linear unit of tear propagation, normally N/m.

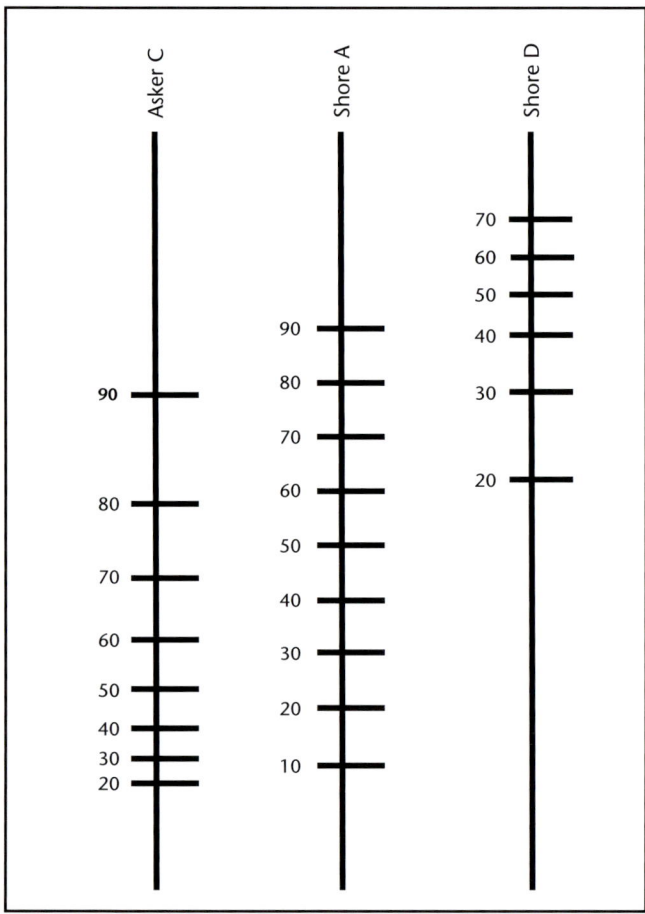

Figure 19-10
Comparison of Shore A, D, and Asker C hardness scales

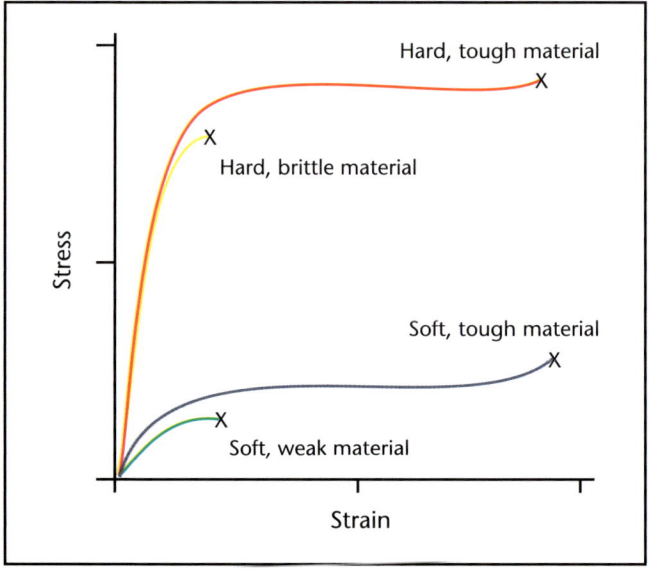

Figure 19-11
Stress strain curves for a variety of materials

Figure 19-12   Configuration of tear test samples

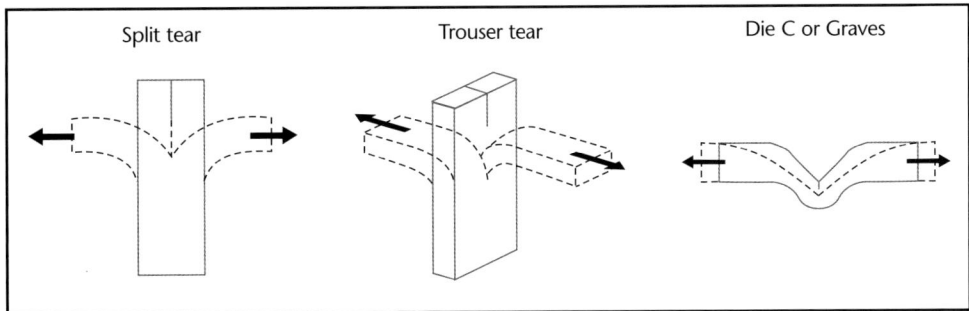

Compression set
The compression set of polyurethane elastomers is determined by measuring the thickness of a sample, compressing it to a specified deflection, holding for a fixed time and temperature then releasing the sample, allowing it to cool and the substance re-measured. The compression set is reported as a percentage of the original deflection with samples usually compressed for periods that range from 24 to 72 hours at temperatures from 23 to 100°C.

Dynamic property testing

Hysteresis
Hysteresis is a measure of the energy loss in a mechanical test and can be measured in tension or compression with the energy to load the sample compared to the energy to unload the sample. Polymer materials will always absorb some of the energy during such a test and a typical trace is shown in Chapter 11, Figures 11-6 and 11-9, where the area within the loop represents the energy lost during deformation. This is a low frequency test with the hysteresis determined at 50 per cent deformation.

Rebound resilience
Rebound resilience is another means of measuring the energy lost during a deformation and recovery of a polymeric material and is a high-speed test that usually involves small deformations. The test methods include the Bayshore falling weight test where the rebound height of the falling weight is measured and reported as a percent of the original drop height. Other rebound tests, where the weight is on a pendulum and the rebound of the pendulum is reported as a percentage, are also used.

Flex fatigue tests
Elastomers find extensive use in applications where the ability of the material to repeatedly flex is exploited. A variety of flex fatigue tests are used in the industry to compare the performance and verify the suitability of the materials. In the Ross test a specimen with a two-millimetre chisel-cut is placed over a steel rod and flexed 90 degrees at a rate of 60 cycles per minute. The test is run for a specified number of cycles and the cut growth is measured. The German Pirmasens test is similar, except the chisel-cut is replaced by a needle punch.

Whole shoe sole tests have also been developed that take into account not only the capability of the material to flex, but also the ability of a tread design to flex without cracking. There are two tests in common use: the Bennewart, Figure 19-13, and the Bata Belt test. In the Bennewart test a sole is repeatedly flexed in a three-point bend for a specified number of cycles, normally 50,000, and the degree or absence of cracking checked. In the Bata Belt test the soles are glued or stitched to a fabric belt that is then cycled around two cylinders of different sizes, the smaller of which stresses the sole, for 50,000 to 100,000 flexes and, again, the level and position of any cracking is noted.

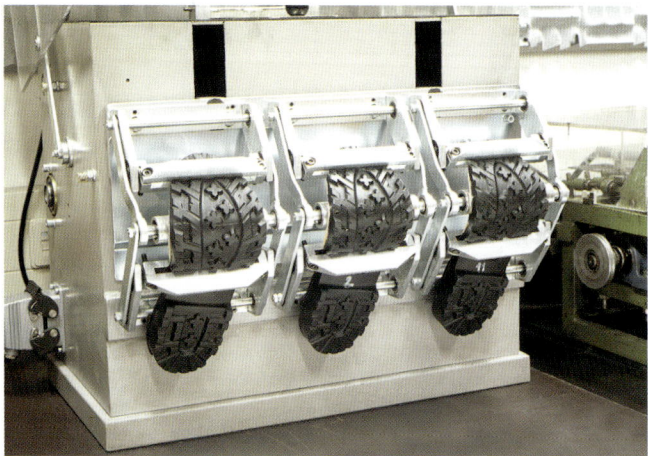

Figure 19-13   Bennewart flex test

Frictional properties

Elastomer materials can be formulated with good frictional properties that are of interest particularly when developing transport belting or shoe soles and the values are determined by either pushing or pulling a sample, loaded with a fixed mass, across a standard surface. The static coefficient of friction is obtained by dividing the horizontal by the vertical force for the sample to just begin moving. The dynamic coefficient of friction is the same ratio, but taken as the amount of force required to maintain the movement at a specific speed. There are several types of test equipment available to determine frictional properties, many of them stemming from the footwear industry and the desire to create a standard for safety with the James test an established test for determining the static coefficient of friction. SATRA (Shoe and Allied Trade Research Association) has developed a dynamic testing device for slip testing that is representative of real wear conditions.

Thermal analysis

Differential scanning calorimetery (DSC) is an important technique in the research and development of elastomer materials and is used to measure the glass transition temperature, T_g, of an elastomer, which is the temperature at which the material begins to behave as a soft viscoelastic solid. For elastomers it is desirable for the T_g to be well below room temperature to ensure good flexible behaviour under ambient conditions.

Dynamic mechanical thermal analysis (DMTA) is another important thermal analysis technique and the test measures the stress of a material created by applying a sinusoidal strain over a broad range of temperatures. This method is used to identify the glass transition temperature of a material as well as the change in mechanical properties with increasing temperature. This technique is particularly important in phase separated polymers because it offers a means of evaluating the degree of phase separation.

Environmental exposure testing

Heat-aged tensile test
Heat ageing an elastomer and then measuring the tensile strength and determining the residual strength is a common accelerated test for polyurethane elastomers and is used as a means of evaluating the long-term durability of a material.

Hydrolysis testing
The resistance of an elastomer to hydrolysis is also a common accelerated test and is determined by checking the tensile strength before and after exposure to a humid environment at elevated temperatures, typically 70 or 80°C at 100 per cent relative humidity (RH) for 0, 7, 14 and 21 days. It is reported either as a percentage change from the original value or else as the tensile strength in MPa.

Weathering testing
Accelerated weather testing is typically carried out using a chamber fitted with an ultraviolet light source, such as a Xenon arc lamp, where the samples are exposed and degradation is measured from the change in colour or a loss of mechanical properties. The test can also be carried out under controlled temperature and humidity conditions, as is the case for a QUV Weatheromoter. However, in many instances long-term actual weathering of samples in various regions is still the preferred method of determining weather resistance.

Abrasion resistance
The resistance of an elastomer to abrasive wear is evaluated by loading a sample, under a standard load, against a specified abrasive material that is calibrated using a standard rubber material. The most accepted test, DIN 53516 Figure 19-14, uses a rotating drum abrader and the abrasion resistance is reported as a loss in weight, volume or thickness of the sample.

Figure 19-14   DIN abrasion test

Oil and solvent resistance
To specify elastomers for a wide variety of applications it is important to determine their resistance to the absorption of common industrial fluids and any resulting change in mechanical properties. Tests are typically conducted by exposure of a sample for a designated period of time and then any change in weight, dimension and in many instances tensile properties are determined. The test liquids that are utilised represent various blends of fuel for automobiles or aircraft, lubricating fluids such as engine oils, hydraulic fluids, and automotive engine coolant.

20. Elastomers for footwear applications

Nick Limerkens

Footwear, like a lot of other consumer products, includes a range of tangible, but not performance-related, properties in its construction. This has always been true for footwear, yet the non-technical tangible factors such as comfort and style have become more critical as increased consumer spending lifts demand for better quality non-utilitarian shoes.

At their most simple, shoes can be only two components per pair – injection moulded PVC sandals for example – or can be complex constructions with 60 or more components per pair. But all shoes can be split into two key areas – upper and sole. All footwear, whatever its function, needs a sole unit and there are two fundamental methods of attaching the sole to the upper – direct moulding or using unit soles.

Direct moulding is where the upper, usually on a metal last, becomes the lid of a three-part mould that consists of a bottom plate, defining the sole pattern, and two side rings that determine the edge profile. A variety of methods are used, but all have the essential feature that the sole is formed and at the same time directly bonded to the upper. With unit soles the moulds and forming processes are much simpler and cheaper than with direct moulding. Once the unit is made it is stuck to the upper using an adhesive. The other advantage, besides cost, is that unit soles can be made from a wide range of soling materials, making this method of production the most popular today.

Sole units were made originally from a single material, even if several components were used, but over the past 30 years increasingly more complex sole constructions have been developed so that now some sports shoe soles can have many components per pair using five to six different materials. However, most sole units are produced as single or dual component construction with one of the advantages of the latter being that it combines a thin hard-wearing outsole with a lower-density midsole material to provide overall lower weight and increased comfort. Another possibility is to produce multi-coloured soles, in either the same or different densities.

The major soling material for all types of footwear was traditionally leather, either on its own or heavily reinforced with nails, with wood also playing an important role. Crepe rubber, cork, reeds and other natural products were also used for soles. Leather, crepe rubber, wood and cork continue to be used as 'natural' soling materials today.

The first synthetic material to be used for soling was vulcanised rubber, introduced in the USA during the 1850s, then spreading to other countries and achieving a major position by the 1930s. The next major change in soling material was the development of flexible PVC, which was introduced in the 1960s. Thermoplastic rubber and polyurethanes at the start of 1970s rapidly followed PVC and the introduction of new soling materials has continued so that now, in the 21st Century, there are over ten primary raw materials available to the shoemaker.

Most soling materials are either thermoplastic, such as PVC, thermoplastic rubber, thermoplastic polyurethanes, which require only to be melted and re-shaped by a mould, or come in sheet form, like leather, EVA, resin rubber, micro rubber and crepe rubber, from which sole units can be cut. Polyurethane is the exception as it is supplied as two liquid components that are reacted in the mould to produce the sole unit.

Whilst the new polyurethane technology provided many advantages and opportunities to the shoemaker and consumer, footwear manufacturers initially had limited experience of the reaction polymerisation required to make polyurethane soles. Today a wide range of equipment, manufacturing processes and material handling procedures are available to take advantage of the great variety of opportunities that polyurethanes offer to the shoemaker for the production of soles.

The over-riding advantage of polyurethane as a soling material is that it maintains its properties as a tough, flexible, hardwearing, durable elastomeric material over a wide density range. With polyurethane it is possible to produce single density outsoles in virtually any fashion style, in any range of thickness and shape over a density range of 350 to 1,200 kg/m^3, although predominantly between 350 and 650 kg/m^3.

Polyurethane is also ideally suited for the production of dual density footwear since systems are available to produce soft, low-density midsole units, of between 280 and 450 kg/m^3 density, that can be combined with thin, tough, hard-wearing outsoles such as polyurethane, thermoplastic polyurethane or vulcanised rubber. Polyurethane midsoles have an excellent combination of toughness, flexibility, good compression, wear-resistance and easier bonding to the outsoles than midsoles based on EVA.

Thermoplastic polyurethanes have had a role in the footwear industry for over 30 years, but mainly in specialist areas such as athletic boot sole plates, Figure 20-1, heel-tips of women's formal shoes, down-hill ski boot shells, Figure 20-2, upper components for sports shoes and several other minor uses. However, this all changed in the early years of the 21st Century when new, softer grades of thermoplastic polyurethane were developed, that opened up the use of the material as an outsole for boots, casual and formal styles.

The application and potential use for thermoplastic polyurethane has been extended by the development of blowing agents that have enabled the density to be reduced to that achievable with the reaction polyurethane systems. To date (2002) thermoplastic polyurethanes are in regular production at a density of 700 kg/m^3, Figure 20-3, and developments are in hand to reduce this to 500 kg/m^3 and eventually to below 400 kg/m^3, at which point it would fully match the density range of the reaction polyurethane systems for outsoles.

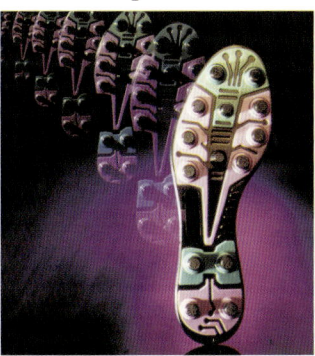
Figure 20-1
Soccer boot plates

Figure 20-2
Ski boot shell

Figure 20-3
Blown TPU outsole

Two-component polyurethanes

There are essentially two competing technologies for the production of reaction polyurethane soling systems with the primary one being based on polyester polyols and the minor one on polyether polyols. These are always used as two components: an MDI-based quasi prepolymer and a polyol blend. In both cases the basic formulations consist of a prepolymer (MDI and polyester or polyether polyol) that is reacted with a polyol blend that besides the polyol contains a chain extender, usually butane or ethane diol, water (blowing agent), surfactants, catalysts and any other additives required. Depending on the polyol formulation and characteristics of the prepolymer, the mix ratio of the two components can vary between 0.8:1 and 1.3:1. The scale and pace of operation requires demould times of one and a half to four minutes. This speed of operation not only determines the type of machine that has to be used, but also the formulation of the polyurethane system has to provide the correct reaction profile.

Polyester technology

The initial development of polyurethane soling systems was based on essentially linear polyester polyols made from adipic acid and glycols (see Chapter 6 for details). These polyols continue to prevail in the polyurethane soling market mainly because polyester-based polyurethane systems have inherently better physical properties than polyether polyol systems due mainly to the inter-chain hydrogen bonding between the ester groups. The hydrogen bonding provided by the polyester systems is difficult to match with a polyether system. This leads to tougher, more wear-resistant products over a wide density range.

However, whilst the ester group contributes to better physical properties it also has the disadvantage of being prone to hydrolytic and microbial attack, both of which are accelerated by heat and humidity. This can limit the use of polyester-based systems in high humidity tropical climates, unless expensive additives are used to reduce the rate of attack. In addition, most polyester systems have poor low-temperature flexibility and this can limit their use in certain cold climates.

These poor low-temperature properties are mainly due to the crystallinity of the polyester soft blocks which have glass transition temperature (Tg) values in the range -10 to +10°C.

The polyester systems can be classified into two main groups, those designed for high performance and those for the lowest cost. In reality there is a continuum of systems that merge into each other with the cheaper style of polyester polyols being blended with the higher performance ones to increase processing tolerance or to improve properties. Typical formulations and properties for all these systems are given in Table 20-1.

Table 20-1   Formulations and properties for polyester systems

	Safety boot	Outsole sport	Midsole sport	Casual shoe	Summer sandal	Platform shoe
Formulation, wt-%						
MDI prepolymer	46.8	42.5	46.2	48.5	56.5	58.3
Polyester polyol (BD/MEG/AA OHv 45)	42.2	50.1	–	42.7	–	–
Polyester polyol (MEG/DEG/AA OHv 56)	–	–	45.4	–	37.4	35.5
BD	6.9	5.9	5.5	7.0	–	–
MEG	–	–	–	–	5.5	5.6
Catalyst	0.3	0.4	0.3	0.3	0.2	0.2
Surfactant	0.2	0.1	0.2	0.2	0.2	0.2
Water	0.2	0.1	0.3	0.2	0.2	0.3
Antistatic agent	1.9	–	–	–	–	–
UV stabiliser	–	–	0.5	–	–	–
Pigment	1.6	0.9	1.6	1.3	–	–
Processing conditions						
Polyol blend temperature, °C		40 – 45 (all formulations)				
Prepolymer temperature, °C		35 – 40 (all formulations)				
Properties						
Density, kg/m^3	600	800	400	550	500	380
Hardness, *Shore A*	65	60	35	55	65	70
Tensile strength, *MPa*	7.5	8.5	4.5	6.5	6.5	6.0
Elongation, %	520	750	500	550	300	200
Abrasion, mm^3	120	75	n/a	150	<250	<250
Ross Flex (-10°C), *kcycles*	150	250	n/a	150	n/a	n/a

High-performance systems

These systems are used for the production of boots, everything from hiking to safety to military, plus high quality casuals and sport shoes. Direct moulding is the major production technique owing to the higher value of the final products, but unit soles are also used. Higher molecular weight polyester polyols, with narrower specification limits on viscosity and acid value, based on adipic acid and longer chain glycols such as butane diol or hexane diol, in conjunction with ethane diol have been developed for this area. The necessity to produce performance items means that quality is more important than speed of production.

Figure 20-4
Safety boots

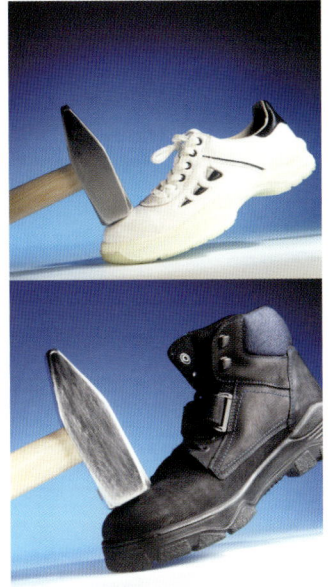

Dual density polyurethane soles are often used in these applications, employing high specification polyester polyols in the outsole, to get maximum wear and performance, whilst using cost effective polyester polyols in the lower density midsole. A typical machine for dual density direct moulding would have 24 stations with one mould per station that with a high-performance system would produce around 2,500 pairs of complete shoes in 24 hours.

Examples of some of the performance footwear that can be produced using these high specification polyester systems are shown in Figures 20-4 and 20-5.

Low-cost systems

These systems are used for the mass production of unit soles for sandals, exaggerated fashion styles such as platforms, slippers and less demanding casual shoes. The key factors required for low-cost soling are:

- Low-cost polyester polyols.
- High productivity.
- Low mould cost.
- Low density.

Figure 20-5
Sports shoe

The four factors are closely interlinked. The cheaper polyester polyols used are still usually based on adipic acid and glycols, such as diethylene glycol and ethylene glycol, but other acids such as 'AGS mixed acids' (Adipic/Glutaric/Succinic) are also used to further reduce costs. The main difference, compared to the 'high-performance' polyesters, is that these lower cost polyols tend to be produced with a lower molecular weight and an increased functionality.

When cost is the major driver it is certain that, whichever soling material is used, unit soles will be made, as the mould costs are significantly cheaper than direct moulding.

Figure 20-6   Sandals, platforms and slippers

When used in conjunction with high MDI content prepolymers it is possible to develop formulations with these polyester polyols that provide unit soles at densities of 350 to 500 kg/m^3 with hardnesses in the range 50 to 80 Shore A and extremely high levels of productivity due to rapid demould times of 60 to 150 seconds.

For instance, a 60-station semi-automatic rotary machine fitted with a pair of moulds per station can consistently produce sandal unit soles with a density of 350 to 380 kg/m^3, a hardness of 70 to 80 Shore A at a rate of around 30,000 pairs every 24 hours.

Examples of some of the cost effective footwear that can be produced using these fit for purpose polyester systems are shown in Figure 20-6.

Polyether technology

There are two main reasons for using polyether polyols in footwear applications; first, the inherent resistance of the ether group to hydrolysis and microbial attack and, second, the ether group provides good flexibility at low temperatures, important for applications such as cross-country ski and winter boots used in cold countries. This is due to the lower Tg of polyethers that is in the range of -60 to -50°C. These two advantages greatly broaden the range of climates accessible to polyether systems.

Another key advantage of polyether polyols is that they are liquid at room temperature, having a lower viscosity than polyester polyols. This leads to easier processing as the system can be operated at a lower initial temperature. The improved processing tolerance and better flow of material in the mould, from the lower viscosity, contribute to improved surface definition and less air entrapment for polyether-based systems. This is especially important for the production of quality fashion units.

The initial development of polyether systems, in the 1970s, was based on modified integral skin foam systems that were blown with CFCs to give a thick skin and a low-density core. High molecular weight, lightly branched polyether polyols were used in conjunction with high MDI content prepolymers, leading to footwear formulations with mix ratios of MDI prepolymer/polyol blend of 1:2. The properties of the final sole units were only acceptable if a thick skin was obtained, in essence producing a dual density sole from a single density process.

Therefore, when CFCs were banned it was necessary to find alternative blowing systems and this led to major developments, during the early 1990s, into the technology of polyether systems. This was required since the alternative physical blowing agents did not produce such clear thick skins and the underlying properties of the polyether polyols were not sufficient to meet the requirements of most applications. Accordingly, a complete re-design of the polyether technology was necessary covering the following three key areas:

- Water blowing.
- Polyether polyol molecular design.
- New prepolymers.

Increasing the percentage of water blowing in the old polyether systems was not satisfactory as it led to higher levels of urea formation, raising the cross-link density, whilst producing thin skins. Therefore, to introduce polyether polyol systems that are fully water blown or using a combination of HFC-134a and water it has been necessary to modify the type of polyether polyol and prepolymer used.

Higher molecular weight, essentially linear, polyether polyols with increased ethylene oxide tipping have been developed for better reactivity and to improve the water-blowing compatibility. New prepolymers with lower isocyanate content, similar to polyester-based prepolymers, have been developed so that the mix ratios, prepolymer/polyol blend, are close to 1:1.

These developments in polyether technology have greatly broadened its range of applications and polyether systems are now used freely for all footwear styles, except work and safety boots, where the inherently poor oil and solvent resistance of polyether polyols can limit their use. Typical formulations and properties of polyether-based systems are given in Table 20-2.

Table 20-2   Formulations and properties for polyether systems

	Midsole sport	Casual shoe	Summer sandal	Platform shoe
Formulation, wt-%				
MDI prepolymer	44.4	48.7	51.2	54.5
Polyether polyol (diol & triol) OHv 30	46.7	43.1	40.3	36.6
Butane diol	0.6	0.5	0.5	0.5
Ethylene glycol	4.3	5.1	5.4	5.9
Catalyst	0.7	0.6	0.6	0.6
Surfactant	0.3	0.3	0.3	0.2
Water	0.3	0.2	0.3	0.3
UV stabiliser	1.0	–	–	–
Pigment	1.7	1.5	1.5	1.4
Processing conditions				
Polyol blend temperature, °C	30 – 35 *(all formulations)*			
Prepolymer temperature, °C	25 – 30 *(all formulations)*			
Properties				
Density, kg/m^3	400	550	400	350
Hardness, *Shore A*	45	55	70	70
Tensile strength, *MPa*	2.5	4.0	3.5	3.0
Elongation, %	350	400	250	200
Tear strength, *kN/m*	15	20	15	12
Abrasion, mm^3	n/a	<150	<250	<300
Ross flex (-10°C), *kcycles*	n/a	>200	n/a	n/a

Hybrid technology

An alternative approach is to combine polyether and polyester polyols into one system. The major advantage of this hybrid technology is the combination of excellent wear resistance with improved hydrolytic and microbial stability. Typically, a polyether prepolymer is used in combination with a polyester-based polyol system.

The two areas where this hybrid technology has found major application are for the production of Wellington boots and sports shoe midsoles; typical examples are shown in Figures 20-7 and 20-8.

Polyurethane offers distinct advantages in waterproof footwear as it keeps the feet warm as well as dry. Figure 20-9 compares the insulation performance of polyurethane to vulcanised rubber and PVC Wellington boots.

Figure 20-7 Wellington boots

Figure 20-9   Insulation properties of Wellington boots versus temperature

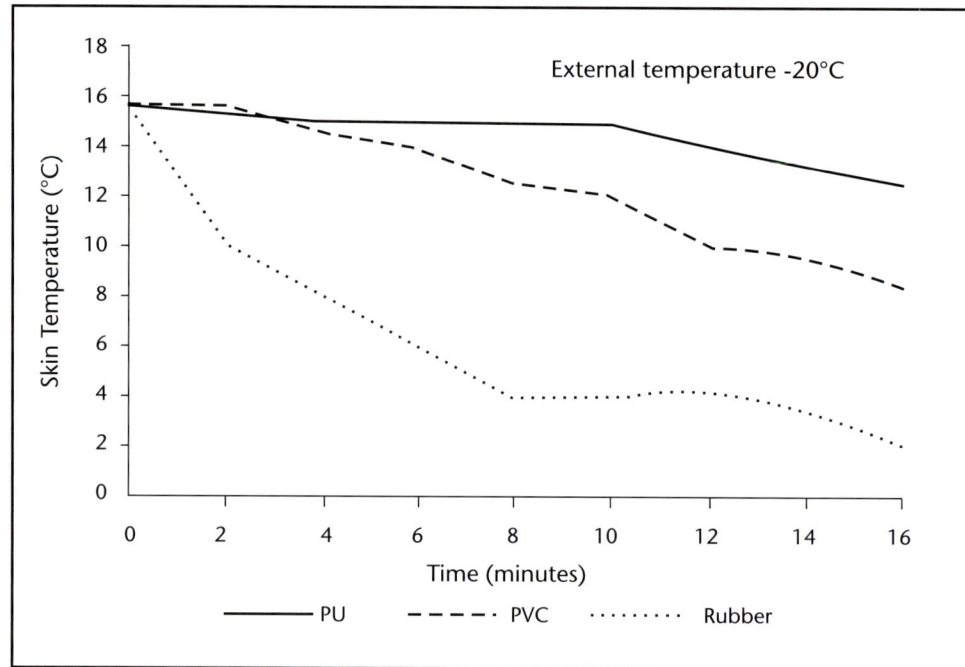

Figure 20-8 Sports shoe midsole

Additives

The basic properties of both polyester and polyether systems can be modified by the use of additives dependent upon the specific requirements of the end application. Some of the areas where additives need to be used are to improve the resistance to UV, to alter the electrical conductivity for work boots and shoes used in the computer industry and to improve hydrolytic and microbial resistance, specifically for polyester systems.

Polyurethane polymers are essentially white when produced and turn yellow with exposure to UV if no additives or colour are added. Pigments or dyes are thus added to change the colour, which is particularly important for direct moulded soles where it is difficult to apply lacquers to the surface after production. The pigments or dyes used can have an effect on the cure and final properties of the polyurethane system and thus need to be selected with care. It is usual to have ranges of pigments and dyes specifically formulated for polyester or polyether systems.

Process technology for two-component polyurethanes

Machinery

All the different types of machines used to produce polyurethane sole units consist of four basic elements – material tanks, pumps, mixing head and mixing screw. The stirred tanks, which maintain the prepolymer and polyol blends at the correct temperature, are connected by heated pipes to and from the mixing head via gear pumps that meter the two components at the correct ratio. The mixing head contains valves to ensure that the two components can be smoothly switched from a recycle mode to dispense mode. Finally the mixing screw ensures that the two components are completely mixed as they are introduced into the mould.

Most modern machines now have mixing heads fitted with extra injection valves, up to four, that can be used to introduce colour or other additives directly into either the polyol blend stream, just prior to mixing, or else directly into the mixed system.

Polyester polyol systems are processed at temperatures of 35 to 50°C to reduce viscosity and to increase the reactivity with the prepolymer typically held at 35 to 40°C. The traditional polyether systems, blown with CFC-11, had to be processed at temperatures around 20°C, necessitating cooling systems on the material tanks and lines to avoid loss of blowing agent. It is only now with the introduction of water blowing that the polyether polyol systems can be processed over a broader temperature range, typically 25 to 35°C.

To obtain best results in terms of reaction characteristics and final properties, the correct amount of polyether or polyester polyol blend must be intimately mixed with the prepolymer. Machine operators carry out routine checks at the start of every shift to ensure the optimum ratio is being maintained and the two components are efficiently mixed.

There are many types of metering and dispensing machines in use for making polyurethane shoe soles, but the majority can be divided into two categories – injection and casting.

Injection machines
Injection machines mix the polyurethane using high speed, 15,000 to 20,000 rpm, helical screws that are usually self-cleaning. This approach is very well suited to the direct moulding process.

Casting machines
With casting machines the two components are mixed at much lower speeds, typically 4,000 to 7,000 rpm, and the mixed material poured into an open mould with the head tracking over the mould contours to precisely place the material. These machines are normally used for unit sole production although it

is possible to produce direct moulded footwear by using a top closure method, where the upper, on its metal last, is brought down on top of the open mould after the polyurethane has been poured. Main Group SpA and Gusbi Off Mecc. SpA produce typical examples of these machines. The relatively simple construction means that many locally-produced alternatives are available.

Moulds

Relatively low pressures are involved in moulding polyurethane sole units, whether by direct moulding or as unit soles, therefore moulds made from cast or machined aluminium are normally used even for long production runs. For prototype testing and shorter runs of unit soles it is possible to use moulds made from metal filled epoxy and metal-sprayed shells backed up with epoxy. It is normal to cast a mould close to its final shape and then to obtain the detail by machining with a final application of handwork to finish off the design.

A single density unit sole mould can consist, in its simplest form, of only two parts, a base cavity plus a lid. Moulds for direct moulding are far more complex usually having at least four moving parts – a base plate (defining the sole pattern), two side rings (determining the side profile) and a last for the upper (the mould lid). If two density sole units are being produced then the number of parts obviously increases and for special effects the moulds can end up with a complex structure and hence a high cost. Typical examples of the moulds used for footwear production are shown in Figures 20-10 and 20-11.

Mould release

Reacting polyurethane footwear systems are an excellent adhesive so to stop the cured soles from sticking to the moulds it is necessary to apply mould release. There are many types, but the commonest are based on a mix of silicone oils that have traditionally been dispersed, at low concentrations in organic solvents such as methyl ethyl ketone and trichloroethane.

However, the legislation restricting the emission of volatile organic compounds (VOCs) into the atmosphere, in force in most developed countries and having a secondary effect in developing countries, has led to the introduction of mould-release agents based on high solids content formulations. This, in turn, has necessitated the development of new spray guns and the use of automatic robotic equipment to apply the release agent. A wide range of release agents is commercially available, tailored for specific products, machines and applications.

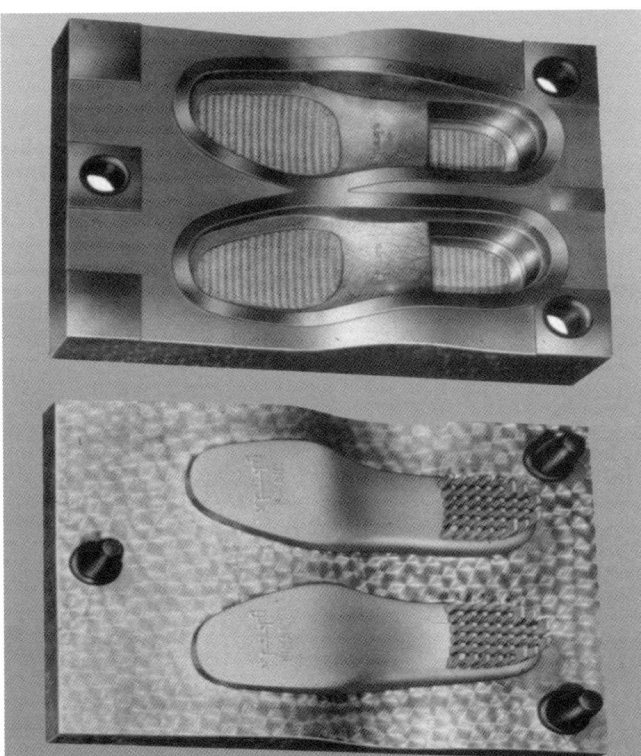

Figure 20-10
Single density unit sole mould

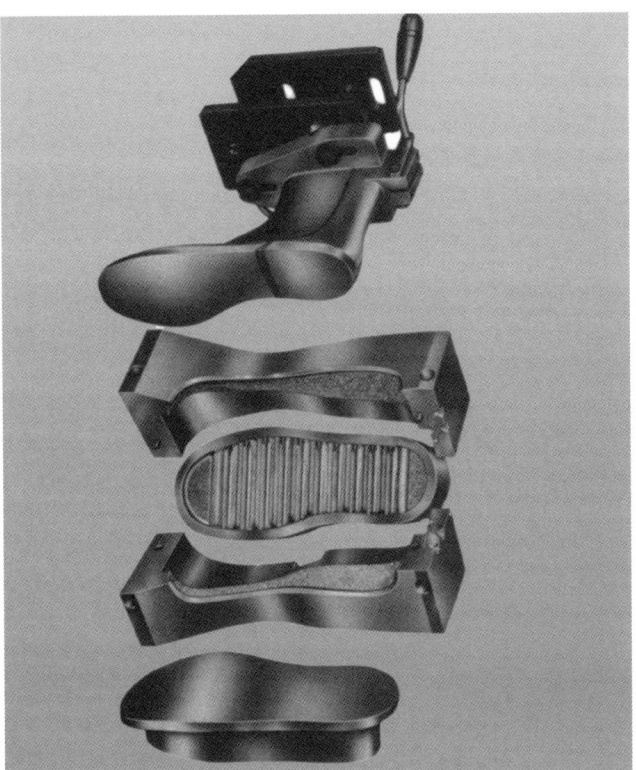

Figure 20-11
Dual density direct moulding mould

Thermoplastic polyurethanes

Thermoplastic polyurethanes have a long history in the shoe industry especially in segments where extreme durability and abrasion resistance coupled with excellent flex life and optimum hydrolysis is required. This mix of properties makes thermoplastic polyurethanes particularly suitable for applications such as sport shoe studded sole plates (football, rugby, golf, hockey), top pieces for the heels of women's court shoes, whilst its excellent cut and scratch resistance combined with low temperature properties make it ideal for ski boots and in-line skate boots.

Other significant applications have been as upper components for sports shoes, such as lace straps, motion control devices, anti-pronation fittings and external heel stiffeners, and air bags in low-density polyurethane midsole units for sports shoes. All of these applications form the core of the traditional market.

The thermoplastic polyurethane moulding compounds that have been used for over 25 years in the footwear industry have been based on MDI and either polyether or polyester polyols with hardnesses in the range 75 Shore A to 85 Shore D. The basic chemistry and production methods used are covered in Chapter 21.

The use of thermoplastic polyurethanes as a general shoe soling material has been limited by the minimum hardness, 75 Shore A, that could be traditionally achieved. This is significantly higher than the 50 to 65 Shore A hardness required to provide flexible comfortable soles for the majority of casual, formal and boot styles. Only in a few applications, such as the thin soles used for women's formal shoes, have thermoplastic polyurethanes been used as a conventional soling material.

Until the late 1990s it was not possible to produce thermoplastic polyurethanes with hardness below 75 Shore A whilst maintaining good processing and acceptable properties. However, that has now changed and softer thermoplastic polyurethanes, in the range 55 to 70 Shore A have been developed. A comparison of the properties for thermoplastic polyurethanes, thermoplastic rubber and PVC is given in Table 20-3.

Table 20-3   Comparison of properties for thermoplastic polyurethanes, thermoplastic rubber and PVC

	TPU Solid	TPU Solid	TPU Solid	TPU Blown	TR Solid	TR Blown	PVC Solid	PVC Blown
Processing conditions								
Barrel temperature, °C	200	175	170	155	170	165	150	145
Demould time, s	45	60	60	90	60	90	60	90
Properties								
Density, kg/m^3	1.20	1.20	1.20	0.75	0.98	0.75	1.25	0.75
Hardness, *Shore A*	–	65	55	55	60	55	60	55
Shore D	65	–	–	–	–	–	–	–
Tensile strength, *MPa*	45	20	15	8	7	5	12	3.5
Elongation, %	400	900	950	750	600	750	450	350
Tear strength, *kN/m*	200	60	45	35	30	25	38	22
Abrasion, mm^3	35	50	75	120	160	300	200	450
Ross flex (-10°C), *kcycles*	>25	>150	>250	>150	>100	>100	>100	>30

One of the first applications for the new soft thermoplastic polyurethanes was as a thin outsole combined with a lightweight polyurethane midsole resulting in very high quality dual density 'combi' soles. There are three major advantages in using thermoplastic polyurethanes as an outsole:

- Design freedom.
- Thin low weight outsoles.
- Durability.

The characteristics of thermoplastic polyurethanes inherently copy even the finest of mould details. Sharper and more varied pattern definitions can be obtained than with normal dual density polyurethane or even thermoplastic rubber. It is now quite feasible using thermoplastic polyurethanes to obtain outsoles with good tactility, crisp sharp look to the pattern and both matt and gloss finishes in the same moulding, simply by optimising the mould design and surface.

The excellent physical properties of thermoplastic polyurethanes enable the production of thin outsoles, 1.5 to 3 millimetres, which leads to lower weight and cost savings, whilst still maintaining good wear resistance.

There are a number of different techniques available to produce these 'combi' soles starting with the conventional method of moulding the thin shell on a standard injection moulding machine and then transferring this shell to a polyurethane machine so that it can be filled with the low-density polyurethane midsole. This is not an efficient process and several companies, such as Klockner Desma Schuhmaschinen GmbH and Main Group SpA, have developed a process of 'injecting/extruding' thermoplastic polyurethanes into an open mould, using a conventional thermoplastic injection moulding head fitted to an existing reaction polyurethane moulding machine so that the midsole can be injected or poured within the same rotation cycle.

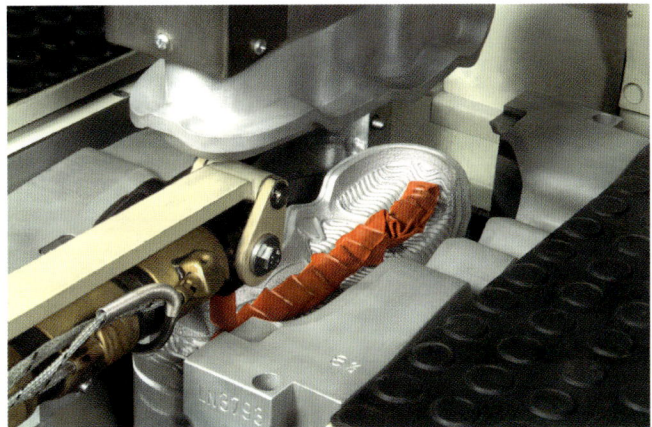

Figure 20-12   *Pouring of thermoplastic polyurethane on Desma machine*

It was necessary to develop the 'open-pour' process as the mould clamping force available on a conventional polyurethane moulding machine is not sufficient to hold the mould closed against the pressure created during the injection of thin outsoles. A typical pour of material from a Desma machine is shown in Figure 20-12.

Many companies have now adopted this 'combi' sole technology and it has been especially popular for the production of work, safety and hiking/trekking boots.

The next development with these softer grades of thermoplastic polyurethanes was to reduce the density, by the addition of a special chemical blowing agent package, in order to make lighter sole units. This opened up the possibility to use the new compounds for the production of single density outsoles for general-purpose casual and formal styles.

The first expanded thermoplastic polyurethane technology was commercialised during 2000 for the production of unit soles and densities as low as 700 kg/m^3 are now achieved in standard production conditions. Lower densities have already been obtained in laboratory tests and it is expected that soles with densities of 400 to 500 kg/m^3 will be in regular production within the next few years.

One of the major advantages of these new blown materials, based on the soft thermoplastic polyurethane grades, is that they can be processed on standard thermoplastic machines to produce single or multicolour unit soles. The unique feature of these materials is that they combine excellent aesthetics with lightness and high physical performance. As with most thermoplastic materials, production scrap and rejects can be recycled.

Whilst good units can be produced using these expanded grades of thermoplastic polyurethane with existing moulds, the best results are obtained by optimising the mould design to exploit the unique combination of aesthetics and performance at low weight.

21. Thermoplastic polyurethanes

Vikram Kapasi

The polyaddition reaction between diisocyanates and diols was first identified in 1937 at the I G Farben laboratories and the first polyurethane elastomer application was a fibre called 'Perlon'. The real breakthrough in commercialising these thermoplastic polyurethane elastomers came when of B F Goodrich presented a paper on 'Virtually cross linked elastomer' in 1957 and the company first marketed the products under the trade name 'Estane'.

The unusual feature of thermoplastic polyurethanes (TPUs) is that they are the one product group in the polyurethane family that is supplied as a fully-reacted product, so that the processor/customer only has to re-shape it into the final form required. All other polyurethane products, other than powder coatings, are supplied as reactive liquids. Thermoplastic polyurethanes can be split into two discrete classes, based on how they are used, since by varying the formulation they can be designed to be processed on conventional thermoplastic equipment or by solvent processing in a range of organic solvents. In the first case they are used for the production of solid components by injection/extrusion whilst in the second they are used as adhesives or coatings.

Thermoplastic polyurethanes are injected or extruded to produce a wide range of products such as footwear, wire and cable sheathing, hoses, tubing, film and sheet, or they are used to produce coated and laminated textiles, protective/functional coatings and adhesives. Details on the coatings and adhesive applications can be found in Chapters 24 and 25 whilst this chapter will focus on the use of thermoplastic polyurethanes for injection and extrusion.

Thermoplastic polyurethanes have a combination of high elongation and tensile strength and so form a bridge between rubber polymers and thermoplastics and in addition their toughness provides excellent abrasion and tear resistance. The type and length of the soft segment leads thermoplastic polyurethanes to have good low temperature flexibility, while the hard segments contribute to temperature resistance. They also can have excellent resistance to oils, fuels, solvents and chemicals.

To produce thermoplastic polyurethanes it is necessary to build a high molecular weight linear long chain polymer that will at the same time be both thermoplastic, implying a relatively low melting point, and yet have the required degree of toughness and high physical strength characteristics needed to meet the performance specifications required. To process thermoplastic polyurethanes requires repeat melt and freeze cycles, using a combination of temperature and pressure, so one essential requirement is that the hard segments can be reversibly melted without thermal degradation. Thermoplastic polyurethanes are supplied as granules or pellets that are converted into end-use items by conventional thermoplastic processing techniques such as injection moulding, extrusion, blow moulding, slush moulding, thermoforming and calendaring.

Applications

Thermoplastic polyurethanes are often described as if they were an industry sector themselves rather than a range of products with applications in a broad range of industry segments. But there is still confusion as their use is classified by process (injection, extrusion), form (film, sheet) or application area (automotive, footwear) and certain companies classify thermoplastic polyurethanes with other thermoplastic products and not as part of the polyurethane family. But all thermoplastic polyurethanes can be categorised as belonging to either the thermoplastic elastomer family, which includes such products as styrene-butadiene-styrene (SBS), SEBS, thermoplastic olefins (TPO), thermoplastic vulcanisates (TPV), co-polyesters (COPE), co-polyamides (COPA or PEBA), or the engineered plastics, which include, polyether-ether-ketone (PEEK), polycarbonate (PC), ABS, polyacetals (PA), as they have high tensile strength and elongation, excellent cut, abrasion and tear resistance, good resistance to solvents, lubricating fluids and fuel oils.

Automotive

Thermoplastic polyurethanes are used in a wide range of automotive parts that can be split into four major areas – interior, exterior, under the bonnet and under car. Most of the applications use thermoplastic polyurethanes with hardnesses in the range 35 to 60 Shore D with softer grades, down to 70 Shore A for specialist areas.

Interior
An important use for thermoplastic polyurethanes is as films flame-bonded to fabric, as a barrier coat for the direct foam-in-fabric process, as the elastomeric quality of the films helps maintain the handle and comfort features of the seat. The recycling properties of thermoplastic polyurethanes are an important aspect for the production of covers for the airbags fitted to the steering column, as the process produces a high level of waste which can be immediately recycled.

A lot of the wiring in the car interior is linked to the engine compartment and, because continuous looms are required, as cables sheathed with thermoplastic polyurethane are increasingly specified, they end up running through into the car interior. There are many other small items in the interior of the car that are made from thermoplastic polyurethanes such as plugs, gaskets, door bearings and gearshift bellows.

Exterior
The two most successful areas are as the external skin of bumpers or shroud elements which are filled with energy absorbing polyurethane foam and collapsible mounting systems, and as glass filled panels where tough elastomeric properties are needed such as in the lower door sections of off-road vehicles.

Another use is as protective body strips down the side of door panels and the ability to design flush fitting windows has been improved by the development of novel window seals based on moulded thermoplastic polyurethane profiles that have the attachment studs already moulded in place. The exterior includes tyres and thermoplastic polyurethanes are used as lightweight snow chains, based on polyether polyols to provide good low temperature performance and hydrolysis/microbial resistance.

Under the bonnet

An aggressive high temperature, oily, damp atmosphere exists under a car bonnet, which has exerted pressure on component suppliers to improve resistance properties and thermoplastic polyurethanes are now available that can be run continuously at 135°C compared to the normal 100°C. This increase in operating temperature has led to a great boost in the use of fuel line connectors, tubing, wiring, shrouds, covers and plugs in the engine compartment. There is also a need for improved transmission belts and high-performance fibre-reinforced thermoplastic polyurethanes are now used due to the high elasticity and good hysteresis.

*Figure 21-1
TPU shrouds*

Under car

The growth in the use of anti-lock braking systems has led to a demand for high-performance cabling, which must have a low risk of failure because of safety, so thermoplastic polyurethanes are used due to the demanding conditions under the car. This is mainly due to their good resistance to common greases and lubricating oils, excellent mechanical/abrasion performance as well as good flexibility over wide extremes of temperature.

Another application for thermoplastic polyurethanes is as maintenance-free flexible bushes in the joints of the steering mechanism and the front and rear suspension systems. The key property is the ability to maintain a consistent size, without swelling, in the presence of grease and oils and polyester-based products with hardness in the range from 55 to 60 Shore D are normally used. Thermoplastic polyurethanes are also used to make shrouds and bellows, Figure 21-1, which are used as flexible protective elements around critical components under the car.

*Figure 21-2
TPU castor wheel*

Engineering

There are virtually endless applications for thermoplastic polyurethanes in the engineering industry due to the extreme versatility of the material, but one of the best known is the injection moulding of tyres, wheels, Figure 21-2, and guide rollers, which find a wide variety of end-uses from shopping trolleys, to low-speed castor wheels for industrial vehicles, to guide rollers for airport luggage conveyor systems.

Thermoplastic polyurethanes are also widely used for the manufacture of gaskets and seals for both hydraulic and pneumatic use when wear resistance, rigidity and resistance to oils/greases are essential and polyester polyol-based materials are used. Small- and medium-sized seals can be produced to a high degree of accuracy by injection moulding and it is possible to produce gaskets with a combination of different hardnesses or in combination with other plastics. Seals can also be directly injected onto metal parts, using metal-to-polyurethane adhesion promoters. These gaskets and seals are widely used in construction machines, forklift trucks, lifting platforms, hydraulic presses, injection moulding machines, hydraulic cylinders, shock absorbers, mechanical pit props, car jacks, rotating piston pumps and pneumatic cylinders, as a few examples.

Polyurethane mineral classification screens have competed with rubber or steel for many years due to their longer life because of not only their high abrasion resistance, but also the good oil/solvent resistance of the polyester and the hydrolysis/microbial resistance of the polyether based-grades. Large units are normally produced using polyurethane cast elastomers, but there is a trend to produce the smaller, complicated, modular units that can be linked together by injection moulding. The screens are used for the classification of materials such as gravel, coal and ores and the separation of dry and wet materials – for instance slotted screens are used for dewatering products in the paper and food industries.

There are many other examples of engineering applications for thermoplastic polyurethanes, but four products illustrate the diversity of thermoplastic polyurethanes. Extruded polyester-based thermoplastic polyurethane profiles are used to fill and protect the V-grooves on cast-iron lift pulley sheaves, separating the sheaves from direct contact with the steel cable wires, leading to a reduction in wear and a doubling in the life of the cable. Thermoplastic polyurethane housings on pneumatic or electric road drills provide a high level of noise reduction and enhanced working conditions. Soft thermoplastic polyurethanes are used to clad electric drills to produce tools that are both comfortable to handle and have high durability and also to make animal ear-tags for agricultural uses.

Footwear

The many existing and new applications for thermoplastic polyurethanes in the footwear industry are fully detailed in Chapter 20.

Medical

The medical market for thermoplastic polyurethanes is small, but has a high growth rate and medical workers depend on the durability and reliability of the products in severe and critical conditions and an example is the seal on the face piece used during artificial respiration.

Thermoplastic film laminated to fabric is used to make pressure infuser bags to deliver intravenous fluids to patients on the way to hospital. The laminated fabrics are radio frequency (RF) welded, which seals them to form a bag that can be pressurised to ensure that the fluids flow when the bag is at any angle. Similar bags are used for compression dressings that pump iced water to injuries to minimise haematosis and swelling. The use of RF-weldable fabrics reduces manufacturing costs and enables any shape to be produced with bladders and fittings included without the need for additional stitching.

Thermoplastic polyurethane tubing is finding wide application in the medical industry, replacing PVC, as its superior properties, such as high burst strength, enable much thinner walled tubes to be produced with the same performance characteristics. Also, the plasticisers in PVC can cause stress cracking in the polycarbonate fittings. Therefore, the total cost of a piece of equipment, including lifetime expectations, can be lower for thermoplastic polyurethanes than PVC.

Pipe, hose and tube

There is no pipe, hose and tube industry since these products are used across a wide range of industries, but there are certain common features with polyester polyol-based thermoplastic polyurethanes with hardness's from 78 to 98 Shore A providing the best balance of properties. Besides the applications already mentioned there are many other areas of interest such as thermoplastic polyurethane pneumatic tubing replacing nylon tubes in automotive plants since the latter is easily abraded, leading to surface roughness that can damage the finished paint films on body parts.

Figure 21-3
Wire-reinforced TPU tubes

The high abrasion resistance of thermoplastic polyurethanes is used when manufacturing large-bore tubing for the pneumatic transport of powders and small particle size abrasive materials. These tubes can be either straightforward extruded pipe or fabricated using a mandrel and continuous welding of a metal wire between two strips of thermoplastic polyurethane; examples are shown in Figure 21-3.

There is a large market for braided hose that is lined with thermoplastic polyurethane as the higher bursting loads of thermoplastic polyurethane leads to a much thinner film than is the case with rubber or PVC-lined hoses. These hoses are made as lay-flat construction with the added advantages of thinner films being that long lengths of hose roll to a smaller diameter and are lighter. The range of applications for this type of hose varies from fire to irrigation to garden, with the construction dependent on the end application. The common feature is that they are normally made from polyether polyol-based thermoplastic polyurethanes since they need to have a life expectancy of at least 10 years when used and stored in damp conditions and with the irrigation and garden hoses exposed to microbial attack.

Wire and cable

Wire and cable, too, is not really an industry sector, but covers a range of products that are linked through related manufacturing routes. The full length of a cable needs to be sheathed with thermoplastic polyurethane without any defects since water or other contaminants could seep in and short out power cables and destroy the integrity of control/ computer connections, examples shown in Figure 21-5. Polyether polyol-based thermoplastic polyurethanes of 78 to 92 Shore A are usually preferred because of their good hydrolysis/microbial resistance, excellent abrasion resistance and low temperature flexibility. Polyester polyol-based thermoplastic polyurethanes can be used, but only where there is no risk of microbial or hydrolytic attack.

Figure 21-4   TPU-sheathed cables

Thermoplastic polyurethanes are not used for high-tension wires, but are widely used as cable sheathing for lower voltages especially where electrical, control or computer cables are used in rough conditions and climates. The service life of such cables can be greatly extended using thermoplastic polyurethane sheathing instead of rubber or PVC.

Film, sheet and calendared articles

Thermoplastic polyurethanes can be made into film or sheet materials with a range of thicknesses using either a blown film or casting process and polyester or polyether polyol-based thermoplastic polyurethanes with hardnesses of 75 to 95 Shore A. Films with a thickness range of 25 to 500 microns can be produced from blown film and from 10 to 1,000 microns by casting, but in this case films thinner than 150 microns require a support carrier such as release paper or rigid plastic film. A calendaring process is used to coat textiles with thermoplastic polyurethanes, which requires special grades and it is possible to apply coats with loadings as low as 50 g/m².

Figure 21-5   TPU-film in laminated windscreens

A special niche application for films is the production of laminated composites such as high-performance windshields for cars, aircraft, Figure 21-5, and trains, security glass windows for prisons, banks and buildings and bullet-proof protective equipment. Aliphatic diisocyanate and polyether polyol-based thermoplastic polyurethanes of 72 to 95 Shore A are normally used. The films, which can vary in thickness from 375 to 1,250 microns, are made by a flat die cast process and used to bond symmetrical composite constructions such as

glass/TPU/PC/TPU/ glass. Aliphatic thermoplastic polyurethane's adhesive property, low temperature flexibility, light stability and plasticiser-free nature makes these ideal materials for bonding to the polycarbonate (PC) interlayer. Some of the end-uses for these films and calendared materials are shown in Table 21-1.

Table 21-1   End-use applications for thermoplastic films

Application	Articles	Key Properties
Automotive	Protective films	Toughness, abrasion resistance, exterior durability
	Adhesive films	Adhesive, low temperature flexibility, hydrolysis resistance
Clothing	Labels, numerals	Adhesive, fast crystallisation, washability
	Apparel	Adhesive, water resistance, light-weight
Fashion	Synthetic leathers	Soft-touch, durability, light-weight, abrasion resistance, elasticity, leather-like feel
Footwear	Bladder, Pumps	Toughness, elasticity, heat sealability, low air permeability
	Uppers	Abrasion, cut resistance, flexibility, low fatigue
Furniture	Kitchen door	Adhesive, fast crystallisation
Inflatables	Air mattress, boats	Elasticity, toughness, low permeability to air, weldability
Industrial Textile	Laminates, coatings	Adhesion, strength, abrasion resistance
	Conveyor belts	Toughness, cut resistance, elasticity, heat sealability,
	Heat seal tape	High crystallinity, adhesion, low temperature activation
Medical	Wound dressing, surgical drapes	Moisture permeability, flexibility, strength

Some application areas such as food contact and medical require compliance with regulations and standards.

Raw materials and properties

Thermoplastic polyurethanes are produced by reacting a diisocyanate with a substantially linear polyether or polyester polyol and a low molecular weight chain-extending diol in either a one- or two-step reaction process and the major raw materials used are listed in Table 21-2.

The majority of thermoplastic polyurethanes are based on MDI, but all the other diisocyanates are also utilised often for specialist niche applications with the aliphatic diisocyanates being used when UV-resistant products are required. By comparison, all the various types of polyols find use in the production of thermoplastic polyurethanes with the polycarbonate polyols being the only niche product.

The MDI-based thermoplastic polyurethanes are, therefore, differentiated mainly on the basis of the soft segment polyol used with the standard polyester and the

Table 21-2   Raw materials used for thermoplastic polyurethanes

Isocyanates	Polyols	Chain Extenders
MDI	Butane diol adipates	1,4 Butane diol
TDI	Ethylene glycol adipates	Ethylene glycol
H_{12}MDI	Hexane diol adipates	1,6 Hexane diol
XDI	Polycaprolactone:	HQEE
IPDI	PTMEG	
	PO/EO polyether polyols	
	Polycarbonate polyols	

Phthalic acid, isophthalic acid, azealic acid, 1,3 propane diol and neo-pentyl glycol are also used to make polyols

propylene oxide/ethylene oxide polyether polyols being the workhorse products. Polyester polyols are selected when the product requirements are a high toughness and oil/solvent/ chemical resistance whilst conversely propylene oxide/ethylene oxide polyols are selected where a good hydrolysis/microbial resistance is specified. Polyester polyols tend to have poor hydrolysis/microbial resistance whilst propylene oxide/ethylene oxide polyether polyols generally have poorer physical properties.

Polycaprolactone polyols have the inherent toughness and resistance of polyesters, but also have improved low-temperature performance combined with a relatively high resistance to hydrolysis. Polytetramethylene ether glycol (PTMEG) has all the polyether benefits, but in addition approaches the polyester polyols in terms of physical properties and has excellent hydrolysis/microbial resistance and low temperature flexibility. Polycarbonate polyesters are a special class of polyester family and are suitable for high humidity exposure applications.

The flexibility, strength and toughness of thermoplastic polyurethanes are related to the modulus of the polymer, which is dependent on the relative amounts and phase separation of the hard and soft segments. The degree of hard segment formation and separation depends not only on the ratio of hard to soft segment, but also on the type and structure of the polyol, the choice of chain extender and on the manufacturing process and reaction conditions. As thermoplastic polyurethanes are manufactured under controlled factory conditions the final polymer structure can be precisely optimised, which is the major reason why they achieve the highest physical property characteristics for a given formulation since if the same components are mixed as a cast elastomer there is less time and control to achieve the required structure. Full details of the implications of phase separation and component structure can be found in Chapter 19.

The physical properties of thermoplastic polyurethanes change only slowly over the normal range of operating temperatures and the glass transition temperature (Tg) of the polyester or polyether polyol soft segment limits the lower operating temperature, which is usually below ambient temperature. The upper operating temperature depends on the softening point of the hard block domain, which occurs at temperatures above 110°C with the precise temperature dependent on the structure.

Other than the type of polyol the main classification of thermoplastic polyurethanes is based on hardness and standard products are available from 70 Shore A to 75 Shore D with lower hardness grades, down to 55 Shore A, traditionally made by compounding with plasticisers, but more recently by modifying the formulation through choice of polyol, chain extender and the control of production process. Typical mechanical properties of thermoplastic polyurethane elastomers based on polyester and polyether polyols and MDI and aliphatic diisocyanates are given in Tables 21-3 and 21-4 respectively.

Table 21-3   Formulation and properties of MDI-based thermoplastic polyurethanes

Formulation, wt-%								
MDI	18.7	25.6	38.9	34.6	48.7	40.8	51.3	48.7
PTMEG (OHv 56)	–	–	–	–	35.4	–	31.6	–
polyether polyol (OHv 112)	74.6	–	51.8	–	–	–	–	–
Polyester polyol (OHv 56)	–	68.3	–	55.4	–	46.6	–	35.4
Butane diol	6.7	6.1	9.3	10.0	15.9	12.6	17.1	15.9
Properties								
Hardness, Shore A	80	80	90	90	–	–	–	–
Shore D	–	–	–	–	55	55	65	65
Density, kg/m^3	1.11	1.19	1.11	1.22	1.14	1.22	1.17	1.22
Tensile strength, MPa	20	30	45	50	53	35	45	37
Elongation, %	430	550	500	600	450	500	350	450
Modulus, MPa 100%	6	5	8	8	12	13	24	25
300%	11	10	17	15	21	20	31	30
Tear strength (Die C), kN/m	63	79	102	96	100	137	145	236
Abrasion, mm^3	35	50	22	65	40	50	35	45

Normal additions: catalyst <0.25% and processing additives <2%

Table 21-4   Formulation and properties of aliphatic-based thermoplastic polyurethanes

Formulation, wt-%						
H_{12}MDI	29.6	31.6	40.0	41.2	48.8	49.1
PTMEG polyether polyol (OHv 112)	66.3	–	50.8	–	37.9	–
Polyester polyol (OHv 56)	–	63.3	–	49.1	–	37.4
Butane diol	4.2	5.1	9.2	9.7	13.3	13.5
Properties						
Hardness, Shore A	75	75	90	90	–	–
Shore D	–	–	–	–	50	50
Density, kg/m^3	1.05	1.10	1.05	1.15	1.06	1.15
Tensile strength, MPa	21	25	54	60	45	48
Elongation, %	520	480	650	600	510	450
Modulus, MPa 100%	4	5	6	6	10	11
300%	10	11	15	17	22	28
Tear strength (Die C), kN/m	75	88	70	77	94	105
Abrasion, mm^3	55	45	44	50	77	70

Normal additions: catalyst <0.5% and processing additives <2%

The formulations of thermoplastic polyurethanes can be modified to provide the structure required for a specific performance and special application requirements such as adhesion to a variety of substrates, high temperature exposure, flexibility at room and low temperatures, good impact resistance at temperatures down to -30°C, low permanent set, hydrolysis and chemical resistance can be achieved by careful selection of the components. The resistance of thermoplastic polyurethanes based on MDI and polyester and polyether polyol to heat, oil and water is shown in Table 21-5.

Table 21-5   Heat ageing and oil resistance of MDI polyester and polyether thermoplastic polyurethanes (87 to 90 Shore A)

	Initial values	% Change			
		Heat	Oil 1	Oil 3	Hydrolysis
Polyester TPU					
Tensile strength, *MPa*	50	-24.0	-26.5	-23.0	-49.0
Elongation, %	600	3.5	5.5	3.5	12.5
Modulus, *MPa* 100%	8	-10.5	-7.5	0.0	-14.5
300%	15	-5.5	-1.5	11.0	-11.5
Polyether TPU					
Tensile strength, *MPa*	45	-24.5	-18.5	-12.5	-28.0
Elongation, %	500	9.0	0.0	21.5	14.5
Modulus, *MPa* 100%	8	6.5	8.0	-3.5	-7.0
300%	17	6.0	14.0	0.0	-5.0

Heat ageing & oil resistance: 70 hours at 100°C; hydrolysis: 14 days at 70°C & 100% RH

The volume resistivity of thermoplastic polyurethanes is normally in the range 10^6 to 10^{13} ohm. cm so they can be classified as electrically insulating, but antistatic and electrically-conducting grades can be produced by the addition of antistatic agents or special grades of carbon black. Thermoplastic polyurethanes based on aromatic isocyanates are susceptible to yellowing and ultimately to degradation when exposed to UV and UV stabilisers and colours are normally added for exterior use both for UV protection, but also aesthetic and fashion reasons. This works well for short-term use, but for high performance applications, aliphatic isocyanate-based products should be used.

Production

There are three production methods currently in use for making thermoplastic polyurethanes:

- Batch.
- Band casting.
- Reactive extrusion.

Batch process

High quality, consistent thermoplastic polyurethane products can be produced using simple batch mix techniques of which there are two variants – hand mix and agitated vessel. The advantage of these methods is that they provide flexibility for the production of a wide range of special and difficult to make grades and the investment in equipment is low. The hand mix technique is the simplest method, mainly used in development work and for specialist products such as fast reacting injection and extrusion grades. The method consists of the following steps.

Weigh 20 kilogrammes of the formulation into an open container using the following conditions:

- First polyol at 70°C, second chain extender at 25°C and then MDI at 40°C.
- Mix vigorously for 30 to 60 seconds.
- Pour into a large casting tray, at 100 to 120°C, under a blanket of nitrogen or into two open, unheated trays.

Once initial reaction is complete the slabs are post-cured in ovens at 120 to 130°C for 24 hours, granulated in water-cooled mills; batches are blended together to ensure consistency and the blended material extruded and chopped into sections followed by final drying and bagging.

The agitated vessel method is typically used for the production of adhesives grades although it can be used for injection and extrusion products and is a scaled-up version of the hand mix process, but based on mixing 100 to 250 kilogrammes in a stirred reactor with the major difference that once the isocyanate is added the product is mixed for 1 to 1.5 minutes before being poured into a stack of trays of about 10-kilogramme capacity. During these operations care must be taken to ensure that all required personal protection is employed.

Band casting

Band casting is a continuous process in which the raw materials are individually fed to a mixing head, fitted with a spreader system, to place a precise stream of mixed material onto a continuous steel or plastic heated conveyor belt. The thermoplastic polyurethane reacts and solidifies before being stripped from the belt and there are two variants of the process: the separate block and the extruder systems.

The difference between the two is that in the first, solid plates of thermoplastic polyurethane are removed at the end of the conveyor belt, thermally treated in an oven, ground to chips and, if necessary, extruded and pelletised. In the second, the reaction mass is transferred directly from the conveyor belt to an extruder for final reaction, homogenisation and pelletising. Both these methods have the disadvantage that the polymer synthesis and pelletising are still two separate operations and also the process is labour intensive as the belts require a lot of cleaning between grades.

Reactive extrusion

In this process the diisocyanate, polyol, chain extenders and, if required, additives such as catalysts and stabilisers are metered in one step into a twin-screw reactor extruder where they mix and react during the transfer down the screw and are then pelletised at a die face. Co-rotating twin-screw reactor extruders, fitted with kneading blocks, have proved particularly suitable for the production of thermoplastic polyurethanes as they meet the following process requirements:

- Good conveying action at low as well as high viscosity.
- Good mixing and homogenising action even at very high viscosity due to the variable configuration of the screws and kneading blocks.
- Short residence time due to the excellent self-wiping effect of the screws as well as the small clearance between screws and barrel to screws.
- Efficient product transfer from screw to screw.
- Good heat transfer from barrel to product.
- Closely defined, effective heating/cooling of product in the reaction area due to individual temperature control of the barrel sections.

These extruders can be used for the continuous production of all types of thermoplastic polyurethanes from soft adhesives, which are made at low reaction speeds, up to the hardest injection or extrusion grades. The short residence time and the relatively high temperature in the twin-screw extruders are the essential differences between this process in comparison to discontinuous synthesis in an agitated vessel, or continuous polymerisation on a conveyor belt.

Formulations, therefore, have to be adjusted and optimised for the extrusion process and due to the speed of production the reaction is rarely fully complete before the die-cutting stage. Dependent on the system, different degrees of cure are reached and the final physical properties are only achieved after several hours of post-production storage. The mix ratio of the components is critical and must take into account the variations in hydroxyl content of the polyols and the isocyanate content of the diisocyanates, since even a relatively small difference will lead to variations in the processability of the thermoplastic polyurethane produced.

To obtain reproducible results, high quality metering systems are required and the continuous metering of the pre-heated polyol, diisocyanate and diol into the extruder is normally done using loss-in-weight feeders which are not only insensitive to viscosity variations, but also allow feedback and recording of the weights actually metered. Additives such as catalysts, stabilisers and antioxidants are usually mixed with the polyol in the feed container and metered together with the polyol, but increasingly they are also injected straight into the extruder barrel, just before the die-cutter.

The pelletising method depends on the viscosity and cure of the material as well as the pellet shape required and there are two basic methods – strand cutting or hot die-face cutting. The former is used for fast-curing products with a high melt viscosity, whilst soft grades and all adhesives are pelletised using the hot die-face cutting process. In strand-cutting, thermoplastic polyurethane strands emerge from the die-head, are cooled in a water bath, dried by an air flow to remove any adhering water and chopped into cylindrical pellets. For hot-die pelletising, thermoplastic polyurethane is extruded through a multi-channel die face, normally fitted with rotary cutter, so that small beads of product are obtained. For adhesives and soft grades, the die head needs to be under water to provide fast cooling, but with harder grades a stream of water is sufficient.

Even though they are directly cooled by water, the pellets still retain a lot of residual heat and to eliminate stickiness, due to lack of cure, the pellets need to reach as high a degree of crystallisation as possible in the cooling water system. A high flow of water through the pelletiser chamber and a low volume of pellets are beneficial, especially for adhesive grades where the process takes several minutes. The residence time is controlled by the speed of the water flow and also by using long pipes, usually coiled, between the pelletiser and separator. However, because of their susceptibility to hydrolysis, particularly for polyester polyol products, the pellets need to be separated from the water and dried, to a minimum moisture content, as soon as possible.

As moisture affects the processing of thermoplastic polyurethane the pellets are supplied either in 25-kilogramme foil-lined thermally-sealed sacks, foil-lined and sealed one-tonne Octabins or in bulk by tanker.

Processing

Compounding

Additives can be added to thermoplastic polyurethanes either during production or just prior to processing. In the latter case, the additives, such as hydrolytic stabilisers, UV-absorbers, fire retardants, lubricants, pigments, biocides, reinforcing fibres and plasticisers, can be added as neat materials, but normally they are produced as master batches using a standard grade of thermoplastic polyurethane, polyester- or polyether-based, to ensure that there are no compatability problems. The master batch pellets are then mixed with the bulk of the thermoplastic polyurethane prior to processing.

Polymer blends

Thermoplastic polyurethanes can be blended with other polymers and as they are polar molecules they are more compatible with other polar polymers. The blend of thermoplastic polyurethane and PVC is an excellent example of where both polymers support each other to achieve specific properties. TPU/ABS, TPU/PC, TPU/Acetal blends are also commercially available. Thermoplastic polyurethanes can also be compounded with non-polar polymers using compatibilising agents and TPU/PP blends are expected to be available soon.

Powder

Thermoplastic polyurethanes can be converted into a powder, by cryogenic grinding, for blending with PVC pellets that are then processed into a TPU/PVC alloy or the powder can be used as a hot-melt adhesive and for textile coating.

Drying

Although it is possible to use thermoplastic polyurethane pellets straight from the bag, all of them are hygroscopic to some extent and will absorb moisture if left exposed to the atmosphere. At room temperature and a relative humidity of 50 per cent they will reach a water content of 0.3 to 0.5 per cent dependent on the grade and the chemical structure of the thermoplastic polyurethane. Those based on polyester polyols are more hygroscopic than polyether types and need drying to a moisture content below 0.1 per cent to ensure satisfactory processing. Insufficient drying can lead to defects in extruded films or moulded articles, which show up as streaks or bubbles. In addition, the moisture can cause polymer degradation due to hydrolysis at the processing temperature.

Drying should be carried out in a desiccant bed dryer, by circulating hot dehumidified air at 60 to 90°C dependent on the grade and the optimum dew point of such air should be around -30 to -40°C.

Injection moulding

Most grades of thermoplastic polyurethane can be injection moulded and the best processing is achieved by screw pre-plasticisation using standard polyethylene type screws with a length/diameter ratio of about 17:1 and a compression ratio of between 2:1 and 2.5:1. Accurate temperature control is essential and typical barrel temperature profiles are shown in Figure 21-6, and typical melt temperatures for different grades given in Table 21-6.

Overheating should be avoided, as significant polymer degradation can occur above 230°C, but insufficient heating can also lead to degradation due to local

Figure 21-6
Temperature profile for injection moulding

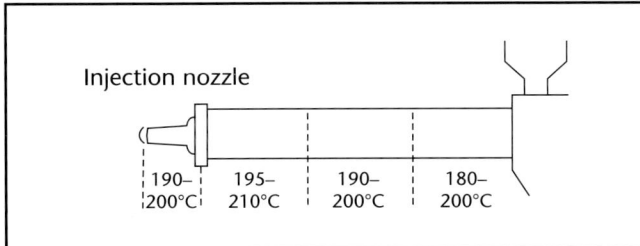

Table 21-6
Typical melt temperatures for injection moulding

Hardness (Shore A)	Hardness (Shore D)	Melt Temperature (°C)
75 – 90	28 – 40	180 – 210
90 – 95	40 – 52	190 – 225
>95	>55	210 – 245

overheating from the high shearing forces created by processing partially melted material. Polyether-based thermoplastic polyurethanes are more sensitive to overheating than polyester products.

Control of the screw and injection speed is also important and the moulds should be controlled in the temperature range of 20 to 30°C. Lower temperatures can be used to help the release of thick mouldings and higher temperatures are sometimes helpful when producing large thin-walled mouldings such as automotive steering rack gaiters. Thermoplastic polyurethane mouldings will shrink by up to two per cent from the mould dimensions and the shrinkage can vary with the cross-section of the moulding.

Cycle times for a given mould vary with the chemical structure of the thermoplastic polyurethane and the processing conditions, but as a general rule caprolactone-based products are faster than polyester, which are faster than polyether.

Extrusion

Thermoplastic polyurethane can be extruded to produce films, sheets, profiles, tubes, hoses, cable sheathings and fibres using three-stage screws with L/D ratios of 24 to 30 and a compression ratio of about 3:1. High-compression screws cause excessive shear heating and are unsuitable. The screw should be driven by an oversized, high-torque motor to ensure uniform rotation of the screw and propulsion of the viscous thermoplastic polyurethane. A breaker plate and screen pack, 80 to 120 mesh filters, should be included to eliminate any contaminants. The melt viscosity varies with the shear rate and the processing controls should include melt pressure and temperature control on the back of the die, in addition to accurate control of the barrel zone temperatures, examples of which are shown in Figure 21-7.

Figure 21-7   Temperature profile for extrusion

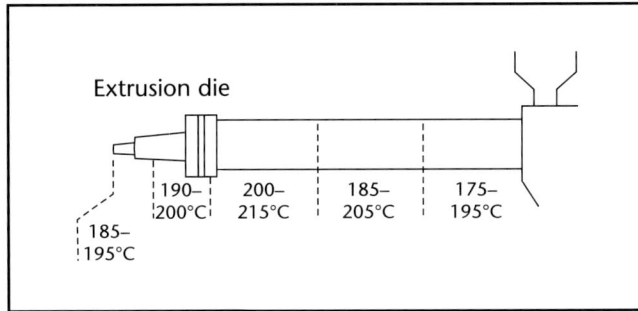

Rotational moulding

Rotational moulding is widely used for the manufacture of hollow articles typically using polyethylene or PVC compounds. Thermoplastic polyurethane is now competing with PVC as the skin for automotive instrument panels (dashboards), so it has been necessary to develop the technology for the new material. The process involves a machine equipped with two arms, each with two roto-moulds that are internally heated with oil to 240 to 250°C. Thermoplastic polyurethane, in the form of powder or micro-pellets, is added to the moulds, pre-treated with mould-release agent, and the moulds are then spun at high speed to evenly spread and melt the material. The total cycle time is around three to five minutes dependent on the grade of thermoplastic polyurethane used. The formulation and melt-flow index of the thermoplastic polyurethane are the major factors controlling the flow of material during melting, but another important factor is the particle size and uniformity of particles to allow uniform melting.

22. Other two-component elastomers

Brian Fogg

The wide diversity of polyurethane chemistry means that there are many other types and applications of polyurethane and polyurea elastomers with the main categories being:

- Cast elastomers.
- Synthetic leathers.
- Elastomeric fibres.
- Integral skin foams.
- Reaction injection moulding (RIM).
- Polyurea elastomers.

Cast elastomers

High-performance cast polyurethane elastomers are produced by manual or automatic casting techniques in which liquid or low melting point components are mixed, using a hot or ambient temperature process and poured into a mould, where they react to form a finished part. This technique was one of the first polyurethane processes developed and from the 1960s cast elastomers have been used for a range of products from soft printing rollers to wheels to mining screens to hard aluminium replacement parts. Cast elastomers can be formulated with a hardness range from 15 Shore A to 85 Shore D and resiliency from as low as 2 per cent up to 85 per cent. Many of these materials maintain their resiliency over a temperature range from 0 up to 150°C. Cast polyurethane elastomers are able to span the hardness gap between conventional natural and synthetic rubber and hard plastic materials and the products have excellent mechanical properties and low tooling costs.

Cast elastomers have found many applications as engineering components because of the low tooling cost in combination with the versatility in processing. One of the major applications is the moulding of tyres for specialty vehicles such as forklift trucks, Figure 22-1. Solid tyres with high load-bearing resistance are moulded directly onto metal hubs. The hub is sandblasted and primed before the hot processed polyurethane cast elastomer is cast into the mould.

Other applications include mining screens, Figure 22-2, conveyor system rollers and pipe linings, all of which take advantage of the excellent abrasion resistance of these materials. Cast elastomers are the material of choice for roller skate wheels, Figure 22-3, due to the abrasion resistance, resilience and ease of pigmentation of the material. Novel applications for cast elastomers include low resiliency gels, which are poured into bicycle and horse riding saddles, as well as comfort applications such as medical seating. Semi-flexible moulds are manufactured from cast elastomers for the production of simulated wood rigid foam parts and for the moulding of epoxy and decorative polyester resins.

Cast elastomers are produced by reacting diisocyanates such as, MDI, NDI, TDI, IPDI, H_{12}MDI and HDI, with a polyether or polyester polyol and a chain extender. The choice of diisocyanate and polyol are dependent on the specified performance of the elastomer with aliphatic isocyanates being used when a light/UV resistant product is required. The simplest method of producing cast elastomers involves mixing the isocyanate, polyol and chain extender in one step before pouring it into a mould where it cures. The resulting product is a lightly cross-linked polyurethane, with high tear and tensile strength properties and this process is known as the 'one-shot process'.

Figure 22-1   Cast elastomer tyres

An alternative method involves the manufacture of a prepolymer by reacting a diisocyanate with a polyol. Full prepolymers with a free isocyanate content of around four per cent require only a chain extender to complete reactions. Quasi-prepolymers are produced by reacting all of the diisocyanate with part of the polyol, to give free isocyanate levels as high as 20 per cent, and this is then mixed with a polyol blend made from the rest of the polyol and the chain extender.

Figure 22-2
Mining screen grid

MDI prepolymers are mainly based on reacting pure MDI with a polyester polyol at temperatures of 50 to 100°C whilst the TDI prepolymers are made from a variety of polyols, the most common being polytetramethylene glycol (PTMEG) having a molecular weight in the range 450 to 1,000. As with all pre-polymerisation reactions the temperature needs to be controlled and the reaction carried out under a blanket of nitrogen. NDI prepolymers are chemically unstable and need to be freshly prepared by reacting NDI at 120°C with most formulations based on polyester polyols. Dependent on the choice of polyol, the viscosity of the prepolymer can be as high as 20,000 mPa.s, requiring processing of the system at elevated temperature to reduce the viscosity.

Figure 22-3   Roller skate wheels

Cast elastomers, obtained by reacting di-functional polyester polyols and chain extenders with diisocyanates, have a polymer structure with a high molecular weight, and a significant amount of hydrogen bonding between the polyol chains in the final elastomer, leading to high levels of mechanical properties. Polyether polyols often have reduced functionality, caused by unsaturation, which leads to chain stopping and lower molecular weight formation in the polymer chain. Formulating techniques can accommodate chain stopping and polyether polyols with low levels of unsaturation are available. However, there is no hydrogen bonding between the chains with polyether polyols so they are inherently weaker than polyesters, but do have inherently better hydrolysis and microbial resistance. The polyols normally used are di-functional with molecular weights of 1,000 to 2,000 with some triols of similar equivalent weight being used to introduce functionality into the formulations. Typical formulations and properties for hot-cure cast elastomers are listed in Table 22-1.

Table 22-1   Typical formulations and properties for hot-processed cast elastomers

Formulation, wt-%	1	2	3	Processing conditions	1	2	3
Prepolymer				NCO Index	106	107	105
MDI	31.5	–	–	Prepolymer temperature, °C	80	90	110
TDI	–	14.5	–	Demould time, *min*	30	30	60
NDI	–	–	23.7	Post cure, °C	110	100	110
Polyester or polyether polyol (OHv 56)	68.5	85.5	76.3	Post cure, *hours*	1–3	1–4	24
				Properties			
NCO content, %	7.8	3.4	6.3	Hardness, *Shore A*	90	90	90
				Density, kg/m^3	1,200	1,200	1,200
Chain extender/100 prepolymer				Tensile strength, *MPa*	25	28	30
Butane diol	7.9	–	6.4	Elongation at break, %	500	450	500
MOCA	–	10.2	–	Tear strength (Die C), *kN/m*	80	45	55
				Resiliency, %	50	60	50

There are many chain extenders used with the major ones being 1,4-butane diol, 1,3-butane diol, neopentyl glycol, hexane diol, trimethylol propane and a range of amines, such as 3,3'-dichloro-4,4'-diaminodiphenylmethane (MOCA) and proprietary mixtures of amines. The amines are mainly used with TDI prepolymers to improve the final properties whilst MDI prepolymers are usually chain extended with diol compounds. MOCA is a known animal carcinogen and should only be used, after consulting suppliers' material safety data sheets, with extreme caution and proper handling procedures.

Processing of cast elastomers requires a long open time to accommodate all the steps of the process such as, mixing, degassing, pouring, flow and levelling, but should be followed by a rapid cure, and minimum post-cure. The original choice of catalyst was mainly based on mercury compounds, but these now have toxicity complications and amines such as TEDA are now used in conjunction with organo metallic compounds based on lead, bismuth, tin, nickel, iron and titanium. The catalysis of cast elastomer systems still presents a problem and work is continuing to optimise solutions.

Several other types of additive can be incorporated into the formulation of cast polyurethane elastomers to modify the properties and improve processing, including the following materials.

The choice of pigments and dyes, used to colour the cast elastomers, can have an effect on the reactivity as well as the hydrolytic and thermal stability of the product. Generally, iron oxide pigments need to be avoided when moulding thick sections where the catalytic effect of iron can induce scorching and embrittlement. Organic pigments are preferred and soluble dyes are often used to produce transparent elastomers.

Plasticisers and extender oils are added to reduce hardness, impart softness, lower the resiliency, balance mix ratios and reduce formulation costs. Many plasticisers commonly used for thermoplastics, such as diisooctyl phthalate (DIOP) and dibutyl phthalate (DBP), can be used as well as petroleum distillates such as aromatic and naphthenic oils.

The processing of cast elastomer formulations can be improved by the addition of silicone-based degassing agents, which reduce surface tension and aid the release of entrapped gas from the reacting mixture, and desiccants, which absorb water and prevent it from being available to react with isocyanate, thus preventing carbon dioxide gassing in the reacting material. Magnesium aluminium silicate, as a powder or as a paste by dispersion in castor oil, zeolite or ion exchange resins are used.

When cast elastomers require bonding to metallic inserts, adhesion promoters can be used to improve the polyurethane-to-metal bond strength. Compounds such as organo-silanes and titanates can be incorporated into the system as adhesion promoters.

If the cast elastomers are going to be used in applications involving friction with electronic or medical facilities then an antistatic additive is often required to reduce the build-up of electrical charge on the component. Many proprietary materials, such as quaternary ammonium salts and zinc compounds are used.

Fillers can be added to cast elastomers to increase the hardness, for reinforcement, to raise the density or to reduce cost. The fillers normally used are, carbon black, titanium dioxide, wollastonite, hammer-milled glass fibre and barium sulphate.

Cast elastomers can be efficiently produced by processes that vary from simple hand mixing of the components to the use of sophisticated machinery. Small parts are frequently produced by carefully mixing degassed components in such a way that air is not introduced, pouring the mix into open moulds where it reacts and cures. Once sufficient cure has been achieved parts are demoulded and post-cured at elevated temperature for several hours in order to achieve the optimum properties. The system components can be processed at ambient temperature, which whilst it provides plenty of time for degassing and mixing, leads to long cure times, or hot-processed using components heated to around 100°C. Hot processing has the advantage of reducing material viscosity and accelerating cure, but limits the mix, final degassing and pour times.

Specialised processing equipment for cast elastomers is available from several manufacturers and is often built to customer specifications and can include tanks for larger moulders to produce the prepolymer and polyol blend directly on the machine. Small moulders normally prefer to purchase two-component systems ready for use. The prepolymer and chain extender, which can be a short chain diol for a total prepolymer system or a polyol blend in the case of quasi-prepolymers, are degassed, normally using a thin film vacuum degassing column, before being added to the machine tanks.

The two components are then accurately metered to a mix-head where they are mixed and poured or injected into a mould. The advantage of the total prepolymer/chain extender system is that by varying the metering ratio it is possible to produce cast elastomers with a range of hardnesses from two components. Typical ratios vary from 100 parts of prepolymer to 10 parts of chain extender up to 50 parts of chain extender. This is not so easy with the quasi-prepolymer/polyol blend systems which tend to have fixed mix ratios closer to 1:1 although systems have been developed where two polyol blends, at the extremes of a hardness range, are pre-mixed in the correct ratio to produce a range of hardnesses. The mix-head is designed to limit turbulence and the introduction of air to the liquid stream and the throughput can vary from a few grammes up to several kilogrammes per second dependent on the size of part being manufactured.

An important niche application for polyurethane cast elastomers is the filling of pneumatic tyres for off-road vehicles such as earthmoving equipment and heavy tractors. The tyres are made puncture-resistant by being completely filled with a slow reacting two-component cold-cure elastomer system. These are based on polyether polyols and either TDI or MDI with a low level of water added to the polyol blend in order to achieve a small amount of foaming, which improves the fill characteristics and provides the tyre with a better pneumatic character. Prepolymer systems are used in order to achieve 1:1 mix ratios, which simplify the injection equipment as then fixed-ratio piston pumps can be used with direct feed from barrels to a static mixer. A typical formulation and properties are given in Table 22-2.

Table 22-2   Formulation and properties for tyre-fill cast elastomer

Formulation, wt-%		Processing conditions	
MDI	15.0	NCO Index	103
Polyether polyol	35.9	Prepolymer/polyol blend ratio	100/100
(MW 6,000)		Prepolymer and	25
Chain extender	3.5	polyol temperature, °C	
Water	0.1	Pour time, *hours*	1 – 2
Catalyst	0.3	Cure time, *hours*	24 – 36
Surfactant	0.2		
Plasticiser	45.0	**Properties**	
		Density, *kg/m³*	1,030
		Hardness, *Shore A*	30
		Tensile strength, *MPa*	2.0
		Elongation at break, %	550
		Tear strength (Die C), *kN/m*	10
		Resiliency ball rebound, %	50
		Compression set (22 hours at 70°C), %	20

With the tyre mounted on the wheel rim, the mixer of the injection machine is connected to the tyre valve stem, a small hole is drilled at the high point of the tyre, for air release, and the tyre is filled. It is then cured at room temperature for 24 to 48 hours before being re-mounted on the vehicle. Vehicles with polyurethane cast elastomer-filled tyres are restricted to a maximum speed of 50 km/hour due to the heat build up in the elastomer caused by hysteresis from repeated compression during use at higher speeds.

Synthetic leathers

Simulated leathers were originally produced by coating textiles with PVC, but the market now predominantly uses polyurethane elastomers as they have better properties, such as a clear finish, bright colours, high delamination strength, breathability, flex-crack, abrasion and moisture resistance. Compared to natural leather, highly consistent, wide, long rolls can be made that are used for applications including garments, footwear, luggage and handbags, car seats and furniture. The major production of synthetic leathers is in China, Taiwan and Italy with a typical production line producing two to four million square meters of finished material a year. Polyurethane can be blended with PVC and used as a hybrid alloy material whilst polyurethane elastomers are also applied as a thin finishing coat to natural leather.

Two production processes are used, based on one- or two-component technologies. The initial manufacturing method was a two-component method that involved reacting an adipate polyester polyol with TDI, forming a high viscosity isocyanate terminated prepolymer – with a molecular weight of 10 to 20,000, which was then directly coated onto a substrate, as shown in Figure 22-4, or onto release paper using a doctor blade. In both cases, the prepolymer was immediately cured by being passed through an atmosphere of primary aliphatic diamine such as ethylene diamine. If a free film is produced it then needs to be laminated to a textile substrate in a subsequent step, the process is shown in Figure 22-5.

Figure 22-4   Direct coating process

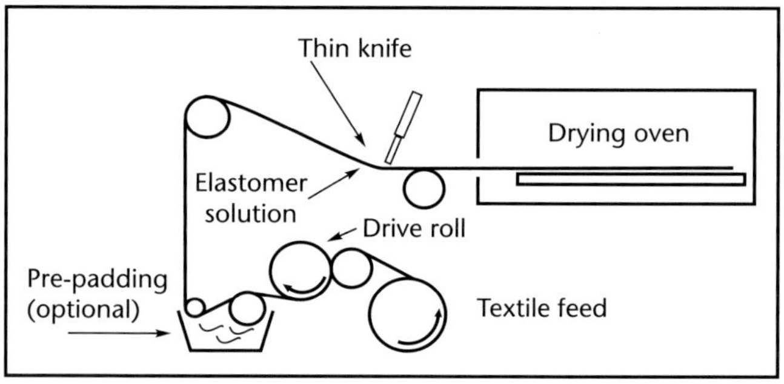

This method produces high molecular weight cross-linked polymer structures with good physical properties and excellent resistance to oils and solvents, but has been modified and it now tends to be based on producing a hydroxy-terminated prepolymer based on MDI and polyester polyols that is cured by reaction with modified liquid MDI or TDI products. The major drawback with the polyester polyol-based products has been their poor hydrolysis resistance, which is catalysed by trace levels of metals or amine catalysts from the raw materials and the urea in human perspiration when worn as garments or shoes. Therefore, hydrolysis stabilisers, based on ethylene diamine tetra-acetic acid (EDTA) or carbodiimides, are normally added to improve the properties and end-use life.

There are two one-component methods: solution in solvents or water-based dispersions. The original method was the reaction of a polyester polyol, pure MDI and chain extenders in a solvent such as the highly polar dimethylformamide (DMF) to produce polyurethane elastomers with molecular weights of 40,000 to 200,000. The high viscosity solution is spread onto a carrier or release paper, the material oven dried and laminated to the substrate. They are not normally cross-linked and so whilst having excellent physical properties have reduced solvent resistance compared to the two-component coatings.

Figure 22-5   Transfer coating process

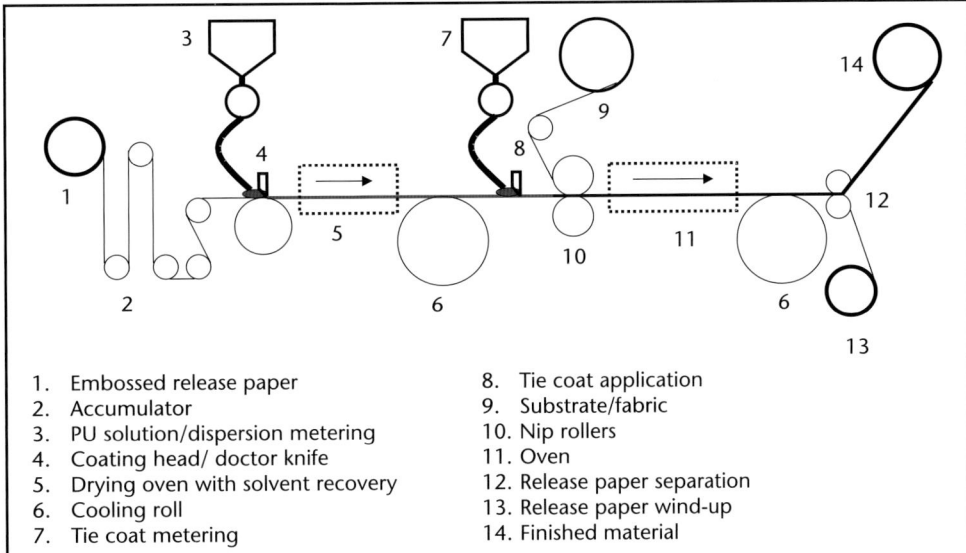

1. Embossed release paper
2. Accumulator
3. PU solution/dispersion metering
4. Coating head/ doctor knife
5. Drying oven with solvent recovery
6. Cooling roll
7. Tie coat metering
8. Tie coat application
9. Substrate/fabric
10. Nip rollers
11. Oven
12. Release paper separation
13. Release paper wind-up
14. Finished material

Polyurethane dispersions in water (PUDs), see Chapter 24 for details, overcome some of these problems, such as solvent resistance, as it is easier to add reactive functional groups to the PUD to increase the cross-link density. The PUDs can also be air frothed to give a low-density interlayer coating or specialty products can be made by abrading the surface to give a simulated suede effect. Poromeric films can also be produced by coating with a PUD that contains water-soluble filler, which after the film is cured can be removed by immersion in water to produce synthetic leather with improved breathability. Controlled porosity is an important property in applications such as upholstery and footwear.

Polyether, polyester and polycaprolactone polyols are now used in all the process methods with the choice being dependent on the application and end properties required. Polyesters provide high-performance, tough elastomers, but have poor hydrolysis and microbial resistance whilst polyethers offer the opposite. Polycaprolactones and PTMEG are used when the application requires high physical properties and excellent hydrolysis and microbial resistance.

Despite the fact that aromatic isocyanate-based products yellow on exposure to light one- and two-component polyurethane coatings based on them perform well in most applications, especially when pigments, anti-UV and anti-oxidant additives are formulated into the system. If a high resistance to UV discoloration is required then it is necessary to use aliphatic diisocyanates such as IPDI, HDI or H_{12}MDI.

Elastomeric fibres

For many years elastomeric fibres, used in garments to provide stretch and fit, were produced from natural rubber, but they have the disadvantage of poor resistance to oxidation, ozone and oils and low abrasion resistance, and they were difficult to colour and make as a fine denier. Polyurethane fibres were first developed and commercialised in Germany during the 1940s, but these were stiff materials and it was the introduction by E I DuPont de Nemours in 1959 of elastomeric polyurethane 'Spandex' fibres that defined the market for these materials. Spandex is now a US Federal Trade Commission approved generic name for such polyurethane fibres.

In comparison to natural rubber, polyurethane fibres generally have the following advantages/disadvantages:

- Improved resistance to oxidation, sunlight, oils, dry cleaning fluid, perspiration.
- Can be dyed using reactive or disperse dyes.
- Inferior stress retention and hysteresis loss.
- Worse resistance to hot detergent solutions.
- Poor resistance to concentrated hypochlorites, but resist the 0.5% solutions commonly found in swimming pools.

Elastomeric fibres are used in many garment applications where stretch, comfort, colour and fine denier are required such as stockings, tights, socks, underwear, swimming costumes and sports garments, Figure 22-6, but these fibres are also used to provide elastomeric properties to shoe upper materials, medical bandages and medical gloves. There is also an expanding market for stretch fabrics for furniture, protective clothing and luggage applications. Products are commonly manufactured using a blend of elastomeric fibre with other synthetic and natural fibres or using covered yarns, made by wrapping nylon, cotton or wool around elastomeric filaments. Most applications use 5 to 20 per cent of elastomeric fibres, dependent on the application, but this can fall to 2 to 4 per cent for sock tops and hosiery as this gives the required properties.

Fig 22-6
Track suit with elastomeric fibres

The elastomeric fibres have a similar structure to thermoplastic polyurethanes, with soft block segments of 1,000 to 3,000 molecular weight and hard segments of substituted polyurethanes or polyureas. The majority of the elastomeric fibre production is based on reacting a prepolymer of pure MDI and a polyester polyol with 1,6-hexane diol as chain extender, but fibres are also made by reacting a prepolymer based on H_{12}MDI and PTMEG with an aliphatic diamine. Titanium dioxide pigment is added to produce a white yarn and UV stabilisers and anti-oxidants additives are incorporated to improve stability.

There are several spinning processes used for the production of elastomeric fibres – dry, wet, reaction or hot-melt extrusion – with the first two being the most important. The dry technique is the most common process and involves spinning the DMF polymer solution in a current of hot air to remove the solvent, which is recovered and recycled. In wet spinning, the DMF solution is extruded into a solvent extraction bath, in which the DMF is soluble, but the elastomeric fibre is not. In reaction spinning, the prepolymer is extruded into a diamine solution to produce a partially cured fibre that is then moisture cured. Hot-melt extrusion spinning avoids the expense of solvents and related recovery systems and allows co-polymer extrusion.

Elastomeric fibres are produced as mono or multi-filaments, with a wide range of thicknesses, and further spun into yarns with the type used being dependent on the final application.

Integral skin foams

The key feature of integral skin foams (ISFs) is that they are blown with a volatile blowing agent, which with a combination of pressure and the correct mould temperature causes a surface densification – the skin – as shown in Figure 22-7. The foam core of the ISFs can be open or closed cell and they are flexible, semi-rigid or rigid dependent on the formulation. ISFs were developed in the 1960s when CFC-11 was introduced, but the phasing-out of CFCs has led to other blowing agents being employed and HFC-141b, HFC-134a and water have been used in recent years. These will need to be replaced by other types, depending on environmental laws. ISFs are used in a broad range of applications some of which are listed in Table 22-3.

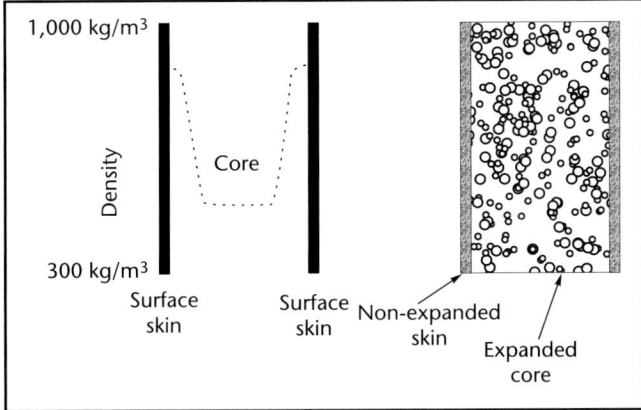

Figure 22-7 Density gradient across a section of ISF

Table 22-3 Some applications for ISFs

ISF type	Automotive and transportation	Furniture and leisure	Miscellaneous equipment
Flexible	Steering wheel	Arm rests	Bicycle seats
	Head rest	Theatre and sports stadium seating	Motorcycle seats
	Airbag deployment door	Pedestal encapsulation	Dunnage
	Handles	Keyboard wrist rests	Sports helmets
	Gear shift knobs	Exercise equipment	Protective equipment
	Miscellaneous interior trim	Pleasure rides	Roll bars
	Bus, train and plane arm rests		Luggage racks
Semi-rigid	Mirror surrounds		Wheel chocks
	Spoilers		Bumpers
	Wheel arch trim		
Rigid	Spoilers	Business machine housings	Filter press plates
	Trim	Computer and ATM housings	
	Sun roof surrounds	Chair arm inserts	

The choice of diisocyanate depends on the modulus and elasticity required, with low-functionality pure MDI-based prepolymers used to achieve good elastic properties and abrasion resistance whilst higher functionality MDIs are used for stiffer polymers. A broad range of polyols is used ranging from high molecular weight propylene oxide/ethylene oxide triols and diols for flexible foams to low molecular weight sucrose-based functional polyols for rigid ISFs. Figure 22-8 shows schematically the range of chemistry used for the production of ISFs and typical formulations and properties are given in Table 22-4.

The main blowing agents are HFCFs and HFCs, since they produce the same effects as CFC-11, and recently cyclo-pentane and iso-pentane have been introduced, but care needs to be taken due to their flammability. Attempts have been made to use carbon dioxide, either introduced as liquid at the mix-head or from the isocyanate/water reaction, but

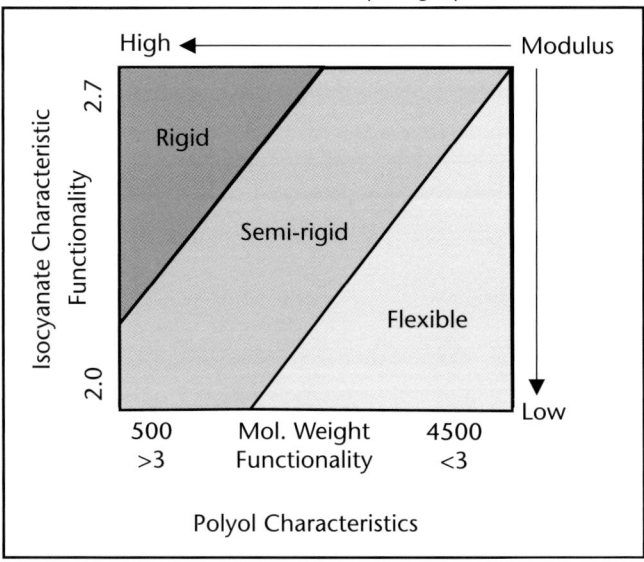

Figure 22-8 The type of ISF manufactured depends on the chemistry employed

Table 22-4 Formulations and properties for ISFs

Formulation, wt-%	Flexible		Semi-rigid	Rigid
	Standard	General purpose		
Polyether polyol triol (MW 6,000)	56.5	–	–	–
Polyether polyol triol (MW 4,500)	–	54.5	–	–
Polyether polyol triol (MW 1,500)	–	–	54.1	–
Sucrose based polyol (MW 500)	–	–	–	58.1
Butane diol	7.6	–	–	–
Ethylene glycol	–	5.2	2.9	3.0
Silicone Surfactant	–	0.1	0.6	0.6
TEDA 33% in DPG	0.6	0.6	0.3	0.1
DMCHA	–	–	0.3	0.5
Water	–	–	0.1	0.2
Pigment paste	2.6	3.1	–	–
Blowing agent	2.6	6.3	5.8	2.1
Modified MDI (NCO 28%)	30.1	30.1	–	35.5
Polymeric MDI	–	–	35.9	–
Processing conditions				
NCO Index	0.98	1.00	1.00	1.05
Cream time, s	4 to 30 (dependent on part size)			
Mould temperature, °C	30 – 40	30 – 40	30 – 50	30 – 50
Demould time, min	1.5 – 5 (dependent on part size)			
Physical properties				
Density skin/core, kg/m^3	1,100/700	900/400	900/400	1,200/900
Tensile strength, MPa	8	1.5	2.5	120
Elongation at break, %	100	75	40	8
Hardness Shore A	85	65	65	–
Shore D	–	–	–	70

with limited success as the result is surface densification rather than true skin formation. For applications requiring UV resistance, or light coloured parts, then in-mould coatings, lacquering or aliphatic diisocyanate-based systems need to be used.

Metal moulds are used for long production runs because they reproduce the fine grain surface effect required and have better temperature control than resin moulds. The latter, made from rigid epoxy or polyester, are normally used for development work or short production runs. Flexible moulds, silicone or polyurethane elastomers, are used to mould parts with undercuts to avoid making complicated multi-piece moulds. The correct mould temperature, 30 to 50°C, is a critical factor for ISF manufacture as above this temperature the skin thickness decreases, while moulds that are too cool lead to surface cure problems. On demould, the ISF cools and contracts and closed cells in the foam will shrink which will result in a permanent deformation unless air is introduced. This is achieved by injecting compressed air, using a hollow needle, or simply puncturing the part. High density ISF does not normally require puncturing.

The pressure generated in ISF moulds can be as high as 1.5 MPa, so construction needs to be robust and a high clamping force is required to prevent moulds opening during injection. To avoid defects it is essential to design in methods for air removal and since most parts have at least five aesthetically visible surfaces, this can be difficult. A variety of techniques are used such as, vents, dummy chambers, avoidance of turbulence at the injection point and injection from the bottom of the part to sweep out air as the polyurethane foam expands.

High-pressure machines are increasingly used as both a way of achieving higher productivity and to eliminate mix-head cleaning solvents from the process. For high volume production, such as steering wheels, it is common for stationary presses with fixed high-pressure mix-heads to be employed with as many as six presses fed from the same high-pressure dispensing unit. Each press is located in a ventilated chamber where mould release and in-mould coating can be applied. For lower volume production, or when a varied range of parts is moulded, then a carousel machine, with manually or automatically clamped moulds, is normally used.

Mould-release agents for ISF are specially formulated to produce a sharp, bright surface finish that can then be easily lacquered with or without removing the mould release. In-mould coatings are used to colour ISFs, especially for steering wheels and arm rests, to produce an aesthetically pleasing durable finish; mould release is still required to remove the part from the mould. The ISF is generally mass pigmented with a colour similar to the coating, which can be added at the mix-head using a series of colour dosing units and it is possible to change colour every shot, although it is more economical to work in campaigns of colours.

Reaction injection moulding elastomers

Reaction injection moulding (RIM) refers to any process where chemicals are mixed, reacted and injected into a closed mould. However, in recent years it has become associated with the production of solid elastomeric parts by the rapid injection of polyurethane, polyurea or hybrid systems using self-cleaning high-pressure machines. Large parts can be moulded and the process is competitive with conventional thermoplastic injection moulding with the advantage of significantly reduced tooling costs. The main applications for RIM have been in the automotive and associated transportation industries through applications such as body panels – bonnets, bumpers, boot lids; reinforced RIM – seat frame and chassis parts; miscellaneous – spare wheel covers, tractor roofs, tractor seats and mudguards.

The modulus of RIM materials can be varied from soft flexible rubber-like elastomers to high stiffness materials. Low viscosity liquids, mixed at pressures of 10 to 20 MPa, fill intricate moulds leading to parts with thicknesses that vary from 1.5 to 10 millimetres. Thicker sections can be produced and whilst they can have better physical properties have the disadvantage of longer demould times, increased weight and lower production rates. Class A, paintable, grained and patterned surfaces can be produced and in-mould coating can be used to achieve a durable surface finish with built-in colour.

Most RIM formulations are based on pure MDI, as prepolymers or carbodimide modified, and propylene oxide/ethylene oxide polyether polyols with the degree of ethylene oxide tipping selected to enhance miscibility and cure. Both diols and triols are used with molecular weight ranging from 600 to 2,000 dependent on the modulus required. Short chain glycols such as ethylene glycol and butane diol are used for chain extension and the reaction accelerated by amine catalysts, such as TEDA, or metal catalyst such as potassium octoate. To improve flow of the components in the mould, blocked catalysts are also used with hindered tin salts preferred. Surfactants that assist nucleation of the polyol component and isocyanate soluble surfactants are employed. Pigments and soluble dyes are often incorporated to colour the part even if post painting is applied. Other common additives include antioxidants and UV stabilisers.

Elastomeric RIM parts are characterised by excellent tensile, tear, chemical, abrasion and impact resistance properties, listed in Table 22-5, which are enhanced by post-curing the moulded part, typically for an hour at 120°C.

Reinforced RIM (RRIM) systems provide increased mechanical properties in comparison to standard non-reinforced material, but the formulations are essentially the same, with the processing equipment changed to accommodate the incorporation of filler. Full details of the chemistry and properties of reinforced RIM products are given in Chapters 28, 29 and 30.

Table 22-5   Properties of elastomeric RIM (4-mm section)

Hardness, *Shore D*	55	64	70
Density, *kg/m³*	1,100	1,100	1,100
Flexural modulus, *MPa* -30°C	660	1,300	1,800
20°C	235	530	1,050
70°C	165	280	425
Ratio of modulus (-30°C/70°C)	4	4.6	4.2
Tensile strength, *MPa*	23	28	32
Elongation, %	260	130	80
Tear strength (Die C), *kN/m*	85	125	145
Heat sag test (10-cm cantilever deformation 1 hour at 121°C), *mm*	3.5	2	1.6

So that parts up to 10 kilogrammes can be injected before the fast reacting materials begin to react, high-pressure machines are used with throughput rates up to 100 kilogrammes per minute and fitted with self-cleaning impingement, straight or L-shaped, mix heads and a 24-millimetre diameter clean-out piston. In the case of larger parts and the extremely fast reacting polyurea systems machines with a throughput of 250 kilogrammes per minute are used. The velocity of the liquid stream as it enters the mould should be the same as the velocity from the mix-head, to avoid surging and turbulence, which can lead to air trap problems.

The mould injection port is normally fitted with a built-in after-mixer, which eliminates lead/lag problems of the two components and improves mix quality. To achieve laminar flow at the entry point of the mould a simple fan, film or tubular gate is used with the choice dependent on the part being moulded. For polyol blends with a viscosity of 1,000 mPa.s mixed with a diisocyanate component with a viscosity between 50 to 100 mPa.s the entry velocity is usually between 2 and 4 metres/second.

As most applications for RIM elastomers are in the automotive industry, the surface finish of the moulds is critical since any surface blemishes will be duplicated in the part. Resin moulds, epoxy or polyester, can be used for prototypes and short runs, but they tend to have a short life that depends on the complexity of the part, as well as the amount of care taken during use. For bulk production, metal moulds are used as they can be produced with a high quality surface finish, have a longer life and offer higher productivity rates.

Machined steel moulds have the greatest durability and are capable of producing up to 250,000 parts, but other metals such as nickel/copper alloy, aluminium and Kirksite are also used. Machined aluminium tends to have a porous surface, leading to problems with the finish, and cast aluminium moulds are preferred as they have lower porosity and perform better. Kirksite moulds are heavy and temperature control is difficult due to the low thermal conductivity, but they perform adequately. To achieve extremely high definition surface finishes a thin coating, usually nickel, can be electro-deposited onto a master to form a shell that is backed up with copper and supported on a fibreglass former. A spray metal technique can be used instead to make the shell. The life of these moulds can be 100,000 parts with the most common form of failure being surface delamination of the shell from the backing.

At the start of production the moulds need heating, but due to the high exotherm of the reactions, it soon becomes necessary to cool them in order to maintain a consistent temperature, which should be in the range 50 to 70°C, and ideally should not vary by more than ± 3°C over the entire surface. To achieve this constant temperature, cooling/heating coils are usually built into the moulds at a consistent depth and spacing. Polyurea RIM systems require a higher mould temperature than polyurethane RIM.

Air needs to escape (vent) from moulds at the same velocity as the mould is filled to prevent air entrapment. Therefore, as the injection time varies from less than one second up to 10 seconds, and the volume of liquid injected can be as high as 10 litres, care has to be taken to correctly size and position the vents. Correctly sized vents will lead to air removal without excessive material loss, but care should be taken to design a mould split line that forms a tight seal since reliance on flash venting is not a reliable technique and leads to the need for flash trimming. Vents that are self-cleaning, through having an inverse conical section, assist demoulding and minimise the risk of damage during mould cleaning.

Mould-release agents for RIM applications are designed to allow post painting, with or without part washing or are specific for use with in-mould coatings. They are temperature sensitive since the waxes used have specific softening points and silicone oils are not used as they can lead to painting problems. During production, a mould release residue builds up with the need for frequent mould cleaning, but this can be reduced by the use of an internal mould-release (IMR) agent. Moulds can be cleaned by the conventional solvent soak method followed by scraping and washing or a solid carbon dioxide blasting system is an alternative.

The incorporation of inserts into parts is relatively easy with the RIM process as the excellent flow characteristics and reactivity of the systems lead to good adhesion to the insert. Thicker sections, around bosses, do not show the same tendency for sink marks that is seen with many thermoplastic injection moulded parts.

Polyurea elastomers

Polyurea elastomers are obtained by reacting polyfunctional amine-terminated polyethers with polyfunctional isocyanates, but the technology is normally limited to di- or tri-functional polyamines reacted with aromatic diisocyanates. The major advantage of these materials is their rapid reaction, fast cure even at low temperatures, which leads to them being used in the injection moulding of RIM applications such as class A surface automotive body parts with cycle times below a minute. The high cost of these materials has inhibited development of the automotive applications, since competitive materials, such as thermoplastics and conventional polyurethane RIM, are cost competitive alternatives with adequate properties.

Application areas where the higher cost of polyurea elastomers is not such a barrier to market growth is in coatings and two-component spray polyurea elastomers are rapidly finding many uses such as:

- Secondary containment.
- Pipeline protective coating.
- Truck bed liners.
- Concrete coatings.
- Mining applications.
- Replacement for PVC skins.

Whilst these materials are true elastomers they are applied as coatings and, thus, their chemistry and properties are detailed in Chapter 24.

23. Introduction to coatings, adhesives, sealants and encapsulants

Jacquin Wilford-Brown

The link between these four applications is the requirement to bond to one or more substrates. There may or may not be an exposed surface of the bonding material (compare a coating with a typical adhesive). The key purpose of coatings is to provide a thin film or layer over a substrate with protective, decorative or specialist properties. Adhesives are used to bond substrates together; sealants and encapsulants can be thought of as similar materials, except that their prime purpose is to hold things apart. For all four areas, technical integrity is of major importance.

The global market for polyurethane-based products in this area is about 2.6 million tonnes split between the key segments as shown in Figure 23-1.

Polyurethanes were introduced to this market segment in the 1950s, a long time after the introduction of alkyd and formaldehyde resins during the early part of

Figure 23-1   Global market share for polyurethane products in coatings, adhesives, sealants and encapsulants

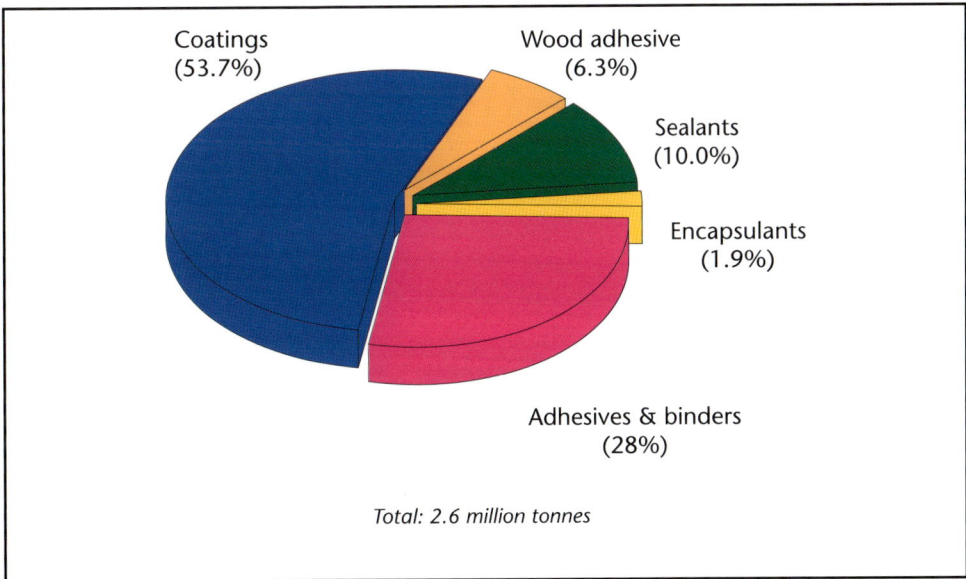

the 20th Century, but at a similar time to the arrival of acrylics and epoxies. It was quickly recognised that polyurethanes would be very useful as adhesives and coatings and their use in paints was quickly accepted and rapidly followed by the manufacture of artificial leather, wood and rubber composites and hot-melt systems. Pressure-sensitive and structural adhesives plus reactive hot-melts soon followed and the arrival of commercial production of the aliphatic isocyanate HDI, in 1967 led to the development of light-stable coatings.

Polyurethane confers many advantages on coatings, adhesives, sealants and encapsulants that include excellent mechanical properties due to the elasticity of the urethane network, which produces tough, flexible, and durable films or glue lines and they can also provide good chemical and solvent/oil resistance. The high reactivity that can be built into polyurethane formulations means that the required polymer structures can be obtained even when components are mixed at room temperature or even lower.

Polyurethane materials are available in a wide variety of types and forms, both liquid and solid, to suit the many different applications of the coatings, adhesives, sealants and encapsulants markets. The key classes and types are listed in Table 23-1.

Table 23-1   Classification of systems

Class	Chemistry (Mechanism)	Types	Applications
Non reactive	TPU & PUDs (Carrier evaporation)	Solvent-borne Water-borne Hot-melt	Contact adhesives
Reactive: one-component	Isocyanates (Moisture cure)	Solvent-borne Solvent-free	Cross-linker Wood binding
	Isocyanate prepolymers (Moisture cure)	Solvent-borne Solvent-free	General purpose adhesives Maintenance coatings Sports floors Sealants
Reactive: two-component	NCO + Amine (Polyurea)	Solvent-free	Pipe coatings Roof coatings
	NCO + Polyol (Polyurethane)	Solvent-borne Water-borne Solvent-free	Auto refinish & OEM coatings Sealants Encapsulants Flexible packaging adhesives Anti-corrosion coatings Synthetic mortar
	Blocked NCO + Polyol (Polyurethane)	Powder Stoving	Coating white goods Can and coil coatings Maintenance coatings
Reactive: other	Urethane alkyds (Oxidation)	Solvent-free	Protective coatings
	Urethane acrylates (Radiation cure)	Solvent-free	Wood coatings

TPU – thermoplastic polyurethane
PUD – polyurethane dispersion

Raw materials

Isocyanates

Most of the isocyanates are utilised in this area with the main aromatic isocyanates being TDI and MDI whilst of the aliphatic isocyanates HDI, IPDI and H_{12}MDI are all used. MDI and its prepolymers/adducts have a low vapour pressure and can be safely handled at room temperature with the same being essentially true for H_{12}MDI. However, TDI, HDI and IPDI have relatively high vapour pressure and toxicity. Therefore, these isocyanates are normally transformed into functional products, which have a lower vapour pressure.

TDI is mostly used as the tri-functional adduct formed from its reaction with trimethylolpropane and whilst TDI-trimer exists, its use is less common. HDI is commercially available as a biuret, trimer and as a mixture of dimer and trimer. The TDI- and HDI-based polymeric products are generally liquids at room temperature with relatively high viscosities, but in order to make them easier to handle they are typically supplied as 50 to 80 per cent solutions in solvents, which have isocyanate values ranging from 8 to 22 per cent. An exception is the mixture of HDI dimer and trimer which is a low viscosity liquid at room temperature. IPDI is used as either the monomer, as solid trimer or as blocked derivatives.

Since aromatic isocyanates yellow on exposure to UV/daylight, for many coatings, and for adhesive applications where light stability is required, aliphatic isocyanates are preferred, the majority of these being HDI or HDI-adducts. In non-light sensitive applications aromatic isocyanates are normally used, not only on cost grounds, but also because they have significantly higher reactivity than the aliphatic isocyanates; see Chapter 5 for further details on isocyanates.

Polyols

A broader range of polyols is used to formulate coating, adhesive, sealant and encapsulant systems than in any other area of polyurethane chemistry and a brief outline of these is given in Table 23-2 with further details in Chapter 6.

The choice of polyol is important in controlling the final properties of the material, with the chemical type, molecular weight, structure and reactivity of the polyol being important especially with regard to flexibility, hardness and chemical resistance. Functional polyols will lead to high levels of cross-linking and, therefore, harder and less flexible films. High molecular weight, low functionality polyols produce more flexible films although there can be an increase in phase separation, which will affect the polymer hardness and toughness. Stiff polymer chains, as a result of the introduction of aromatic rings in the main chain or bulky groups along the chain, generally give polymers with higher glass transition temperatures.

Table 23-2   Key polyols used for coating, adhesive, sealant and encapsulant applications

Type	Examples	Comments
Polyether	PO PO/EO PTMG	Generally lower viscosity and lower cost. Good acid/alkali resistance. PTMEG has higher performance, viscosity and price.
Polyester	Mainly polyadipates (adipic acid & various glycols) but also polycaprolactones	Polar polyester groups tends to help adhesion. Good flexibility and solvent resistance. Better weatherability than polyethers, but generally more viscous and higher cost. Can be amorphous or crystalline.
Acrylic	Acrylic polymers based on hydroxy ethyl acrylate (HEA) and/or hydroxy ethyl methacrylate (HEMA)	Used in high performance coatings to improve gloss, hardness, colour and durability. Also used to increase water compatibility.
Alkyd	Typically based on phthalic acids, glycerol/TMP and fatty acids	Used where solvent resistance and durability is not critical, and where vegetable oils are widely available; very low cost.
Hybrids	Acrylic/polyesters	Mainly acrylic.
Others	Castor oil based polyols Polyhydroxybutadiene	Used where hydrophobicity is important.

Other ingredients

A wide range of other components is needed to allow these coating, adhesive, sealant and encapsulant formulations to meet their required specifications and, more importantly, to control the ease of processing. Some of the commonly used materials are listed in Table 23-3. This list is by no means exhaustive, but these secondary products have all been discussed in detail earlier, mainly in Chapters 9 and 10.

Surface interactions

All coating, adhesive, sealant and encapsulant formulations share one common key characteristic in that all of them are required, in one way or another, to interact with surfaces. The dominance of this feature is what distinguishes these materials from all other areas of polyurethane applications and key factors controlling this interaction are:

- Surface wetting and spreading.
- Controllable reaction rate.
- Strong adhesive bond.

Table 23-3 Other ingredients used in coating, adhesive, sealant and encapsulant formulations

Type	Examples	Comments
Chain extenders	Butanediol Hexanediol HQEE derivatives	To control the formation of the hard segments in the polymer structure.
Catalysts	Tertiary amines Organometallics	Used to control the rate of the reactions.
Solvents	MEK Toluene Ethyl acetate	Dry organic solvents can be used to reduce viscosity.
Pigments	TiO_2 Carbon black Mica	Can be both decorative or functional. Conductivity. Anti corrosion.
Fillers	Calcium carbonate Clays Silicas Organic and polymeric materials Fibres including glass and carbon	Used to provide bulking at low cost. Increased modulus. Silicas/clays can be specially treated to help dispersion and are used as thixotropic agents. All must be very dry.
Flow aids/ rheology modifiers	Silicones	Improve appearance of the final film. Can be difficult to use without introducing other problems.
Foam control/ bubble reduction	Mineral based – hydrophobic silica + surfactant Silicone based – contain hydrophobic oils	Trapped air from mixing, or carbon dioxide formed on curing can cause bubbles in thicker coatings/adhesive films. Additives work by lowering the surface tension.
Moisture scavengers	Molecular sieves Oxazolidones	May be needed in reactive systems to ensure the desired NCO-OH reaction takes place, rather than NCO-water.
Miscellaneous	Antioxidants Light/UV stabilisers	Dependent on the application.

In principle, two ultra clean and flat solid surfaces could be joined together, but in practice contamination by other species, including atmospheric oxygen and the irregularity of real surfaces at the microscopic level, means that liquid products are required to bridge between the two surfaces. Therefore, the ability to rapidly and easily wet the surface or surfaces being covered or glued is a critical feature for all these systems, which by inference need to be liquids at the application temperature. Once the surfaces have been wetted then it is essential that the liquid system solidifies at the required rate to produce a uniform final layer of coating or adhesive, or moulding. Finally, none of this is relevant unless the system has the ability to form a sufficiently strong adhesive bond to the surfaces being coated, glued, sealed or encapsulated that is stable and durable in the environment of use.

Reactions at surfaces and interfaces

Wetting and spreading

A liquid will only wet a surface if its surface tension is lower than the critical surface tension of the solid, if not, then the liquid will remain as discrete droplets as shown in Figure 23-2. The chief effect of variations in surface tension is to control the ability of a material to spread over a surface. The polyurethane components used in coating, adhesive, sealant and encapsulant formulations generally have low surface tension and will wet most substrates effectively, except for very low surface tension solids such as polyethylene and polytetrafluoroethylene (PTFE).

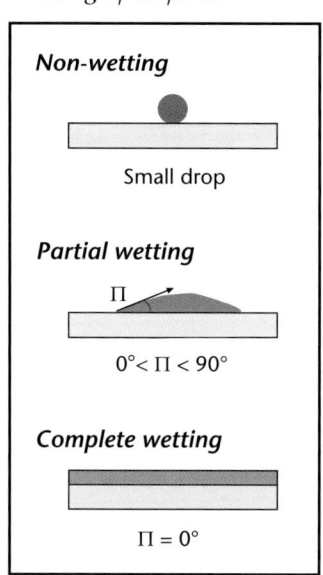

Figure 23-2
Wetting of surfaces

The extent of spreading across a surface, including the microscopic flow, depends additionally on the surface roughness and porosity, the chosen application conditions, the cleanliness of the surface and the viscosity of the liquid. The rheology of the systems can be adjusted, by the choice of raw materials and additives, to give the necessary flow during the time allowed for curing. For porous substrates internal wetting and penetration into the pores can be important, for example in wood binders and adhesives. Obviously, the flow properties and the cure properties of a system in practise are interrelated: as the system cures, the viscosity increases and the extent of flow is reduced. The flow properties also determine the choice of handling and application equipment for these isocyanate/polyurethane systems.

Surface reactions

There are five classical modes of adhesion, all of which can give very good adhesion and there is often significant debate about which mechanism is dominant:

- Primary bond formation – covalent or ionic links.
- Secondary bond formation – hydrogen bonds, Van der Waals interactions.
- Mechanical interlocking.
- Electrostatic forces – between positive and negative charges.
- Diffusion interface – where the resin and the substrate are compatible and both diffuse across the interface.

Another important element in adhesion is the type and condition of the surface being bonded to as, for instance, a weak surface layer on the substrate can severely reduce the overall strength. Clearly, good surface preparation can improve adhesion and either cleaning the surface, to remove contamination, or priming or roughening, to activate the surface, should provide a good surface for chemical or mechanical bonding.

For isocyanates and polyurethanes mechanical and/or physical bonding to the surface can be quite strong, especially if the surface is sufficiently rough to allow 'keying'. Where the polyurethane film can be formed adjacent to the surface, then the good intimate contact will enable this physical process to take place and also allow Van der Waals-type bonding to occur.

However, often some form of chemical bonding or reaction takes place as the majority of surfaces have strongly bound active sites, such as active hydrogen (-NH or -OH) or chemisorbed water groups. Isocyanates can react with these forming covalent bonds or polyurethane can hydrogen bond with them, in both cases generating strong adhesive bonds.

Solidification and bonding

Once the surface(s) are covered, the isocyanate or polyurethane material needs to solidify or cure to form the final film or joint. A wide variety of hardening mechanisms are possible – from physical freezing of a high molecular weight polymer, via solvent removal, to a chemical reaction – depending on the exact system. The system can therefore be chosen to fit the curing profile desired. For example, moisture-curing prepolymers are used in rubber crumb binders to make sports tracks, to give a one-component system with relatively long working times followed by cure at ambient outdoor conditions. In contrast, polyurea systems are designed to cure very rapidly even at low temperatures once the components are mixed together. Polyurethane systems have a particular advantage for thermally sensitive surfaces, such as plastics, because of their ability to cure at ambient or low temperatures.

The overall strength of the system depends on the bulk and surface properties of the coating or adhesive, those of the substrate, and the interface between the two. Polyurethane systems generally have good performance because of the strength of the network formed, and its ability to flex with the substrate and dissipate stresses. Typically, systems are formulated to have a modulus high enough to withstand the expected stresses, but low enough to have some ductility. Examples of failure modes are shown in Figure 23-3.

If there is a sharp boundary between the polymer and substrate, then an interface is formed, but if there is no clear boundary then the region of interpenetration is known as an interphase. Failure can be through the bulk material, at the boundary layers or in the interphase region and with many failures small amounts of polymer matrix or substrate are left on either side.

Test methods

A large variety of tests are performed on both the isocyanate/polyurethane formulations and the final coating, adhesive, sealant or encapsulant products. Typical tests for liquid systems and components include density, flashpoint, non-volatile content, contact angle measurement (to determine the wetting of surfaces) and, for hot-melts and thermoplastic materials, the thermal stability and softening point. On the final products these vary from elementary property measurements, such as fracture energies, to tests designed to give a measure of a

Figure 23-3   Adhesive/cohesive failure modes

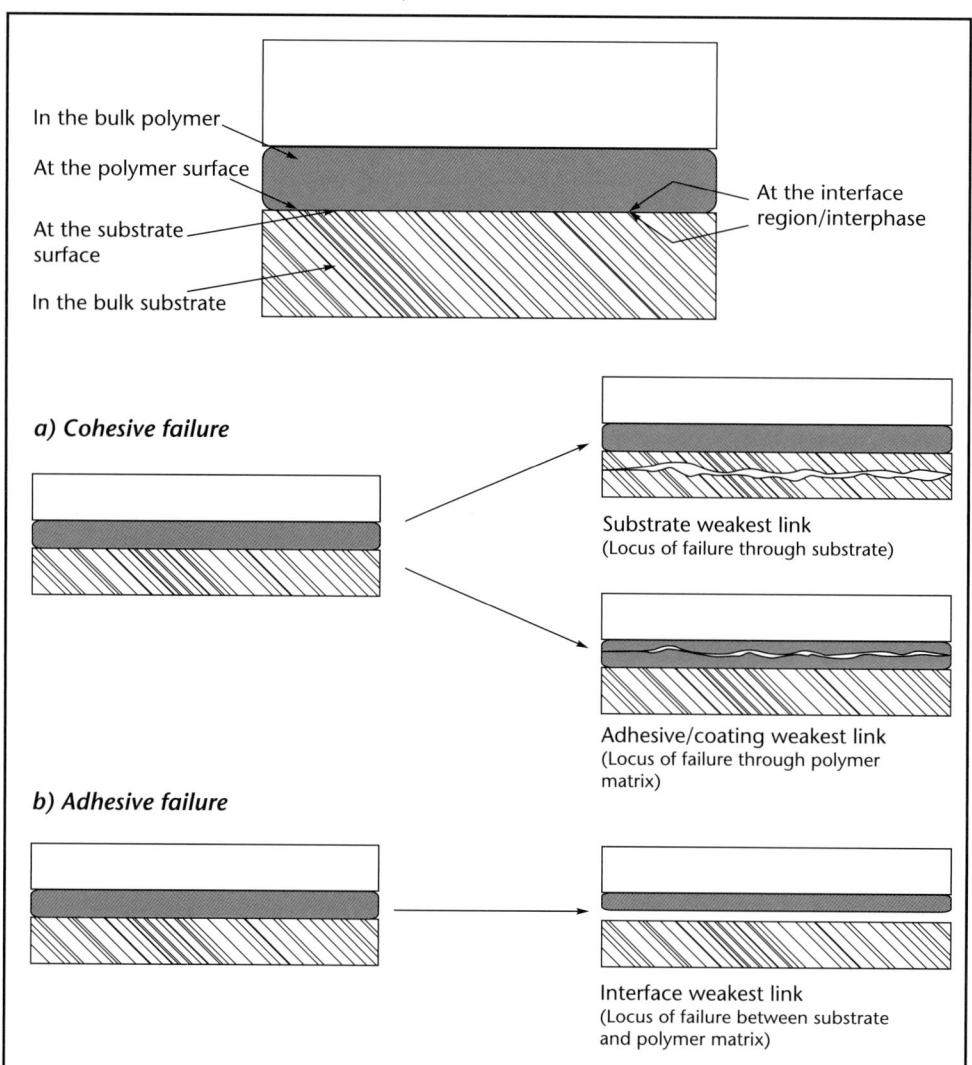

property of relevance for the specific application such as gloss, abrasion resistance, chemical resistance and composite joint strength. This is by no means an exhaustive list and some of the key tests in this area are given in Table 23-4.

Properties of basic materials

Some tests are specific to the isocyanate/urethane system, for example determining the isocyanate content of components used for paints, varnishes and adhesives, but the main properties of interest for the basic materials are flow, storage stability and cure profile.

Flow properties
The usual measure of the flow properties is viscosity, or the ratio of stress over shear rate, which represents how far a material will flow under an imposed force.

Table 23-4   Standard test methods

Type of test	ISO	European EN
Basic materials		
Density of adhesives		EN 542:1994 & 543:1994
Density of paints and varnishes (2 parts)	ISO 2811:1997	
Determination of isocyanate content in adhesives		EN 1242:1998
Flashpoint		EN 456:1991 & EN 924:1994
Freeze-thaw stability		EN 1239:1998
General test methods for polyisocyanate resins in paints and varnishes	EN ISO 11909:1998	
Non-volatile content	EN ISO 3251:1993	
Softening point (thermoplastics)		EN 1238:1999
Surface wetting		EN 828:1997
Thermal stability (hot-melts)	EN ISO 10363:1995	
VOC content (2 parts)	ISO 11890:2000	
Viscosity		
Brookfield viscosity for liquids, emulsions, dispersions	ISO 2555:1989	
Cone and plate viscosity (high shear)	ISO 2884-1:1999	
Flow cup	ISO 2431:1993	
Rotational viscosity for liquids, emulsions, dispersions	ISO 3219:1993	
Viscosity of adhesives		EN 12092:2001
Cure		
Gel-time of thermosetting powders	ISO 8130-6:1992	
Pot-life of adhesives	ISO 10364:1993	
Surface dry (Ballotini)	ISO 1517:1995	
Through-dry of coatings	ISO 9117:1990	EN 29117:1992
Sample preparation		
Adhesive/metal preparation	ISO 4588:1995	
Adhesive/plastic preparation	ISO 13895:1996	
Film thickness	ISO 2808:1997	
Surface preparation and coatings application methods	EN ISO 4618-3:1999	
Temperature/humidity conditioning	ISO 3270:1984	
Mechanical properties		
Abrasion resistance, paints and varnishes (3 parts)	ISO 7784:1997	
Bond strength, engineering plastics	ISO 15509:2001	
Creep properties under static load	ISO 15109:1998	
Cross-hatch test	EN ISO 2409:1994	
Fatigue properties in tensile shear	EN ISO 9664:1995	
Gloss	EN ISO 2813:1994	
Impact for paints and varnishes (falling weight)	EN ISO 6272:1994	
Impact for structural adhesives		EN 29653:1994
Leather & footwear adhesives		EN 1392:1998
Mandrel bend, conical	EN ISO 6860:1995	
Mandrel bend, cylindrical	EN ISO 1518:2000	
Peel of a flexible-to-rigid assembly (2 parts)	ISO 8510:1990	EN 28510:1993
Peel resistance of structural adhesive bonds		EN 1464:1994

Standard test methods (continued)

Type of test	ISO	European EN
Mechanical properties (continued)		
Pencil hardness	ISO 15184:1998	
Shear behaviour of structural adhesives (2 parts)	ISO 11003:2001	
Tensile lap shear strength of adhesives		EN 1465:1994
Wood adhesives, shear strength (tensile)	ISO 6237:1987	
Wood adhesives, non-structural		EN 205:1991
Wood adhesives, shear strength (compression)	ISO 6238:2001	
Wood adhesives, structural (4 parts)		EN 302:1992
Weathering		
Artifical weathering	EN ISO 11341:1997	
Continuous condensation	EN ISO 6270-1:1998	
Corrosion resistance, salt spray	ISO 7253:1996	
Cyclic corrosion (2 parts)	ISO 11997:1998 & 2000	
Natural weathering of exterior wood coatings (5 parts)		EN 927:2000
Solvent resistance (2 parts)	EN ISO 2812:1994 & 1999	
Water vapour transmission	EN ISO 7783-1:1999	

Specific viscometers are designed to measure different types of systems, from low viscosity isocyanates, polyols or polymers in solution to highly viscous polymer melts, under different conditions of shear or stress, to assess how easy it will be to pump materials. At moderate concentrations and for relatively simple liquids, the viscosity is independent of shear rate, and the liquid is Newtonian. However, for concentrated polymers, filled systems and most products used in this area, the viscosity does depend on shear rate, non-Newtonian behaviour, and only the apparent viscosity can be measured. All viscosity measurements should, therefore, indicate the instrument used and/or the shear conditions, as well as the temperature of the measurement.

The time-dependent behaviour of systems can also be important as directly after a change in shear condition, the viscosity of materials can change either slowly or quickly. For example, after an adhesive is forced out of an applicator at high shear, does it continue to flow, or does the viscosity increase? When a coating is applied to a surface, can it be rolled or brushed to give a smooth film? The viscosity can be measured with a viscometer, but it is also important to use a range of practical tests to determine the precise handling performance of individual systems.

For many coating systems, a 'cup' measurement is used in which the time for a given amount of a coating system to flow through a cup with a standard orifice is measured. This is a simple and portable measurement, but can give reproducibility problems for reacting systems.

Storage stability
The storage stability, or shelf-life, of products is another important criterion and depends on factors such as the physical separation of components and whether chemical reactions continue to take place in the stored material, either via internal processes or through external contamination. Pigmented or filled systems can settle and some systems may de-mix. Freeze-thaw stability can be important, especially for exterior use systems, and for systems with free isocyanate levels the stability limit is determined by the need to protect these products against chemical reaction with adventitious or atmospheric water.

Cure profile
Knowledge of the cure profile is important in determining the processability of a system and there are a variety of methods available to measure it, all of which involve measuring the change of a property versus time:

- Cure kinetics.
- Viscosity or modulus build-up.
- Industry specific tests.

The reaction exotherm can be measured and used to determine cure kinetics by differential scanning calorimetry (DSC), and the reaction of specific groups, such as isocyanate, can be followed using fourier transform infra-red spectroscopy (FTIR), which works well for both thin films and bulk materials. FTIR is described in Chapter 9.

One technique for following the build-up of viscosity, or modulus, during reaction is the scanning vibrating needle curemeter (RAPRA Technology Ltd) that follows the reaction profile, from mix to final gel-point, through monitoring the frequency and amplitude of vibration of a needle immersed in the reacting mass. In early stages of the cure, the resonance amplitude is a measure of the viscosity of the resin; after the gel point, it is related to the resilience and network formation. The resonance frequency shifts at the gel-point, and provides a measure of the build-up of the storage modulus.

Cure profiles are often indirectly determined by looking at the change of a key property with time/temperature and dependent on the application, this could be bond strength, surface adhesion, solvent resistance and so on. A common test involves a mechanically-driven needle which is placed in contact with a wet cast film on a supporting plate, as the film cures, the needle slides through the cast film leaving a trace in the resin. The start-of-open-trace 'SOT' and the end-of-open-trace 'EOT' are measured as the 'gel time' and 'cure time' respectively. This test is often performed inside a humidity/climate chamber to mimic typical application conditions.

Others
Many tests are not standardised, but relevant to the use of the material, such as: tack time, touch-dry time, gel time and cure time with the pot-life/working time being a measure of how long a system can be used in a reproducible way.

Sample preparation

For any coating, adhesive, sealant or encapsulant system, the preparation of the surface or joint and the conditioning of the sample can significantly affect the properties of the final article. This is true not just for testing of properties, but also for using the materials in practice. Wherever possible, samples should be tested under conditions related to those in service.

It is important for all surfaces to be degreased, cleaned and freshly prepared if possible, to prevent problems from contamination or irreproducibility of surfaces. Specific preparation may depend on the exact substrate and guidelines exist for adhesives applied to plastics and metals.

For coatings, the thickness and uniformity of the coating can be critical and standard applicators are used to produce controlled coatings, often on standardised substrates. The properties of the coating or adhesive on the substrate can then be assessed, but if the properties of the coating/adhesive alone are required then a freestanding film can be prepared by removing the cured film from the surface.

In most cases, the sample must be kept under controlled conditions of temperature and humidity to ensure precise control of the drying and cure processes.

Adhesion and other properties are measured according to well-defined procedures, with control of the sample geometry to ensure tests are performed at standard thickness and configuration under controlled stress conditions.

Final product properties

Strength and adhesion

Mechanical properties are often the most significant in determining whether a particular product can be used in a given application and there are a number of methods for assessing the strength of adhesion to a substrate, depending on whether the substrate is rigid or flexible. Common tests are illustrated in Figure 23-4. Lap shear tests are used both to assess the adhesion to a substrate and the cohesive strength of an adhesive, whilst peel and blister tests are used to measure the adhesion to the substrate. For simple cases, the peel force is a direct measure of the tear energy.

Tests are designed to apply a well-defined stress to the bond, and to minimise the amount of energy that is absorbed by deformation either of the substrate or the coating/adhesive, so that the

Figure 23-4   Adhesion testing formats

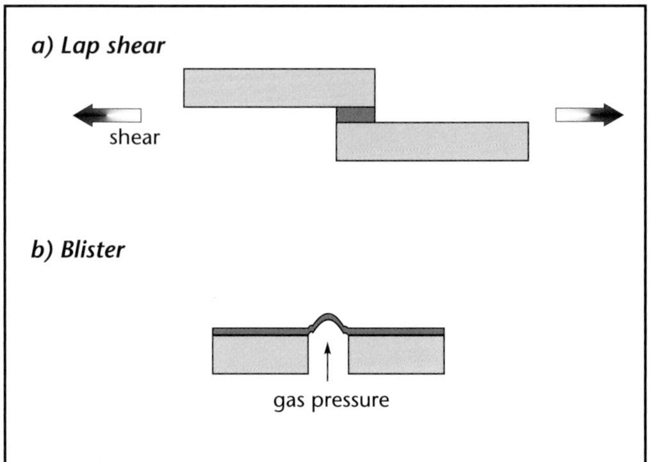

a) Lap shear

b) Blister

c) Peel

Figure 23-5   Single lap joint testing

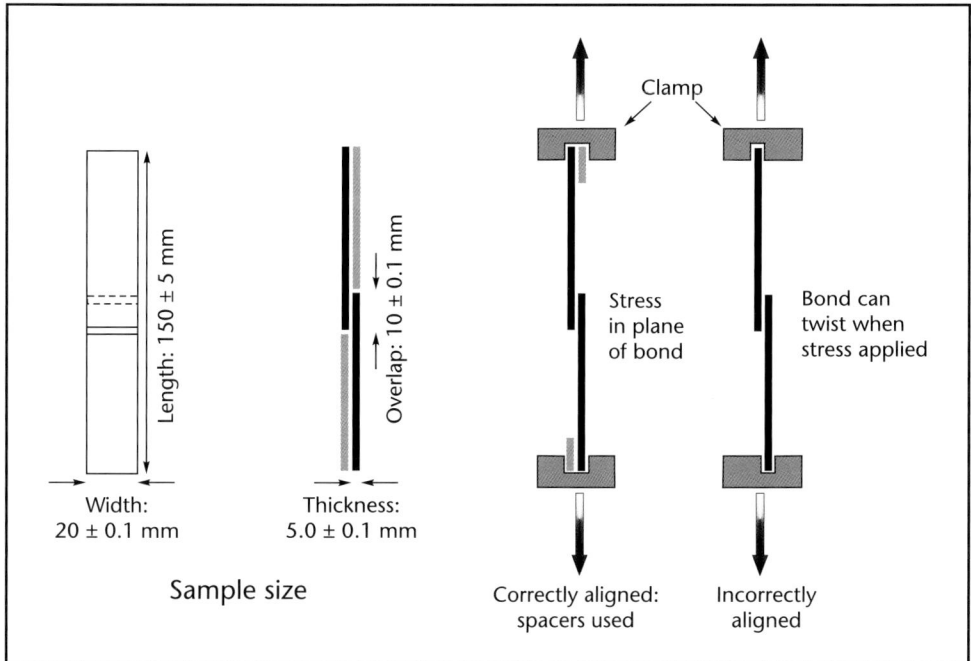

strength measured is the strength of the bond formed. The sizes and shapes of the bonds and joints are precisely controlled, so that stresses are applied accurately and the results can be compared. For single lap joints, care must be taken to ensure that the joint is correctly aligned and gripped with spacers in the testing machine, so that the stress applied when the bond is pulled is pure shear and does not twist the sample, Figure 23-5. Typical measurements of the strength of the bond are normalised stress at failure, modulus and elongation.

A selection of the relevant standards to cover the multiplicity of mechanical tests required to evaluate coatings, adhesives, sealants and encapsulants is listed in Table 23-4.

More empirical methods are also used to test for adhesion, such as the cross-hatch and mandrel bend tests, which are, respectively, used for films on rigid and flexible substrates. In the cross-hatch test pressure-sensitive tape is applied and removed over a series of cuts that have been made in the film, simulating scratches or damage to the coating. Depending on the exact test, the cuts may be linear, or cross- or straight-hatched. The tape used is appropriate for the level of adhesion the coating will need in practise and the level of damage from repeated applications of the tape is assessed. In the mandrel bend test, the coating on a flexible substrate is rolled around either a cylindrical or a conical mandrel of smaller and smaller size to produce tighter bends and the cracking of the coating measured. As well as giving a measure of coating adhesion, this test simulates what might happen to a coating when in use on a substrate that is bent.

Hardness

The surface hardness of a coating is a measure of its resistance to localised deformation and is often used as an indicator of the wear properties. Most tests use an indentor, either a durometer or a pencil and with the durometers a number of scales are used such as the Shore A and Shore D. The relationship between these two scales is provided in Chapter 19. Pencil hardness is a popular test in the paint industry as it is very cheap, simple and quick. A range of calibrated pencils is individually, from the hardest to the softest, pressed onto the surface at 45 degrees and pushed away. Any lines of damage to the coating can easily be seen and the quoted pencil hardness is the hardest pencil which does not mark the surface. Although very quick, this test can be operator dependent and, therefore, it is difficult to obtain reproducibility.

Abrasion

Abrasion resistance depends not only on the surface hardness, but also on the frictional properties and it is normally measured by the loss in weight of a film during a standardised abrasion process in which a sample is rotated against a surface with a specified roughness. A technique commonly used in the industry is the Taber® abrader, Figure 23-6. Specialised tests can be required dependent on the application, for instance a method to measure stone chip resistance for automotive coatings.

Figure 23-6 Taber® abrader

Visual properties

The visual appearance of the final product is important, especially for coatings and pigmented systems, which should normally cover or hide the substrate. A typical measurement involves the determination of the gloss level of a coating and this is measured by comparing the relative intensity of light reflected by the coating. Systems can be very high gloss, as for automotive applications, or very low gloss, as for architectural or protective coatings. Gloss measurements at different angles can vary with surface roughness, and with the type of substrate or under-layer.

Weathering

Weathering or exposure properties are especially important for exterior and/or structural applications of coatings, adhesives, sealants and encapsulants. The effects may be aesthetic, yellowing or chalking of materials due to UV degradation, or a reduction of mechanical properties of the coating or the assembled joint. Materials can be exposed to the natural weather in a range of environments and the results assessed. However, this is not always practicable and a range of accelerated weathering tests is also used to simulate extreme weather conditions. Examples include salt-spray cabinets, 50°C humidity cabinets, UV light exposure and thermal cycling.

In addition to exposure testing, the permeability to water vapour or liquid is often used as a measure of the durability of a material since the ingress of water is often the key step in the deterioration of properties.

For adhesive applications the test must assess the response of the complete joint as the durability of the substrate is also important. This is especially true with wood adhesives, where the wood may swell and degrade on weathering, adding to the stresses on the adhesive bond. Cyclic exposure tests are frequently used to simulate and speed up the measurements.

Chemical and corrosion resistance

The resistance of a material to specific chemicals, such as acids, alkalis, oils and fats, is usually measured by timed exposure at fixed temperatures followed by inspection and measurement of properties. Solvent resistance can be assessed by rubbing the surface of a coating with the desired solvent and evaluating the damage after a specified number of rubs.

Environmental issues and future trends

In all of these applications, environmental issues are becoming increasingly important as customers are demanding reduced atmospheric pollutants, volatile organic compounds and hazardous air pollutants (HAPs), as part of clean air regulations and desires. Polyurethane systems are already being widely adopted for these reasons, examples being moisture-curing systems with no added solvent and reactive hot-melts. In addition to solvent restrictions, higher energy efficiency, lower cure temperatures, faster curing and better material utilisation, thinner films, and reduced material waste are increasingly important. Many of the new developments in binders, adhesives and coatings are in these areas. A brief qualitative comparison of the main routes to environmentally-friendly polyurethane systems is given in Table 23-5.

Table 23-5   Environmental impact of systems

System type	Advantages	Disadvantages
Solvent-borne (one- and two-component)	Very high performance	Solvent emissions
Solvent-borne high solids	Reduced emissions Conventional equipment	Can be difficult to apply without defects Lower molecular weight/functionality hence poorer properties in general
Solvent-free moisture-cure	Good performance (cross-linked network) 1K, robust cure No solvent	Cure depends on water availability Bubble formation
Water-borne (one-component)	Much reduced emissions One-pot system	Need cross-linking Some quality problems (wetting, water sensitivity)
Water-borne (two-component)	Much reduced emissions Can give high quality coatings	Some quality problems (wetting, water sensitivity, NCO Index) Limited use
Powders/hot-melts	No emissions Powder over spray can be recycled	Thick layers High curing temperatures
Radiation-cured	No emissions Energy efficient	Capital cost of equipment Not suitable for all shapes

24. Coatings

Berend Eling

Polyurethane coatings were originally defined as products made from polyisocyanates and polyols but now a broader definition is used that includes all systems based on a polyisocyanate whether the reaction is with a polyol, a polyamine or with water. Consequently a polyurethane coating may contain urethane, urea, allophanate and biuret linkages. Since their introduction in the 1950s polyurethane coatings have grown rapidly because of the highly versatile chemistry and the excellent properties including toughness and abrasion resistance, combined with flexibility, chemical resistance and good adhesion.

The polyurethanes coating industry, a global market of around 1.4 million tonnes in 2000, can be divided into three broad market segments, which are industrial products (44 per cent), construction and architectural (36 per cent) and speciality/maintenance (20 per cent). The main application areas for industrial coatings are automotive, aircraft, electrical insulation, electronics, textile, leather, wood products, coil coatings, appliances, metal furniture and machinery. In construction the major applications for coatings are the protection of metalwork and sealing floors and roofs whilst the decorative architectural applications mainly involve wood varnishes and pigmented enamels. Speciality/maintenance coatings are used mainly in automotive refinish, anti-corrosion, flooring, roofing and decking.

There are four broad types of polyurethane technology used in the coating industry, the first three of which are reactive systems with the fourth category covering all systems with no isocyanate reaction during final application:

- Two-component systems consisting of a polyisocyanate and a polyol or polyamine that are mixed just prior to application with cure generally taking place at ambient temperature (53 per cent).
- Oven-curing polyurethanes use similar materials to the first category except that a blocked isocyanate is used to provide a storage stable one-pack mix with the polyol or polyamine; the isocyanate is then de-blocked at the stoving temperature and reacts. This includes powder coatings (10 per cent).
- Moisture-cure polyurethanes are one-component, high molecular weight, low free-isocyanate content prepolymers that cure by the reaction with moisture from the environment to form urea linkages (4 per cent).
- Non-isocyanate reactive systems, such as thermoplastic polyurethane-based lacquers, aqueous polyurethane dispersions, urethane oils and alkyds, and radiation-cured polyurethanes already containing urethane or urea linkages and where there are no further isocyanate reactions during application (33 per cent).

The reactive polyurethane coatings are generally cross-linked, due either to branched polyols and/or isocyanates, or through the formation of allophanate and biuret. Cross-linking, whilst increasing hardness and abrasion resistance, improving the resistance to water, solvents, weathering and temperature can, if too high a level is used, lead to decreased flexibility.

Materials selection

Isocyanates

There are five isocyanates commonly used in coating formulations, the production and properties of which are described in Chapter 5:

- MDI – 46 per cent.
- Aliphatics – 37 per cent (H_{12}MDI, HDI, IPDI).
- TDI – 17 per cent.

Aliphatic isocyanates have a lower rate of reaction and form softer coatings than aromatic isocyanates when used in comparable formulations and are used where UV or light stability is required, for example in top-coats, and in water-borne systems. HDI and HDI adducts are the most widely used, with H_{12}MDI finding use in water-borne systems. For applications, which are not light sensitive, such as primers, and heavily pigmented maintenance coatings, then aromatic isocyanates are often preferred for economic reasons.

Polyols

There are three main types of polyols used in coatings technology, with hydroxyl values in the range 30 to 500:

- Acrylics.
- Polyethers.
- Polyesters.

Acrylic polyols are amorphous, random copolymers prepared by the radical polymerisation of esters of acrylic and methacrylic acid plus vinyl monomers such as styrene with hydroxyl groups introduced through the use of hydroxylated methacrylates, such as hydroxyethyl methacrylate (HEMA). Three or more different monomers are normally used in the synthesis of acrylic polyols with the choice dependent on the manufacturing process, performance required and cost.

The production and properties of the polyester and polyether polyols are described in Chapter 6.

Acrylic and polyester polyols are generally preferred for harder coatings with better weatherability. Care has to be taken in comparing the coating performance of systems based on the various types of polyol as it is strongly dependent on the level of branching and the hydroxyl value of the polyol. Nevertheless, some general trends can be found that are applicable to all the isocyanates commonly used and these are shown in Table 24-1.

Table 24-1   Coatings performance versus polyol type

	Acrylic	Polyester	Polyether
Application conditions	Equal	Equal	Equal
Viscosity	Medium	High	Low
Appearance	Excellent	Very good	Good
Hardness	Hard	Medium	Soft
Brittleness	Fair	Excellent	Excellent
Gloss retention	Excellent	Fair	Poor
Solvent resistance	Excellent	Fair	Fair
Salt water spray resistance	Excellent	Very good	Good

Amines

The amine compounds used in coating technology are polyoxyalkyleneamines, basically amine-tipped propylene oxide/ethylene oxide copolymers, and amine-terminated chain extenders, such as diethyl toluene diamine (DETDA) or isophorone diamine (IPDA). Full details of the production and properties of the polyoxyalkyleneamines are provided in Chapter 6.

Solvents

Solvents have traditionally been added to reduce the viscosity of components, in order to improve processing; they should not react with isocyanates and need to have a water content of less than 500 ppm. These limitations do not apply to the non-reactive polyurethanes.

It is common practice to use a mixture of at least three solvents in order to dissolve every component in the system or form a stable emulsion. The solvents evaporate at different stages during film formation, some quickly, to avoid excessive sagging and dripping, whilst during the final stages slow evaporation is required to provide acceptable levelling and adhesion and to release stresses in the coating. Solvents such as esters, ketones, ether-esters and polar aromatic and aliphatic solvents are split into the following three groups:

- Fast – boiling point under 100°C.
- Medium – boiling point 100 to 150°C.
- Tail or heavy – boiling point greater than 150°C.

There is a drive to reduce the solvent content of all formulations in every application due to the legislation that limits the levels of volatile organic compounds (VOCs) that can be emitted in some parts of the world. This is a major issue for aliphatic polyurethane spray coatings as these systems are generally based on high molecular weight polymers that need a high solvent level to obtain processable systems. This legislative pressure has led to the development of low viscosity polyol and isocyanate components leading to lower solvent usage (and also water-borne systems). Systems that contain more than 60 wt-% or 70 vol-% non-volatiles are termed 'high solids' coatings. Because the amount of solvents required is strongly dependent on the application, the impact of recent legislation on VOCs is application specific.

Technology of reactive coatings

The reactive polyurethane coating systems consist of many sub categories that are defined, along with typical formulations and properties in Table 24-2.

Two-component polyurethane coatings

Two-component systems, also known as two pack, twin pack or 2K, were the first polyurethane coatings developed and are still by far the most important products. The primary reaction is of isocyanate with polyol. These systems are characterised by a 'pot-life' since once the isocyanate and polyol are mixed they start to react, leading to an increase in viscosity and eventually gelation. The effective use time of the system is dependent on formulation, rate of reaction and cure time. An important feature of these types of reactive polyurethane systems is that curing occurs at low temperatures, producing coatings with excellent properties.

Low-temperature curing is not only important for reasons of processing and cost, but it also allows coatings to be applied to heat sensitive materials such as plastics and to large items, like aeroplanes or floors that cannot be heated because of their size. Full cure is normally reached after several days. In general it can be said that two-component polyurethane systems are used in those applications where the highest quality of finish is required combined with the ability to dry at relatively low temperatures.

Two-component polyurethane systems can be further divided into three categories: solvent-borne, water-borne and solvent-free.

Solvent-borne two-component polyurethane coatings
The major use for solvent-borne coatings is in automotive and aviation applications especially for refinish, where polyurethanes have largely replaced the traditional nitrocellulose and acrylic lacquers. The cross-linked

Table 24-2   Formulations and properties for reactive polyurethane coatings

		Solvent-borne	Water-borne	Solvent-free	Polyurea	Solvent-borne stoving	Water-borne stoving	Powder coating	Moisture cure
Components		2	2	2	2	2	2	2	1
Technology		NCO + OH	NCO + OH	NCO + OH	NCO +NH$_2$	Blocked Iso + OH	Blocked Iso + OH	Blocked Iso + OH	NCO + H$_2$O
Application		Clearcoat	High gloss top coat	Floor coating	General purpose	Coil coating	Gloss top coat	High gloss white	Zn primer for steel
Formulation, wt-%									
MDI	Mid functionality	–	–	19.5	–	–	–	–	–
MDI prepolymer	NCO 15 to 16%	–	–	–	50	–	–	–	6.1
HDI	Trimer	11.0	33	–	–	–	–	–	–
HDI	Blocked	–	–	–	–	15	8.7	–	–
IPDI	Blocked	–	–	–	–	–	–	17	–
Castor oil, modified	OHv 165	–	–	34.8	–	–	–	–	–
Saturated polyester polyol	OHv 66	–	–	–	–	20	–	–	–
Isophthalate polyester polyol	OHv 82	–	–	–	–	–	–	56.5	–
Amine-tipped polyol	MWt 2,000	–	–	–	32	–	–	–	–
Acrylic polyol	OHv 145	26.2	–	–	–	–	–	–	–
Acrylic polyol	OHv 43	–	21	–	–	–	30.7	–	–
Chain extender	Amine	–	–	–	13	–	–	–	–
Catalyst	Organo-metallic	0.002	0.007	–	–	–	–	–	–
Additives *		0.8	0.8	7.9	–	0.1	3.4	1.5	2.6
Pigments/fillers **		–	23.4	37.8	5	34.9	22.4	25	78.2
Solvents		62.0	–	–	–	30	–	–	13.1
Water		–	21.8	–	–	–	34.8	–	–
Processing conditions									
NCO Index		105	150	105	105	100	100	100	n/a
Solids, %		38	78	100	100	70	65	100	87
Cure temperature, °C		80	Ambient	Ambient	Ambient	350	150	180	Ambient
Cure time, *min*		30	360	1,440	1	1	30	20	1,440
Thickness (system), *mm*		0.075	0.1	1 – 5 1 coat	0.5	0.02	0.04	0.05 1 coat	0.075 – 0.1 1 coat
Properties									
Density, *kg/m^3*		1,050	1,300	1,450	1,100	1,600	1,400	1,250	3,000
Hardness, *Pencil*		H	H	H – HB	H	4H	3H	H	HB

* UV/HALS stabilisers, rheology control, defoamers moisture scavengers etc.
** Titanium dioxide is the main pigment. Zn dust is used in the steel primer. BaSO$_4$ is used in floor coating.

polyurethanes have much better mechanical and durability properties plus they have the advantage that they can be applied at high solids loading, have a high dry film thickness per coat, good colour matching, are UV stable and have excellent gloss retention.

In automotive refinish applications, Figure 24-1, polyurethane coatings are used as primers and solid colour or clear topcoats. The systems are based on acrylic polyols with a trimethylolpropane-TDI adduct normally added to the primer formulation whilst the top and clear coats are based on aliphatic isocyanates such as HDI-trimer. Coatings are applied at about 60 per cent solids, the isocyanate index is usually close to 100 and dibutyltin dilaurate catalyst can be used to accelerate the drying time. In addition, UV absorbers are normally added to maximise colour retention in service. The systems need to be designed to cure at temperatures from ambient up to 70°C. The coatings are spray applied using a single feed gun so a reasonable 'pot life' is important, which limits the catalyst level in the formulation. The presence of a polyisocyanate-curing agent means that suitable breathing apparatus must be worn during application of the coating to protect the health of the applicator.

Figure 24-1   Car refinish

Similar products are also used in automotive original equipment manufacturers coating applications, especially as clear coats for 'clear over base' systems, but also in pigmented systems. The formulations are based on HDI and IPDI-trimers or mixtures of the two, acrylic polyols and catalysts followed by stoving for around 30 minutes at about 80°C. Because the coatings are applied within a factory environment dual feed spray guns become commercially viable. The increasing use of plastics in automotive components has accelerated the use of two-component coating systems because of their ability to cure at low temperatures, thus preventing deformation of the plastic components as a result of high cure temperatures. They are favoured because of their water white colour, high gloss and 'wet look' appearance and high levels of durability.

Another major application area for solvent-borne two-component coating applications is in topcoats for aeroplanes. Aliphatic isocyanates, such as HDI-biuret, are used in conjunction with polyester polyols, or blends of polyester and acrylic polyols. Polyurethane coatings are the most commonly used in this application area because of their ability to cure at ambient temperatures whilst giving excellent abrasion, corrosion, moisture and UV resistance and an excellent resistance against 'Skydrol' (butyl phenyl phosphate), which is used as hydraulic fluid in aeroplanes.

Water-borne two-component polyurethane coatings
The development of water-borne polyurethane coatings has been driven by the need to reduce the emission of volatile organic compounds even further than was feasible with high solids systems and their introduction has been helped by the fact that they are a close match, in both performance and appearance properties, to the solvent-borne systems. They are now widely used on metal

substrates in automotive, transportation, machinery, furniture and protective steel coating applications whilst their high flexibility also enables them to be used on plastic and wood substrates.

These two-component water-borne systems are based on dispersible isocyanates and polyols, such as polyacrylates or polyesters that are emulsifiable or soluble in water. The main isocyanate used is HDI trimer, but IPDI trimer is also used, as is allophanate modified HDI. Aromatic isocyanates are generally not used because of their high reactivity with water. The isocyanates, which are hydrophobic, can be used directly or made easily emulsifiable by partially reacting them with a hydrophilic polyol.

Hydrophobic isocyanates normally require high-shear mixing to obtain a fine dispersion in water and to prevent incompatibility problems that can lead to film defects. Hydrophilically-modified isocyanates emulsify in water even under low-shear conditions, giving fine and uniform dispersions, but the water-compatible functional groups can lead to softer films with poorer humidity resistance. Most water-borne coating formulations however, still need some solvents, currently around 10 per cent, to help the polymer form a homogeneous film. The coatings are normally cured at ambient temperature, but using temperatures up to 80°C can increase the rate.

To obtain a good coating it is essential that the isocyanate-polyol chain extension reaction occurs as normal, but this is compromised by the reaction of the isocyanate with water, forming urea, which in a water-borne system is unavoidable. Therefore to ensure adequate polymer network formation an excess of isocyanate, isocyanate index from 120 to 300, is used to compensate for the isocyanate consumed in the urea reaction.

The carbon dioxide formed as a consequence of the urea side reaction is not a problem for thin films, but tends to form blisters and micro foam with thick layers. This problem can be minimised through optimisation of the raw material components, modifying the formulation and changing the application technique.

Solvent-free two-component polyurethanes
Solvent-free two-component polyurethane coatings have obvious advantages, regarding lower emissions of volatile organic compounds, and are predominately used in the building industry where they are applied either as sealants and membranes or as coatings to roofing and flooring for car parks, chemical production facilities, such as dye works, flooring for industrial buildings and reservoirs. They are also used in the construction sector as anti-corrosion coatings for steel roofing, scaffolding, ramps, bridges and storage tanks. They also have the advantage that the surface finish and texture can be easily modified, Figure 24-2.

Most systems are based on MDI, plus MDI oligomers, polyether or oil-modified polyester polyols, chain extenders and catalysts. Low viscosity raw materials are required since a high level of filler is used both to keep down cost and to

Figure 24-2   Solvent-free two-component polyurethanes

The aesthetic appearance and performance of two-component polyurethane systems can easily be adapted to specific requirements, as shown in the adjacent photograph, with all the coatings applied to hardboard. The bottom strip is a non-filled, non-pigmented formulation and has the appearance of a glossy varnish, whilst in the next strip the resin has been sanded in to give anti-slip properties. The third strip is a filled system containing red pigment, whilst the final strip has been given a textured surface by adding a thixotropic agent and using a special spray technique.

provide the required performance characteristics. All fillers and pigments must be dried in order to reduce the formation of carbon dioxide that can lead to pinholes and craters and to help reduce this moisture scavengers are often included in the formulation.

The pot life of these systems ranges from 2 minutes to 2 hours, dependent on the formulation and level of catalyst. They cure at temperatures between 5 and 30°C and can be applied at high humidity. The slower coatings can be hand applied, using brush, roller or putty knife techniques, whilst the faster systems are normally spray applied. Solvent-free systems are typically applied at thicknesses of 0.5 to 10 millimetres, compared to solvent-borne coatings, which are limited to thicknesses of 0.05 to 0.25 millimetres.

Formulations can be varied to produce coatings that range from flexible and elastomeric to hard and tough, both with crack-bridging properties, and excellent resistance to aqueous salt, dilute acid and alkaline solutions, heating oil and with low water absorption. The hard and highly cross-linked systems also show very good chemical resistance.

Highly filled two-component polyurethane systems are used as synthetic mortars. Sometimes they are referred to as polyurethane-bound concrete. Such systems may contain up to 80 wt-% of filler material consisting of cement and different types of sand with various particle sizes. Synthetic mortar can be used for concrete repair and as foundation material for heavy equipment. It is also used as a coating for industrial flooring applications. In these applications the coating typically has a thickness between 4 and 10 millimetres. The coatings are gas permeable and provide excellent wear and slip resistance.

Spray elastomer coatings are used for industrial applications such as belting for mining, pipe linings for handling abrasive slurries and as protective membranes in building. The combination of toughness and high elasticity with the capability of forming seamless membranes is particularly useful in buildings where spray elastomers are used to seal insulated flat roofs.

Polyurea two-component coatings

Polyurea spray systems are normally based on MDI or, if an aliphatic isocyanate system is required, on IPDI or TMXDI with the polyol being a polyoxyalkyleneamine or an amine-terminated chain extender, such as diethyl toluene diamine (DETDA) or isophorone diamine (IPDA). The reaction between amine and isocyanate, to form polyurea, is extremely rapid so no catalyst is required. It can also take place at very low ambient temperatures, allowing application at well below 0°C. The polyurea coatings are more stable, when held for prolonged periods at high temperatures, than conventional polyurethanes because of the urea linkage and the absence of catalyst whilst the increased crosslink density also means that coatings have a high resistance to humidity, heat and cold.

Due to the fast reaction it is essential to ensure good mixing and to achieve this the two components are heated to 70 to 80°C, to lower the viscosity, a high-pressure impingement mixing spray gun is used and the formulation adjusted so that the mix ratio, by volume, is 1:1. This results in isocyanate prepolymers with an isocyanate content of about 15 per cent and spray-applied systems with gel times of 3 to 15 seconds.

Polyurea spray coatings are developing their own application fields, due to the fast curing behaviour and specific properties, which is broader that that of the conventional polyurethane spray coats. Polyurea is an excellent choice to coat pipes, freight ships, industrial floors, Figure 24-3, automotive truck-beds, Figure 24-4, train wagon liners and roof coatings, as they need to be put back into service as soon as possible and any post cure is impossible or too expensive.

Figure 24-3   Industrial floor coating

Figure 24-4   Automotive truck bed with polyurea spray

Oven-curing or stoving systems

'Oven-curing', also known as 'stoving', coatings are based on blending a blocked isocyanate with a polyol to form a pseudo one-component mix that is stable at room temperature. When this is heated to its activation temperature the isocyanate unblocks and reacts with the polyol forming the coating. The free blocking agent then either remains in the coating or is removed by evaporation.

There are many aromatic or aliphatic blocking agents, but they all contain one active hydrogen, usually of the form -NH, -NOH or -OH. In the coatings industry caprolactam is the most commonly employed blocking agent, but butanoxime, 3,5-dimethylpyrazole, phenol and 2-ethylhexanol are employed. Another method of blocking is to form uretidinedione or dimer links.

Blocked IPDI is used in most coating formulations, but all the other aliphatic and aromatic isocyanates can be used and have found specific uses in various applications. The unblocking temperature is dependent on the type of blocking agent and isocyanate employed and can vary from 100 to 200°C. Aromatic isocyanates have much lower unblocking temperatures than aliphatic isocyanates, when using the same blocking agent and the addition of catalyst lowers the temperature and reduces the cure time.

As these coatings have to be cured at relatively high temperatures they can only be applied to substrates that do not deform in the heating/cooling cycle and in practice they are predominantly used for coating metal. The coatings can be applied either as solvent-borne, water-borne or as a powder.

Solvent-borne oven-curing coatings

The dominant use of solvent-borne oven-cured systems is for the high-speed coating of continuous coils of steel and aluminium that are further processed by profilers to fabricate panels for the construction, appliance and transportation industries with typical examples being commercial vehicle and caravan cladding. Therefore, the coatings need to be highly flexible. A primer and topcoat are typically applied in a one-machine operation, with TDI commonly used for the base coat whilst the topcoat is produced using an IPDI trimer. With HDI-based systems both base and topcoat use the same isocyanate. The blocking agent selected not only sets the stoving temperature and cure rate, it also affects the yellowing resistance, adhesion and high build facility of the coating. The polyols used in these applications are predominately polyester polyols, acrylics and phenolics, and must be non-yellowing. The solvents must have higher boiling points than those used for conventional two-component systems.

Special grades of these solvent-borne oven-cured systems are used for coating copper or aluminium wires, Figure 24-5, that can vary in configuration from fine, 0.001 millimetres, to wide, 10 millimetres. The formulations are based on blocked aromatic isocyanates and either hydroxy-terminated polyurethanes or highly-branched polyester, polyesterimide or polyamideimide polyols with the thermal resistance increasing from 170°C for the first to 210°C for

Figure 24-5   Wire coating

polyamideimide. The components are dissolved in aromatic solvents such as xylene or mixtures of esters and aromatic compounds at 10 to 55 per cent solids.

The coating is applied by a continuous method, running at 5 to 1,000 metres per minute, using either a horizontal or a vertical design, in which an excess is first applied then metered with wiping devices, dies or felts. The wire is then passed through an oven at 400 to 750°C so the cure time is only a few seconds or much less. Thus, in order to produce defect-free films the coatings can only be applied in thin layers and to obtain the required film thickness for insulation the wires must be coated several times, typically 6 to 20.

Water-borne oven-curing systems

Water-borne oven-curing systems are used predominantly in the appliance and automotive industries mainly as primer or base coats, to provide corrosion and stone chip protection in the latter case. The coating systems are based on dispersions of blocked aromatic isocyanates, usually modified with hydrophilic components to give them good emulsifying properties, and polyester polyols or modified epoxies. The coating is normally applied using cathodic electro-deposition, sometimes referred to as EPD coating or e-coat, and after drying the isocyanate is activated on stoving, leading to reaction and cross-linking with the polymers in the formulation.

Powder coatings

Polyurethane powder coatings, that are electrostatically applied and heat cured, are used to coat metals with major applications in construction – wall panels and window frames – and general products – appliances, metal furniture, lamp housings, garden/farm equipment and bicycles.

IPDI, blocked with caprolactam, is normally used in conjunction with polyester or acrylic-based polyols with the proviso that all the ingredients for the formulation have to be solids. The ingredients are dry-blended then melted and further mixed using an extruder at temperatures of 90 to 140°C, which, as it is close to the curing temperature, means it is imperative to complete the mixing quickly. The melt is then force cooled, granulated and sieved to obtain the required mix of particle sizes.

Several techniques exist to apply the powder, but the most commonly employed method is electrostatic spraying where the charged powder adheres to the earthed substrate, which upon heating to temperatures of around 180°C for about 10 to 20 minutes melts, flows and cures. The films have excellent mechanical properties and high wear resistance, and when aliphatic isocyanates are used also excellent UV resistance. Another advantage is that thick coatings can be produced in one pass.

One-component polyurethane coatings

One-component moisture-cure polyurethanes are widely used for maintenance and repair because of the ease of application and their excellent mechanical performance. Typical examples are coatings for: steel constructions, such as bridges, Figure 24-6, and cranes, wood floors in bowling alleys, primers and sealers for concrete and in synthetic mortar.

Figure 24-6   Steel bridge

These coatings are prepolymers, liquid at ambient temperature, based on reacting MDI, TDI or HDI with polyether or polyester polyols and are linear or slightly branched with an isocyanate content between 3 and 16 per cent. The primary cure reaction is between isocyanate and water, to form urea, with the rate and extent of the cure reliant on the availability of water, so it is dependent on both humidity and temperature. Cure usually takes place from the outside inwards, and is controlled by the diffusion rate of water through the curing polymer.

At higher temperatures and levels of humidity these systems may react so fast that the carbon dioxide formed causes defects, blisters and pinholes, especially with thicker coatings. This can be avoided by reducing the viscosity of the prepolymer through the addition of solvent, generally less than 40 per cent so they are still counted as high solid systems, but also by applying thinner coats.

The prepolymers on their own produce clear lacquers, but most applications require pigmented or filled systems and it is important that all additives are essentially water-free. The water level of the prepolymers can also be controlled by the use of water-scavenging additives such as monoisocyanates – p-toluenesulfonyl isocyanate, oxazolidines or ortho esters – triethylorthoformate, which react with water, or by adding molecular sieves, which physically absorb the water. In addition de-foamers or de-aerators can be added. These additives not only prevent foam formation and viscosity increase during manufacture and storage, but also aid the release of bubbles in the curing stage.

They can be applied by brush, roller or by airless spraying techniques. Damp surfaces are an advantage rather than a disadvantage and there is no temperature restriction of the substrate provided there is no risk of ice formation. Successful application at temperatures as low as -7°C has been reported, but to ensure satisfactory drying the minimum relative humidity should be above 30 per cent.

These one-component systems have the advantage that no metering and mixing is required, they are storage stable and the second component, water, will almost always be sufficiently available at the conditions of applications. Properly formulated coating systems have shelf lives of up to about six months.

The final properties of the coating depend both on the original prepolymer formulation and the urea links formed on cure but the hardness increases as the functionality and the isocyanate value of the prepolymer increases. Moisture-cure films are hard, glossy and show high abrasion resistance and exhibit a high resistance to water and chemicals. Furthermore, they show good adhesion to a variety of substrates, such as wood, steel and concrete. These properties make moisture-cure systems excellent materials for anti-corrosion coatings. In these applications the paints are filled with corrosion protecting pigments such as zinc dust, zinc phosphate, lead silicochromate and micaceous iron oxide. Often the formulation contains coal tar pitch to make the coating more water repellent. For re-coating work on metal a three-coat system would typically be applied, consisting of a base coat followed by a first and then a second topcoat.

Technology of non-reactive coatings

The essential feature of the non-reactive polyurethane systems is that they all contain fully formed polymers, containing urethane or urea linkages, but no free isocyanate groups.

Solvent-borne lacquers

Solvent-borne lacquers are made by forming or dissolving in solvents high molecular weight linear thermoplastic polyurethanes, made from aromatic or aliphatic isocyanates, mainly MDI and IPDI, plus polyester or polyether polyols and chain extenders. Because of the high molecular weight of the polymers (about 100,000) only relatively low solids contents can be used, typically 5 to 10 per cent. The lacquers are applied using conventional techniques such as brush or spray and film formation is by physical evaporation of the solvent.

These films, which dry rapidly, are extremely flexible and elastic and have remarkable resistance to mild solvents. The major application is as topcoats, clear or pigmented, for flexible substrates such as leather, fabrics, shoe soles and integral skin foam. Another key application is in the electronic industry where they are used as coatings for magnetic tapes when polyester polyol-based thermoplastic polyurethanes, modified with sulphonate or phosphate groups to bind and hold magnetic particles in the film, are used. Through adjusting the molecular structure films can be produced with minimal frictional resistance, high durability and good particle orientation.

Polyurethane dispersions

Polyurethane dispersions (PUDs) are fully-reacted polyurethane systems produced as small discrete particles, 0.01 to 0.1 micron, dispersed in water to provide a product that is both chemically and colloidally stable, which only contains minor amounts of solvents and thus emit very little volatile organic compounds. PUDs are based on aliphatic – IPDI or H_{12}MDI – or aromatic – MDI or TDI – isocyanates, modified polyether and/or polyester polyols, chain extenders, catalysts plus additives to modify the coalescence, flow, thickness, coagulation and defoaming properties.

PUDs are used in many application areas to coat a wide range of substrates – for example wood lacquers for flooring and furniture, leather finishing, vinyl upholstery topcoats, plastic coatings, printing inks, automotive base coats and glass fibre sizing.

PUDs are produced in conventional stirred reactors fitted with distillation equipment. The first step in the manufacture of an anionically-stabilised PUD is to prepare a prepolymer from isocyanate, polyol (containing either carboxylate or sulphonate side chains) and chain extenders in a water-miscible solvent such as acetone. The reaction product is an isocyanate-terminated polyurethane or polyurea with pendent carboxylate or sulphonate groups. These groups can be converted to salts by adding a tertiary amine compound, which, as water is added to the prepolymer/solvent solution, disperses the prepolymer in the water. The next step in the synthesis is the reaction of the remaining isocyanate groups with more chain extender or a cross-linker or a mixture of both. The solvent is then stripped, leaving the water-borne polyurethane dispersion with only a low solvent content. The critical factor is achieving a fine enough particle size of the fully reacted polyurethane so that it maintains a stable dispersion once the solvent is removed. The final dispersions contain 35 to 50 wt-% of dispersed particles.

Alternatively, a low molecular weight hydrophilic prepolymer can be chain extended at the same time as the aqueous dispersion is formed, providing that the isocyanate reacts preferentially with the amine rather than the water. In each case, the final polymer is a mixture of urethane and urea groups. The majority of PUDs are made from the slower-reacting aliphatic isocyanates.

Hybrid systems, especially urethane-acrylates, are also increasingly used. Simple blending of the two individual dispersions can be used, but the film properties are better if a mixed synthesis is used, giving a continuous phase on drying with final film properties typical of polyurethane on its own.

PUDs can be applied by a range of techniques – such as brush, spray, dip, curtain – and the aqueous dispersions form films by a coalescence process in which the individual particles are forced together, as water is lost during drying, so that the particles deform and eventually fuse together. The process is dependent on a number of parameters with a faster rate obtained from a small particle size, low polymer glass transition temperature, increasing temperature and the addition of a coalescing agent to achieve sufficient flow and fusion of the particles.

Cross-linkers, such as isocyanate, aziridine, carbodiimide, melamine and epoxy, can be added just prior to application to improve the performance of the coating, but then the blends have a pot life and usually need temperature activation in order to achieve full cure.

Urethane oils and alkyds

Heating natural oils, for example linseed, soybean or safflower, with polyols such as glycerol or pentaerythritol converts them to diglycerides and monoglycerides. These hydroxyl-containing oils can then be reacted with isocyanates, typically TDI or IPDI, to form long chain hydroxyl-terminated polymers that can be dissolved in solvents. These can be applied by brush or spray and after solvent evaporation the film cross-links by air oxidation of the unsaturated groups in the original oils, with the reaction promoted by metal catalysts.

Urethane alkyds consist of an alkyd resin in which some of the dibasic acid, usually phthalic acid, has been replaced by a diisocyanate. In the synthesis the ester links are formed in the usual way, by polycondensation, and the isocyanate added to react with the remaining hydroxyl groups to form urethane linkages. Urethane alkyds are basically similar to ordinary alkyds and may be drying or non-drying. The terms 'urethane alkyd', 'urethane oil' and 'uralkyd' are often used to describe urethane oils as defined above so the urethane alkyds should be called 'urethane-modified alkyds'.

Urethane oils and alkyds give coatings with high abrasion resistance, and have better mechanical properties and weathering resistance than unmodified alkyds, but not nearly as good as the other reactive urethane coatings. They find their main outlets in varnishes for floors, windowsills and boats, general use in undercoats and for industrial maintenance finishes.

Radiation curing

Urethane acrylate coatings are one-component, low viscosity, 100 per cent solids products that are easy to apply and can be rapidly cured by ultraviolet (UV) or electron beam (EB) energy sources at room temperature. Aromatic urethane acrylates are used in wood, paper, plastic and ink coatings whilst aliphatic urethane acrylates find applications where non-yellowing is important such as PVC floor tiles and continuous flooring. The UV curable systems are also used in adhesives, sealants and potting compounds.

The basic urethane acrylate oligomer is produced by reacting a prepolymer, obtained from the reaction of a diisocyanate and a polyether or a polyester polyol, with a stoichiometric amount of a hydroxyl-containing acrylate such as hydroxypropyl acrylate. The molecular composition of the oligomer can be varied to achieve the properties required in the finished coating.

Urethane acrylate oligomers are normally blended with some acrylate monomer, such as tripropylene glycol diacrylate or trimethylolpropane ethoxylate acrylate, as a reactive diluent and a photoinitiator, for UV curing. The photoinitiator, benzophenone is a typical example, absorb UV energy, producing free radicals, that then initiate the cross-linking through the acrylate groups. By comparison, the impact of the energy from EB is generally powerful enough to initiate free radical polymerisation without the need for photoinitiators.

The cure times of these urethane acrylates can be extremely rapid, of the order of milliseconds, whether using UV or EB curing and can be set up to have instant start and shut down capabilities and uses compact equipment that can often be retrofitted to other plants. The main difference between UV and EB curing is that EB energy can penetrate opaque or thick films whilst UV curing is limited to clear or thin films.

25. Adhesives

Berend Eling

Polyurethane adhesives, which vary widely in composition, are used in many application areas due to their outstanding properties, their simple and economical processing and their high strength. They account for about eight per cent of the global adhesives market, at around 530,000 tonnes, excluding their use as binders for wood and other materials. Polyurethanes are a major element in the high value reactive adhesive category because of their versatility and moderate pricing. Other reactive adhesives are epoxy resins, and to a lesser extent, modified acrylic, cyanoacrylates and radiation-cured systems. The market segments in which polyurethane adhesives find most use are construction 31 per cent, flexible packaging 27 per cent, footwear 17 per cent, woodworking 17 per cent and transportation including assembly 8 per cent.

Polyurethane adhesives are normally defined as those adhesives that contain a number of urethane groups in the molecular backbone or where such groups are formed during use, regardless of the chemical composition of the rest of the chain. Thus, a typical urethane adhesive may contain, in addition to urethane linkages, aliphatic and aromatic hydrocarbons, esters, ethers, amides, urea and allophanate groups.

A common factor for all polyurethane adhesives is that they cure to produce essentially thin films, used to bond two similar or dissimilar surfaces together, with the correct type of polymer structure for the end application.

Types of adhesive technology

Polyurethane adhesives can be divided into two main classes: non-reactive and reactive. In both cases, the aim is to put a thin continuous layer of high molecular weight polyurethane between the two surfaces to be joined. Non-reactive adhesives are based on high molecular weight polymers, which are applied as solvent-borne, water-borne or as hot-melts. Film forming occurs through evaporation of the solvent or water for the first two whilst hot-melts are applied at high temperatures and films form upon cooling.

Reactive adhesives are supplied as one- or two-component systems or as hot-melts. The one-component reactive systems, based on low isocyanate prepolymers are usually moisture-cured, whilst the prepolymers for two-component systems are reacted with a mix of polyol and chain extender. The types of technology are summarised in Table 25-1.

Table 25-1   Types of polyurethane adhesives

Type	Form at room temperature	Curing or film-forming mechanism
Non reactive		
Solvent-borne	Liquid	Physical evaporation of solvent
Water-borne	Liquid (dispersion)	Physical evaporation of water
Hot-melt	Solid	Physical cooling
Reactive		
One-component	Liquid	Chemical cure, NCO + moisture
Two-component	Liquid	Chemical cure, NCO + polyol
Reactive hot melt	Solid	Physical cooling + chemical cure, NCO + moisture
Cross-linker	Liquid	Chemical cure, NCO + active H

For all types of adhesives there is a variation in the build-up of strength with time, Figure 25-1, and the relationship between application and failure temperature, Figure 25-2. The variety of strength characteristics is used to obtain specific effects in different adhesive applications.

Figure 25-1   Strength development of adhesives

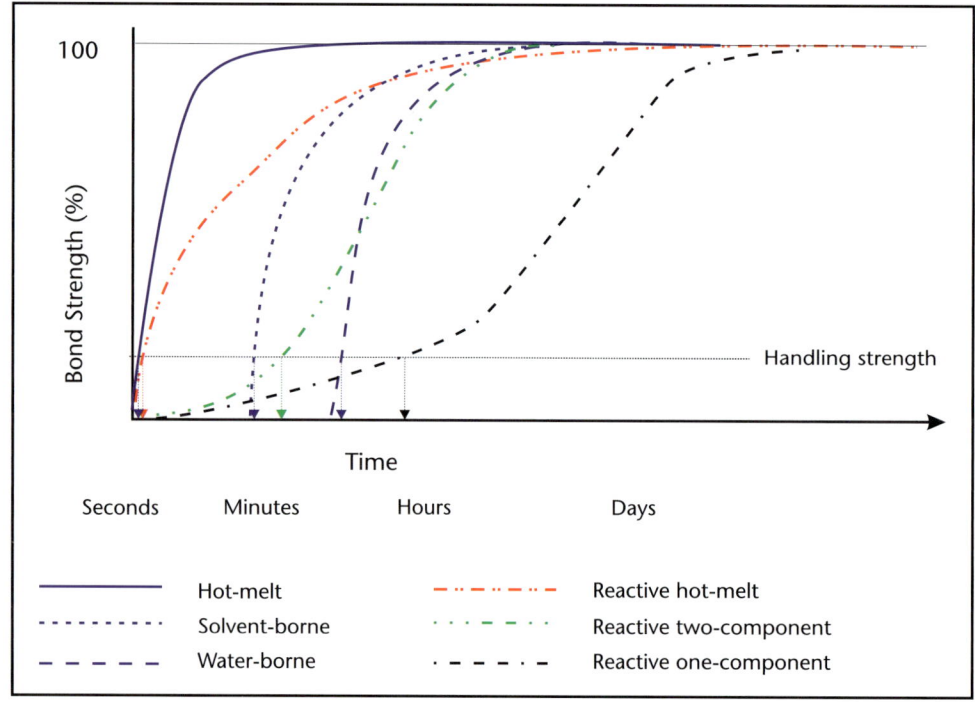

In Figure 25-1 bond-strength is normalised to 100 per cent for all the adhesives and a line drawn to represent the strength at which the adhesive bonds can be handled. Typically hot-melt adhesives have a short melt time, but then develop strength rapidly as the compound cools/crystallises after the bond is formed.

One- and two-component reactive systems have a typical S-shape strength build-up, directly related to the isocyanate reactions taking place and the solvent-borne and water-borne products have a distinct drying time before film forming starts.

Figure 25-2 compares the application temperature of the various adhesives and the retention of bond strength as a function of temperature. Hot-melts are applied at temperatures of 140 to 220°C, but fail at temperatures well below this. Conversely reactive hot-melts are applied at temperatures of 100 to 120°C, but fail at much higher temperatures. The solvent-borne, water-borne and the one- and two-component reactive systems are all applied at room temperature with the former two having similar strength retention to that of the hot-melt whilst the latter achieve the highest failure temperature due the degree of chemical cross-linking introduced.

Figure 25-2   Effect of temperature on the strength of adhesives

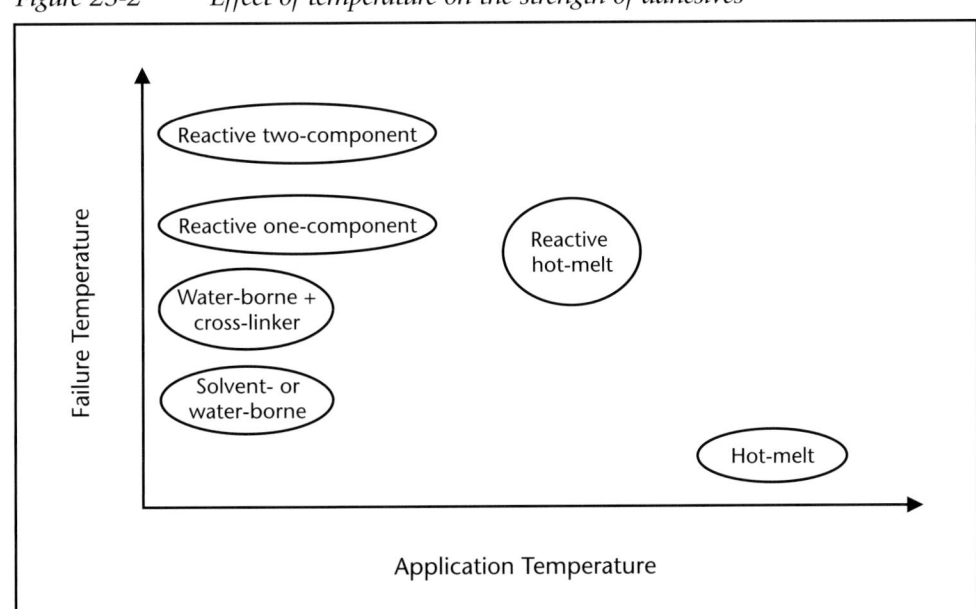

The many diverse applications for which adhesives are used need a range of parameters to meet their required specification and function with key factors being the degree of cross-linking in the final polymer film and its hardness. For all adhesives the concentration and type of hard segments affects the stiffness/hardness of the polymer. The degree of cross-linking is dependent on the type of adhesive with hot-melts, solvent- and water-borne products having no or a low level of cross-linking since otherwise they would not be thermoplastic or soluble in solvents. However, cross-linkers can be added to the non-reactive solvent or water-borne systems to improve their strength retention, as illustrated in Figure 25-2 for a water-borne system.

Cross-linking reactions for the reactive hot-melts and the one- and two-component reactive systems occurs, due to the excess of isocyanate present, during the reaction phase following application, leading to the formation of urea, allophanate and biuret cross-links.

Materials selection

A broad range of raw materials is used in the production of polyurethane adhesives and the key components are listed in Table 25-2.

Table 25-2   Polyurethane raw materials

Isocyanates	Polyols	Solvents	Fillers/plasticisers	Cross-linkers
MDI	Polyester polyols	Methyl ethyl ketone	Calcium carbonate	MDI/polymeric MDI
TDI	PO polyether polyols	Acetone	Fumed silica	MDI trimer
H_{12}MDI	PO/EO polyether polyols	Ethyl acetate	Carbon black	TDI adducts
HDI	Castor oil polyols	Toluene	Titanium dioxide	IPDI trimer
IPDI	Polybutadiene polyols	Xylene	Phthalates	HDI biuret, trimer

TDI is used to manufacture low viscosity prepolymers for flexible substrates, but its volatility and special handling precautions have limited its use. MDI is used where high tensile strength, toughness and heat resistance is required and its lower volatility makes it easier to handle. Not only is the pure 4,4'-MDI isomer used, but also mixtures of 2,4'- and 4,4'-MDI and polymeric MDI. MDI is also modified, to produce liquid products using several methods such as reaction with a polyol to make a prepolymer, formation of uretonimine groups and mixtures of 2,4'- and 4,4'-MDI.

Aliphatic isocyanates are used for those applications which require resistance to yellowing due to exposure to UV light. This is not a concern for most applications as substrates shield the adhesive from sunlight. Commonly used aliphatic isocyanates are H_{12}MDI, HDI and IPDI. The latter two have a relatively high vapour pressure and are not commonly used on their own, but instead as IPDI 'trimer' and in the case of HDI as 'biuret', 'trimer' or 'dimer' modifications.

MDI accounts for 75 per cent of the isocyanates used in polyurethane adhesives, TDI for about 20 per cent with the remainder being a mix of aliphatic isocyanates.

Both polyester and polyether polyols are used, with the structure of the polyol playing a large part in the properties of the final polyurethane adhesive. The molecular weight and functionality are the main factors, but the structure is also important. Polyethers, both propylene oxide and propylene oxide/ethylene oxide, are resistant to hydrolysis and are generally low in viscosity and have excellent wetting properties. Their major drawback is that the cohesive strength of the adhesive is relatively low, leading to low bond strengths. Polyester polyols combine high levels of adhesive and cohesive properties, but they are more viscous and therefore more difficult to handle. Also, polyester polyols have worse hydrolysis resistance than polyethers. Polyester polyols, dependent on their composition, can be amorphous or crystalline. When highly flexible bonds, with a good resistance to hydrolysis are required then castor oil and polybutadiene polyols are often used.

Solvent-borne cross-linkers, which are aliphatic or aromatic isocyanates with functionality greater than two, are added to solvent-borne thermoplastic polyurethane and polychloroprene adhesives at about five per cent. The cross-linkers react with the polymer chains to form chemically cross-linked polymers, which raises the softening point of the adhesive film and improves the resistance to grease, oil and solvent. Another effect of the isocyanate is its reaction with groups in the substrate surface, thereby increasing adhesion. However, the addition of cross-linker means the system becomes two-component with a pot-life typically between 6 and 48 hours for solvent-borne thermoplastic polyurethane systems.

Water-borne 'emulsifiable' cross-linkers, based on aromatic or aliphatic isocyanates that are generally non-ionically stabilised, are added, at levels of 5 to 15 per cent, to water-borne polyurethane and latex adhesives and the high growth of the former product over the past five years is mainly due to their introduction. The pot-life of these systems is strongly dependent on the type of isocyanate, with aromatics reacting much faster with water than aliphatics. For a given isocyanate the temperature and the pH control the reactivity with water. Optimum pot lives are obtained at conditions close to neutral or slightly acidic with the temperature kept as low as technically feasible. They work in the same way as the solvent-borne cross-linkers.

There is a broad choice of solvents, used in both the solvent-borne systems and to reduce the viscosity of one-component reactive adhesives, but the commonest are methyl ethyl ketone, ethyl acetate, acetone and toluene. The major criterion is that they must be dry to avoid reaction with any excess isocyanates and to avoid blisters forming as solvent-borne films dry.

Fillers are often added to adhesive formulations to improve physical properties, control the rheology or simply lower the cost. The most commonly used are calcium carbonate, talc or clay. Fumed silica or carbon black are used where non-sagging properties are needed. The use of pigments such as titanium dioxide is rare as most adhesive bonds are between two surfaces and essentially out of sight. If a filler or pigment is used, then they must be carefully dried before addition and a moisture scavenger is also frequently added as well. Plasticisers, such as esters, can also be used to lower viscosity or improve the low temperature performance, but generally result in poorer adhesion.

Non-reactive polyurethane adhesives

These are all essentially high molecular weight linear thermoplastic polyurethanes, unless a cross-linker is added. Cure is by solidification, leading to the crystallisation of the polymer; with the rapid rate leading to the fast generation of properties with excellent values due to the high molecular weight of the polymer structure. At room temperature these products are supplied as solids, hot-melts, or liquids, solvent-borne and water-borne.

Solvent-borne adhesives

Solvent-borne adhesives are produced by dissolving high molecular weight thermoplastic polyurethanes, made from MDI and usually an adipate polyester polyol with a crystalline structure, in solvents. Dependent on the molecular composition of the polyester, the melting point of the crystalline segments ranges from 40 to 90°C and above this temperature the polymer is essentially amorphous, whilst below it is partially crystalline. The crystallisation temperature and the crystallisation rate are strongly dependent on both the molecular weight of the polyester polyol and the level of hard segments. Polyester polyols made from glycols with an even number of carbons give fast crystallisation rates, which lead to good initial 'green' bond strength development. The solution viscosity of the adhesive is related to both the molecular weight of the thermoplastic polyurethane and the solvents used.

The footwear industry is a high-volume application for these adhesives where they are used for bonding soles to uppers as they bond most materials, have high levels of grease and plasticiser resistance and good 'green' strength leading to rapid production rates. For polyurethane and PVC soles it is only necessary to either roughen the bonding margin or solvent-wipe it, before applying adhesive, but for vulcanised and thermoplastic rubber and EVA soles a more aggressive pre-treatment is required to activate the surface. For rubber compounds a chlorinating agent needs to be used whilst for EVA even more complex methods are required, often using UV activated compounds. The key steps in sole bonding, which typically uses 10 grammes of dry film thermoplastic polyurethane per pair, are listed below and shown in Figures 25-3 to 25-6:

- Apply adhesive solution to shoe upper.
- Apply adhesive solution to sole unit.
- Force dry or allow 7 to 30 minutes for solvents to evaporate.
- Activate adhesive by flash heating upper and sole unit at 50 to 80°C.
- Spot bond sole to upper.
- Press for 30 seconds at 0.2 MPa.

*Figure 25-3
Application of adhesive to shoe upper*

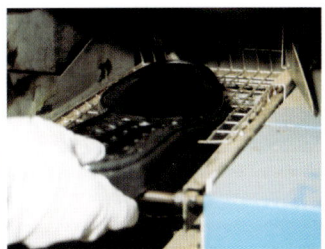

*Figure 25-4
Flash activation of sole*

*Figure 25-5
Spot bonding of sole and upper*

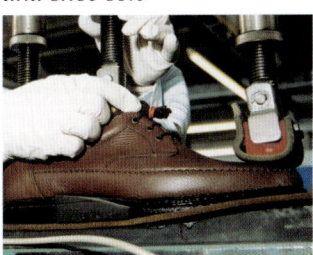

*Figure 25-6
Press bonding of shoe upper and shoe sole*

Solvent-borne cross-linkers are frequently added to solvent-borne thermoplastic polyurethane adhesives to improve the strength and resistance of the bond, but as mentioned earlier this turns a one-component adhesive, where the only concern is to control the loss of solvent, into a two-part adhesive with a limited pot-life. However, for certain applications this is necessary in order to achieve the required performance and processing speed.

The solvent-borne adhesives are prepared in closed mixers at a temperature below 40°C or, in mixing equipment fitted with a condenser, at temperatures as high as the boiling point of the solvent. Methyl ethyl ketone is the normal solvent used although ethyl acetate, acetone and toluene are also employed and often mixtures of these solvents are utilised. The choice of solvent is important, because it determines both the viscosity and the drying time of the adhesive.

Typically, about 50 per cent of the solvent required for the final adhesive solution is charged to the mixer and the thermoplastic polyurethane granules added in stages, to avoid coagulation, to the stirred mixture. The shear forces generated, especially when using high shear mixers, can raise the temperature as high as the boiling point of the solvent. The remaining solvent is added over time to control the viscosity and temperature of the mixture. The dissolution time depends on the process conditions and can be 24 hours using low-temperature equipment or two hours for reflux conditions. The solutions can be used as made, but may also be compounded with, for instance thixotropic additives added to alter the rheological properties. The final adhesive solution will, at a 15 to 20 per cent loading of thermoplastic polyurethane, have a typical viscosity of 4,500 mPa.s.

Solvent-borne adhesives are, however, under increasing pressure due the fact that as they are used the solvents evaporate, releasing volatile organic compounds into the atmosphere thus potentially damaging the environment and creating poor working conditions. Equipment to capture and recycle the volatile organic compounds is now commonly used in developed countries or non solvent-borne adhesives are used.

Water-borne adhesives

Water-borne polyurethane adhesives are fully reacted, high molecular weight polyurethanes dispersed in water (PUDs – polyurethane dispersions) that only contain minor amounts of solvent and thus have a very low emission of volatile organic compounds. Because of their low solvent level they are replacing solvent-borne adhesives in many areas.

The main applications for water-borne PUD adhesives are in the construction and packaging industries and in the manufacture of foils and textile laminates. Their use in the footwear industry is growing due to the environmental pressure on the solvent-borne adhesives. They are also used to a small extent in consumer/do-it-yourself applications, in the automotive industry and as patching compounds for plywood. Water-borne polyurethanes are normally roller or spray applied and heating is required to speed up the drying process.

The production of adhesive PUDs is based on the same technology used for the preparation of coating PUDs, as described in Chapter 24, with the main difference being in the basic polyurethane formulation. Adhesive PUDs need to be based on the same range of raw materials, already defined in this chapter, that are used for solvent-based or two-component adhesives. Again the particle size is a critical factor in the subsequent use of the PUD as it controls the coalescence rate of film formation as the water is driven off. The solids content of the final PUD adhesives is normally in the range of 35 to 50 per cent, but the viscosity is significantly lower than solvent-based adhesives with 15 to 20 per cent solids. However, the higher solids content compensates for the lower viscosity so that the coating weights obtained for PUDs and solvent-based adhesives are very similar.

The mechanism of film formation is similar to that for coating PUDs in that the carrier water has to be eliminated, usually by forced drying, followed by the coalescence of the PUD particles to form a coherent adhesive film. The difference, however, is that the adhesive film, on the coated substrate, then requires heat activation in order to achieve a bond with a second coated substrate or material.

Commercial PUDs can be classified on the basis of two characteristics. These are the internal stabilisation mechanism – anionic, cationic, or non-ionic – and their chemical composition – notably the type of isocyanate, aliphatic or aromatic, and polyol, polyether or polyester. Most adhesive PUDs are based on H_{12}MDI or IPDI, with some TMXDI. The type of stabilisation chosen determines the usable pH range for the PUD.

PUDs can be used as made, but are also used as mixtures with latexes of other polymers, such as natural rubber, polychloroprene or polyvinylacetate, since blending allows optimisation of the cost versus performance requirements of the adhesive. PUDs can be used as one-pack adhesives. However, as they normally contain linear, or only slightly cross-linked, polymer networks they do not compete technically with one- or two-component reactive adhesives. Cross-linking will normally improve the mechanical properties and this can be obtained by adding suitable water-borne reactants using isocyanate chemistry and special isocyanates have been developed that can be dispersed in water.

PUDs have a good bonding performance with many materials, including PVC substrates, although their adhesion to rubber materials is generally less satisfactory, but this can be increased through halogenation. The water-borne PUD adhesives that are used in sole bonding applications are usually produced from crystalline polyesters and their processing is very similar to that previously described for solvent-borne adhesives. For leather and polyurethane or PVC-coated fabrics it is essential to roughen the surface to ensure optimum bonding. Problems can be experienced with PUD adhesives when they are used with materials having certain surface treatments or with high oil content leather.

Hot-melt adhesives

Hot-melt thermoplastic polyurethane adhesives are similar, in terms of chemical composition and physical behaviour, to those used to produce solvent-borne adhesives. These hot-melt adhesives are usually supplied as granules and the solid material is either heat extruded or melted in holding tanks. On a smaller scale, 'glue sticks' which are melted in hot guns at the point of use are also used. The molten hot-melt is applied to the substrates, which when brought together under pressure rapidly form an adhesive bond on cooling. The application is readily automated, using special hot-melt machinery or a hot-melt gun and they are used to bond paper, board, wood, metals, plastics and for laminating fabrics to foams. The high temperatures of application can cause substrate deformation and because of their linear structure they have poor property retention at high temperatures. Their ability to adequately wet surfaces can also be poor as they have a high molecular weight and thus are highly viscous.

Hot-melt thermoplastic polyurethane adhesives can be formulated using amorphous or crystalline polyesters and control of the composition gives a range of melt viscosities from 50,000 to 250,000 mPa.s, whilst varying crystallisation rates and 'open tack time' ranging from less than five seconds to more than five minutes makes these adhesives suitable for a variety of assembly operations.

These hot-melt adhesives can also be produced as filament webs, free films or as films on release paper. Several adhesive producers specialise in the manufacture of these adhesive films using extrusion equipment. The hot-melt films are available in a range of weights from 25 to 150 grammes/m^2. To produce a bond the adhesive is placed between the substrates, the laminate heat activated at temperatures above the melting point of the crystalline polyester segment of the thermoplastic polyurethane and pressed under specified press conditions to fuse the layers together.

Hot-melt films combine low heat activation with high viscosities at the softening point and a benefit of hot-melt films is that when properly processed they do not penetrate porous substrates such as cloth or foam.

Reactive polyurethane adhesives

Reactive adhesives, that can be supplied as one- or two-component systems or as hot-melts, are applied as a liquid, which cures and hardens by chemical reaction. They make good adhesives for a number of reasons:

- Low viscosity and effective at wetting the surface of most substrates.
- Interact with the substrate through polar interactions such as hydrogen bonding.
- Easily penetrate porous substrates due to their low molecular weight.
- Stiffness, elasticity and cross-linking can be tailored to suit specific needs.

One-component adhesives

One-component polyurethane adhesives are prepolymers with low isocyanate values, produced by reacting an excess of isocyanate with a high molecular weight polyol; they are cured by reaction with water that is either added or, more commonly, present in the atmosphere. For successful curing, it is important that moisture can diffuse into the bond and if atmospheric moisture is needed, then the joint must be designed to allow water diffusion, or at least one of the substrates must be porous. MDI is the preferred isocyanate, on both reactivity and cost terms, although TDI is used and so are aliphatic isocyanates for special applications and both polyether and polyester polyols are used.

To produce a prepolymer with adequate storage stability and with satisfactory strength, elasticity and resistance, the functionality of both polyol and isocyanate and the level of free isocyanate needs to be carefully balanced. A prepolymer with too high a free isocyanate content for an application can lead to the formation of hard and brittle adhesive films due to excessive water reaction. Conversely, with too low a free isocyanate content, the adhesive may be too soft for most applications and be more suitable for sealing applications. One-component adhesives for structural applications need a degree of functionality so that the final adhesive film is a cross-linked polymer to prevent creep and increase water, oil and solvent resistance. Branching can be introduced through the polyol or by using polymeric MDI or functional MDI adducts.

One-component polyurethane adhesives have a long pot-life, if contact with water is eliminated and they do not need special mixing equipment as two-component adhesives do. They are commonly applied as 100 per cent solids, but solvents can be added, to reduce the viscosity and catalysts, to increase the reaction rate of free isocyanate and water.

Figure 25-7
Free isocyanate content versus type of application for one-component moisture-cured adhesives

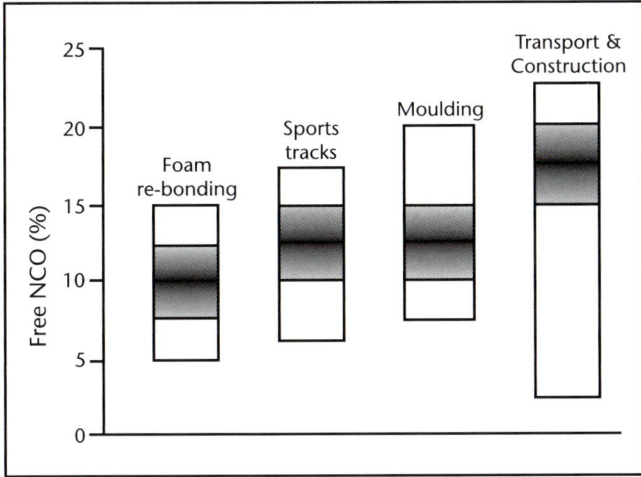

The three major applications for these moisture-cured prepolymer adhesives are:

- Rubber crumb bonding.
- Foam re-bonding.
- Construction and transport.

The ranges of free isocyanate content used in the applications of one-component moisture-cured adhesives are shown in a simplified way in Figure 25-7.

Rubber crumb

Moisture-cured polyurethane adhesives are used to bind re-milled rubber waste, or rubber crumb, and the composite materials then find applications as floor coverings, including playgrounds, running tracks, Figures 25-8 and 25-9, and also for garden tiles, drainage layers, vibration-absorbing bases and as sound and heat insulating layers.

Figure 25-8   Application of top layer

To manufacture these polyurethane composites the rubber granules are coated with prepolymer, transferred to a mould or press, where the mix is compressed and cured, or for sports/play surfaces, laid directly at the building site by processes similar to those employed in road construction for the laying of asphalt. The properties of the composites are dependent on the particle size of the rubber granules, the constitution and the amount of the polyurethane one-component adhesive and in case of the moulding process also on the degree of compaction during curing.

Figure 25-9   Rubber crumb running track

Prepolymer adhesives used for the wet-lay process for sport flooring applications usually have an isocyanate value of about 10 per cent and are produced from linear or slightly-branched polyether polyols and MDI, a mixture of MDI and TDI, or TDI alone, with the market trend moving towards use of all MDI systems. The reactivity of such systems can be adjusted by balancing the 2,4'- and 4,4'-MDI content of the prepolymer. In the wet-lay process typically 15 to 25 wt-% of prepolymer is added to the rubber crumb and cured in place by moisture from the environment.

For moulding applications the prepolymer adhesives have isocyanate values of about 15 per cent, with functionality greater than two, and 7 to 10 wt-% is mixed with the rubber crumb. Typically, 0.5 per cent water is added to the prepolymer/rubber crumb mixture to ensure thorough cure and amine catalysts can be added to speed up the cure rate.

Foam re-bonding

Moisture-cured polyurethane prepolymer adhesives are also used to bond scrap foams and the details are outlined in Chapter 14. The prepolymers used are based on TDI or MDI and flexible polyether polyols and usually have isocyanate contents of 10 to 15 per cent. To reduce the cure time, steam is injected through the mixture of foam and prepolymer; catalysts can also be added.

Construction and transportation

Moisture-cured adhesives are also employed in the construction and transportation industries to bond porous materials such as wood, which can contain some humidity, with each other or with non-porous substrates such as metal or plastic sheets and closed-cell rigid foam PVC, polystyrene or polyurethane panels.

Main areas of application are in the manual or automated manufacturing of sandwich constructions such as partition walls, doors, insulation panels for cool stores, factory building panels and side panels for caravans, as shown in Figure 25-10. The various components of the construction provide strength, heat insulation and sound dampening properties. These adhesives are also used to laminate aluminium foil to plasterboard.

Figure 25-10   Side wall of a caravan

Ranges of special moisture-cured prepolymer adhesives have been developed for these applications, as harder and stronger bonds with a higher cohesive energy and low levels of deformation under constant load are required. Therefore, the prepolymers have a higher isocyanate content, from 15 to 20 per cent, contain some branching and have open times that vary from a few minutes to an hour.

The moisture-cured polyurethane adhesives are cold applied using doctor blades, roll coaters or spray equipment. In order to obtain complete cure in a reasonable time span a water mist is sprayed onto the substrate prior to or during the assembling stage and the slight foaming provides some gap filling properties. To ensure full development of high mechanical strength and to avoid blister formation the cure is performed under pressure, and heat is normally used to accelerate the cure. Other applications in this area are covered in Chapters 26 and 27.

Two-component adhesives

Two-component polyurethane adhesives consist of a prepolymer and polyol blend that when accurately mixed react to provide bonds with excellent mechanical properties. For continuous applications the two components are mixed using automated equipment whereas for discontinuous processes an appropriate quantity for the application is mixed. This 'pot' of adhesive then has a 'pot-life' that is related to the reactivity of the system and the temperature/humidity of application. Whilst continuous mixing provides consistent bonds, care has to be taken with the discontinuous process as the increase in viscosity over time, due to the cure reaction that starts once mixing takes place, affects the penetration depth of the adhesive, which may alter the consistency of the bond. A balance is thus required between a long pot-life and speed of final cure.

Two-component polyurethanes are used in many adhesive applications in the construction and transportation industries, especially for lamination processes, but as they cure by reaction of the two components, they can be used to bond moisture impermeable substrates such as plastic-to-plastic and plastic-to-metal. For instance, construction panels, manufactured in a continuous process, can be made by bonding steel, aluminium or foil stressed skin materials to polyurethane or polystyrene foam, mineral wool or other insulating cores.

The fast cure speeds that can be obtained with two-component polyurethane adhesives are used not only for laminating applications, but also on assembly lines, which require rapid quick fixing of parts, especially under conditions where no heat can be applied. Another advantage is that thick glue-lines can be applied providing gap filling as well as bonding and, unlike the moisture-cured adhesives, there is no foaming to interfere with the adhesive properties.

A further advantage of the two-component polyurethane adhesives is that the mix ratio of the two components can be adjusted to vary the properties within an isocyanate index range of 90 to 150. An isocyanate index above 100 is used to increase the hardness of the adhesive and to improve bonding to the substrate, since a lot of substrates, such as wood, paper or glass have hydroxyl groups on the surfaces that can react with the excess isocyanate. Isocyanate indexes below 100 are sometimes used to produce softer and more flexible adhesives.

The two-component polyurethane adhesives are normally based on MDI and TDI with the functionality usually increased to greater than two either by adding polymeric, in the case of MDI, forming uretonimine groups or, in the case of TDI, through the production of an adduct with a functional compound such as trimethylolpropane. Aliphatic isocyanates are used in those cases where discoloration of the bond-line due to exposure to UV light would cause problems. A wide variety of polyether and polyester polyols is used with molecular weights of 300 to 4,500 and functionalities of two to four.

Two-component polyurethane adhesives developed for the construction industry are usually cross-linked and amorphous in nature with the cross-links introduced through the use of branched polyols and/or isocyanates or through the use of an isocyanate index in excess of 100. The use of low molecular weight polyols and the high levels of chemical cross-linking prevent phase separation. As a result, these adhesives have a high modulus and hardness, with few elastomeric properties, and the bonds have good tensile, shear stress and peel strength properties, especially at high temperatures. In general, the resistance of the bond to high temperature, water, solvents and ageing increases with increasing levels of cross-linking. More elastic bonds can be obtained by using lower levels of cross-linking and higher molecular weight polyols.

These adhesives can be applied as 100 per cent solids, but if the viscosity is too high for a particular application it can be reduced by the addition of low levels of solvents, such as acetone, methyl ethyl ketone, ethylacetate, toluene and xylene. Fillers, such as calcium carbonate, talc or clays, are added to control the rheology and to reduce shrinkage and costs. Fumed silica is used as a thixotropic agent to prevent the adhesive penetrating porous materials like paper, textile, leather and concrete. Solvents and fillers need be dry to prevent foaming, but besides careful drying the addition of moisture scavengers, such as zeolites in the form of powder or paste, is recommended. Flow agents (cellulose derivatives and vinyl acetate co-polymers) and film forming additives (hydroxyl-terminated polyurethanes) can be added to solvent-based systems. Whilst catalysts are not normally required to achieve satisfactory cure times for MDI or TDI systems, those based on aliphatic isocyanates, which are intrinsically slower, generally do need catalysis.

If the adhesive has a long pot-life then it can be applied manually by using brushes, rollers, notched trowels, coating knives, roll coaters or by casting or spraying. Fast-reacting systems, however, have to be applied using meter-mix-dispense units and static mixers are adequate for low-volume application, but dynamic mixers are required for larger volumes.

An example of the use of a flexible two-component polyurethane adhesive is in the lamination of packaging films, Figure 25-11. There are two application methods. The first uses a solvent-borne adhesive system that is applied to the substrate by roller with the solvent then removed in a vented oven and lamination completed using a heated nip roller. The second method, now used on most lamination lines, applies a solvent-free system using a roller coater.

Figure 25-11
Adhesive bonded laminate packaging film

The amount of adhesive required varies from one to four grammes/m^2 dependent on the type of laminate and higher levels may be required for textile, unevenly printed boards and for sterilisable laminates. A related application in the print industry involves the lamination of a transparent plastic film to printed paper or board to produce the glossy protective finish required for articles such as book covers, table mats and playing cards. Polyurethanes make excellent packaging/lamination adhesives because of their good adhesion, heat and chemical resistance and fast cure. They can also be designed to obtain Food and Drug Administration (FDA) approval and, therefore, can then be used for food packaging.

Reactive hot-melt

Reactive hot-melt adhesives combine the advantage of polyurethane hot-melts and one-component polyurethane adhesives as they provide fast initial strength build-up, show excellent mechanical property retention at high temperatures, but can be applied at intermediate temperatures. They have found application in the bookbinding, woodworking, packaging, automotive and construction industries. In each case, either the substrate is porous, or the joint must be designed so that moisture can diffuse into the adhesive and complete the cure.

They are essentially prepolymers made by reacting a polyol, or mix of polyols, with about a two molar excess of isocyanate, usually MDI. To produce a material solid at room temperature it is necessary to ensure that the polyol, or at least one of those in a mix, has a crystalline structure. The adhesive is applied at temperatures 30 to 50°C higher than the melting point of the crystalline polyester chain segment to achieve a low viscosity, which leads to good wetting. After application of the reactive hot-melt to a substrate it cools and crystallises quickly to a solid rapidly providing bonds with good green strength properties so that components can be quickly handled. The further build-up of properties comes from the free isocyanate groups, which subsequently react with atmospheric moisture. Reactive hot-melts can contain fillers and tackifiers with the fillers added to improve creep resistance, to increase the viscosity and to modify the rheological properties whilst tackifiers are used to adjust the viscoelastic properties.

The adhesive must be protected from moisture prior to use and this can be achieved by supplying it in discrete portions, in moisture impermeable foil packaging, to be utilised in one lot. Reactive hot-melt technology is applicable to continuous assembly line operations, using an automated dosing unit, but can also be applied manually using hot-melt guns.

Hybrid systems

In addition to the main classes of polyurethane adhesives described above, there have been a number of attempts to combine the advantages of polyurethanes with other adhesives in hybrid systems. Urethane acrylic systems range from the use of water-dispersible isocyanates to cross-linked acrylic emulsion polymers through the use of acrylic polyols to co-polymerised resins. Urethanes can be used to toughen vinyl-terminated acrylic polymers for improved impact resistance and polyurethanes with acrylic functionality are also used in anaerobic or radiation-cured adhesives, again to increase toughness. Urethanes can also be combined with epoxy chemistry, using amine-curing systems. This can lead to products with fast curing, good adhesion and high temperature tolerances for structural and heavy-duty applications.

26. Wood adhesives

Chris Phanopoulos

Wood is an abundant resource and the source of the natural polymers cellulose, lignin and hemi-cellulose. Despite the profusion of trees and the fact that wood is a renewable resource, ever-efficient utilisation is required since there are increasing demands on the use of land, diminishing the locations and area available for forestry and wood harvesting. Also, although there are effective ways of increasing tree growth rates, they remain a relatively slow growing crop. Additionally, environmental, ecological and social pressures increasingly limit the ability to utilise certain types of large diameter hard wood or broad leaf deciduous trees and forests.

Figure 26-1
OSB panel close-up

Methods that will replace solid lumber applications, reduce the amount of waste and increase the utilisation efficiency of wood are constantly being sought. One major area has been the gluing of pieces of wood to produce composite structures and this has progressed from the early products like the traditional plywood and chipboard panels to such new materials as wafer board, oriented strand board (OSB), Figure 26-1, medium density fibreboard (MDF), Figure 26-2, and structural engineered lumber such as glulam, Figure 26-3. The main types of engineered wood composites are listed in Table 26-1.

Figure 26-2
MDF panel close-up

As the use and types of wood composites developed, increasing demands were put on resin quality both in terms of production of boards (panels) and final performance. Traditionally, urea-formaldehyde (UF) resins have the largest volume of the market because of their low cost and high cure speed. However, they can only be employed in interior quality composites as they have poor hydrolysis resistance. The hydrolysis of UF resins results in the loss of bond integrity and leads to the production of formaldehyde. Melamine modified urea formaldehyde (MUF), melamine formaldehyde (MF), melamine urea phenol formaldehyde (MUPF), phenol formaldehyde (PF), and phenol resorcinol formaldehyde (PRF) all provide increasing resistance to hydrolysis and a reduction of formaldehyde emissions.

Figure 26-3
Glulam close-up

However, in the same order resin costs increase dramatically, but for exterior applications, which require hydrolytic resistance, these resins are used and have a large market share. By the late 1950s the rapid expansion of these types of resin was over and since then any developments have concentrated on improving the performance of existing technologies in terms of processing tolerance, production speed, utilisation rate and the final board properties.

Moisture-cured isocyanate-based resins were first introduced in the 1960s and immediately established themselves due to their ability to easily and rapidly penetrate and bond to most types of wood, for their excellent hydrolysis resistance and extremely low vapour release. The grade of resin required varies dependent on the type of wood in the composite being manufactured and the end performance requirements. Generally, the smaller particle composites such

Table 26-1   Main types of engineered wood composites

Composite name	Abbreviation	Primary wood type	Wood dimensions (mm)	Types of resins employed	Description	Applications
Finger-jointed lumber	FJ	Lumber or an EWP	Depends on elements	PRF, PF, PUR, PVAc (non structural)	Elements joined into long 'continuous' lengths	Glue laminated lumber, any solid lumber application
Glue laminated lumber	Glulam	Lumber or jointed lams	50 mm wide	PRF, MUPF, PUR	Lams glued together, grain parallel to long axis, can be 3 m wide, 2 m deep, 30 m long	Arches, spans, roofs
Plywood	PW	Veneer sheets	2 mm thick, 1,500 mm wide & long	PF, UF	Veneer sheets (plys) stacked with grain direction alternating	Floor sheathing, shearwalls packaging, furniture
Laminated veneer lumber	LVL	Veneer sheets	2 mm thick, 1,500 mm wide & long	PF, PUR	Veneer sheets (plys) stacked with grain direction parallel, continuous length, 40–50mm thick	Flanges for I-joists, headers, beams
Oriented strand board	OSB	Strands/flakes	100 x 40 x 0.4 mm	pMDI, PF	Strands glued into panels	Roof, floor and siding panels; webstock for I-beams
Medium density fibreboard	MDF	Fibres	Individual wood cells up to a few mm length	pMDI, MUF, UF	Fibres glued into panels	Window frames, cabinetry, furniture
Particleboard	PB	Chips	Up to 10 mm long	pMDI, UF	Chips glued into panels	Cabinetry, furniture

Does not include wood products that are made from these, such as I-joists.

as MDF, particleboard (PB) and OSB utilise polymeric grades of MDI with some product differentiation due to processing or composite requirements. The processing method for MDF requires isocyanate resins with increased affinity for water such as emulsifiable MDI (eMDI).

Composites assembled from large pieces of wood such as glulam, finger-jointed (FJ) and laminated veneer lumber (LVL) cannot be bonded with polymeric MDI other than under selective conditions and therefore prepolymers, based on polymeric MDI and polyether polyols are used. These formulated resins (PUR) are also used because of the specific processing conditions required for these composites, long open assembly times for glulam, high tack for LVL and room temperature cure for I-beams, which consist of glulam flanges and OSB webs. They are also used as they have good gap filling properties and improved resistance to creep.

The use of isocyanates continues to grow rapidly and they now have a significant market share, especially for exterior applications.

The choice of resin type varies from region to region reflecting, to some extent, the types of wood composites made and the final applications. There are large differences in the relative amounts of the different types of composites for North America and Europe as shown in Figure 26-4.

Figure 26-4   Market share of types of wood composites (2000)

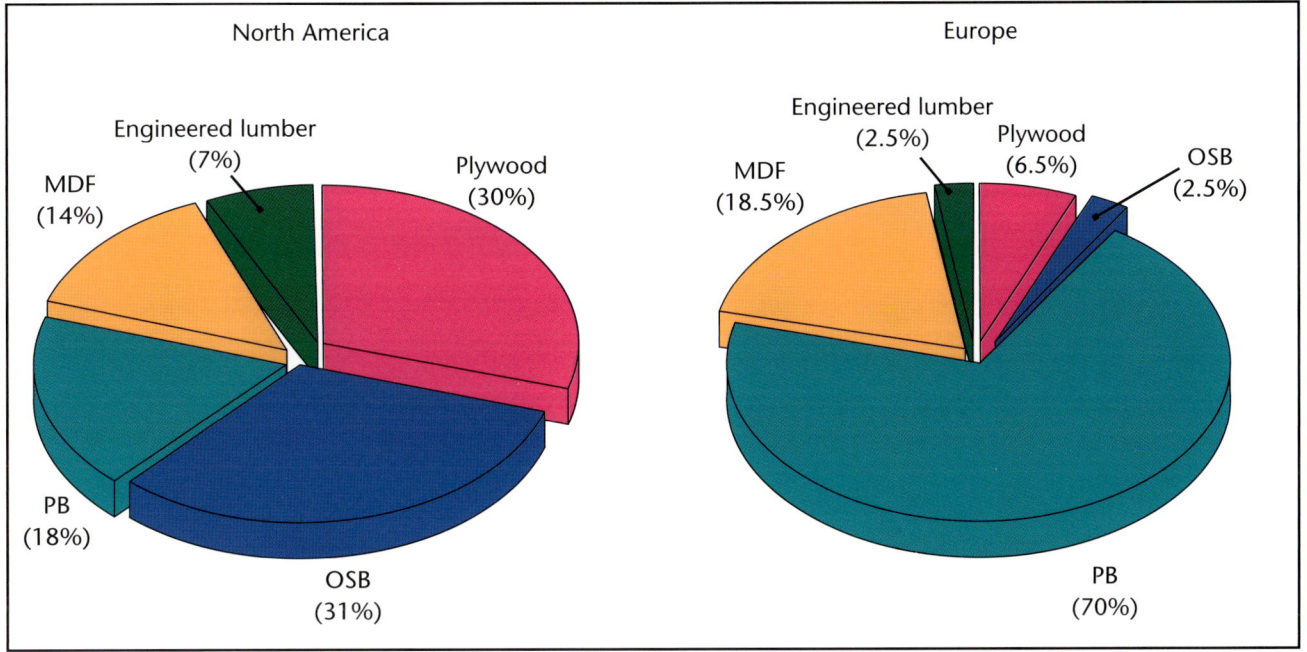

Mechanisms of adhesion

The various wood composites involve different manufacturing techniques, but there is a common mechanism of adhesion for isocyanates and wood that can be outlined by looking at the bonding of solid wood and, in particular, OSB. A good adhesive should wet and spread over the surfaces, cure to a solid material, and form a strong, durable bond. Isocyanates can be applied to the surface of the wood particles, fibres, chips, strands, veneers or lumber by spraying, pouring or rolling, dependent on the type of composite being produced.

Isocyanates wet wood reasonably well and uniformly and then spread and penetrate the wood with the extent of penetration strongly dependent on the species of wood, the moisture content, the type of surface cut, whether it is early or late wood and the age of the surface. It also depends on the viscosity, surface tension and reactivity of the resin used as well as the temperature and to some extent the relative humidity.

It has been determined that a penetration depth of at least 0.25 to 0.3 millimetres is required to provide good adhesive strength; this distance is slightly greater than the typical depth of the surface damage of sawn wood. Isocyanate-based resins have been found to penetrate wood to a depth of one to two millimetres. This ability to penetrate into the wood surface and produce an 'interphase' distinguishes isocyanate from formaldehyde adhesives, which form a surface layer that is less than 0.15 millimetres deep. This ten-fold difference in penetration between isocyanate and formaldehyde resins is a key factor in their relative moisture tolerance and thickness swell, the degree of expansion of the wood composite on exposure to moisture.

In many traditional gluing applications, such deep penetration is seen as an inefficient use of the resin and even results in starved gluelines which do not adhere particularly well. In the case of isocyanates, however, resin penetration has been determined to be of critical importance.

The amount of resin at any given spot of wood will also impact the depth to which the resin can penetrate. It has been found that increasing the resin availability initially increases the depth of penetration, but eventually this levels off and any further addition has no effect. Increases in resin will, however, increase the spreading of the resin and increase the volume of wood which is penetrated. Hence, in applications such as OSB where the resin is sprayed onto the strands, it is important that some of the resin droplets or the cumulative effect of combined droplets are large enough to achieve the required minimum penetration.

Penetration of resin beyond the 'critical' 0.3 millimetres level is of benefit with regard to additional strength, but more specifically with regard to thickness swell. Increasing the depth of penetration increases the volume of resin-modified-wood and this contributes to the excellent thickness swell performance characteristics of isocyanate bonded wood. In OSB, those droplets not sufficiently large to penetrate beyond the critical depth are not wasted as the spread of resin across the macroscopic and probably the internal wood surfaces is critical to the reduction in thickness swell due to moisture ingress.

Recently, a more detailed analysis of the penetration of resin by x-ray microscopy shows that the resin enters and interacts with the wood in a very specific manner preferentially locating itself in one of four regions. First, some of the resin remains at the wood-to-wood interface and in fractured cells. Secondly, the resin runs into the ray cells, which are exposed on the macroscopic wood surface. Thirdly, the resin can be seen at some distance from the wood-to-wood interface in the lumen of the large vessels. It is most probable that the resin accesses the lumen of the cells exposed on the surface since the grain angle will not always be perfectly parallel to the macroscopic surface of the wood. Having entered the lumen of the vessels, the resin can flow along the inner cell walls some distance. The resin appears to wet the inner cell wall of the vessels rather well, but does not always fill the cavity.

Finally, it can also be seen that the resin can pass from the lumen of one vessel to that of an adjacent vessel by passing through the connecting pits. Resin does appear to fill the free space of the pits and does wet the primary cell walls in these areas. From the x-ray microscopy studies there is some tentative evidence that in fact the resin reaches a fifth location within the wood, namely inside the cell walls.

This is the mode of isocyanate bonding for most wood composites in which 'solid' wood particles are employed. However, the situation with wood fibres, as in MDF, is slightly different. In the fiberisation process, the wood is cleaved in the middle lamella and so the isocyanate, which is subsequently added, contacts a lignin-rich surface.

The complete bonding process involved has not yet been fully investigated, but x-ray micrograph images do show that the different nature of the wood fragment surface is significant in how the isocyanate is interacting. In these images it can be seen that isocyanate is wetting the fibre surfaces and either bridging fibres directly or in combination with lignin making agglomerate-type bridging structures. Additionally, the isocyanate can be seen as thin layers partially covering the surfaces of the fibres, but not binding fibres together.

Types of wood composites

Oriented strand board (OSB)

Process technology
Logs are usually soaked in ponds to soften the wood then debarked and sliced or stranded into thin wood elements, approximately 0.3 to 0.8 millimetres thick and 7 to 15 centimetres long by two to five centimetres wide. The aspect ratio of the strands is one of the key characteristics of OSB.

The strands are separated into two streams for the face and core of the final board. If a PF resin is being used on its own then both face and core material need to be dried to an average moisture content of about four per cent as PF resin cannot tolerate a high moisture level. However, the usual procedure in the USA is to use a PF or PF/MDI resin for the face and a polymeric MDI resin for the core as the latter works well with the 12 per cent natural moisture level of undried wood. The elimination of the drying step, offering savings in terms of energy and time plus a reduction in the volatile organic compound emissions, is a major advantage for isocyanate resins.

The strands are blended in rotating drums with the appropriate resin and other materials such as waxes, used to improve water resistance. A typical level for MDI resin is between 1.5 and 8 per cent compared to 6 to 12 per cent for PF resins and wax levels are usually around one per cent.

The resin-coated strands are transferred to a mat lay-down system where, typically, three layers of strands are laid. The bottom face is laid with the strands, hence the grain, orientated parallel to the conveyer direction, a core layer is then laid at 90° and finally the top face is laid with the same orientation as the bottom face.

The mats are conveyed into a heated press, traditionally multi-daylight static presses, Figure 26-5, heated to 170 to 240°C where the mats are pressed to the required thickness. The heat activates the reaction of the isocyanate with the water, steam and wood components. For isocyanate resins, core temperatures of 110 to 120°C are needed to achieve the desired reaction. Controls on the pressure cycle are required to accommodate decompression of the wood and to accommodate any spring back of the board due to relaxation. The press is opened and the boards are removed, edge trimmed, conveyed to cooling racks and then cut to the desired dimensions.

Figure 26-5
Multi-daylight press during installation

Increasingly, continuous presses, Figure 26-6, are being used in which a mat of resin-coated strands is added to the bottom one of the two steel belts that move through the press where pressure is normally applied from one side only. To ensure that the required thickness is achieved the pressure and temperature of the belts is controlled in zones. The pressed board is cut to length on-line, cooled, stacked and processed as for product from static presses. These presses can be 2 to 3 metres wide, 30 to 40 metres in length, with line speeds up to 180 metres per minute. Continuous presses allow closer production control and can process higher moisture content wood, but obviously are best suited to making large quantities of the same thickness/ grade material. They have improved heat transfer during pressing, therefore a faster cycle time and produce boards with improved properties such as lower thickness swell, better dimensional stability, better thickness tolerance and potentially offer less loss of material on cutting.

Figure 26-6   Continuous press

Since polymeric MDI is a strong adhesive not only for wood-to-wood bonding, but also for wood-to-metal bonding, it can cause sticking of the mat to the press platens and so cannot easily be used in face layers. Therefore, MDI is typically used in the core layer, with either a PF resin, in North America, or MUPF resin, in Europe, in the face layers. However, the use of MDI in the face as well as the core layers leads to a panel with better dimensional stability, so solutions to this problem have been found through a range of alternative technologies to provide release:

- Wood floor facings that are subsequently sanded off.
- Thin UF bonded outer facings that are subsequently sanded off.
- External release agents such as soaps and waxes that are sprayed either on the press platens or on the mat.

Increasingly, the most commonly used approach is the application of external agents which with good engineering controls works well, especially on continuous presses.

Resins and properties of OSB

For most applications, standard grades of polymeric MDI are used for OSB, but there still remains the problem of release and the most attractive approach would be a 'self-releasing' isocyanate – a resin, which does not adhere to the press platens, maintains wood-to-wood adhesion and all the expected isocyanate properties. Such self-release systems have been developed through controlling the chemical reactions whilst maintaining all the mechanical, aesthetic and environmental performance normally associated with MDI resins.

Increasing the rate of cure for isocyanate resins is another key development as this leads to an increase in production capacity or a reduction in energy consumption through pressing the panels at lower temperatures.

Isocyanates with improved cure speeds have recently been commercialised, offering reductions in cure times of several seconds per millimetre of board thickness. This is equivalent to a 10 to 20 per cent lowering of the press time, allowing either a reduction in press temperature or a faster press opening time at the same temperature. This is obviously of benefit in those manufacturing sites where the production rate is limited by the press capacity.

The properties of boards depend upon the following factors:

- Quality of wood.
- Density of board and density profile.
- Resin loading and type.
- Orientation of the wood grain in the board.
- Thermal and moisture history of board.

These factors introduce a degree of variability to the properties of panels, particularly under production conditions, so it is not always possible to produce precise and reproducible figures for the performance levels of particular properties and ranges are usually quoted. However, laboratory conditions are easier to control and a comparison of the relative performance of PF and isocyanate resins in the face of an OSB board are given in Table 26-2, which includes the standards used for these tests. This shows that MDI has better properties than PF, especially regarding the modulus of rupture and wet internal bond strength and this is even more pronounced at a lower board density.

Other performance characteristics such as core and face strengths for tongue-and-groove joints, nail-ability, nail and screw holding strengths, durability, creep, fungal and insect resistance and fire resistance can be important properties dependent on the application. If biological and fire resistance properties are important they are usually obtained by adding the appropriate additives. Fungal infestation is strongly dependent upon board swelling due to moisture, so as isocyanate bonded boards show a delayed on-set of swelling and generally swell less than PF bonded boards, fungal infestation is slower and to a lesser degree.

Table 26-2   Properties of PF and MDI OSB boards

Performance characteristic	PF face resin		Isocyanate face resin		European standard (OSB3)
Resin loading, %	4.2	4.2	4.1	4.1	
Thickness, mm	18	18	18	18	
Density, kg/m^3	608	590	608	590	
Modulus of rupture, MPa	25	18	33	30	20
Modulus of elasticity, MPa	4,100	3,000	4,700	4,300	3,500
Internal bond strength, MPa					
dry	0.62	0.32	1.02	0.81	0.32
wet aged (1087-1)	0.20	0.04	0.28	0.27	0.13
Thickness swell (24 hours), %	26	34	13	12	<15

All panels made with an MDI core, pine wood and pressed under laboratory conditions, so unoriented with no major/minor axis. OSB3 is higher specification than OSB2. Source: Huntsman.

OSB applications

Oriented strand boards are used for a number of applications, including, roofing, Figure 26-7, walling (siding), flooring, Figure 26-8, I-beam webbing, hoardings, and packaging. In the USA, by far the biggest use is in domestic house construction. Increasingly, such composites are also being employed, for similar applications, in Europe.

In most of these markets, OSB competes mainly with plywood, as both are structural boards, with layers of wood with perpendicular grain. Since OSB is cheaper to produce and does not require the use of large diameter trees, to peel the veneers used for plywood, it could be expected that OSB would have eroded the market for plywood. However, what seems to have happened is that OSB has grown in line with the growth in demand for boards without decreasing demand for plywood, which itself continues to grow, although rather slowly.

Figure 26-7   OSB roofing

Figure 26-8   OSB flooring

Medium density fibreboard (MDF)

Process technology

Wood from a wide range of sources can, and often is, used in the production of medium density fibreboard, but the principal source is fast growth young trees. Industrial waste wood, both raw lumber and composite wood and even domestic waste wood can be used.

Initially the wood is reduced to rough chips of a few centimetres in length that are soaked and heated, at seven to eight bar and 150°C, then conveyed into a defibrator, also called a refiner. This typically consists of two discs, counter-rotating at about 1,500 rpm on the same axis, with radial grooves that become smaller towards the edges. The steamed chips are fed into the centre and pushed out through the smaller grooves by centrifugal force, the residence time being about a minute. At a temperature of 150°C, the lignin is quite soft and the mechanical action of the discs essentially cleaves adjacent cells from each other by breaking the connecting middle lamellae, producing individual wood fibres. This defibration step and the quality of the fibres produced have a very significant impact on the final board properties.

From the defibrator, the fibres are discharged directly into the 'blowline', essentially a long pipe, which 'blows' the fibres from the defibrator to the dryer. Resin, wax and water are added to the fibres in the blowline and the turbulent flow gives good mixing and dispersion. The blowline can be from 5 to 30 metres long and widens from a narrow entrance of 40 millimetres to an exit diameter of 1.5 metres. The internal pressure is 0.6 to 1.0 mPa and the temperature varies along the length from about 180 to 190°C at the inlet to 50 to 100°C at the outlet. The damp fibres are then dried to a moisture content of 10 to 14 per cent.

For isocyanate-based resins, either standard polymeric MDI or emulsifiable MDI grades can be used, with the latter providing some processing and performance benefits. The emulsifiable resin is pumped to a static mixer where it is emulsified in water and is then pumped into the blowline. It has been found that isocyanates can be added anywhere along the length of the blowline without detriment to the final product performance. Typically 3 to 7 wt-% of resin is added. Due to the nature of the wood particles, release of the board from the press platen surfaces is usually easier than for OSB. Even so, internal release aids, such as Montan wax or polyethylene waxes, are often employed. These waxes are added to the fibres in the blowline usually at a level of approximately one per cent based on wood weight.

The dried fibres are kept in a fibre bin and transferred to a lay-down or mat former, which produces a loose mat of fibres between 230 and 610 millimetres thick, dependent on the board thickness required. This is conveyed first to a pre-press and then to the main press. The majority of MDF is made by continuous production, with the mat-forming, pre-pressing and press directly linked. The main press can be 30 to 40 metres long, with typical platen temperature control of 170°C, in feed and exit, to 205°C at the centre. Application of pressure is usually controlled so that a relatively uniform density profile or a smooth u-shaped profile, without discontinuities, is obtained. Press times are similar to those used in OSB and decompression steps are used at the end of the press cycle to avoid explosive decompression and consequent blows in the boards.

For lower capacity plants, multi-opening presses are used, where after pre-pressing, the endless fibre mat is divided into individual lengths, which are loaded into the press openings. For thicker boards, a stacking station can load two or more mats on top of each other before they are loaded into the press.

After pressing, the hot boards are cooled in a star-cooler, then trimmed, sanded and finished as required.

Resin and properties of MDF

Urea formaldehyde (UF) and melamine modified urea formaldehyde (MUF) are the most commonly used resins, but boards made with UF can only be used for interior applications as they have poor water resistance. However, UF resin is cheap, the cure speed high and the melamine modification improves the quality and performance of the boards as well as reducing formaldehyde emission levels, but at a higher cost and slower cure speed. Isocyanate resins have much better moisture properties, do not contain formaldehyde, but take longer to cure with press speeds about 20 per cent lower than UF and 10 per cent lower than MUF. Typical properties for MDF boards made using UF and MDI resins are given in Table 26-3.

Table 26-3   Properties of MDF boards made from UF and MDI resins

Performance characteristic		Standard values (humid use)	UF resin	MDI resin
Resin loading, %			12	6
Panel density, kg/m^3		650 – 800	750	750
Press factor, s/mm			10	12
Thickness, mm		12 – 19	16	16
Modulus of rupture, MPa		24	36	46
Modulus of elasticity, MPa		2,400	3,200	3,220
Internal bond strength, MPa	dry	0.75	1.00	1.82
	wet aged (V-100)	0.12	0.02	0.40
Thickness swell (24 hours), %		<8	8	3.7

Source: Huntsman. See Table 23-4 for test methods.

Polymeric MDI can be used successfully to make MDF, but it is difficult to efficiently mix the small quantities of viscous liquid with the high surface area wood fibres. By comparison, UF resins have a lower viscosity and are added at 12 per cent. The use of emulsifiable MDI brings a double processing improvement as the eMDI can be easily mixed with the water added to the blowline and the resulting added volume of material is easier to mix with the fibres ensuring better dispersion. There is also evidence to suggest that this results in an improved product performance especially regarding internal bond strength, which shows a 20 per cent increase for the dry value and 30 per cent after humid ageing when compared to polymeric MDI.

Some of the properties of UF and MUF-bonded boards can also be improved by hybrid resin technologies, in which UF/MDI and MUF/MDI systems are employed. This technology is increasingly used with significant success.

MDF applications

MDF boards have excellent machinability properties and they can be cut, profiled, engraved and otherwise shaped and decorated with a well-defined finish. Many of the applications of MDF boards employ this characteristic and this has ensured that they are used extensively in DIY applications.

Applications include furniture, cabinets, shelving, door panels and frames, window frames and signposts. The specific application defines the thickness required as well as the type and quantity of resin needed. For instance, garden furniture and kitchen cabinets require low thickness swell values and a range of thicknesses whilst standard internal door panels are thin (two to three millimetres) and do not require significant resistance to swelling. Signposts and shop front signs require both low thickness swell and low linear expansion. MDI is commonly used for exterior grades.

MDF can also be used to make moulded parts since after the fibres are blended with resin, they can be used in moulding applications to make complicated shapes and parts with well-defined and finished surfaces.

Engineered lumber (EL)

The term 'Engineered lumber' is a generic name for a range of composite wood products all with the key feature that they are structural load-bearing components and are often made by bonding other wood composites together. The range of products includes glulam, laminated veneer lumber (LVL), oriented strand lumber (OSL), timber strand or parallam and I-beams. Such structures can be tens of metres long, often achieved by finger-jointing (FJ) individual pieces together. For this reason, finger-jointing applications are also considered as engineered lumber.

EL composites compete with solid sawn wood with the advantage that they are much lighter and longer lengths can be produced than is possible with sawn timber. The properties of EL are also less variable since the composite structure overcomes the natural variation of sawn wood. The overall conversion of tree to product is higher for EL than for solid lumber; the final product yields up to 75 per cent from the log compared to 40 per cent for traditional means.

The increase in properties combined with the fact that EL can be pre-fabricated into large timber frames, offering time and labour savings, has resulted in a large market growth.

Process technology

To manufacture an EL assembly wood, usually freshly cut to the required dimensions, is dried to a specified moisture content and resin applied, typically 200 to 400 grammes/m² for PRF resin and 90 to 150 grammes/m² for polyurethane resins. A second piece of wood or wood composite is then brought into contact with the resin coated piece and the two pieces, or more in the case of glulam, are pressed together. The pressing step can be very short, a few seconds in the case of I-beams or relatively long, several hours in the case of glulam. Generally, no heat is applied and the components are required to cure at room temperature, or only slightly elevated temperatures.

For I-beams, Figures 26-9 and 26-10, grooves in the flanges, solid wood or LVL, are cut before resin is applied then tapered edge OSB or plywood panels, the webs, are pushed into the resin-coated grooves of the flanges. Since I-beams need to be of specified length both flanges and web stock are finger-jointed previously or in situ. In the case of PRF resins the loadings are relatively large and cure is comparatively slow and so overflow occurs, resulting in unsightly resin flow lines, due to the deep red colour of PRFs, over the web and flange. This can cause adhesion between touching beams in the stack. Isocyanate-based resins, which are wood coloured, cure faster and require lower volume usage and so generally suffer less from these problems.

Figure 26-9   I-beam close-up   *Figure 26-10   I-beam in roof structure*

Glulam, Figure 26-11, is prepared by conditioning to a specific moisture content 3 to 15 metre-long pieces of wood before applying resin to the first freshly prepared surface. A second piece of wood is then placed on the glueline and its upper surface coated with resin and the process repeated for as many layers of wood as required. Again, finger-jointing is used to extend the lengths and the pieces are stacked in such a way as to control the orientation of the growth rings with respect to the thickness of the stack. The upper and lower pieces are arranged so that the growth rings curve away from each other. The stack of lumber is then mounted in a simple press and clamped for up to 24 hours.

Figure 26-11   Glulam in roof structure

Owing to the nature of the wood substrates, the resins need good gap-filling performance and this is more important for glulam than for I-beams. PRF resins achieve this by the inclusion of fillers, such as wood dust and modifications to the formulation. The loading of resin is slightly excessive to fill the gaps, where they occur, so again this can lead to resin flow lines on the outside of the assembly.

Resins and properties of EL

Two-component isocyanate-based adhesives are used commercially as they provide a high degree of control of component viscosities, stabilities, reactivities and performance and good processability. However, such resin systems do require good mixing at the specified ratio and require constant monitoring during production. One-component moisture-cured urethane-based adhesives are also used, relying on the moisture in the wood, atmosphere or applied as a spray for cure. Application and processing is easier as the resin does not need mixing and a longer shelf life is obtained.

PRF resins are commonly used in these applications and where only limited load bearing performance is required, polyvinyl alcohol (PVA) and acrylic-based resins are also used. Modified polyvinyl alcohol and acrylate resins, using isocyanate-based materials to improve performance, have met with commercial success. Some grades of these latex adhesives with added isocyanate cross-linkers are even used successfully in structural applications. The amount of isocyanate addition varies, but is typically 5 to 15 per cent, based on the weight of the principal adhesive. The poly-functional isocyanate additives, usually polymeric or emulsifiable MDI, increase the cross-link density of the major polymer, thereby increasing strength and durability.

Most applications of composites, such as glulam and I-beams (other examples include parallam, laminated strand lumber and composite strand lumber) are load-bearing. For this reason, they are subject to stringent codes and standards. Standards are set on a national level, with each country requiring a particular set of performance tests. These tests range from initial strengths and bending moduli of assemblies to long-term endurance and creep testing. Tests can include glueline thickness assessments and aggressive short-term hydrolytic ageing assessments.

In Europe, the German test institute, Forschüngs-und Materialprüfungsanstalt, Baden-Württemberg (FMPA) are seen as the critical evaluators of resins for structural applications and although national tests need to be passed, passing of the FMPA tests is usually seen as a strong indicator of success.

EL applications

Structural applications for these types of composites include among others, roof trusses, floor joists, support columns and cross-beams. In the USA, LVL is mostly used as flange material for I-beams (60 per cent) with the remainder in structural beam applications such as garage door headers. I-beams are primarily used in floor joist and roof rafter applications in North American platform construction. In Asia, the main consumption is of glulam, especially in Japan, for traditional post and beam house construction. In Europe, glulam is the dominant composite, used in commercial and industrial applications such as beams, arches and columns for large-span buildings such as gymnasia, auditoria and exhibition halls.

Other types of wood composites

There are a number of other types of wood composites, including particleboard and plywood, that have not been covered here, but ranges of isocyanate-based resins are available for these composites in competition with the other resins. Each composite has a particular set of production and use performance requirements and the only peculiar aspect to both particleboard and plywood and to certain LVL production processes is the need for the resin to provide tack to 'fix' the components of the composite in position prior to pressing. Historically, this has proven very difficult with isocyanates, but recent developments have shown that this is now achievable.

27. Sealants and encapsulants

Wolfgang Pille-Wolf

Sealants and encapsulants form the lesser-known part of the application area known as coatings, adhesives, sealants and encapsulants (CASE). Nevertheless, they are still significant in terms of volume with a global market of 260,000 tonnes for sealants and 50,000 tonnes for encapsulants.

Sealants

The need to protect joints against environmental impact is common to a wide range of industries in which components or constructions are assembled from a variety of different materials. In order to prevent the migration of moisture, water, oils and other contaminants into or out of such an assembly a material is required, which adheres tightly to both surfaces, forming a joint that is stable under all environmental conditions. Sealants have to perform under a wide range of climatic conditions, need superior and durable adhesive properties, must stay flexible enough to follow the movements of the component without stressing the joint and there should be no loss of adhesion even when different materials or large dimensional parts are used or where expansion and contraction with temperature can vary widely.

Therefore, sealants are usually soft and flexible elastomers with a low modulus, high hydrophobicity and oil, gas and petrol resistance. Any class of chemical can be used as a sealant material provided that it conforms to these requirements. Initially, coal tar or naturally occurring bitumen was used to seal the joints in timber-clad ships, wax for sealing wine corks and then kneadable putties, based on air-drying vegetable oils, were introduced as the use of window glass became more widespread.

The first concrete roads were built in the middle of the 19th Century without transverse gaps and it was rapidly realised that expansion joints were necessary, filled with an elastic material, to prevent premature damage of the road surface due to water ingress and subsequent freeze/thaw cycles. Initially bitumen-based products or rubber profiles were used, but the disadvantages were the lack of resilience of thermoplastic material such as bitumen and the lack of adhesion of the rubber profiles.

The first elastic, rubbery polymers used as sealants in the construction industry were based on two-component liquid polysulphide systems and had the advantage that they could be applied at low temperature and had increased movement capability. Their good chemical and weathering resistance meant that, despite their initial high price, they rapidly replaced oil- and resin-based putties. Further developments led to the silicones and subsequently polyurethanes becoming the major sealants and now besides these high performing sealant raw materials there are only limited specialist uses for materials such as acrylic water-based dispersion polymers and butyl rubber. It was during the 1960s that the first polyurethane-based sealants, both two- and one-component, were first commercialised.

Applications

The major application for sealants is in the construction industry, which represents 64 per cent of the market, whilst the next largest market is in automotive where direct glazing of windscreens has a 23 per cent share whilst insulated glazing and other transport areas each have 6 per cent and other applications account for the final 1 per cent.

Formulations and properties of sealants

Polyurethane sealants are available in a range of systems designed to meet specific processing and performance requirements:

- Two-component sealants.
- One-component sealants.
- Direct glazing.
- Foam-in-place.

Typical formulations and properties of these four types of polyurethane sealants are listed in Table 27-1.

Two-component polyurethane sealants

The initial use of polyurethane-based sealants, during the 1960s, was in joints exposed to abrasive wear such as roads and a typical example is a concrete expansion joint, Figure 27-1.

The initial systems were based on polyester polyols, but these were replaced by polyether polyols in order to increase the hydrolytic stability and to lower the viscosity. All the systems are based on MDI, either low functionality prepolymers with relatively low isocyanate values or modified polymeric MDI with a higher functionality, the choice dependent on the final mechanical performance required. Castor oil-based polyols, with long hydrocarbon chains, are still used extensively due to their superior hydrolytic and oxidation resistance. Other additives used in the formulations are mineral fillers, colourants, plasticisers, thixotropic agents, adhesion promoters and solvents. One of the major advantages of moving to the lower viscosity polyether polyols was that higher mineral filler ratios could be used, leading to significant cost reductions.

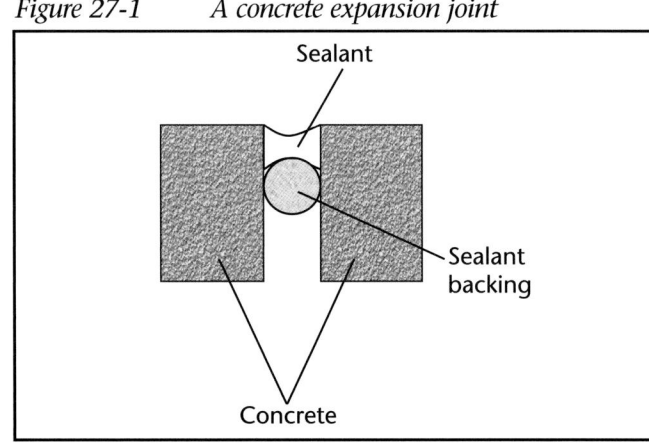

Figure 27-1   A concrete expansion joint

Table 27-1   Typical properties of sealants

	1-component	2-component	Direct glazing	FIPG
Formulation, wt-%				
MDI, Pure/liquid pure	9.0	13.0	6.5	33.3
PPG 2000	52.0	–	–	–
Polyether polyol (OHv 35)	–	16.0	30.0	61.6
Chain extender	–	–	–	1.1
Plasticiser	12.0	20.0	40.0	–
Catalyst	–	0.5	0.1	0.3
Adhesion promoter	–	–	3.7	0.3
Surfactant/wetting agents	0.8	0.5	–	–
Filler	25.0	49.0	–	2.0
Carbon black	–	–	18.4	–
Thixotropic agent	1.2	1.0	–	0.4
Water	–	–	1.4	0.9
Properties				
Density, kg/m^3	1,300	1,350	1,150	200
Hardness, *Shore A*	35	45	40	5
Viscosity (23°C), *mPa.s*	Non-sagging	20,000	Non-sagging	Non-sagging
Tack-free time, *min*	30 – 180	30 – 60	20 – 60	10 – 20
Cure time, *days*	3 – 7	2 – 4	0.25 – 1.0	0.25
Modulus, *MPa* 50%	0.3	–	0.9	–
100%	0.5	0.7	2.0	0.3
Tensile strength, *MPa*	1.5	3	5	0.45
Elongation, %	800	750	700	200

The advantage of the reactive two-component polyurethane sealant systems is that they can be applied at low temperatures and the variability of the polyurethane chemistry means that the formulations can be tailored to the specific needs of an application.

Tensile strengths between two and five MPa and elongation of more than 500 per cent are achievable with good adhesion to a large variety of substrate. Sometimes, however, the application of a primer is required in order to improve the weathering resistance of a joint.

A special case is insulated glazing, which requires extremely low gas permeability of the polymer sealing the gap between the two or three glass panes. Initially this area was dominated by polysulphides, but polyurethane systems based on polybutadiene polyols are now much more common due to their easier and robust application by automated equipment.

Moisture-curing (one-component) sealants

The introduction of two-component polyurethane sealants had been a step improvement for the construction industry in sealing horizontal joints, due to tolerant mixing ratios and their self-levelling nature, but despite the availability of thixotropic systems it was not easy to seal vertical joints. These were usually filled using caulking guns to apply putties and other one-component sealants and application guns for two-component polyurethane sealants were either non-existent or less convenient to handle.

One-component polyurethane sealants, based on MDI and cured with water or the humidity in the air, had been known since the 1940s, but they had low hydrolytic and weathering resistance and too high a modulus and poor elongation plus the storage stability for a marketable product was very difficult to achieve.

The introduction of systems based on TDI and high molecular weight polyether polyols in the mid-1960s led to products that had the processing and properties required by the construction industry. As TDI has a lower reactivity than MDI, the risk of premature gelation was considerably less and the use of cartridges made from aluminium or high-density polyethylene gave a storage life of over six months. The rheology of these one-component sealants is very important since the material has to be applied to vertical joints, so non-sagging is the key basic requirement. However, a low viscosity is also required as the sealants have to be applied using hand-operated caulking guns, Figure 27-2.

Figure 27-2   Application of sealant

Although the required thixotropic characteristic can be achieved with additives such as mineral fillers, colourants, plasticisers, thixotropic agents, adhesion promoters and solvents, the addition of partially gelled PVC plastisols, as a major component, has been found advantageous for creating the necessary sag resistance. All the additives used need to be thoroughly dried and the storage stability can be further improved by the addition of moisture scavengers such as methyl orthoformate or toluene sulfonyl isocyanate. Stripped TDI prepolymers are now used in order to meet environmental concerns.

The critical processing features of one-component sealants are the skin formation time, which depends on the humidity and the temperature, and the through cure time that is related to these, but, in addition, to Fick's law of diffusion. Owing to the need for high elongation and low modulus the prepolymers for these sealants tend to have a low functionality so that the molecular weight build up, during cure of the sealant by cross-linking, is slow.

The low reactivity of TDI, whilst an advantage for storage stability, leads to typical skin formation times of about an hour with a cure rate of less than one millimetre per day at 25°C and 50 per cent relative humidity. MDI-based systems have much faster cure rates with skin formation times of 15 minutes and cure rates of four millimetres per day, however, the storage stability is decreased. To formulate storage stable MDI-based construction sealants hydrophobic plasticisers, acidic mineral fillers and non trimerisation catalysts need to be used. The storage life also depends on limiting the side reactions of the isocyanate group that can generate urea, allophanate, biuret and isocyanurate groups all of which can lead to molecular weight increases.

Carbon dioxide is produced during the moisture cure reaction, but only in small quantities so that its concentration is low enough for it to diffuse through the cured skin. However, if the cure speed is too high or if there is a high isocyanate content then the carbon dioxide build-up can lead to the formation of gas bubbles in the cured sealant. These need to be avoided as they can fill up with water during the lifetime of the seal leading to deterioration and premature breakdown. The lower the isocyanate content the less carbon dioxide will be formed, but it cannot be too low since this then affects storage stability and application viscosity. A low cross-link density of the prepolymer and polar plasticisers and fillers decrease the tendency to bubble formation.

Since these sealants are not load-bearing and are not used in structural applications then the most important features are not strength, but their flexibility, resilience, weatherability and durability.

Direct glazing
Polyurethane direct glazing now dominates the global automotive industry where 80 to 90 per cent of all cars are fitted with windscreens using this technology. It is also used in buses, trucks, trains, ships, yachts and telephone cabins, with the European market in 2000 being approximately 15,000 tonnes.

The American automotive industry started changing the method of fastening windshields and rear windows in cars after more stringent requirements for crash testing were introduced in the early 1960s. A three-component elastomeric polysulphide, used to bond windows into aeroplanes, initially replaced the original preformed rubber profiles, which did not provide any structural strength. This polysulphide sealant not only improved the car body stiffness, by bonding the glass to the metal, but also prevented water leakages, another major problem of car design at the time. However, the system was too complex for the cycle times of the car industry.

One-component moisture-curing polyurethane sealants soon replaced the polysulphides and consist of up to four different components for the conditioning and pre-treatment of the surfaces to be bonded. The exposure of the polyurethane sealant to sunlight during use would lead to rapid degradation unless precautions were taken to reduce the effect and although the light has to pass through single or double layers of glass, with thickness between five and eight millimetres, this does not filter out all UV radiation and other steps are required.

Because the joint is a structural element of the car its failure could have catastrophic consequences. Therefore, the application of all the components is tightly controlled and automated in production, with the design principle of the joint shown in Figure 27-3.

Figure 27-3   Joint design for car windscreen

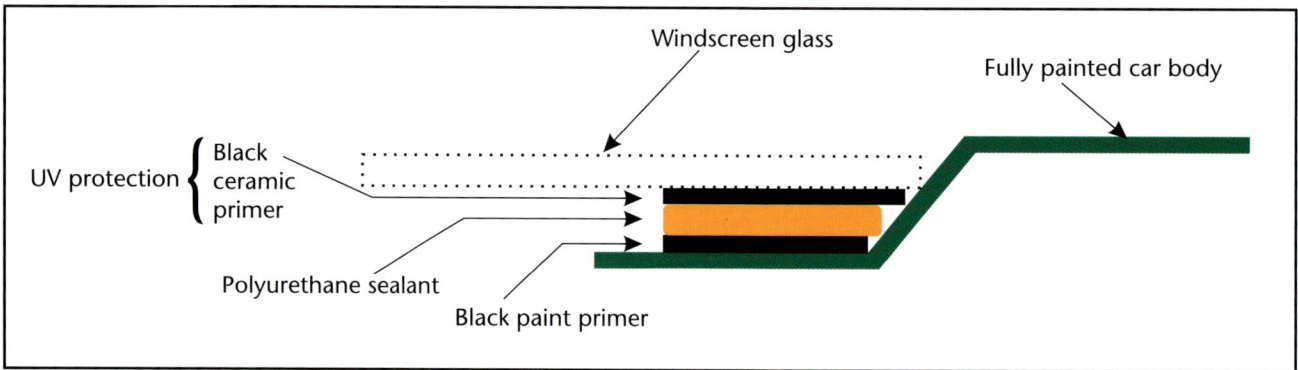

The following steps are involved in preparing the joint:

- The glass is treated with methanol or ethanol containing small quantities of silanes or titanates to both remove traces of grease or other impurities and to prime the surface.
- A black-pigmented glass primer, a low melting pigmented lead oxide glass, is then applied to the glass by screen-printing.
- A black primer is applied to the car body.
- Finally the bead of polyurethane sealant is applied to the car body and the windscreen carefully positioned.

The black primer is applied to the car body because it will be coated either with freshly-cured lacquer topcoats containing additives to reduce wetting and absorption of water and grease, or the middle paint layer, which can also contain residues such as SBS polymers that have a detrimental effect on adhesion.

The sealant has always been applied by robots and initially the various cleaners and primers were hand applied, but nowadays these operations are also fully automated and only the insertion of the glass is still done manually. The curing behaviour of the one-component polyurethane direct glazing sealant has to suit the cycle time of the car manufacture. Similar compounds are used for replacement windscreens, when application is by hand-gun.

The rheological demands on the polyurethane sealant are quite significant as 700 to 900 grammes have to be applied within a minute in the form of a triangular bead with a base of not more than 10 millimetres, but a height of 30 millimetres with no sagging or movement. Also, the green strength of the one-component sealant, the build-up of which is affected by the severe shear forces created during the pumping of the high viscosity material, is not sufficient to support the weight of the glass, so supports are used to stop it gliding out of position in the frame.

A critical processing feature of the direct glazing bead is that it must build up enough strength to meet the first roll test, which is usually three hours after glass insertion. At high temperatures and high relative humidity the sealant can form blisters whilst at low temperature and low relative humidity, typically seen in winter, the cure speed can slow dramatically.

The formulations, based on MDI, use functional prepolymers with the level of functionality and isocyanate content adjusted to provide a balance between viscosity and cure time. The type of filler is critical and china clay is often used as it provides additional strength whilst substantial amounts of carbon black are added to improve the tensile, shear and modulus properties and to provide a black sealant. When using silica gel, acidic grades need to be used to improve the storage stability. The modulus is typically 0.8 to 1 MPa at 50 per cent elongation in order to provide stiffness to the car frame.

Some car manufacturers specify a maximum conductivity of the sealant as a high level of carbon black can affect the field strength of radio antennas, which are sometimes encapsulated in the sealant, and also contact corrosion may occur with aluminium body shells.

The introduction of two-component or hot-cured systems has helped overcome the problem of variable cure seen with the one-component systems and has provided a lot more stability and decreased the susceptibility to climatic conditions. These newer systems have cure rates of more than five millimetres per day and are formulated to provide high modulus material of 2.5 MPa at 50 per cent elongation, which further stiffens the car body.

A final key factor of the sealant is the ease with which a windshield can be replaced following damage and the level of tear resistance can be set so that removal is easier. The freshly cut sealant surfaces are usually easy to adhere to, but an activator containing silanes is often used to improve the adhesion.

The polyurethane sealant direct glazing application, with very few cases of joint failure in use, has helped to open the door for other structural adhesive applications in the automotive industry.

Foam-in-place gaskets (FIPG)
The variety of different parts that needed sealing in automotive construction meant that traditionally a large number of rubber profiles, seals, inserts and gaskets had to be stocked, but these are increasingly being replaced by directly applied polyurethane sealants.

Complex seals can be laid down robotically and due to the ease of reprogramming it is possible to accommodate part variation and rapidly adapt to the introduction of new models. Since the requirements are non-structural and the seal is joining a variety of materials the sealant is usually foamed to improve resilience and provide a softer seal. The polyurethane sealant can be foamed using water to produce carbon dioxide or by using a physical blowing agent. A typical example is shown in Figure 27-4.

Figure 27-4 Finished part with a sealant bead already applied

Encapsulants

Bitumen, waxes and phenolic compounds were initially used to encapsulate electrical components to protect them against environmental impact. From the 1950s epoxy resins became available and were used due to their high strength, chemical resistance and excellent adhesion. Silicones and polysulphides were also used as encapsulants from this date.

Polyurethanes are also electrical insulators and their use as potting or casting resins for the protection of electrical and electronic parts has grown substantially over the years. Cast polyurethanes are used for insulators, transformer housings and junction boxes, and for sensitive equipments such as sensors, switches and complete integrated circuit boards. Polyurethane encapsulant systems can be formulated to have a low reaction exotherm, low viscosity and to vary from gels to hard and tough materials. Their electrical properties are comparable to the epoxy resins, so can be used in similar applications.

Applications

Until the early 1960s power cables were joined by feeding the two cables into a cast-iron box, which provided protection and earth continuity, where the conducting cables were soldered together and, finally, the joint was insulated with bitumen. Following the development of polyurethane encapsulants in the 1960s it was possible to replace the cast iron by moulded thermoplastic boxes since the new two- or three-component systems, which were normally hand mixed on site, provided sufficient strength to the joint assembly. A typical example is shown in Figure 27-5.

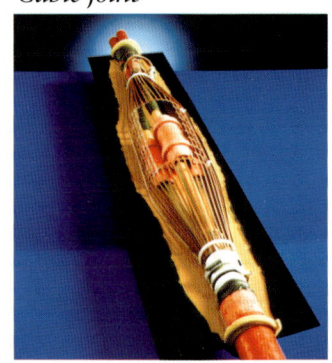

Figure 27-5
Cable joint

The protection of capacitors and inductors requires polyurethane encapsulants with particularly low cure temperature since thermal stress can damage the electrical properties of the component. On the other hand, the growth in the utilisation of electronic components and sensors in the automotive industry has led to very demanding temperature shock tests in which parts are shock heated from -40 to +150°C and vice versa. Polyurethanes are also used widely in sonar sensors due to their acoustic density match to sea-water.

*Figure 27-6
Potting of electronic components*

With the miniaturisation of condensers and inductors the flow properties of the polyurethane are becoming ever more critical in order to avoid air entrapment whilst enabling small gaps to be filled and ensuring wetting of the thin wires, Figure 27-6.

Formulations and properties of encapsulants

The majority of the polyurethane encapsulant formulations are based on MDI. Dependent on the final application this can range from polymeric MDI, with or without modification, to prepolymers based on pure MDI with varying degrees of functionality. The addition of the lower reactivity 2,4-MDI is used to control the reaction rate, the viscosity build-up and to reduce the exotherm. Typical formulations and properties are given in Table 27-2.

Table 27-2   Formulations and properties for polyurethane encapsulants

	Cable joint encapsulant	Soft encapsulant	Industrial assemblies
Formulation, wt-%			
MDI	28.0	20.0	17.5
Castor oil	64.0	–	–
Polyether polyol	–	58.0	36.8
Chain extender	–	5.0	–
Plasticiser	–	16.5	–
Catalyst	0.3	0.1	0.2
Surfactant	–	0.4	0.5
Filler	–	–	40.0
Dessicant	7.8	–	5.0
Properties			
Density, *kg/m³*	1,200	1,100	1,700
Hardness, *Shore A*	50	20	70
Tensile strength, *MPa*	5	2	15
Deflection under load, °C	45	55	70
Glass transition temperature, °C	10	0	40
Dielectric constant, (20°C and 100 Hz)	4	5	5
Volume resistivity (20°C), *ohm.cm*	10^{14}	10^{12}	10^{11}

Initially polyester polyols were used, but it soon became clear that the hydrolysis resistance was insufficient for the application. They were replaced by polyether polyols, which not only had much better hydrolysis resistance but also had a lower viscosity that gave the mixed system a longer open time and better wetting of more complex surfaces. The dielectric constant of polyether polyol-based encapsulants is also better.

Ricinoleate polyols, based on castor oil, have even better properties than the polyether polyols, as their long carbon chain is very hydrophobic, so reducing the sensitivity to attack by moisture. However, they have poor compatibility with MDI and on mixing there is a rapid formation of separate phases, which turns the initial transparent liquids into opaque or translucent materials. An increased level of 2,4-MDI improves the compatibility with these polyols resulting in more transparent castings. Transparency is required in order to be able to see the electronic components in repairable products. To adjust the hardness and strength of the castor oil-based polyurethane encapsulants they are modified with rosin or terpene resins and also combined with polyether polyols.

Polyurethanes based on ricinoleate polyols have a much lower dielectric constant, as low as three at 100 Hz, than polyether-based polyurethanes and the volume resistivity is as much as two orders of magnitude higher.

Encapsulants based on polybutadiene polyols have proven their superiority for applications in which the combination of resistance to temperature shock and very low moisture absorption is required. Due to their very low glass transition temperature, -70°C, no internal stress is generated over a wide temperature range – an important feature for pressure-sensitive electronic components.

Since materials with low electrical conductivity also tend to have a low thermal conductivity then, if the dissipation of the reaction heat is critical the addition of mineral fillers can become necessary as they absorb heat. The first mineral filler used was mica because of its contribution to electrical insulation and other mineral fillers are quartz or aluminosilicate. Aluminium hydroxide is used in the formulation of self-extinguishing encapsulant materials.

28. Introduction to polyurethane composites

Alan Hamilton
Stephen O'Nein

The continuing demand for lower cost production achieved by more efficient processing and lighter weight components leads to the need for higher performance and one of the ways this can be achieved is through the use of composite materials. Polyurethane composites are predominantly thermoset resins that are reinforced with a variety of fibres or minerals. The advantages of strength, stiffness and lightness offered by polyurethane composites coupled with novel design techniques have been fuelling the growth of new applications in transportation, construction, marine, electrical and consumer products.

Fibre reinforced composites have become established for two main reasons, their physical properties and their processability. The key advantages are:

- Tensile strength to density ratio up to six times greater than steel or aluminium.
- Modulus (stiffness) to density ratio of 3.5 to 5 times that of aluminium or steel.
- High fatigue endurance and ability to absorb higher impacts.
- The ability to be strengthened precisely where required.
- Easily formed into complicated component shapes; joints and fasteners can be eliminated.
- Excellent corrosion resistance.

Conventional polymer composite technology is based on reinforcing thermoset resins such as polyester, vinyl ester, and epoxy resins. Polyurethane is increasingly used because it offers specific advantages in terms of processability, material properties and speed of manufacture.

The most commonly used fibres are glass, aramid and carbon; the optimum diameter for all of them is 15 to 20 microns. Filaments of these materials are produced by extrusion processes followed by drawing to the required diameter, with a further pyrolysis step required in the case of carbon fibres. The fibres are subsequently grouped into bundles of filament called a roving in the case of glass and aramid, and a tow in the case of carbon. The fibres have a coating, sometimes known as sizing, applied to protect the surface from corrosion, oxidation, mechanical damage and to help the filament to adhere to the matrix material. All these fibres can be produced with different properties, customised towards their end use.

General properties of composites

Composites by their very nature are heterogeneous materials and it is difficult to carry out stress analysis on parts made from them. In a composite the fibres are the main load-bearing component and the tensile strength of the composite is dependent upon the fibre length with the material properties decreasing, as the fibres get shorter. The fibre type and the way it is formed also affect the final tensile strength of the composite component.

Also, flaws caused by the poor distribution of fibre, air bubbles or polyurethane rich areas can have significant effects on the properties of the composite structures. Because of this, frequent stress testing of final components is necessary. Although composites offer considerable advantages over other materials, when considering the material properties and ease of manufacture, their production and subsequent use has to be applied with care. The material properties can also change with time thus introducing a perceived variation part-to-part.

The major variation in strength of a composite component comes from the placement of the fibres with the key variables being: distribution, direction and density of loading. Another way component strength can be affected during production is shown in Figure 28-1 where the problem is due to poor bonding between the fibre and the resin, whilst Figure 28-2 shows a composite with good bonding.

Figure 28-1   Flaws in fibre reinforced composite   *Figure 28-2   Good fibre reinforced composite*

 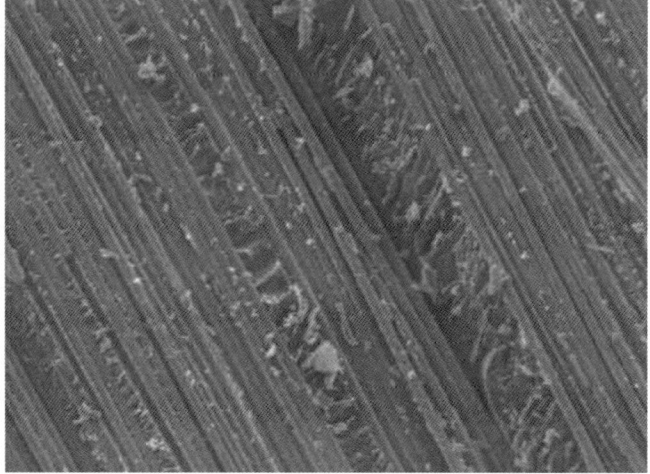

The poor bonding can be caused by problems between the polyurethane, glass and the sizing or the introduction of 'micro bubbles' due to poor venting of the tool. The difference in properties can be considerable. For example, the impact strength of a 'micro bubbled' part can be reduced by around 40 per cent.

Design considerations when working with composites

New materials need innovative design concepts to fully utilise their benefits. The common issues for all materials are related to stress analysis and detailed design points, but the effects are accentuated for composites due to the heterogeneous nature of the material.

Stress analysis

- Stress is introduced by non-uniform thickness of the component thus keeping uniform wall thickness is critical in part design.
- Press-fit and snap-fit joints need to be considered as a source of stress in the final product due both to the fit during final assembly and also possibly because of stress built up during manufacture. In the latter case this is often due to the fact that the snap-fit joint connection on the part acts as a large heat sink in the component.
- If a part is specified to withstand repeated assembly and disassembly then cyclic loading tests should be designed to meet expected life figures and a minimum of 1,000 cycles should be built in.
- With a part that is going to experience high long-term loading, creep failure data on test samples should be used in the design calculations to avoid stress cracking, to maintain tightness of joints and to ensure part functionality.
- Warpage and bending of parts after they have been demoulded can be induced by internal stresses and even well designed parts have some warpage.

Detailed design

- Fibre reinforcement, oriented in a favourable direction, increases the strength of a component.
- Where non-uniform sections are unavoidable the thickness should be increased gradually and not in a step change.
- For parts with large spans stiffeners, such as ribs and gussets, should be used to increase strength, prevent distortion and minimise weight.
- Increasing the rib height and/or decreasing the spacing between the ribs also improve performance.
- Proper design of the ribs is important to prevent sink marks on the part and in areas where sink marks are expected, ribs of a thickness up to 75 per cent of the adjoining walls thickness can be used, but to further avoid sink marks they should not exceed 30 per cent of the local area.

- To aid removal of the part from the tool all ribs should have proper draft angles and generally a draft angle of 1° is used, but it can range from 0.5° to 1.5° depending on the depth of the rib. These draft angles also apply to all other vertical walls in the component.
- With highly polished moulds the composite can stick to the mould leading to the formation of micro cracking on the surface of the finished part.
- To avoid stress concentrations all corners and areas where the width and thickness changes should be designed with a smooth, generous radius.
- A minimum radius of 0.5 millimetre is recommended in all corners, also on the root of ribs and larger radii should be used when possible in the ratio of radius:thickness greater than 0.6:1.
- This combined with draw angles aids the flow of material during the production process and part removal from the tool.
- With plastic-to-plastic surface joints it is necessary to consider stop surfaces to prevent screw over-tightening or to limit the depth of engagement of two matching taper surfaces.
- The design should take account of the expected exposure temperature, including the range of change, but not be too 'over-engineered' and the design also needs to consider differential expansion around inserts in the part.

Raw materials

Polyurethane systems

A broad range of polyurethane formulations is used to produce polyurethane composites varying from the high-speed, high-density RRIM process, based on amine tipped polyols, to the low-density, 250 to 600 kg/m^3, slower-curing LD-SRIM process using mainly functional modified sucrose polyols. However, there are two constants for the different production methods – all of them are based on MDI and are processed using high-pressure machines. Details of typical formulations and the properties obtained when glass filled are provided in Chapter 29.

Form and types of glass fibre

Glass fibre is used as continuous strands or cut into various lengths. Fibres with varying aspect ratios are used to produce several forms of product such as chopped strand mat, woven strand mat, chopped glass and the aspect ratio for cut fibres can vary from 50:1 to 5,000:1, as the length changes from 1 to 60 millimetres compared to an average diameter of 15 microns.

E-Glass
This is the most inexpensive and commonly used where the 'E' designation stands for 'electrical' because of its superior insulation properties. E-Glass has reasonable tensile strength, 3.5 GPa and relatively low modulus, 73 GPa.

A-Glass
This is the second most frequently used fibre and is comparable in price with E-Glass and is made from recycled glass, mostly light bulbs. It is commonly used in continuous strand mat.

S-Glass
Is an aerospace grade of glass where the 'S' stands for 'strength' as it is stronger, tensile strength 4.3 GPa and has a higher modulus, 87 GPa, than E-glass. 'S' is a trade name of Owens-Corning Inc. and other glass fibre manufacturers with comparable products use designations such as 'Te-glass' and 'R-glass.'

C-Glass
This is a corrosion resistant type, and is usually used as a surfacing veil cloth on outer surfaces of laminates, or against tool surfaces to protect the laminate from corrosion.

Continuous fibre
This is used for filament winding and pultrusion and as the feed for the spray chopped fibre process.

Woven glass cloth
Produced by conventional textile weaving methods in virtually any variation and thin cloths can make laminates of high tensile strength and modulus.

Woven roving (WR)
Is used in conjunction with chopped strand mat to provide bulk and directional strength. Glass fibres are arranged at set angles to each other so that their orientation provides balanced strength in the directions required.

Chopped glass
Chopped glass, produced in short (1 millimetre) through to long lengths (60 millimetres), is used mainly in reaction injection moulding (RIM) and spray lay-down processes.

Chopped strand mat (CSM)
Supplied in roll form, CSM is a mat of randomly chopped strands held together by a light binder. It provides uniform strength in all directions.

Mineral reinforcements

Mica is often used for applications where improved surface finish is required whilst barium sulphate and talc are used to increase density when a high damping performance is specified.

Carbon fibre

This is a relatively new reinforcement material for polyurethane composites and growth will depend on the production of a low-cost continuous carbon fibre. Carbon fibres are predominantly made by the pyrolysis of polyacrylonitrile fibres but lower cost pitch-based carbon fibres are being commercialised.

Natural fibre

Hemp and sisal are used, in the form of strands or woven mats, but as both are natural products there is always a problem ensuring consistent strength and moisture level.

Processing techniques

For full details of the processing techniques used for the production of polyurethane composites see Chapter 29.

Test procedures

Mechanical testing, non-destructive testing and field evaluation of polyurethane composites is performed according to industry standards such as ASTM (American Society for Testing Materials), ANSI (American National Standards Institute) and ISO (International Organization for Standardization). The choice of test method will depend upon the final performance objective of the composite. The tests – similar to those for any material to determine properties such as physical strength, high temperature properties, creep properties, ageing effects and surface properties – are used to check material characteristics, to control quality in production and to check product durability. The critical mechanical tests for composites are impact resistance and flexural modulus, which respectively, measure the ability of a component to withstand a sharp blow and determine its stiffness.

The Izod impact test (ISO R180) is normally used for polyurethane composites although others such as the Charpy test (ISO R179) and falling dart or weight tests (ASTM D3029) are also used. In the Izod test the impact energy required to break a standard sized specimen is determined by increasing the fall height of a fixed weight pendulum until the sample breaks, normally using a fresh sample each time. Pendulum impact tests are used for solid composite samples, whilst the falling weight or dart tests are usually used for foam composites.

The flexural strength and modulus of composites are usually measured using a three point loading test such as ISO 1209 where a standard-sized bar is bent at a speed of 10 ± 2 millimetres per minute. If fracture occurs before 20 millimetres of deflection has occurred the breaking load is recorded, whilst if the sample has not broken by 20-millimetre deflection then the flexural stress at a given strain is recorded. The flexural modulus can also be derived from the stress-strain curve as well.

The major difference with composites, compared to other polyurethanes, is that there is a critical need to identify defects in parts by non-destructive means since composites can fail at a defect at much lower loadings than the bulk material can withstand. As with metals, several non-destructive methods can be used – radiography, ultrasonics, mechanical impedance and thermography.

Radiography relies on the penetration of radiation through the specimen and variations in the structure of the composite will affect the absorption of the radiation, highlighting any defects. It is difficult to obtain good results, as it is difficult to produce a good contrast on the film. This restriction on the process limits the minimum size of voids that can be detected within the composite. Because carbon has much lower radiation absorption than glass, the direction of carbon fibres is undetectable within the matrix. With radiation a two-dimensional image is produced and it is nearly always possible to identify a flaw. This makes it a powerful inspection method for non-carbon fibre composites that is more accurate than the ultrasonics method.

Ultrasonic testing is the most widely used non-destructive test method and is based on the transmission of ultrasonic waves, in pulses, from a piezo-electric crystal. The acoustic impedance in the material affects the pulse and it is this effect on the reflected pulse, which the process measures. Defects have a large effect on the pulse and thus can be detected. The best results are obtained with good contact between the head and the component so a common technique is to immerse the test piece in water. There are two methods in use, the 'A' scan and 'C' scan. The key difference is that the 'C' scan requires a more complicated operating system, but it portrays the results pictorially in plan form, that is in two dimensions. The 'A' plan gives a one-dimensional readout from an oscilloscope, only displaying the flaw in one plane.

In mechanical impedance inspection a localised forced oscillation is applied and its effect measured through the use of an electronic transmitter and receiver assembly. The piezo-electric transmitter converts electrical signals into vibrations and the effect of these vibrations is observed. Defective areas respond differently to the bulk of the material, but the process is most sensitive to defects near the surface.

Thermography is a simple process to use, but it only shows the general shape of defects and its clarity is limited compared with ultrasonics. The process is based on directing a sinusoidal thermal wave onto the surface of the composite. Part of the wave will penetrate the specimen and part will be reflected from the surface and internal defects. The reflected waves are out of phase so interference patterns emerge, thus enabling a pattern of the defects within the composite to be produced.

29. Polyurethane composite technology

Stephen O'Nien
Mike Anderson

The technology used to combine fibre reinforcement with a liquid polyurethane system varies depending on the properties of the product that is produced, the complexity of the final part and the number of items to be produced. There are currently five methods being used to produce polyurethane composite materials. Together these are able to meet the demands of the disparate parts being made by this technology:

- SRIM, P4 and LD-SRIM.
- RRIM.
- Sprayed chopped fibre.
- Filament winding.
- Pultrusion.

SRIM, P4 and LD-SRIM

Structural reaction injection moulding (SRIM) is a process by which a preformed glass mat is placed into a mould that is then filled with an appropriate polyurethane system. SRIM can be either produced in a closed or open mould process, with the latter generally preferred as it reduces the damage to the glass mat near the injection point, Figure 29-1. There are several steps involved in the process.

The cover material, such as a vinyl sheet or preform, is placed in the bottom cavity of the mould and for complex parts this cover stock is vacuum-formed directly in the mould. Then the reinforcing mat, pre-cut into a shape that resembles the profile of the finished part, plus any attachment clips is placed into the mould. With the open mould process the polyurethane system is poured over the glass matt and the mould lid immediately closed whereas with the closed process the polyurethane is injected into the closed mould held in the clamp unit.

Figure 29-1
Glass mat being placed in mould

With the injection process the glass fibre mats need several key properties for the process to work successfully. The matting must be strong enough to resist the polyurethane moving it out of place as it flows through the mould from the injection points; it also needs to be strong enough to maintain structural integrity and resist breaking due to the injection pressures and the forces produced by the viscosity increase during curing. Additionally, the mat needs to be open enough to avoid air being trapped in the finished part.

A variation on the SRIM method is the P4 process, which has been developed for the production of large structural parts with the key advantage that the chopped glass insert can be produced automatically. The process consists of spraying chopped glass, produced from rovings fed through cutters, over a former, which has vacuum suction to hold the glass in place. Once this first layer of glass is in place a second layer of chopped or continuous glass fibre combined with a dry, powder binder, is sprayed on top. Random or oriented distribution of these fibres can be used to obtain the desired final part properties. The upper part of the mould is now applied to compress the preform and the powdered binder is melted to bond the glass assembly together by the use of hot air injected through the suction ports in the former.

The former is then replaced by the other half of the mould and a reaction injection moulding (RIM) polyurethane system injected into the glass mat. All this can be performed in a cycle time of approximately four minutes. The use of equipment designed for full automatic production enables a large flexibility to be built into the process making complex geometries relatively easy to produce. The elimination of trimming by using automation and the use of the cheap binder system for large parts means that P4 offers a favourable economic solution to the use of thermoformable glass mats in a RIM process.

Polyurethane RIM chemistry has been formulated to provide the long injection times needed for making complex parts, whilst still maintaining fast demould times of 45 to 120 seconds. The key changes include improvements in flow of the polyurethane through the glass mat, changes in reactivity profiles to allow filling of complex moulds whilst still achieving fast cycle times, improvements in internal mould release capability and significant improvements in cure time.

Internal mould release (IMR) technology has been developed in the past few years so that moulders only need to apply an external release agent once a week, leading to significant cost saving and increases in productivity of 10 to 20 per cent. Furthermore, a good IMR system provides a clean tool surface, significantly reducing downtime to clean moulds, as well as improving the consistency and quality of the parts.

Another variant on the SRIM process is the use of a blown polyurethane system to provide parts with a lower density – the LD-SRIM process. The main variation compared to SRIM is that water is added to the polyol blend of the polyurethane system to produce lighter weight parts. Therefore, it is possible using the LD-SRIM, P4 and SRIM processes to produce a complete range of parts where the weight and performance characteristics are varied as required by the end use of the component.

RRIM

The essential element of the RRIM process is that the reinforcing material, which can be either mica or hammer milled or chopped glass strand, is added to the polyol blend storage tank. Recirculation needs to be adequate to prevent settling of the reinforcing material and several other problems emerge both due to the increase in viscosity and roughness that they introduce.

Current chemistry limits loading levels to around 20 per cent as, above this level, the polyol blend becomes too viscous for good mixing with the isocyanate. The problem with the more viscous systems has led to the use of alternative metering systems. These are based on single-action dosing cylinders or displacement lances, that are driven at an accurately controlled speed, with a typical throughput of eight kilogrammes per second for RRIM systems.

Because the polyol blend contains the reinforcing material it is very abrasive and hardened cylinders, non-standard metering units and special mixing equipment is required. There are major problems with erosion caused by the introduction of fillers, which are most acute around the jet mixing orifice area, where the velocity is highest. This leads to a need to monitor the wear of the mixing orifices and change them regularly so that the quality of the mixing remains good.

Polyurethane reinforced with mica, 1.5 millimetres long compared to glass fibre at normally 25 millimetres, is now used for exterior body panels such as bumpers and other outside panels as it provides superior surface qualities. Glass reinforcement can achieve the same strength, but not the finish as it always has a slightly rough surface.

Sprayed chopped fibre

The key element of the sprayed chopped fibre process is that glass fibre is chopped, as required, above the mixing head of a RIM polyurethane machine and intimately mixed and sprayed into a mould, Figure 29-2. This is a relatively new technology which has only become possible with the development of specialised dispensing machines.

Continuous glass roving is fed into a chopper positioned above the mixing head and the chopped glass is drawn into the mixed polyurethane by a Venturi air feed system. Intimate mixing of the two materials is achieved by feeding the fibre in through the centre of the main cleaning bore plunger on an L-shaped head of a RIM high-pressure polyurethane machine. This plunger is also designed to aid the mixing of the fibre and the polyurethane within the main bore prior to its exit into the oscillating air jets that fan out the stream of composite material as it leaves the head. The amount of glass added to the polyurethane can be varied during the course of a single spray pattern.

The process has the advantage that it eliminates glass mat or preform handling, decreases cycle-time and improves the work environment. Other advantages of the technology are that the glass reinforcement is only placed where it is required and not into areas that do not need the strength or are going to be trimmed. For instance when moulding an interior door panel then areas such as map pockets, speaker grill areas, and door handle regions can be sprayed without glass saving 5 to 10 per cent material in every part. By comparison, if a glass mat is used, as in the SRIM process, then the entire part must be moulded with the high glass reinforcement.

Figure 29-2   Sprayed chopped fibre process

Formulations have been developed for the process where the chemistry has a 'snap cure' reactivity profile which provides the long spray times needed for making large parts, whilst still maintaining fast demould times of 90 to 120 seconds.

The challenge is to increase glass loadings past the original design maximum of 30 vol-%, equivalent to about 50 wt-%, in order to further improve the property envelope of the composite.

Filament winding

The key advantage of filament winding is that continuous fibres can be laid in a specific direction, with a high density of fibre reinforcement, producing parts that are strong and stiff in one direction and flexible in another.

Dry glass fibre is fed through a bath of liquid polyurethane with a pot-life of 20 to 120 minutes, which is worked into the fibre by a system of breaker bars within the dip tanks and the wet fibre is wound around a mandrel that is either removable or remains as part of the final composite, Figure 29-3. A key parameter is fibre tension since to obtain the optimum properties the fibre tension must be kept equal on all of the fibres laid. The key to doing this and thus controlling the part quality depends on the equipment used to lay the fibre onto the part, which is controlled through a roving delivery system.

Figure 29-3   Filament winding

As the mandrel is rotated, the 'feedeye' of the roving delivery system reciprocates along the length of the mandrel, producing a controlled roving action and a steady laydown of the fibre. With complex parts and the need for multi directional strengths a high degree of machine flexibility is required. This production process, with its tight winding of densely packed fibres, produces a compact laminate as the part is produced with glass loadings of 50 to 60 wt-% and wall thicknesses that can be easily varied from two to four millimetres.

Once the composite is fully wound the whole assembly is cured in an oven at 65°C for 30 minutes. After curing, the product is either left on or removed from the mandrel dependent on the type used. The complete process is highly automated and efficient.

Pultrusion

Pultrusion is an automated process that produces continuous, thousands of metres-long, dense high-fibre composites that are straight and of constant cross-section. Glass roving is used to give the unidirectional strength, but fabrics can be used to add off-axis fibres and provide some multi-directional strengthening.

The production process is similar to that of filament winding, in that fibres are pulled through a polyurethane bath with the key difference being that the entire cross-section is pulled at once. This requires a large number of glass fibre spools with a good feed control mechanism to prevent tangles. Almost any type of fibre can be used, but glass fibre is by far the most common.

Similar mechanisms to those in filament winding are used to ensure the fibres are fully wetted and the wetting system also serves the purpose of flattening and spreading out the individual rovings. The coated fibres pass into the mould, which maintains the fibre alignment, compresses the composite matrix to the desired volume and cures the composite. The polyurethane system needs to have a fairly low viscosity to achieve good impregnation, but as the process is fast, a short cure time is desirable as a full cure is required during the time the pultrusion is in the mould.

The whole process is pulled through from the finished product end of the pultrusion line using a set of padded clamps, Figure 29-4. Two sets of grips are used, with one pulling while the other travels back to its initial position. Finally, the material moves through cutting tables where the product is cut to the required lengths.

Figure 29-4   Pultrusion profile

Mould technology

The moulds used for composite technology can be made from epoxy, metal-filled epoxy or metals with the choice dependent on the number of parts required, the quality of finish, the production rate and the possible need for modifications. Epoxy moulds are flexible and easy to modify, but this advantage is offset by many problems such as, poor thermal conductivity, short life due to surface degradation, inferior surface finishes compared to metal moulds and problems with poor mould release. All of this restricts the use of epoxy tools to prototypes and short runs.

Metal moulds can be made from steel, aluminium and zinc alloys or by using beryllium/copper or nickel/copper shells backed up with either epoxy or metal. Moulds made from machined tool steel, finished by polishing or by chromium or nickel plating, provide both good durability and surface finish. Other materials such as aluminium are less durable, but have the advantage of better thermal conductivity and are lightweight. Steel moulds generally pay for themselves when production runs exceed 10,000 parts and upwards of 200,000 parts can be turned out on a set of such moulds.

The design of the injection points and the speed of injection determine how the RIM polyurethane flows and fills a mould. To measure these flow patterns it is necessary to monitor both the heat generated and the pressure created within the mould at various points during the production process. From these measurements the effects of velocity, mould temperature and formulation changes on viscosity build-up and mould filling can be studied whilst the increase in temperature provides an indication of the reaction rate.

Whilst there are heat losses from the outer surfaces of the component to the mould, requiring mould heating to prevent the edges from cooling too rapidly, the overall reaction is so rapid that little or no heat transfer from the bulk of the component through the mould walls occurs. This enables simple modelling of the flow in the mould, assuming adiabatic conditions, to be carried out using programmes that predict flow patterns and pressure build up in moulds so helping with mould design. These programmes help predict the flow of material in moulds, how flow fronts build up and are also useful in avoiding air entrapment.

To reduce flash the mould surfaces should be free of flaws and be a tight fit whilst to prevent air entrapment provision should be made to allow air to escape as the mould fills. Alternatively, a vacuum can be applied to the mould, which speeds up filling and removes air. The extraction process is normally designed into the mould system with strategically positioned ejector pins that push the part out of the mould.

Process properties and subsequent finishing of the product, such as painting, generally determine the choice of release agent, either water or solvent-based, and moulds need to be kept clean of all flash and any build-up of release agent.

Formulations and properties

The polyurethane chemistry used for all the methods of composite production is very similar and typical formulations are given in Table 29-1.

The results of testing composites from the different methods are provided in Table 29-2 and compared to steel and filled ABS.

Table 29-1   Formulations of RIM composite systems

Formulation, wt-%	SRIM	LD-SRIM	Chopped fibre	RRIM
Glycerol-based polyol (OHv 650)	32.1	22.6	31.3	–
Modified sucrose polyol (OHv 575)	3.6	9.7	–	–
Amine-terminated polyol (OHv 56)	–	–	–	42.8
Low viscosity cross-linker	–	–	1.6	–
Monoethylene glycol	–	–	–	8.5
Catalyst package	0.7	1.0	0.5	1.3
Water	0.1	0.5	0.4	–
Surfactants	1.8	3.2	1.1	–
MDI	61.8	63.1	65.0	47.5
Processing conditions				
NCO Index	110	115	105	105
Isocyanate, °C	25	25	25	30
Polyol blend, °C	30	30	30	30
Cycle time, s	90	120	180	60

Table 29-2   Properties of RIM composite products

	SRIM	LD-SRIM	Chopped fibre	RRIM	Steel	ABS
Glass loading, wt-%	50	20	47	23	n/a	20
vol-%	30	4.2	29	10.4	n/a	9.4
Density, kg/m^3	1,450	550	1,590	1,170	7,750	1,220
Flexural modulus, GPa	12.5	1.6	10.8	1.0	–	6.0
Tensile modulus, GPa	11	1.5	11.8	–	207	6.2
Tensile strength, MPa	185	28	185	30	331	76
Elongation at break, %	1.8	2.1	2.0	40	37	2.0
Flexural strength, MPa	340	49	370	–	–	107
Notched Izod impact, kJ/m	1.5	–	29.5*	2.1	n/a	1.2
Coefficient of linear thermal expansion, $mm/(m.°C)$	0.017	0.24	0.018	0.22	0.122	0.022

* Un-notched Izod

It is clear that glass-filled polyurethane composites have a role to play versus steel and ABS both in terms of stiffness, impact resistance and toughness. However, it is still relatively early days in the development of polyurethane composite systems and further improvements will be identified.

Choice of process

The choice of which of the five methods to adopt for a particular application depends on 10 parameters that between them dictate the choice of process technology. These are:

- Component strength.
- Ability to define direction of physical properties.
- Scale of manufacture.
- Ability to automate.
- Complexity of shape.
- Size of part.
- Consistency.
- Flexibility.
- Relative costs.
- Speed of production.

In certain applications all these parameters are called into play whilst in other cases only one or two are of key importance. Each of the production methods that can be used to produce polyurethane composites has it own advantages and disadvantages. Some are more applicable to stronger components, some to mass production etc. In selecting the production process, many factors need to be considered, not least the costs.

Although it is not easy to define the one specific best solution in all cases, Table 29-3 gives some idea of the range of each process, the limitations and advantages.

Table 29-3   Assessing optimum process of manufacture for polyurethane composites

	Ranking 1	Ranking 5	Manual layup	SRIM & LD-SRIM	RRIM	Chopped fibre	Filament winding	Pultrusion
Strength	Low	Very high	4	2,5	1	2.5	4.5	4
Directional properties	Hard	Easy	3	1	1	1	4.5	1.5
Scale production	One off	Large	1	5	5	4	4.5	3
Automation	Manual	Fully	1	5	5	4	3	4.5
Complexity of parts	Simple	Complicated	5	3	2.5	3	4	1
Size of parts	Small	Large	4	3	1	3	2	1.5
Consistency of production	Poor	Good	1	5	5	3	5	5
Flexibility of process	Low	High	5	3	3.5	3	4.5	1
Cost	Low	High	1	2	3	2	4.5	4
Speed of production	Slow	Fast	1	5	5	4	2	4.5

To use the table as a guide the following simple process should be followed. First, identify the top two or three key parameters and rank these. Then, work through these parameters using the diagram to identify the possible alternative processes that best suit the requirement for each case. Repeat this for each parameter until the best process is identified.

For example, consider the following situation:

- *Top priority is repeatability and quality of production process.*
  The table gives the possibility of several processes: RRIM, SRIM, pultrusion, filament winding and P4 process.
- *The second priority is mass production.*
  The table shows all the processes are compatible except filament winding. So this is removed from the list of options.
- *The third priority is to have a strong part.*
  Of the processes that meet the first two requirements pultrusion is the best for this and SRIM and the P4 process would probably be satisfactory.
- *The final requirement is a complex shape.*
  Then pultrusion is very poor for this as can be see from this line on the table, so the best selection for the process would probably be SRIM or the P4 process.

This process does not give a definitive answer and it does not allow the optimisation of all relevant criteria, but it helps to identify the most important factors and gives some idea of where to start to define the correct production process.

30. The use of polyurethane composites in automotive applications

Mike Anderson

Cars manufactured prior to the 1970s were constructed predominantly of steel and thus tended to be heavy with high fuel consumption. However, the rise in the price of oil during the late 1970s meant that consumers looked for more fuel-efficient cars. To improve fuel consumption, weight had to be taken out of cars and plastics and engineered composites started to be used in place of steel due to their lower weight and high strength to weight ratios.

According to studies by the US Department of Energy, about 75 per cent of a vehicle's fuel consumption is directly related to factors associated with weight. Heavier vehicles use larger engines, bulkier drive trains and a more massive chassis. Consequently they require more energy to accelerate, decelerate and overcome rolling resistance. Besides energy costs, environmental demands are leading the push towards lower weight vehicles as improved fuel economy leads to reduced carbon dioxide emissions and lowered air pollution. Recent studies have shown that a 40 per cent reduction in overall weight would improve average fuel consumption to 80 miles per gallon (3.5 litres per 100 kilometres). Polyurethane composites have played and will continue to play a large role in automotive weight reduction.

The key areas where these composite materials are finding automotive applications are in door panels, sun shades, package trays, headliners, seatbacks, floor pans, load floors and some exterior areas.

Automotive interiors

The common composites in automobile interiors are made with polypropylene (PP), acrylonitrile butadiene styrene (ABS), compressed wood and glass-reinforced polyurethanes. The main advantages and disadvantages of these are shown in Table 30-1. This table shows that glass reinforced polyurethanes offer low tooling costs, high strength-to-weight ratio, low densities and the potential for part integration of components into one piece.

Low-density structural reaction injection moulding (LD-SRIM) composites are successfully displacing PP and ABS thermoplastics, as well as wood fibre composites in a wide variety of interior trim substrates. The main benefits of this technology are high strength-to-weight ratios, which enable the production of low weight parts and significant capital and tool savings due to the low moulding pressures required. The development of new, lower density, polyurethane systems allow LD-SRIM to compete more favourably with other plastics in applications such as interior door panel substrates, package trays, sun-shades, seat pans and backs, instrument panel retainers, consoles, and load floors. The polyurethane system improvements that have facilitated this include faster demould times and enhanced internal mould release capabilities.

Table 30-1   Comparison of common interior structural materials

Material	Advantages	Disadvantages
Polypropylene	Low price Moderate density Mouldable	Lower strength High moulding pressures High moulding temperatures High levels of reinforcement not attainable High tooling costs
ABS	High strength-to-weight ratio Moderate density Mouldable	Low reinforcement level High moulding pressures High moulding temperatures High tooling costs
Wood Substrate	Low price High strength-to-weight ratio Moderate density	No complex shapes Gluing of separate pieces High forming temperatures Very high water absorption
Steel	Very high strength Proven processability	Very high density Corrodes No complex shapes Welding, grinding, polishing
Polyurethane	High strength-to-weight ratio Low density Tailored to requirements Low pressure Low cost tooling Ability to consolidate parts without secondary moulding operations	Higher price per part for >150,000 parts/mould

The biggest advantage that glass-reinforced polyurethane has over thermoplastics is in weight reduction. A number of interior trim parts, such as door or quarter panels, do not have high load-bearing requirements, therefore, materials with high flexural modulus, such as ABS, are over engineered for these applications. Glass-reinforced polyurethane composites can provide up to a 50 wt-% reduction compared to thermoplastic composites Table 30-2.

Table 30-2   Savings attainable using glass reinforced polyurethane composites

Material	ABS	PP	Wood Substrate	LD-SRIM
Panel volume, cm^3	1,250	1,250	1,250	1,250
Density, kg/m^3	1,060	970	1,000	550
Reinforcement, wt-%	–	10 (talc)	–	20 (glass)
Flexural modulus, MPa	2,000	1,500	3,800	1,750
Part weight, g	1,325	1,213	1,250	680
Weight savings v ABS, %	0	8	6	48

Significant work has been done by the thermoplastics industry in recent years to exploit the high modulus properties of PP and ABS to make parts thinner, thereby reducing weight. Glass-reinforced polyurethanes can be made with a higher modulus simply by increasing the glass level and modifying the chemistry to keep the overall density constant providing the potential to make thinner, lighter weight composites.

An alternative way to save weight is to make a thicker, lower density part with increased glass loading when an additional 20 per cent weight saving can be achieved over conventional glass-reinforced composites. This is based on the fact that a part's stiffness is directly proportional to its modulus and to the cube of its thickness. Therefore, small increases in thickness can lead to large increases in stiffness as is shown in Table 30-3. The lower density formulation has a decreased modulus, but due to the increased thickness, it has a similar stiffness, as measured by the heat distortion temperature.

Table 30-3   Weight savings attainable using thicker parts

Property	Composite 1	Composite 2
Density, kg/m^3	550	300
Glass, wt-%	20	25
Heat distortion temperature (at 1.8 MPa), °C	70	70
Flexural modulus, *MPa*	1,750	700
Panel thickness, *mm*	2.5	3.8
Panel volume, cm^3	1,250	1,875
Panel weight, *g*	690	560
Weight saving, %	–	19

The primary reason that thicker parts have not been moulded in the past is that increased thickness has lead to substantially increased cure times. However, recent advances in polyurethane chemistry minimise processing issues associated with moulding thicker parts. This has allowed systems to be developed with overall densities of 250 to 600 kg/m^3 at acceptable mould cycle times.

Door panels

Door panels were one of the first automotive interior parts where composite materials made significant inroads, Figure 30-1. The load-bearing part of the door panel was originally made of steel, but this was significantly over-engineered for the application and polyurethane composites offered a significant weight reduction, while still meeting the functional performance requirements.

Polyurethane composites also offer the opportunity to reduce the complexity of door panels. Panels made from steel consisted of many smaller pieces that were stamped and then welded together whilst with composites, the entire door panel can be made in one piece. In addition, the post-assembly vinyl adhesion step can be eliminated by placing the vinyl skin in the mould and forming the composite onto the skin.

LD-SRIM parts can be moulded at pressures of less than 0.5 MPa compared to the 3.5 to 7 MPa needed for thermoplastics, so moulds for LD-SRIM parts can be lighter and consequently are usually only a quarter of the cost of thermoplastic moulds. In addition, the presses required for moulding LD-SRIM parts are smaller and less than half the price of those used in thermoplastic moulding.

Figure 30-1   Door panel

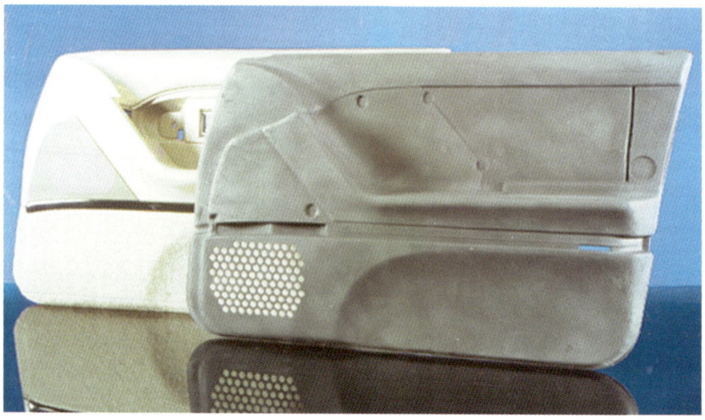

Glass-reinforced polyurethane was selected by an American automotive manufacturer for the door panels of a truck, requiring the production of 3.5 million parts a year. Traditionally, glass-reinforced polyurethanes would not be considered for such a high production volume. However, because of recent process and chemistry improvements, glass-reinforced polyurethanes are now a viable option for these higher volume programmes.

Sun-shades

Sun-shades slide under the sunroof window to block the sun, Figure 30-2, and polyurethane composites are now being used because of the weight savings and processing advantages. Traditionally sun-shades were made of high density sheet moulding compound (SMC), but as with steel in other non-load bearing applications, parts made from SMC are over-engineered and overweight. The same functional part can be made with glass-reinforced polyurethane composites at a half to a third of the weight of SMC.

Figure 30-2   Sun-shade

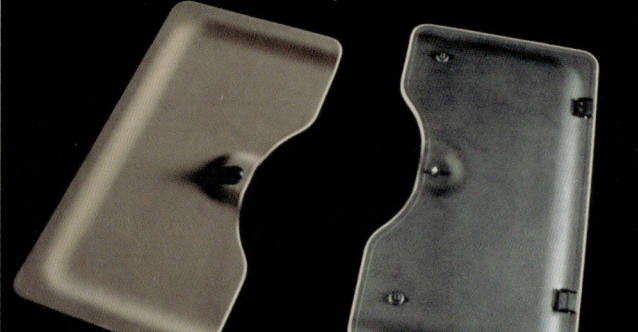

The major customer complaint with sun-shades is that they can slide closed when a car comes to an abrupt stop. Because glass reinforced polyurethane composite sun-shades are significantly lighter, they have lower inertia when a car suddenly brakes, and less of a tendency to slide forward. They also are able to meet very tight dimensional tolerances of ±0.5 millimetres in a one-meter part. This is required for preventing light leakage around the sun-shade and for maintaining a good fit with headliners.

Glass-reinforced polyurethane composite sun-shades can be moulded at densities between 500 to 1,100 kg/m^3 with 10 to 20 per cent glass and are produced in a similar process to door panels, but a cloth is used instead of using a vinyl cover stock. Current technology leads to demould times as short as 60 seconds.

Package trays

Package trays are the ledges between the back seat and the rear window, Figure 30-3, and glass-reinforced polyurethane composites are used for weight reduction (compared to pressed wood fibre and other alternative materials). They are recommended for their load-bearing capabilities, especially for holding the speakers for the advanced car sound systems that are being mounted in the boot (trunk).

*Figure 30-3
Package tray*

Headliners

Headliners cover the ceiling of the car interior, to provide head impact protection for passengers. They have traditionally been manufactured by taking a piece of flexible polyurethane slabstock foam, applying glass fibre mat to both sides of the foam, spraying an isocyanate binder on the glass and shaping the part in a heated mould. A fabric layer is then bonded onto the part to make the finished headliner. These headliners, in their simplest form, have provided acoustical and visual functions. Any requirements for head impact protection have required energy absorbing parts to be moulded and bonded to the headliner. The final headliner is then glued to the metal roof.

Recent trends are towards more modular headliners, which also support light fixtures, consoles, and air ducts. Indeed, structural headliners are being developed that will provide support for the roof of the vehicle, reducing the size of the steel crossbeams, thereby saving weight. Both these trends have led to the development of headliners with higher load-bearing properties than traditional polyurethane headliners.

LD-SRIM systems have been developed that meet these head impact and structural requirements and headliners can now be moulded directly in one piece. Moulders are able to make complex shapes that vary in thickness from 4 to 24 millimetres within the same part. The LD-SRIM mouldings have densities of 100 to 150 kg/m^3 with 10 per cent glass and demould times as short as 60 seconds, making them competitive with traditional polyurethane headliner processes.

Seatbacks

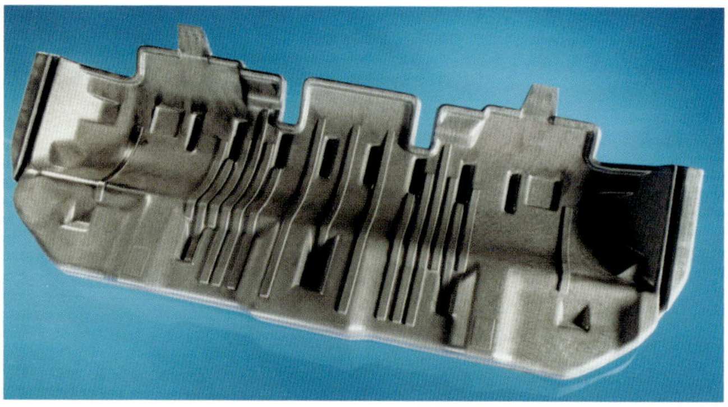

Seatbacks and pans supply the structural support required in automotive seating, Figure 30-4 and Figure 30-5, and they need to be made of materials strong enough to withstand crash impact forces. Polyurethane composites are ideally suited to this application because of their high strength to weight ratio. They have the added advantage that simpler parts can be made, eliminating the need to cut and fasten several steel and aluminium parts together. It has been possible to achieve 30 to 50 per cent weight savings using polyurethane LD-SRIM composites, while significantly reducing the number of parts required, leading to major manufacturing savings.

Figure 30-4   Seat frame

Figure 30-5   Seatback

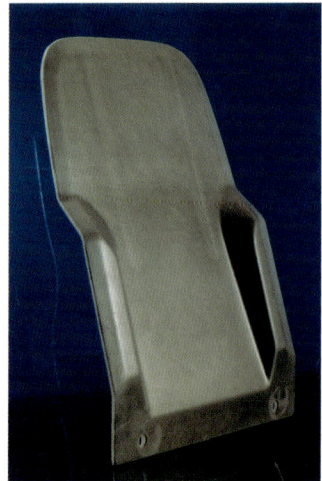

Load floors and floor pans

Load floors bear the weight of the passengers, the seating and floor consoles, and are a relatively new area for polyurethane composites. When designing load floors, it is important to perform finite element analysis (FEA) to ensure proper material selection and part design. FEA analyses assist the design engineer in assessing whether or not a part will meet vehicle safety requirements, without resorting to costly trial and error evaluation.

Automotive exteriors

Glass- and carbon-fibre reinforced polyurethane composites are ideal for replacing steel in exterior structural applications, because of their high strength-to-weight ratios. These parts offer the greatest potential for weight savings in cars and trucks.

The load deflection analysis of a flat plate with supported ends can be used, provided the modulus of elasticity of the materials is known, to determine the thickness of various materials to maintain a constant stiffness. Table 30-4 shows the theoretical weight savings that various materials can attain compared to steel. The values show the calculated thickness of various materials compared to a three-millimetre, 30 vol-% glass-reinforced polyurethane composite. They show that the glass-reinforced polyurethane composite has the potential to be 44 per cent lighter than a 1.1-millimetre steel sheet. Because carbon fibre has a higher modulus of elasticity and lower density than glass, carbon fibre-reinforced polyurethane composites have even greater potential for weight savings.

Table 30-4   Theoretical weight savings of some structural materials

Material	Flexural modulus	Thickness (mm)	Density (kg/m^3)	Weight 1 m^2 section (kg)	Weight saving relative to steel (%)
SRIM 30 vol-% glass	11	3.0	1,600	0.48	44
SRIM 40 vol-% carbon fibre	46	1.9	1,400	0.27	69
Aluminium	69	1.6	2,700	0.43	50
Steel	210	1.1	7,800	0.86	–

The actual weight savings achieved are often less than the theoretical maximum. However, weight reductions of 20 to 30 per cent have been achieved in going from a steel to a glass fibre-reinforced polyurethane composite pick-up truck bed, Figure 30-6. Various composite resins were investigated, with the most viable other candidate to polyurethane being vinyl ester resin in an SMC process. However, it was determined that parts made with the polyurethane resin were more chip resistant and had a greater retention of properties after long-term humid exposure tests. In a series of weathering tests, vinyl ester composites showed a 50 per cent loss in many key design requirements, while polyurethane composites showed only a 10 per cent loss.

Figure 30-6   Truck bed and tailgate

Two-year field studies, including some highly aggressive environments, showed that the glass-reinforced polyurethane composite was the best candidate to meet the durability requirements. The composite material provided excellent scratch and dent resistance, was rust proof and maintained structural integrity in severe environments.

As steel truck beds have a tendency to rust, there is a large automotive after-market for plastic bed liners to cover the steel boxes. These bed liners offer scratch, chip, and dent resistance, but unfortunately, the steel boxes have an even greater tendency to rust when covered with the plastic bed liners. Polyurethane composite boxes will not rust and can give the same benefits as the steel box and plastic bed liner combination at lower cost.

Another advantage of the reinforced polyurethane composites is that large parts can be made in one step, which eliminates the stamping, welding, grinding and polishing steps currently used with steel boxes. By using glass-reinforced polyurethane the American automotive manufacturer was able to go from four steel panels to one composite part for making the truck bed.

For a part as large as a full-size pick-up truck bed, the significantly lower moulding pressure requirements of the LD-SRIM process, 2.4 MPa compared to 8.5 MPa for SMC, results in tremendous capital and tooling savings for presses as well as for moulds. Glass-reinforced polyurethanes can be up to 15 per cent lighter than SMC parts. All these factors make glass-reinforced polyurethane composites economically competitive in low volume production runs.

Two main hurdles restrict glass- or carbon fibre-reinforced polyurethane composites from replacing steel and SMC in exterior vehicle parts; the cost of manufacturing and the difficulty in attaining visually acceptable, 'Class A' surface finish.

The current truck box is made using the P4 process in which a glass mat preform is first made in a press, this is cured in an oven, then placed over a mould and polyurethane injected behind it in another press, with the whole process being very energy and capital intensive.

Recent developments have focused on combining the glass and polyurethane in the mix-head and laying them down in the mould with a robot to save the costs of making a glass preform. The equipment for sprayed chopped fibre technology for use in door panel applications has been developed and is commercially available. Work is continuing to scale this up from glass levels of 20 wt-% to the greater than 50 wt-% levels needed for exterior body parts. Polyurethane systems have been developed specifically for this application that offer excellent glass wet-out, rapid cure and demould. In addition, this technology has the ability to control the levels of reinforcement in the areas where it is most needed.

Truck boxes are being produced with 50 wt-% glass-reinforced polyurethane composites, but are grained and painted and do not have a true 'Class A' surface finish. Films have been developed for glass-reinforced polyurethane exterior parts for the SMART car roof and for agricultural exterior parts, but the surface finish, quality, and look of these parts are not as critical as traditional automotive exterior bumpers (fenders), side panels and bonnets (hoods). Unlike mica-reinforced polyurethanes, which contain very small milled fibres, good surface finish parts cannot be attained with high levels of glass mat or chopped fibre at this time. However, significant development work is ongoing in this area.

Future trends

While there continues to be development in all areas of polyurethane applications, it is in the composite area, which is still at an early stage of its cycle, that the more dramatic future changes are likely to be seen. Polyurethane composites are only at the start of their growth curve and nobody really knows which direction they will take.

Seat pans and load floors are two areas where polyurethanes can offer weight savings in vehicles. The most benefit will be attained by full utilisation of computer modelling to design the proper chemistry and part for the application. By performing an FEA on each part in the design phase, exact reinforcement requirements can be determined.

This is especially important for the sprayed chopped fibre process, where precise stresses can be accounted for by altering the amount of fibre being injected. By combining FEA and 'tailored' materials in the design phase, large weight and cost reductions can be achieved.

According to figures obtained in 1999 for the American Metal Market, a typical automotive vehicle contains 111 kilogrammes of plastic and composites while steel and iron account for at least 906 kilogrammes. The potential to save weight by using fibre reinforced polyurethane composites is tremendous.

To further replace steel, a number of technical obstacles remain. These require developments in the following areas:

- Economically acceptable manufacturing processes for high volume applications.
- Ability to meet all functional requirements, particularly crash integrity and long-term durability.
- Rapid and reliable methods of joining composites to steel and other dissimilar materials.
- Significantly less expensive carbon fibre.
- Cover-stock films that will enable 'Class A' finishes to be produced.

Composites have been recognised for delivering mechanical and thermal performance, weight savings, part consolidation and overall economic savings in low volume production runs, less than 150,000 parts per year per mould. The challenge is to deliver the same value in high volume runs.

The current focus is to make 1.5-millimetre thick, 40 vol-% carbon fibre polyurethane composites economically, which will help the automotive companies to meet their weight reduction targets of 40 per cent.

Appendix 1   Calculations

The amount of isocyanate required to react with the polyol and any other reactive additives is calculated in order to obtain the chemically stoichiometric equivalents. This theoretical stoichiometric amount of isocyanate may then be adjusted upwards or downwards by varying the isocyanate index, dependent on the polyurethane system, the properties required of the polymeric product and known effects such as the scale of manufacture and the ambient conditions.

Hydroxyl value

The hydroxyl value (OHv) of a polyol, sometimes called the hydroxyl number, is determined by acetylating the polyol with pyridine and acetic anhydride and then titrating the excess acetic anhydride with standard KOH solution, measuring the difference between a blank solution and one containing polyol.

The OHv can then be defined as the weight of KOH in milligrams that will neutralise the acetic anhydride capable of reacting by acetylation with one gramme of polyol. The molecular weight of KOH is 56.1.

Per cent hydroxyl

An alternative way of expressing the hydroxyl content of a polyol is as the per cent hydroxyl groups present in the structure and this value is obtained from the following formula:

$$\text{Per cent hydroxyl (\%OH)} = \frac{\text{OHv} \times (\text{MWt OH}) \times 100}{56.1 \times 1{,}000} = \frac{\text{OHv} \times 17 \times 100}{56.1 \times 1{,}000} = \frac{\text{OHv}}{33}$$

Acid value

The acid value (Av) of a polyol, sometimes called the acid number, is determined by measuring the difference between titrating, with a standard KOH solution, a blank solution and one containing polyol.

The Av is defined as the weight of KOH, in milligrammes, that neutralises the acid in one gramme of polyol.

Molecular weight of a polyol

The molecular weight of a polyol is determined from the OHv and the Av as follows:

$$\text{Molecular weight polyol} = \frac{\text{Functionality} \times 56.1 \times 1{,}000}{(\text{OHv} + \text{Av})}$$

Equivalent weight of a polyol

The equivalent weight of a polyol is defined as the portion of the molecular weight that contains one hydroxyl group and is defined as:

$$\text{Equivalent weight} = \frac{\text{Molecular weight}}{\text{Functionality}} = \frac{56.1 \times 1{,}000}{(\text{OHv} + \text{Av})}$$

Equivalent weight of chain extenders (hydroxyl- and amine-terminated)

The molecular and equivalent weights of a range of common hydroxyl- and amine-terminated chain extenders are given in Table A1-1. Their chemical structures are shown in Chapter 10.

Table A1-1   Functionality, OHv and molecular/equivalent weights for a range of chain extenders

Chain extender	Molecular weight	OHv (mg KOH/g)	Functionality	Equivalent weight	Weight of diisocyanate (g/100 g)	
					TDI	MDI
Ethylene glycol	62.07	1,808	2	31.04	280	401
Diethylene glycol	106.12	1,057	2	53.06	164	235
Propylene glycol	76.09	1,474	2	38.05	229	329
Dipropylene glycol	134.17	836	2	67.09	130	186
1,4-Butane diol	90.12	1,245	2	45.06	193	278
1,6-Hexane diol	118.17	949	2	59.09	147	212
1,4-Cyclohexandimethanol	144.21	778	2	72.11	121	174
Hydroquinone dihydroxyethyl ether	198.21	566	2	99.11	88	126
1,2,4-Butanetriol	106.12	1,586	3	35.37	246	354
Trimethylol propane	106.12	1,586	3	35.37	246	354
Glycerol	92.09	1,827	3	30.70	284	407
Water	18.02	6,230	2	9.01	968	1,389
Ethanolamine	77.09	1,456	2	38.55	226	325
Diethanolamine	105.14	1,601	3	35.05	248	357
Triethanolamine	149.19	1,128	3	49.73	175	252
m-Phenylene diamine	108.14	1,038	2	54.07	161	231
Diethyl toluene diamine	178.27	629	2	89.14	97.7	140
Dimethylthiotoluene diamine	214.34	523	2	107.7	81.2	116

Isocyanate value

The isocyanate value is the weight percentage of reactive isocyanate (NCO) groups in an isocyanate, modified isocyanate or prepolymer and is determined using the following equation, where the molecular weight of the NCO group is 42:

$$\text{Isocyanate value} = \% \text{ NCO groups} = \frac{42 \times \text{Functionality}}{\text{Molecular weight}} \times 100$$

$$= \frac{4{,}200}{\text{Equivalent weight}}$$

The isocyanate values and molecular and equivalent weights of a range of isocyanates are given in Table A1-2. Their chemical structures are shown in Chapter 5.

Table A1-2   Isocyanate values and molecular/equivalent weights for a range of isocyanates

Isocyanate	Molecular weight	Isocyanate value (%)	Functionality	Equivalent weight
TDI	174.16	48.2	2	87.08
MDI	250.26	33.6	2	125.13
NDI	210.19	40.0	2	105.10
IPDI	222.30	37.8	2	111.15
$H_{12}MDI$	262.35	32.0	2	131.18
HDI	168.20	49.9	2	84.10
PPDI	160.13	52.5	2	80.07
CHDI	166.18	50.5	2	83.09
TMXDI	244.30	34.4	2	122.15
HDI trimer	504.60	25.0	3	168.20
MDI trimer	750.78	16.8	3	250.26
IPDI trimer	666.90	18.9	3	222.30
TDI-TMP adduct	628.60	20.0	3	209.53

Isocyanate index

The isocyanate index for a formulation is a measure of the excess isocyanate used relative to the theoretical equivalent amount required and is defined as:

$$\text{Isocyanate index} = \frac{\text{Actual amount of isocyanate used}}{\text{Theoretical amount of isocyanate required}} \times 100$$

An isocyanate index of 105, sometimes written as 1.05, indicates that there is a five per cent excess of isocyanate.

Worked examples

The mix ratio of the components for a given formulation is obtained by calculating the amount of isocyanate required to react with the isocyanate active components, either based on a 100 parts of polyol or with all the components expressed as percentages. The data required for the calculation are the isocyanate value of the isocyanate and the hydroxyl value, residual acid value and water content of the polyol and other reactive additives.

Example 1

How much TDI is required to make a foam blown with 4.3 parts of water per 100 parts of a polyester polyol having an OHv of 55 mg KOH/g and an acid value of 0.4 mg KOH/g, at an isocyanate index of 105? The method of working out equivalent weights has been outlined above and all the calculations required to derive the amount of TDI are provided in Table A1-3.

Table A3-3   Calculations for a TDI foam formulation

Formulation	Equivalent weight	pbw	Equivalents pbw method	Polyol blend (%)	Equivalents polyol blend method	Overall formulation (%)
Polyester polyol	1,013	100	0.099	94.5	0.093	63.1
Amine catalyst		0.25		0.2		0.2
Tin catalyst		0.25		0.2		0.2
Silicone surfactant		1		0.9		0.6
Water	9	4.3	0.478	4.1	0.452	2.7
Total other components		105.8	0.576	100.00	0.545	
Effect of NCO index			0.605		0.572	
TDI 80/20	87.08	52.7		49.8		33.3
Overall total		158.5		149.8		100.0
NCO index	105					

Polyester polyol OHv = 55 mg KOH/g, Av = 0.4 mg KOH/g (Equivalent weight = 1,013)

The key point of the calculations is to determine the number of equivalents of hydroxyl compounds and them to multiple this by the isocyanate index before further multiplying the number by the equivalent weight of the isocyanate. The ratio of TDI to the other components is worked out in three ways, all of which are used in the polyurethane industry, based on 100 parts of polyol, 100 parts of polyol blend and an overall formulation including the TDI. These give TDI weights of, respectively, 52.7, 49.8 and 33.3 but are in fact all the same formulation expressed differently.

Example 2

How much polymeric MDI (31.5% isocyanate value) is required to react with 100 pbw of a polyether polyol with an OHv of 28 mg KOH/g, an acid value of 0.01 mg KOH/g and a water content of 0.01 per cent, blended with 6.0 pbw of ethylene glycol and 4.0 pbw of *m*-phenylene diamine chain-extending agents, at an isocyanate index of 98? The results of the calculations are shown in Table A1-4.

Table A1-4   Calculations for a polymeric MDI foam formulation

Formulation	Equivalent weight	pbw	Equivalents pbw method	Polyol blend (%)	Equivalents polyol blend method	Overall formulation (%)
Polyether polyol	2,003	100	0.050	85.1	0.042	62.3
Ethylene glycol	31.04	6	0.193	5.1	0.165	3.7
m-Phenylene diamine	54.08	4	0.074	3.4	0.063	2.5
Amine catalyst		0.3		0.3		0.2
Tin catalyst		0.1		0.1		0.1
Silicone surfactant		1		0.9		0.6
Water	9	0.1	0.011	0.1	0.009	0.1
Blowing agent		6		5.1		3.7
Total other components		117.5	0.328	100.0	0.279	
Effect of NCO index			0.322		0.274	
Polymeric MDI	133.3	42.9		36.5		26.7
Overall total		160.4		136.5		100.0
NCO Index		98				

Polyester polyol OHv = 28 mg KOH/g, Av = 0.01 mg KOH/g, water content = 0.01% (Equivalent weight = 2,003)

Example 3

What is the optimum mix ratio for a shoe sole system based on a pure MDI prepolymer, with an isocyanate content of 19.6 per cent, and a polyol blend that consists of two polyols, polyester polyol A OHv 45 mg KOH/g, acid value 0.3 mg KOH/g (52 per cent) and polyester polyol B OHv 75 mg KOH/g, acid value 0.4 mg KOH/g (30 per cent), blended with butane diol (14.2 per cent), water (0.5 per cent), amine catalyst (0.5 per cent), silicone surfactant (0.3 per cent) and pigment (2.5 per cent), run at an isocyanate index of 103? The results of the calculations are shown in Table A1-5.

In practice the optimum cure for such a system is identified by taking the theoretical best mix ratio, in this case 98.8 parts of the MDI prepolymer to 100 parts of the polyol blend, and evaluating the cure of the foams at this ratio and then plus and minus two and four per cent. With the improved analytical techniques used, which allow precise characterisation of all components, then it is normal for such a system to be mouldable close to its theoretical optimum with deviations occurring mainly due to variations in machine and operating conditions.

Table A1-5   Calculations for an MDI prepolymer water-blown shoe sole formulation

Formulation	Equivalent weight	Polyol blend (%)	Equivalents polyol blend	Overall formulation (%)
Polyester polyol A	1,238	52.0	0.042	26.2
Polyester polyol B	745	30.0	0.040	15.1
Butane diol	45.06	14.2	0.315	7.1
Water	9	0.5	0.050	0.2
Amine catalyst		0.5		0.3
Silicone surfactant		0.3		0.2
Pigment		2.5		1.3
Total polyol blend		100.0	0.447	
Effect of NCO index			0.461	
MDI prepolymer (19.6% NCO)	214.3	98.8		49.7
Overall total		198.7		100.0
NCO index	103			

Polyester polyol A OHv = 45 mg KOH/g, Av = 0.3 mg KOH/g (Equivalent weight = 1,238)
Polyester polyol B OHv = 75 mg KOH/g, Av = 0.4 mg KOH/g (Equivalent weight = 745)

Appendix 2   Conversion factors

Length

1 inch	=	25.4 mm
1 mm	=	0.03937 inches
1 cm	=	0.3937 inches
1 metre	=	39.37 inches

Mass

1 pound	=	0.454 kg
1 kg	=	2.203 pounds

Volume

1 in^3	=	16.387 cm^3
10^3 in^3	=	0.016387 m^3
1 ft^3	=	0.028316 m^3
1 cm^3	=	0.061024 in^3
1 m^3	=	35.316 ft^3

Density

1 lbs/ft^3	=	16 kg/m^3
1 lbs/ft^3	=	0.016 g/cm^3
1 g/cm^3	=	1,000 kg/m^3
1 g/cm^3	=	62.4 lb/ft^3

Temperature

0°C	=	32°F
0°F	=	-17.8°C
100°C	=	212°F
100°F	=	37.8°C

To convert °F to °C subtract 32 and multiply by 5/9.
To convert °C to °F multiply by 9/5 and add 32.

Pressure

1 lb/in^2 (psi)	=	0.0703 kg/cm^2
1 kg/cm^2	=	14.22 lb/in^2 (psi)
1 N/m^2	=	1 Pa
10^3 N/m^2	=	1 kPa
10^6 N/m^2	=	1 MPa
1 lb/in^2 (psi)	=	6.895 kPa
1 kPa	=	0.145 lb/in^2 (psi)
1 kg/cm^2	=	0.09807 MPa
1 MPa	=	10.197 kg/cm^2

Tear strength

1 lbs/in	=	0.1752 kN/m
1 lbs/in	=	0.1786 kg/cm
1 kg/cm	=	5.60 lbs/in
1 kg/cm	=	0.9807 kN/m

Viscosity

1 P (poise)	=	0.1 N s/m^2

Thermal conductivity

1 kcal/(m h °C)	=	1.1630 W/(m. K)
1 cal/(cm s °C)	=	418.68 W/(m. K)
1 Btu/(ft h °F)	=	1.7307 W/(m. K)
1 Btu in/(ft^2 h °F)	=	0.1442 W/(m. K)
1 Btu in/(ft^2 s °F)	=	519.22 W/(m. K)

Energy

1 Btu	=	1.054 kJ
1 kcal	=	4,187 kJ
1 kJ/kg	=	0.4299 Btu/lb

Dielectric strength

1 kV/in	=	39.370 kV/m

Appendix 3 Physical properties of isocyanates

Aromatic isocyanates

Property	2,2' MDI	4,4' MDI	2,4 MDI	Polymeric MDI	2,4 TDI	2,6 TDI	TDI 80/20	TDI 65/35	NDI
Physical state (room temperature)	Solid	Solid	Solid	Liquid	Liquid	Liquid	Liquid	Liquid	Solid
Colour	White to pale yellow	White to pale yellow		Fawn to dark brown	Colourless to pale yellow				White to yellow
Odour	None. Pungent at high temperatures	None. Pungent at high temperatures		None to aromatic at room temperature	Characteristic, sharp, pungent				Pungent
Molecular weight	250.3	250.3	250.3	~450	174.2	174.2	174.2	174.2	210.2
Density, kg/m^3									
20°C					1,218	1,227			
25°C	1,188	1,230		1,239	1,214				
40°C		1,190	1,192	1,224					
50°C									
Melting point, °C	46	40	35	0	21.8	18.2	14.0	8.5	130
Boiling point, °C	142	161	152						
0.07 kPa									
1.33 kPa					120		120	120	190
101 kPa		314			251		251	251	263
Heat of fusion, J/g		101.6							
Viscosity, $mPa.s$									
20°C				400	3.2		3.2	3.2	
30°C				200	2.6		2.6	2.6	
50°C		4.7							
Flash point, °C (Cleveland open cup)		212 – 214		210 – 230			135		155
Fire point, °C (Cleveland open cup)				220 – 250			143		
Specific heat, $J/g/°C$									
25°C		1.38		1.49			1.56	1.58	1.20
40°C									
Vapour pressure, Pa									
25°C		6×10^{-4}		$<6 \times 10^{-4}$			3.33		0.4
40°C		2.5×10^{-3}		$<2 \times 10^{-3}$			11.1		
50°C									1.7
70°C		1.3×10^{-1}		$<1 \times 10^{-1}$			86.5		6.7

Synonyms for MDI
4,4'-diisocyanato diphenylenemethane
diphenylmethane diisocyanate
4,4'-diphenylmethane diisocyanate
p,p-diisocyanato diphenylmethane
methylene diphenyl diisocyanate
isocyanic acid, 4,4'-methylene diphenyl ester
methane, bis(4-isocyanato phenyl)

Synonyms for TDI
toluene diisocyanate
tolylene diisocyanate
methyl phenylene diisocyanate

Synonyms for NDI
naphthalene diisocyanate
1,5-diisocynato naphthalene

Aliphatic isocyanates

Property		HDI	IPDI	H$_{12}$MDI	PPDI	CHDI	m-TMXDI	p-TMXDI
Physical state (room temperature)		Liquid	Liquid	Liquid	Solid	Solid	Liquid	Solid
Colour		Colourless	Colourless	Colourless	Colourless	Colourless	Colourless	White
Odour		Pungent			Pungent	Pungent		
Molecular weight		168.2	222.3	262.35	160.1	166.2	244.3	244.3
Density, kg/m^3	20°C	1,047	1,062		1,441			
	25°C	1,047		1,070			1,050	1,090
Melting point, °C		-67	-60	78 – 82	94 – 95	59 – 62	-10	72
Boiling point, °C	0.07 kPa	96 – 110						
	1.33 kPa	127	158	155 – 166	110 – 112	117 – 120	150	150
	101 kPa	255			260	260		
Viscosity, $mPa.s$	20°C	4.5	15				9	
	25°C	3						
	80°C							8
Flash point, °C (Cleveland open cup)		130 – 140	100	>200	>99			
Vapour pressure, Pa	20°C	7	0.04	0.13				
	50°C		0.93					
	100°C						67	53

HDI   1,6-hexane diisocyanate
IPDI   Isophorone diisocyanate
H$_{12}$MDI   4,4'-diisocyanato-dicyclohexylmethane
PPDI   Phenylene diisocyanate
CHDI   Trans-cyclohexane-1,4-diisocyanate
TMXDI   Tetramethylxylene diisocyanate

Index

Figures and Tables are indicated by *italic page numbers*, main entries by **bold page numbers**.
Abbreviations: "CASE" means "coatings/adhesives/sealants/encapsulants", "PU" is "polyurethane" and "TPU" is "thermoplastic polyurethane".

abrasion resistance
  coatings 360
  elastomers 299, *304, 307, 312*
  test methods 299, 360
  thermoplastic polyurethanes 323
absorption hazards 41
acetals, formation of 123, *124*
acetone, conversion to isophorone 64, *68*
acid rain potential, improvements 27
acid value (polyols) 447
acidity, isocyanates 81
acoustic absorption foams 190, 226
  testing of 185, *186*
acoustic performance test 186
acrylic polyols, in CASE formulations 350, 364, 365, 367
acrylic resins, in wood composites 407
acrylonitrile–butadiene–styrene (ABS), properties compared with PU composites *433, 438*
additives **151–168**
  in CASE formulations *351*, 382, 383, *411, 417*
  in elastomers 308, 333–334, 337
  see also adhesion promoters; anti-hydrolysis agents; anti-microbials; antioxidants; anti-static additives; blowing agents; catalysts; chain-extenders; cross-linking agents; fillers; fire retardants; lubricants; pigments; surfactants; ultraviolet stabilisers; viscosity reducers
adhesion mechanisms 352
  in wood composites 397–399
adhesion promoters 163, 334
adhesive/cohesive failure modes *354*
adhesive strength
  adhesives/coatings, test methods 358–359
  rigid foams, factors affecting 255
adhesives **379–394**
  applications 1, 7, 17, 19, 384
  cross-linking in 381–382
  and fire risks 61
  hybrid systems 394
  in manufacturing 3

materials selection **382–383**
non-reactive 379, **384–388**
reactive 155, 379, **388–393**
strength
  development with time *380*
  temperature effects *381*
  test method for 358–359
types **379–382**
weathering tests 360
in wood composites 407
see also non-reactive...; reactive adhesives
adhesives/sealers/binders, market share *12, 13*
advantages of PUs 3, 246, 348
aeroplane *see* aircraft applications
ageing tests
  flexible foams 187–188
  see also heat ageing
A-glass fibre 423
air bubble nucleation, in flexible foam production 198, 205
air pollution 28–29
  improvements 27
aircraft applications
  coatings 6, 366, 368
  propellers **283**
  seat cushions 57
airflow properties, flexible foams 184–185
aldol condensation reactions 123
aliphatic isocyanates 63
  applications 83
  in CASE formulations 349, 364, 382, 391
  in elastomers 286
  light-stability of PUs 83, 87, 126, 349
  physical properties *455*
  production of 64, *67–68*, 88
  reactivity of isocyanate groups 83
  solvents for 77
  TPUs based on *323*
  trimerisation of 119
  see also CHDI; H$_{12}$MDI; HDI; IPDI
aliphatic polyester polyols *90*, 107, **108–109**
alkanes

as blowing agents 133, 250, *251*
  see also hydrocarbons
alkyd coatings 155
alkyd polyols, in CASE formulations 350
allophonates, formation of 117
aluminium trihydrate (ATH), as fire retardant *161*, 162
Americas (North/Central/Latin)
  base chemical consumption data 13, *14*
  population growth *10*
  supply/demand balance 13, *14*, 15, 16
amine-based polyols *90*, 101
amine catalysts 137, **140**
  applications *140*, 152, 153, *154*, 195, 343
  factors affecting reactivity 141–143
  preparation of **137–138**, *139*
  reaction mechanisms 141–143
amine-terminated polyether polyols *90*, 106
  applications 365, *433*
ammonium polyphosphate, as fire retardant *161*, 162, 207
analysis
  of isocyanates 80–82
  of polyols 112
aniline, production of 63, *65*, 71–73
aniline–formaldehyde condensation 63, *65*, 73–76
  side-reactions 76
anionic (base-catalysed) polymerisation, of propylene oxide 94, 97
annealing, of elastomers 291, *292*
anti-corrosion coatings 18
anti-hydrolysis agents 166
anti-microbials 167
anti-oxidants 165
anti-static additives 163, 334
appliances **245–256**
  coatings for 372, 373
  design 246–249
  ecolabels for *33*
  energy labels for 32, *33*, 252
  recycling of 38

457

applications 1, *2*, 6–7, 16–21
  market share *16*
  see also automotive...; coatings...; construction...; footwear...; furniture...; thermal insulation applications
aquatic ecotoxicity, improvements 27
aquatic oxygen demand, improvements 27
aromatic isocyanates 63
  in CASE formulations 349, 364, 382, 391
  dimerisation of 118
  in elastomers 286
  light-susceptibility of PUs 83, 86, 125, 167, 324, 349
  physical properties 455
  production of 63–64, *65–66*
  reactivity of isocyanate groups 83, 86
  TPUs based on *323*
  trimerisation of 119
  see also MDI; NDI; PPDI; TDI; TMXDI; TODI
aromatic polyester polyols *90*, 107, **110–111**, 232
Asia–Pacific
  base chemical consumption data 13, *14*
  population growth *10*
  purchasing parity power per capita *11*
  supply/demand balance 13, *14*, 15, 16
Asker C (hardness) scale *296*
asphyxiants 62
ASTM standard test methods
  elastomers *295*
  flexible foams 179, 183, 186
  rigid foams *241*
atomic force microscopy (AFM), flexible foams 175
autoclave ageing test, flexible foams 187
auto-ignition/oxidation, in foam manufacture 50, 215
automotive applications 5, 7, 17, **190–193, 217–218, 316–317**
  bumpers (fenders) **217–218**, 316, 429
  CASE formulations 393, 413–415, 416
  coatings used 6, 366, 367, 368
  composites used **437–445**
  door panels 439–440

exteriors **217–218, 316–317, 442–444**
  flexible foams used **190–193**
  floor pans 442
  headliners 441
  instrument panels (dashboards) 217, 329
  interiors **190–192**, 217, 316, 329, **437–442**
  load floors 442
  market share *16*
  package trays 441
  seatbacks 442
  seating **190–192**
    comfort considerations **191–192**
    process technology **199–201**
    raw materials and formulations 104, **195–196**
    seat design 191
  semi-rigid foams used **217–218**
  sound insulation **192–193**
  sunshades 440
  TPUs used **316–317**, 329
  truck beds 443–444
  under-bonnet (under-hood) 317
  under-vehicle 317
  weight savings 442–443, 445
  windscreens/windshields 320, 413–415
aviation *see* aircraft applications

Baker mechanism (for amine catalysts) 141
ball rebound test 185
band casting, TPUs produced by 325
barium sulphate 162, *166*, 204, 423
base chemicals
  environmental burdens due to production 26–27
  global consumption 13–14
  handling of 44–46, 47
  hazards associated 42–44
  production capacity 14–16
  storage and transportation of 47–49, 82
  see also isocyanates; polyols
Bata Belt (flex) test 298
batch process, slabstock foam manufacture 215
Bayer, Otto 113, 315
Bayshore falling weight (rebound resilience) test *295*, 297
Bechamp process 71
bedding applications
  test methods 177
  viscoelastic foams used 220
  see also mattresses

Bennewart flex test 298
biocompatible PUs 1
bis(N,N-dimethylaminoethyl)ether (BDMAEE)
  as catalyst *140*
  reaction kinetics 147, *148*, *149*, *150*
  reactivity 142
  production of 138, *139*
bis(dimethylaminoethyl)-methylamine, production of 137, *138*
1-bis(3-(dimethylamino)-propyl)amino-2-propanol, as catalyst *140*
biurets, formation of *66*, *67*, 77, 78, *79*, 116–117
blends, PU/PVC 327–328, 336
blister (adhesion) test 358
blocked isocyanates 79, 114, 115, 123, 371
blowing agents **127–136**
  carbon dioxide (from isocyanate–water reaction) 115, 127, 135, 204, 280, 282, 283, 416
  chemical **134–135**
  chlorofluorocarbons 23, 127, 131, **131**, 204, 250, *251*, 261
  factors affecting thermal conductivity of foam *236*, 251
  factory emissions 29
  fire/explosion hazards 59
  flammability *131*, *132*, 133, *133*, *134*, 136
  for flexible foams 127, 134, 135, 201, 204
  gas thermal conductivity *233*, *236*
  hydrocarbons **133**, 250, *251*, 261, 282
  hydrochlorofluorocarbons **132**, 250, *251*, 261, 273, 274, 280
  hydrofluorocarbons **132–133**, 250, *251*, 261, 274, 280, 282
  for integral skin foams 339, 340
  liquid carbon dioxide 134, 201, 204, 214
  for rigid foams 250, **261**, 273–274, 280
  selection criteria **135–136**, 233
    cost effectiveness 233
    ease of handling 233
    environmental considerations 135–136
    feasibility factors 136
    gas thermal conductivity 233
    performance 136
    service-conditions behaviour 233

water–isocyanate reaction 127, 135, 204, 280, 282, 283
  see also carbon dioxide; CFCs; HCFCs; HFCs; hydrocarbons
blowing catalysts 137, *140*
  reaction kinetic studies 147–150
boardstock **259–265**
  blowing agents for 261
  conveyor in lamination process 263–265
    'Early Contact' process 265
    fixed-gap laminator 264
    floating-platen laminator 264
    functions 263
    'Inverse Laminator' 265
  cutting and stacking operations 265
  facing materials 262
  lay-down methods 262–263
    'gas poker' method 262–263
    liquid stream process 262
  line speeds vs board thickness 265
  manufacture 263–265
    compared with sandwich panel manufacture 270
  raw materials and formulations 260–261
  mixing of 262
boiling point(s), isocyanates *455–456*
bookbinding applications 393
bottoming-out effect, in flexible foams 184
BS standards/test methods, fire-propagation test 56
bubble stabilisation 156–157
build-up ratios, for polyols 95
building materials
  environmental declarations 33–34
  fire regulations 59–61
  see also insulation panels
bulk storage, of base chemicals 48
buoyancy applications 283
1,4-butanediol, as chain-extender *164*, *448*
1,2,4-butanetriol, as chain-extender *448*
tert-butyl hydroperoxide, in manufacture of propylene oxide 91, *92*
butylated hydroxytoluene (BHT), as anti-oxidant 165

cable sheathing applications 6, 320
calcium carbonate filler 162, *166*, 204, *351*, 383
calendared materials 320

carbamates, formation of 114
carbamic acid 115
carbamoyl chlorides, formation of 76, 121, *122*
carbodiimides **120–121**
  as acid scavengers 166
  formation of 78, 120–121, 135, *146*
  effect of catalysts 120–121
carbon dioxide
  atmospheric concentration *32*
  as blowing agent 115, 127, 130, 134, 135, 204, 214, 280, 282, 283, 416
  from isocyanate–water reaction 115, 127, 135, 204, 280, 282, 283, 416
  liquid 134, 201, 204, 214
  outgassing from rigid foams 238
  physical and environmental properties *131, 236*
  reduction targets 31
  sources of emissions *30*, 257
carbon fibre
  as filler 166
  as reinforcement 419, 424
carbon fibre reinforced PU
  composites 424, 442
  future trends 445
  weight savings *443*
carbon monoxide, toxicity 62
carousel plant
  moulded flexible foams 199
  rigid foams 255, 256
carpet backing foams **227–228**
carpet underlay
  polyols used 104
  recycled material in 37, 227
cars
  recycling/re-use of 37–38
  see also automotive applications
CASE sector *see* coatings/adhesives/sealants/encapsulants
cast elastomers 6, 285, **331–336**
  additives in 333–334
  applications 331, 335
  catalyst(s) used *140*, 333
  market share 12
  polyol(s) used 107, 332–333
  process technology 334–335
  raw materials and formulations 332–334
casting machines
  moulded foams 199
  shoe soles 310

castor oil based polyols 111, 218, *382*, 411, 418
catalysts **137–150, 151–156**
  amine catalysts 137, **140**
    applications *140, 152, 153, 154,* 195, 343
    factors affecting reactivity 141–143
    preparation of **137–138**, *139*
    reaction mechanisms **141–143**
  in CASE formulations *351*, 392
  for cast elastomers *140*, 333
  for coatings 367
  for flexible foams *140*, **151–152**, 195, 197
  in isocyanate production 71, 72, 73
  organometallic catalysts 137, **140**
    applications *140, 154*
    preparation of **139**
    reaction mechanisms **143–145**
  in polyol production 94, 97–98
  reaction kinetic studies **145–150**
  for rigid foams *140*, **152–153**
catalytic hydrogenation 63, 64, *65, 66, 69,* 71–73
cationic (acid-catalysed) polymerisation, of propylene oxide 97
cellular structure, flexible foams 175
CFCs *see* chlorofluorocarbons
C-glass fibre 423
chain-extending agents 163
  in CASE formulations *351*
  chemical structures *164*
  as cross-linkers 126
  in elastomers 286, 333
  equivalent weights *448*
  functionality *448*
  hydroxyl value *448*
  molecular weights *448*
char-formation additives 162, 211
Charpy impact test 424
CHDI *see* 1,4-cyclohexane diisocyanate
chemical blowing agents **134–135**
chemical control regulations 39–40
chemical reactions
  isocyanate production 63–77
  polyol production 93, *106*, 108
  PU production 1, 113–123
chemical resistance 361
chimney effect, in fires 52, 58
China
  fire regulations 60
  footwear production 19
  population growth *10*

China (*continued*)
    purchasing parity power per capita 11
chipboard *see* medium-density fibreboard; particleboard
chlorofluorocarbons (CFCs)
    as blowing agents 23, 127, 131, 250, *251*, 261, 306
    effect on ozone layer 23, 127
    phasing-out of *128*, 250, 261, 282, 339
    physical and environmental properties 131, *236*
chlorohydrin process, propylene oxide produced by 89, 91
2-chloropropane, physical and environmental properties *132*
chopped fibre process **429–430**
    advantages 430, *434*, 445
    applications 430, 444
    formulations used *433*
    limitations *434*
    properties of composites *433*
chopped glass fibre 423
chopped strand mat (CSM) 423
climate change 31
clothing applications 1, 6
coating applications 1, 6, 18
    market share 12, *13*, 16
coatings 363–377
    abrasion resistance, test methods 360
    applications 17, 363
    automotive applications 17
    catalysts used 154–155
    chemical/solvent resistance, test methods 361
    for fabrics and textiles 6
    hardness, test methods 360
    isocyanates and derivatives used 80, 364
    market share *13*, *347*, 363
    moisture-cured systems 363, 373–374
    non-reactive 363, **374–377**
    one-component systems 363, **373–374**
    oven-curing/stoving systems 363, **371–373**
    polyols used 364–365
    raw materials and formulations **364–366**, *367*
    reactive 363, **366–374**
    solvents used 365–366
    two-component systems 363, **366–370**

types 363
visual properties 360, 374
weathering tests 360
see also non-reactive coatings; reactive coatings
coatings/adhesives/sealants/ encapsulants (CASE) **347–418**
    additives used 350, *351*
    advantages of PUs 348
    catalysts used 153–156
    classification of systems *348*
    environmental impacts **361–362**
    isocyanates used 349
    market share 12, *13*, *347*
    polyols used 349, *350*
    raw materials **349–350**, *351*
    solidification and bonding in 353
    surface interactions **350–353**
    test methods **353–361**
        environmental exposure testing 360
        final-product properties 358–361
        raw-materials properties 354, 356–357
        sample preparation for 358
    see also main entries: adhesives; coatings; encapsulants; sealants
cobalt-based catalysts 154, 155
cold-cure moulded foams
    process technology **199–201**
    raw materials and formulations 152, *170*, **194–197**
combined heat-and-power systems 27, 30
combustion-modified high-resilience (CMHR) foams 57, *209*
comfort factor, flexible foams 180–181
Community Awareness and Emergency Response code 39
competitive materials 7, 17, 21
composite wood products 18–19, **395–408**
    fire risks 58
composites **419–445**
    advantages 419
    design considerations **421–422**
    factors affecting choice of process 434–435
    formulations *433*
    future trends 22, 444–445
    general properties **420–421**
    materials **422–424**
    moulds used 431–432
    non-destructive testing of 425
    process technology **427–435**

filament winding 430–431
LD-SRIM process 428
P4 process 428
pultrusion 431
RRIM process 429
sprayed chopped fibre process 429–430
SRIM process 427–428
properties *433*, *438*, *439*
    compared with other materials *433*, *438*
test methods for **424–425**
see also wood composites
compression force deflection (CFD), flexible foams 180
compression hardness
    microcellular elastomers, test method 295
    technical foams, typical values *219*, *221*, *222*
compression set
    calculations 183
    test methods 181–182, 297
    typical values, flexible foams *195*, *197*
compression strength
    rigid foams
        test method 242
        typical values *251*, *260*, *269*, *274*
computer-controlled operations
    cutting of foam 216
    moulded foams 199
concrete expansion joint 409, *410*
cone calorimeter 58
construction industry applications
    CASE formulations 19, 390, 391, 393, 395–418
    fire risks 58–61
    manufacture of products, raw materials and formulations **258–259**
    market share *16*
    rigid foams 7, 19, **257–271**
    see also boardstock; sandwich panels; wood adhesives
contact adhesives 348, 384–385
continuous glass fibre 423
continuous processes
    carpet backing 227–228
    laminating of boardstock 259–260, 263–265
    laminating of foam-backed textiles 224, *225*
    laminating of packaging film 392–393

laminating of sandwich panels
    269–270, 391
pipe foam insulation 277
pultrusion (composites) 431
slabstock foam production
    212–214
thermoplastic PU production
    325–327
wood composites production 400,
    403–404
conversion factors 453
'cook-down' (in polyol manufacture)
    95
corporate average fuel efficiency
    (CAFE) regulations 190
cost advantages, PUs 3
cream time 230, *231*, 250
    comparison of blowing agents *251*
    comparison of catalysts *147*
    typical values
        integral skin foams *341*
        rigid foams *251, 260, 269, 274,*
            280
creep testing, flexible foams
    181–182, 184
cross-hatch (adhesion) test 359
cross-linking agents 126, 163, *165*,
    383
cross-linking of PUs 120, 126
    in adhesives 382
    in coatings 364, 366
cup-flow viscosity test method 356
cure profiles, CASE products 357
cure times
    measurement of 357
    typical values
        coatings 367
        flexible foams 200, 202, 214
        sealants 367
        semi-rigid foams 218
curing agents 163
    see also chain-extending agents;
        cross-linking agents
cushioning
    slabstock foam used 203
    viscoelastic foams used 220
cyclic ureas, formation of 77, 78, *79*
cyclo-addition reactions *122*, 123
1,4-cyclohexane diisocyanate (CHDI)
    63, *449, 456*
1,4-cyclohexanedimethanol,
    as chain extender *164, 448*
cyclopentane
    as blowing agent 133, 250, 340
    physical and environmental
        properties *134, 236*

DADPM *see*
    diaminodiphenylmethane
damping tests, flexible foams 185
decontaminant solutions 45, 48
demould times
    integral skin foams *341*
    moulded flexible foams *195, 197,*
        200, 202
    rigid foams 254, 255
    RIM composites 428, 440, 441
    semi-rigid foams 218
density
    composites *433, 438, 439, 440, 441*
    conversion factors 453
    elastomers
        determination of 294, *295*
        typical values 5, 285, 302, *304,*
            307, 312
    expanded/blown TPU 302, *312,*
        313
    flexible foams
        determination of 179
        moulded foams *170, 195, 197*
        slabstock foam *170,* 203, *206,*
            207, 208, 209
        typical values 5, *170*
    integral skin foams *341*
    isocyanates 455–456
    microcellular elastomers 5, 302,
        304, 307, 312
    PUs 3
    rigid foams 5, 152, 229, *251, 260,*
        269, 274
        determination of *241*
    RIM elastomers *343*
    sealants and encapsulants *411, 417*
    thermoplastic PUs *312,* 323
    wood composites *402, 404*
density–stiffness range of PUs 4
design considerations, composites
    421–422, 442
Desma TPU extrusion/moulding
    machine 313
dialkyltin
    dicarbonates/dialkylthiolates
        as catalysts *140*
        reaction mechanism 143–144,
            145
    production of 139
diamines
    molecular structure of isomers 75
    production of 63, 65, 73, *74*
diaminodiphenylmethane (DADPM)
    as initiator in polyol production
        101
    molecular structure of isomers 75

production of 63, 65, 73, *74*
1,4-diazabicyclo[2.2.2]octane
    (DABCO)
    as catalyst *140*
    production of 137, *138*
    see also triethylenediamine
dibutyltin(IV) dilaurate (DBTDL)
    as catalyst *140,* 145, *154,* 195
    production of 139
dibutyltin(IV) dilaurylmercaptide
    as catalyst *140,* 145
    production of 139
dichlorodifluoromethane
    as blowing agent 131
    physical and environmental
        properties *131*
Die C (tear strength) test 296, *297*
dielectric strength
    conversion factors 453
    typical values for encapsulants *417*
diethanolamine, as chain extender
    448
diethyl ethyl phosphate (DEEP), as
    fire retardant *161,* 162
N,N-diethylcyclohexylamine
    (DECHA), reactivity as catalyst 142
diethylene glycol, as chain extender
    164, 448
N,N'-diethylpiperazine, as catalyst
    140
diethyl toluene diamine (DETDA), as
    chain extender 163, *164,* 365, *448*
differential scanning calorimetry
    (DSC) 175, 298, 357
diisocyanates
    hazards 42–43
    see also isocyanates
4,4'-diisocyanatodicyclohexyl
    methane (H_{12}MDI) **87**
    in CASE formulations 349, 375,
        382
    equivalent weight *449*
    isocyanate value *449*
    market share 63
    physical properties *456*
dimensional stability
    rigid foams, typical values *280*
    test methods *241,* 243
dimethyl terephthalate (DMT)
    polyester polyols 110
2-(2-(dimethylamino)ethoxy)ethanol
    (DMAEE)
    as catalyst *140*
    production of 138, *139*

461

2-((2-dimethylaminoethoxy)-
    ethylmethylamino)ethanol, as
    catalyst *140*
N,N-dimethylaminoethyl N,N-
    dimethylaminopropyl ether
    (DMAEDMAPE), reactivity as
    catalyst *142*
N,N'-dimethylbenzylamine, as
    catalyst *140*
N,N-dimethylcyclohexylamine
    (DMCHA)
    as catalyst *140*, *153*
        reaction kinetics *147*, *148*
        reactivity *142*
        production of *138*
N,N-dimethylethanolamine (DMEA)
    as catalyst *140*, *152*, *153*
    production of *138*
N,N'-dimethylpiperazine, production
    of *137*, *138*
dimethylthiotoluenediamine, as
    chain-extender *448*
dimorpholinodiethyl ether (DMDEE),
    as catalyst *140*, *154*
dinitrotoluene
    hydrogenation of *64*, *66*
    production of *64*, *66*, *70–71*
diphenylmethane diisocyanate (MDI)
    see methylene diphenyl diisocyanate
dipropylene glycol, as chain extender
    164, *448*
direct-glazing sealants **413–415**
    formulations *411*, *415*
    properties *411*
Distribution code *39*
district heating/cooling systems,
    insulated pipes for *30*, *273*, *276*,
    277
domestic appliances *see* appliances
double-glazing applications *411*
double metal cyanide catalysts,
    polyols manufactured using *97–98*
Draka/Petzetakis (slabstock foam
    production) process *212*
drum stock, storage and transport of
    47
dry spinning, for elastomeric fibres
    339
dual-density shoe soles *305*, *310*, *311*
dual-hardness cushions *191*, *200*
durability tests, flexible foams
    180–186
durometer (hardness) test *295*, *360*
    comparison of scales *295*, *296*
dust, from rigid PU foam processing
    50

dyes *167*
    in cast elastomers *334*
    in flexible foams *206*
dynamic mechanical spectroscopy
    (DMS), flexible foams *175*
dynamic mechanical thermal analysis
    (DMTA), elastomers *289*, *291*, *298*
dynamic property testing
    elastomers *297–298*
    flexible foams *183*, *184*, *186*

E-coat *372*
ecolabels *32–33*
E-glass fibre *422*
elasticity modulus, wood composites
    402, *404*
elastomeric coatings *6*, *286*
elastomeric fibres *6*, *286*, **338–339**
    advantages/disadvantages *338*
    applications *338*
    production of *338–339*
    spinning of *339*
elastomers **285–345**
    abrasion resistance
        effect of polyol types *292*
        test methods *295*, *299*
        typical values *304*, *307*, *312*
    additives used *308*
    applications *1*, *285–286*
    compression hardness
        effect of molecular weight of
            polyols *294*
        test method *295*
    compression set, test methods *295*,
        297
    cross-linking in *288*
    density
        test methods *294*, *295*
        typical values *285*, *302*, *304*, *307*
    flex fatigue resistance
        test methods *295*, *297–298*
        typical values *304*, *307*, *312*
    hard blocks in polymer structure
        287
        effect of annealing *291*, *292*
        effect of chain extender type *289*
        effect of isocyanate type *289*
        packing/stacking of *287*, *288*
        properties affected by *289–291*
    hardness
        factors affecting *288*
        test methods *295*, *295*
        typical values *286*, *304*, *307*
    hysteresis characteristics
        effect of hard block content *290*

        effect of molecular weight of
            polyols *294*
    interphase areas *287*, *288*, *291*
    market share *12*, *13*
    process technology *309*
    raw materials and formulations
        303–308
    rebound resilience
        effect of hard block content *290*
        effect of molecular weight of
            polyols *294*
        test methods *295*, *297*
    soft segments in polymer structure
        287
    structural effects on properties
        286–294
    surfactants used *160*
    tear strength
        effect of hard block content *290*
        effect of polyol types *292*
        test methods *295*, *296*, *297*
        typical values *290*, *304*, *307*, *312*
    tensile strength
        effect of hard block content *290*
        effect of polyol types *292*
        test methods *295*, *296*
        typical values *290*, *304*, *307*, *312*
    test methods **294–300**
        dynamic properties *297–298*
        environmental exposure testing
            299–300
        frictional properties *298*
        mechanical properties *294–297*
        thermal analysis *298*
    weathering resistance, effect of
        polyol types *292*
    see also cast...; microcellular...;
        microporous...; RIM...; spray
        elastomers; thermoplastic
        polyurethanes
electrical goods
    recycling of *38*
    see also appliances; refrigerators and
        freezers
electron beam (EB) radiation,
    urethane acrylate coatings cured
    with *377*
electrostatic spraying (of powder
    coatings) *373*
Employee Health & Safety code *39*
emulsifiable MDI, in wood
    composites *396*, *403*, *404–405*, *407*
encapsulants **416–418**
    applications *416–417*
    formulations *417*
    market share *409*

humid ageing test, flexible foams 187
humidity, effect on flexible foam hardness 204–205
Huntsman Polyurethanes
　environmental improvements 26–27
　expanded TPU technology 313
hydrocarbons 133
　as blowing agents 133, 250, *251*, 261, 282
　physical and environmental properties 133, *134*, *236*
　as propellants 282
hydrochlorofluorocarbons (HCFCs) 132
　as blowing agents 132, 250, *251*, 261, 273, 274, 280, 340
　phase-out of 129–130, 250, 274
　physical and environmental properties *132*, *236*
　as transitional replacements for CFCs 128–129
hydrofluorocarbons (HFCs) **132–133**
　as blowing agents 132, 250, *251*, 261, 274, 280, 282, 339, 340
　physical and environmental properties *133*, *236*
　as propellants 282
hydrogenation processes **71–73**
　liquid slurry phase process 72
　liquid–vapour slurry phase process 72
　vapour-phase fixed-bed processes 72–73
　vapour-phase fluidised-bed process 73
hydrolysable chlorine content of isocyanates
　analytic method 80–81
　typical values 83
hydrolysis resistance
　elastomers 299
　flexible foams 187
　TPUs 324
hydrolysis stabilisers 166, 337
hydroperoxidation, propylene oxide produced by 91–93
hydrophilic foams 208, **223–224**
　applications 224, 284
　characteristics 223
　formulations 223–224
hydroquinone dihydroxyethyl ether (HQEE), as chain extender *164*, 286, 289, *448*

N-hydroxyethyl-N,N′,N′-trimethyl bis(aminoethyl)ether (HETMBAEE), as catalyst 147, *147*, *148*, 150
hydroxyl content, polyols, effect of ethylene oxide 100
hydroxyl-terminated polybutadienes 90, 112
hydroxyl value
　chain-extenders *448*
　polyols
　　calculations 447
　　meaning of term 98
　　typical values 111, *304*, *307*, *323*, 364
N″-hydroxypropyl-N,N,N′,N′-tetramethyliminobispropylamine (HPTMIBPA), as catalyst, reaction kinetics 147, *148*
hyper-soft foams 208
hysteresis
　elastomers, typical values *290*
　flexible foam, test method 180

I-beams 406
　applications *406*, 408
ignitability
　of slabstock foams 211
　tests 55–56, 60, 188
impact resistance tests, composites 424
impedance tube test 186
impregnated foams 226
in-line mixing
　flexible foams 198
　rigid foams 253–254, 262
incineration 36
indentation force deflection 181, 190
indentation test, flexible foams 180
India, population growth 10
indoor air quality 34
industry structure 7–8
　future trends 8
ingestion hazards 41
inhalation hazards 41, 42
　from fire effluents 62
initial hardness factor, flexible foams 181
injection moulding
　cold-cure foams 199
　elastomeric shoe soles 309
　TPUs 6, 328–329
insulated pipe applications **273–278**
　joining of pipes 276
　see also pipe insulation applications
insulation (building) panels 19
　environmental declarations 33–34

manufacture
　energy requirements 26
　polyols used 110
insulation materials **257–258**
　benefits 257
　comparison of various building materials 258
integral skin foams (ISFs) 306, **339–342**
　applications 17, *339*
　blowing agents used 339, 340
　process technology 341–342
　properties 340
　raw materials and formulations 339–340, *341*
Intergovernmental Panel on Climate Change (IPCC) 31
intermediate bulk containers 48
internal mould release (IMR) technology 428
International Isocyanate Institute, gaseous emissions study 28
ion viscosity data, catalysis reaction kinetics studied using **150**
ionisable chlorine, isocyanates 81
IPDI see isophorone diisocyanate
irrigation tubing 319, *320*
irritation hazards 41, 43, 50
ISO standards/test methods
　basic-property determination 179, *179*, *241*, *255–256*, *295*
　for CASE products *355–356*
　durability/comfort tests *179*, 180, 181, 183, 185, *241*, *295*, 355
　for elastomers *295*
　fire tests 55, 58
　for flexible foams *179*
　for rigid foams *241*
isobutane
　as blowing agent 250
　physical and environmental properties *134*, *236*
isocyanate derivatives 79–80
　see also prepolymers
isocyanate index 449
isocyanate value 80, 449
　listed for various isocyanates *449*
isocyanates **63–88**
　analysis of 80–82
　　acidity 81
　　gas chromatography 82
　　hydrolysable chlorine 80–81
　　ionisable chlorine 81
　　isocyanate value 80
　　viscosity 81

raw materials and formulations **196–197**
test methods 177
future trends, factors affecting 21–22, 444–445

gas chromatography, isocyanates 82, *84*
gas thermal conductivity 235–236
and blowing agents 233
relationship with molecular weight 236
gaskets and seals 318, 415–416
gel formation, chemistry 113, *114*
gel permeation chromatography
polymeric MDI 84, *85*
polyols 112
gel products 223, 332
gel time 230, *231*, 250
comparison of blowing agents 251
measurement of 357
typical values, rigid foams 251
gelation catalysts 137, *140*
reaction kinetics compared *148*, 149
Gibbs–Marangoni effect 157
glass fibre
as filler 166
as reinforcement 422–423
glass fibre reinforced PU composites
automotive applications 437–445
formulations 433
properties 433
raw materials 422–423
weight savings 443
glass transition temperature
elastomers 291, 293
encapsulants 417
flexible foams 175
TPUs 322
Global Health and Safety initiative 40
global warming 31
global warming potential (GWP) 135
of various blowing agents *131*, *132*, *133*, *134*
gloss measurements, coatings 360
glue laminated lumber (glulam) 395, *396*, 405
applications 396, *407*, 408
process technology 406–407
glue stick guns 387
glycerol, as chain extender *448*
glycerol-based polyether polyols 101
applications 232, 433

glycols, as chain extenders 164, 286, 289, 343, *448*
Graves (tear strength) test 296, *297*
greenhouse gases 31
atmospheric concentrations *32*
guarded hot plate method, thermal conductivity determined using 242

H-point test method 181, *182*
halogen-based fire retardants 160, *161*, 162
handling procedures 44–46
hardness
coatings, test method 360
comparison of durometer scales 295, *296*
elastomers
test method 295
typical values 286, *304*, *307*, *312*, 331
flexible foams
test method 180–181
typical values 190, 207
integral skin foams 341
RIM elastomers 343
sealants and encapsulants 411, 417
TPUs 312, 316, 317, *323*
hardness–density maps, slabstock foams 207, *209*
hazards
manufacturing 49–50
raw materials 42–44
types 41
HCFCs *see* hydrochlorofluorocarbons
HDI *see* hexamethylene diisocyanate
health surveillance programme 44
heat ageing
elastomers 295, **299**
flexible foams 187
TPUs 324
heat flow meter, thermal conductivity determined using 241–242
heat of fusion, aromatic isocyanates 455
heat loss, in buildings 278
heat transfer, factors affecting 233–236
hemp fibre 424
hexamethylene diamine 64
hexamethylene diisocyanate (HDI) 87
in CASE formulations 349, 368, 372, 373, 382
chemical structure *67*, 87
equivalent weight *449*

isocyanate value *449*
market share 63
physical properties *456*
production of 64, *67*, 77
trimer *449*
1,6-hexanediol, as chain extender *448*
HFCs *see* hydrofluorocarbons
HFEs, as blowing agents 134
high-density flexible foams 5, 227–228
high-density rigid foams **284**
high ethylene oxide polyether polyols 103, 223
high-frequency welding 208, 225
high-load-bearing (HLB) foams, formulation 208
high-performance cast PU elastomers 331
high-performance PU soling systems 304–305
high-pressure mixing 198, 253–254, 262, 271
high-resilience (HR) foams
cell-size control 205
combustion-modified (CMHR) 57, 209
hardness–density maps 209
manufacture
process technology 210
raw materials and formulations 100, 103, *140*, *170*, 172, 194, **208–211**
market share 208
porosity 185
properties 209
stress/strain (hysteresis) curves 178, *181*, 205
hip-point (H-point) test method 181, *182*
H_{12}MDI *see* 4,4'-diisocyanato dicyclohexylmethane
hose applications 6, 319, *320*
hot-cure moulded foams
process technology **202**
raw materials 103, 152, *170*, **197–198**
hot-melt adhesives
non-reactive adhesives 380, 381, **387–388**
reactive adhesives 380, 381, **393**
hot-melt extrusion spinning, for elastomeric fibres 339
hot-melt films/webs 387–388
hot-water cylinders, insulation of **282–283**
housekeeping procedures 45

flame-retardancy 55
see also fire retardants
flammability, blowing agents 131, 132, 133, *133*, *134*, 136
flash point, isocyanates 455–456
flex fatigue tests
　elastomers **297–298**
　　typical results *304*, *307*, *312*
　flexible foams 183
flexible foams 4, 5, **169–228**
　ageing tests 187–188
　airflow properties 184–185
　alternative materials 21
　applications 1, 4, 5, 17, 189–193, 203
　cell-size control 195, 198, 205
　cellular structure 175
　chemistry 171–174
　consumption data *20*
　density range(s) 5, *170*, *195*, *197*, 203, *206*, 207, *208*, *209*
　dynamic testing of *177*, 183–184
　　frequency ranges 177–178
　environmental performance 188
　fire-propagation tests 56–58, 188
　fire retardants in 162, 208, 211
　fire risks 51–55
　flame-retardancy 55
　foam stability during processing 195, 205–206
　functionality/performance tests 177–188
　high-density 5
　historical development 169–171
　ignitability tests 56–58, 188
　in integral skin foams 340
　low-density 5
　manufacture
　　carbon dioxide as blowing agent 127, 134, 135, 201, 204
　　catalysts used 140, **151–152**, 195, 197
　　hazards associated 49–50
　　isocyanate emissions 29
　　raw materials and formulations 103, 109, *170*, **193–198**
　　surfactants used **158–159**, 195
　market share *12*, *13*, 169, *170*
　morphology 175–177
　polyols in 103, 109, *170*, 194, 195, 197
　properties *195*, *197*, *206*, *208*, *209*
　raw materials and formulations **193–198**, **204–211**
　rise-height development over time *174*

stages of foam development 172–174, *173*
　cell opening 172–173, *173*
　temperature–time plot *174*
　tensile strength, test methods 179–180
　testing of **177–188**
　　ageing tests **187–188**
　　basic properties **179–180**
　　conditioning of samples prior to 178
　　durability/comfort tests **180–186**
　　environmental tests 188
　　fire tests 188
　　standards/test methods listed *179*
　see also moulded foams; slabstock foam
flexural modulus/strength
　composites
　　test method 424
　　typical values *433*
　RIM elastomers *343*
floral foam applications **283–284**
flotation foams 283
flow aids, in CASE formulations *351*, 392
flow properties, CASE formulations 354, 356
fluorinated ethers, as blowing agents 134
FMPA tests, structural wood composites 408
foam-control additives 158–159, 195, 206, *351*
foam-in-fabric technology, for automotive seating 191, 199, 200–201
foam-in-place gaskets (FIPG) **415–416**
　formulations *411*
　properties *411*
foamed PUs 4–6
　formulation worked examples 450–451
　see also flexible foams; moulded foams; rigid foams; semi-rigid foams; slabstock foam; technical foams
food packaging applications 393
food preservation 245
footwear applications 5, 6, 7, 19, **301–314**
　adhesives for 7, 384–385, 386
　catalysts used 155, 156
　direct moulding methods 301, 310, 311, 313

formulation(s)
　polyester systems *304*, *452*
　polyether systems *307*
　worked example 451–452
　market share *16*
　polyols used 109, 160, 303
　production locations 19
　surfactants used 160
　TPU used **311–314**
　traditional materials 301
　two-component PUs for
　　hybrid polyol technology 307–308
　　polyester polyols 303–305
　　polyether polyols 306–307
　　process technology **309–311**
　see also shoe soles
forklift truck tyres/wheels 331
formaldehyde
　condensation with aniline 63, *65*, 73–76
　emissions from urea–formaldehyde materials 34
formulations
　worked examples 450–452
　see also under individual PU types, e.g. adhesives; coatings; elastomers; encapsulants; flexible foams; rigid foams; sealants; semirigid foams; thermoplastic polyurethanes
Fourier transform infrared (FTIR) spectroscopy, reaction kinetics studied by 146, **148–149**, 357
freezers *see* refrigerators and freezers
frictional properties, elastomers, test methods 298
fuel consumption, by automobiles 437
fumed silica, as thixotropy control agent 383, 392
functionality
　chain extenders *448*
　isocyanates 83
　　typical values 196, *449*
　polyols
　　factors affecting 109, 110
　　meaning of term 98
　　typical values 110, 111
fungal infestation resistance, wood composites 401
furniture applications 4, 5, 7
　fire-propagation tests 56–58
　fire risks 51–54
　ignitability tests 56–58
　market share *16*
　moulded foams 193, 196

properties *417*
end-of-life phase, environmental impacts 35–38
end-of-rise time (for rigid foams) 230–231, *231*
energy, conversion factors 453
energy-absorption foams
  flexible foams 177
  semi-rigid foams 217, 219
  testing of 177
energy absorption tests
  flexible foams 177, 180, 185
  semi-rigid foams 219
energy consumption
  by automobiles 437
  labels 32, *33*
  in manufacturing processes 25, *26*
    improvements 27
energy efficiency, contribution of PU foams 23, 30–31, 246
energy recovery, from waste 36
engineered lumber (EL) composites **405–408**
  applications *396*, 408
  process technology 406–407
  resins used *396*, 407
  testing of 407–408
  see also glue laminated lumber; I-beams; laminated veneer lumber
engineering applications 6, **317–318**
environmental considerations 23–24, 188, 261
environmental declarations 33–34
environmental exposure testing
  CASE products 361
  elastomers **299–300**
  flexible foams **187–188**
environmental impacts
  CASE products 361–362
  end-of-life phase 34–38
  production phase
    PU products 28–30
    raw materials 26–27
  use phase 30–34
environmental improvement strategy, objectives 26–27
EO *see* ethylene oxide
EPD coating 372
epoxy curing agents 106
equivalent weights
  chain extenders *448*
  isocyanates *449*
  polyols 448
esterification 125
ethanolamine, as chain extender *164*, *448*

ethyl benzene hydroperoxide, in manufacture of propylene oxide 91, *92*
ethylene amines, amine catalysts produced from 137, *138*
ethylene glycol, as chain-extender *164*, *448*
ethylene oxide
  oxidation of 124
  production of 8
ethylene oxide/propylene oxide polyether polyols 99–100, *170*, *171*, 322
ethylene oxide tipped polyols 99, 100, *170*, *171*, 189, 209
N-ethylmorpholine, production of 138, *139*
Europe–Middle East–Africa region
  base chemical consumption data 13, *14*
  population growth *10*
  purchasing parity power per capita *11*
  supply/demand balance 13, *14*, 15, *16*
European standard test methods
  CASE products 355–356
  elastomers 295
  flexible foams 179
  rigid foams 241
European Union
  Directives, on waste management 37–38
  energy consumption labels for appliances 33, *252*
  fire classification system 60–61
exfoliating graphite 162, 211
exothermic reactions
  in foam manufacture 49, 50, 215, 230, 232
  isocyanate reactions 113, 114, 115, 119, 230
  polymerisation of propylene oxide 94
extrusion processes
  for elastomeric fibres 339
  for pipe foam insulation 277
  reactive, for TPU production 326–327
  for TPUs 6, 313, 329

fabrics 1, 6
factory emissions 27, 28–29
falling dart/weight (impact) test 424
falling weight tests
  for energy absorption of foams 185

  for impact resistance of composites 424
  for rebound resilience of elastomers 297
Farka mechanism (for amine catalysts) 141
fatigue tests, flexible foams 183
feedstock recycling 37
ferric acetylacetonate (catalyst) *140*, *154*, 156
fibre-reinforced composites 1, **419–445**
  advantages 419
  design considerations **421–422**
  general properties **420–421**
  process technology **427–435**
  raw materials 419–420, **422–424**
  test methods for **424–425**
fibres 6
filament winding (of composites) **430–431**
  advantages and limitations 434
  formulations used 433
  properties of composites 433
fillers 165–166
  in CASE formulations 351, 383
  in cast elastomers 334
  as fire retardants 162
  in flexible foams 204
film materials 320–321
  end-use applications 321
filtration applications 203, 226
finger-jointed (FJ) lumber *396*, 405
finite element analysis (FEA) 442
fire effluents 61–62
fire losses, US data 53, *54*
fire-reaction tests 51–52, 55–58
fire regulations 59–61
fire-propagation tests 56–58, 60, 188
fire-resistance tests 52
fire-retardant foams
  catalysts used *140*
  polyols used 90, 111–112
fire retardants 55, **160–162**
  for flame lamination of foams 207, 225
  in flexible foams 162, 208, 211
  synergetic effect of bromine and phosphorus 160, *161*
fire risks 51–53
  flexible foam 53–58
    during manufacture 49–51
    test methods 56–58
fire safety engineering 62
flame lamination of flexible foams 207, 224–225

463

'blocked'/hindered 79, 114, 115, 123, 371
in CASE formulations 349
chemical structures 65–69, *288*
cyclo-addition reactions *122*, 123
decontamination of drums 48
dimerisation of 118–119
in elastomers 286
factory emissions 28–29
hazards 42–43
heating of drums 47
insertion reactions *122*, 123
light-stability of PUs 83, 87, 126, 349
market split *63*
nucleophilic addition reactions 123
occupation hygiene monitoring for 46–47
physical properties *455–456*
product characteristics 82–87
production of 7, **63–79**, 87–88
production capacity 14, 15, 16
purification of 78–79
reactions with
  acid anhydrides *122*, 123
  alcohols 114, 115
  amines 116
  carboxylic acids *122*, 123
  cyclic carbonates *122*, 123
  dialkylmalonates *122*, 123
  hydroxyl compounds 114–115
  isocyanates 118–121, 135
  oximes 121, *122*
  phenols 114
  phosgene *122*, 123
  propylene oxide *122*, 123
  urea 116–117
  urethanes 117
  water 115
reactivity 83, 113–114
resonance forms *113*
solvents for 77
storage and transportation of 47–49, 82
trimerisation of 119–120
trimers 119, *120*, 449
see also CHDI; H$_{12}$MDI; HDI; IPDI; MDI; NDI; PPDI; TDI; TMXDI; TODI
isocyanurate catalysts *140*, 145
isocyanurates
  formation of 119, *120*
    reaction mechanisms 145, *146*
ISOPA (European Isocyanate Producers' Association) 24
  life-cycle inventory studies 25

isopentane
  as blowing agent 133, 250, 340
  physical and environmental properties *134*, *236*
isophorone, production of 64, *68*
isophorone diisocyanate (IPDI) **87**
  in CASE formulations 349, 371, 373, 382
  chemical structure *68*
  equivalent weight *449*
  isocyanate value *449*
  market share *63*
  physical properties *456*
  production of 64, *68*
  trimer *449*
  volatility *43*
Ixol B251 (fire retardant) *161*, 162
Izod impact test 424
  typical values for composites *433*

James (friction) test 298
Japan
  population growth *10*
  purchasing parity power per capita *11*
  standards/test methods *179*, 181
  supply/demand balance *15*

Kyoto Protocol 31

lambda value 233
  calculations 234
  component parts 234
  determination of **241–242**, 252
  gaseous conduction 235–236
  inverse 234
  radiative transfer 235
  solid conduction 235
  typical values for rigid foams *251, 280*
  see also thermal conductivity
laminated composites 320
laminated PU foams 224–225
laminated veneer lumber (LVL) 396, 405
  applications 396, 408
laminating
  of boardstock 259–260, 263–265
  of foam-backed textiles 224, 224–225, *225*
  of packaging film 392–393
  of sandwich panels **269–270**, 391
lance injection techniques
  pipe insulation 275
  sandwich panels 271
lap shear (adhesion) tests 358, 359

LD-SRIM *see* low-density structural reaction injection moulding
lead-based catalysts 137, *140*, *154*, 155, 156
leather replacement material 336–338
length, conversion factors *453*
life-cycle approach 23, *24*
life-cycle assessment 24–26
life-cycle considerations 23–38
life-cycle inventory studies 25
light-stability of PUs, effect of isocyanate 83, 86, 87, 125, 167, 324, 349
liquid carbon dioxide, as blowing agent 134, 201, 204, 214
local extract ventilation 44
low-density flexible foams 5
low-density rigid foams 4, 5, **284**
low-density structural reaction injection moulding (LD-SRIM) 428
  advantages and limitations *434*
  applications 437, 440, 441, 444
  formulations used *433*
  properties of composites *433*
low-temperature applications, footwear applications 306, *308*
lubricants 166
lung function tests 44

magnetic tapes, coatings for 375
mandrel bend test 359, *360*
Mannich bases 101, *102*
Mannich polyols 101
market **9–22**
  diversity 8
  European
    direct-glazing sealants 413
    wood composites *397*
  factors affecting growth 10–13
  global/world-wide
    CASE products *347*, *363*, *379*, *409*
    flexible foams *170*
    propylene oxide *9*, 89, *91*
    PU base chemicals *14*, *63*, 89, 107
    PUs *9*, *13*, *14*, *16*, *347*
    rigid foams *229*
  split by end-use applications *16*
  split by region *14*
  split by technology *13*, *347*
market share, wood composites *397*
mass, conversion factors *453*
mattresses 4, 203
  fire risks 52, 53, 54

mattresses (*continued*)
   flex fatigue testing of 183
Maxfoam (slabstock foam production) process 213
MDA *see* methylene dianiline
MDI *see* methylene diphenyl diisocyanate
MDI-based PUs
   growth in market 12
   in wood composites 18, 396, 401, 404, 407
MDI-based TPUs 323
MDI polyether foam, formulations 451
mechanical impedance inspection of composites 425
medical applications 1, **318–319**
medical surveillance programme 44
medium-density fibreboard (MDF) 395, 396, **403–405**
   applications *396*, 405
   process technology 403–404
   properties *404*, 405
   resins used *396*, 404
melamine, as fire retardant/filler 161, 162, 166, 211
melamine–formaldehyde (MF) resins, in wood composites 395
melamine modified urea–formaldehyde (MUF) resins, in wood composites 395, 404
melamine urea phenol formaldehyde (MUPF) resins, in wood composites 395
melting point(s), isocyanates 455–456
mercury-based catalysts 137, *140*, *154*, 156, 333
methane, atmospheric concentration 32
methylchloroform, as blowing agent 134
1,1′-methylene bis(2-chloroaniline) (MOCA) 333
methylene chloride
   as blowing agent 130, 131
   physical and environmental properties *131*
methylene dianiline (MDA)
   conversion to H$_{12}$MDI 69
   production of 63, *65*, 77
methylene diphenyl diisocyanate (MDI) **83**
   in CASE formulations 349
      adhesives 382, 388, 391, 396

      coatings 364, *367*, 370, 373, 374, 375
      encapsulants *417*
      sealants 410, *411*, 415
      wood composites 400, 401, 404
   chemical structure 65, 83, *288*, 289
   in composites 433
   dimerisation of 118–119
   emissions from foam production 29
   emulsifiable (eMDI) 396, 403, 404–405, 407
   equivalent weight 449
   in flexible foam production *170*, 171, 172
      moulded foam *170*, 189, 193, 194, 195, 196
      slabstock foam *170*, 203, 205, *209*, 210
   foam formulation worked example 451
   functionality 449
   isocyanate value 449
   life-cycle inventory data 25
   light-stability of resulting PUs 86
   market share 63
   molecular weight 449, 455
   physical properties 455
   production of **63–64**, *65*, 73–76
   purification of *65*, 78–79
   reactivity 83, 113, *114*, 115
   solvents for 77
   storage and transportation of 82
   trimer 449
   in viscoelastic foams 220
   volatility 43
   in wood composites 396, 401, 404, 407
   see also MDI-based PUs; polymeric MDI
N-methylmorpholine, reactivity as catalyst 142
N-methylpiperidine, reactivity as catalyst 142
mica, as reinforcement 423, 429
microbial attack
   protection against 167
   testing for resistance 188
microcellular elastomers 285
   applications 5, 17, 19
   compression hardness, test method 295
   density range 5
   effect of hard block content on properties 289–291
   manufacture, catalysts used *140*

   tear strength *290*, 296, *297*
microporous elastomers 6
microscopy, flexible foams studied by 175
millable PU gums 286
mine stabilisation applications **284**
mineral classification screens 318, 331, *332*
mineral fillers 166, 351
mineral reinforcements 423
model systems, catalysis reaction kinetics studied using 147–148
moisture-cure adhesives 348, 379, 388–390
   applications 389–390, 407
moisture-cure coatings 154, 348, 363, 373–374
   applications 373
   formulations *367*
   properties 374
moisture-cure foams 281–282
moisture-cure resins, in wood composites 395, 403, 404–405, 407
moisture-cure sealants **412–413**
   formulations *411*
   properties *411*
moisture scavengers 351, 374, 383, 412
molecular weight
   chain extenders 448
   isocyanates 449, 455–456
   polyols 448
Montreal Protocol 127, 128
   amendments 128
   effect on CFC/HCFC production 130, *131*
   phase-out of CFCs *128*, 250
   phase-out of HCFCs *129*, 250
morpholine, amine catalysts produced from 137–138, *139*
mould-release agents 200, 310, 342, 344
   see also internal mould release
moulded foams **189–202**
   applications 5, **189–193**
   cold-cure process 189
      catalysts used 152, 195
      process technology **199–201**
      raw materials *170*, 189, **194–197**
   compared with slabstock foams 189
   hot-cure process 189
      catalysts used 152, 197
      process technology **202**
      raw materials 103, *170*, 189, **197–198**

manufacture
  crushing after demoulding 49, 200
  process technology **198–202**
  raw materials and formulations 103, *170*, 171, 189, **193–198**
  market share *170*
  properties *195*, *197*
  raw materials and formulations **193–198**
  sound-insulation systems for automobiles 192–193

1,5-naphthalenediisocyanate (NDI) **86**
  chemical structure 86, *288*, 289
  equivalent weight *449*
  isocyanate value *449*
  market share 63
  molecular weight *449*, *455*
  physical properties *455*
natural fibre, as reinforcement 424
NDI *see* 1,5-naphthalenediisocyanate
nitration
  of benzene 63, *65*, 70
  of toluene 64, *66*, 70–71
nitrobenzene
  hydrogenation of 63, *65*, 71, 72–73
  production of 63, 70
    energy efficiency of plant 27
nitrogen, physical properties *236*
nitrous oxide, atmospheric concentration 32
non-destructive testing, composites 425
non-reactive adhesives 379, **384–388**
  hot-melt adhesives *380*, *381*, **387–388**
  solvent-borne adhesives *380*, *381*, **384–385**
  water-borne adhesives *380*, *381*, **386–387**
non-reactive coatings **374–376**
  PU dispersions **375–376**
  radiation-cured coatings **377**
  solvent-borne lacquers **374–375**
  urethane oils and alkyds **376**

occupation hygiene monitoring 46–47
Occupational Exposure Limits (OELs) 44
  for blowing agents *131*, *132*, *133*, *134*
oil-modified coatings 155, 376
oil pipelines, insulation of 273, 278

oil resistance
  elastomers 300
  TPUs 315, 317, *324*
one-component foam (OCF) **281–282**
one-component (moisture-cure) adhesives 379, **388–390**
  applications 389–390, 407
  development of strength with time *380*
  effect of temperature on strength *381*
one-component (moisture-cure) coatings 363, 373–374
  applications 373
  formulations *367*
  properties 374
one-component (moisture-cure) sealants **412–413**
  formulations *411*
  properties *411*
one-shot process
  for cast elastomers 332
  for flexible foams *170*
open-cell rigid foams 5, **239–240**
organometallic catalysts 137, *140*
  applications *140*, 155, 156
  production of **139**
  reaction mechanisms 143–145
organotin catalysts
  applications *140*, 155
  production of **139**
  reaction mechanisms 143–145
oriented strand board (OSB) *395*, *396*, **399–402**
  applications *396*, 402
  process technology 399–400
  properties 401, *402*
  resins used 18, *19*, *396*, 401
oven-curing coatings **371–373**
  formulations *367*
  solvent-borne *367*, 372
  water-borne *367*, 372
overpacking
  moulded flexible foams 199
  rigid foams 254
oxygen, physical properties *236*
oxygen index test 58
ozone-depletion potential 128
  of various blowing agents 128, *131*, *132*, *133*, *134*
ozone layer, depletion of 23, 31, 127

P4 RIM process (for composites) 428
  applications 444
packaging, recycling/re-use of 37

packaging film, lamination of 392–393
packaging foams **221–223**
  formulations 221–222
  process technology 222
  testing of 177, 185
particleboard *396*, 408
peel (adhesion) test 358
pencil (hardness) test 360
pendulum impact tests 424
pendulum rebound resilience test *295*, 297
pentabromodiphenyl ether 55
pentamethyldiethylenetriamine (PMDETA) *140*
N,N,N',N'',N''-pentamethyldipropylene triamine, as catalyst *140*
pentanes
  as blowing agents 133, 250, 261, 273, 340
  physical and environmental properties *134*, *236*
per cent hydroxyl, calculations *447*
perfluorocarbons, as blowing agents 134
performance-based building codes 61
Perlon fibres 315
personal protective equipment 46
phase-separation of PUs 126
  in elastomers 288, 293
    analysis 298
PHD polyols *see* Polyharnstoff Dispersion polyols
phenol–formaldehyde (PF) resins, in wood composites 395, *402*
phenol resorcinol formaldehyde (PRF) resins, in wood composites 395, 406, 407
phenyl isocyanate, in reaction kinetics studies 147–148
N-phenyldiethanolamine, as chain extender *164*
p-phenylene diisocyanate (PPDI) 63, *449*, *456*
m-phenylenediamine, as chain extender *164*, *448*
phenylmercuric propionate (catalyst) *140*, 154, 156
phosgenation 63, 64, *65*, 68, 69, **76–77**, *66.67*
  alternatives to 87–88
  side-reactions 76–77, *122*, *123*
phosgene 77, 87
phosphorus-based fire retardants 160, *161*, 162

469

photochemical smog formation,
    factors affecting 132–133, 135
PHT-4 Diol (fire retardant) 162
physical properties
    isocyanates 455–456
    PUs 3, 4
pigments and dyes 167
    in CASE formulations 351
    in cast elastomers 333–334
    in flexible foams 206
pillows 208, 220
PIPA polyols see polyisocyanate
    polyaddition polyols
pipe applications 319
pipe insulation applications 30,
    273–278
    continuous moulding method 273,
        277
        formulations and foam properties
            274
    discontinuous moulding methods
        273, **274–276**
        formulations and foam properties
            274, 276
        joining of pipes 276
    spray foam method **277–278**
        formulations and foam properties
            274
PIR foams see polyisocyanurate foams
Pirmasens (flex fatigue) test 297
Planibloc (slabstock foam production)
    process 213
plasticisers 167
    in CASE formulations 383
playground surface applications 389
plywood 396, 408
PMDETA catalyst 153
    reaction kinetics 149
PO see propylene oxide
Pollution Prevention code 39
polyacrylonitrile/polystyrene see
    styrene–acrylonitrile (SAN) polymer
polybutadienol-based materials 382,
    418
polycaprolactone polyols 90, 110,
    322, 337
polycarbonate polyols 90, 111, 322
polyester polyols **107–111**
    aliphatic 108–109
    applications 107, 303–305
    aromatic 110–111
    in CASE formulations 350, 364,
        383, 391
    compared with polyether polyols
        in elastomers 303
        in flexible foams 205, 207

degradation of 125
    in elastomers 303–305, 332
    in flexible foams 170, **206–207**
    footwear applications 303–305
    linear/lightly branched aliphatic
        108–109, 303, 306
    market share 108
    polycaprolactones 110
    polycarbonate polyols 111
    production of 109
        raw materials for 108, 303
    in synthetic leathers 336–337
    types 90, 107
    viscosities 109
polyether polyols **93–107**
    acid decomposition of 123, 124
    amine-terminated 90
    in CASE formulations 350, 364,
        382–383, 411, 417
    compared with polyester polyols, in
        flexible foams 205, 207
    degradation of 123–125
        anti-oxidants used 165
    in elastomers 306–307, 332
    in flexible foams 170, 197, 203,
        207–208
    footwear applications 306–307
    life-cycle inventory data 25
    modified 104–105
    molecular weight distribution 98,
        99
    production of
        catalysts used 94, 97–98
        cyclic ethers used 89
        from ethylene oxide + propylene
            oxide 99–100
        from propylene oxide 93–99
        from tetrahydrofuran 106–107
        initiators used 94, 101
        side reactions 96–97
    structure 98
    in synthetic leathers 337
    types 90
polyethers, amine-terminated 106
polyethylene terephthalate (PET)
    based polyols 111
Polyharnstoff Dispersion (PHD)
    polyols 105
    in flexible foams 170, 194, 204,
        211
polyisocyanate polyaddition (PIPA)
    polyols 105
    in flexible foams 170, 194, 204,
        211
polyisocyanurate (PIR) foams 110,
    232

fire resistance 59
polymer polyols 104, 171
    in moulded foams 170, 194, 197
    in slabstock foams 204, 210
    styrene level in 104
polymeric MDI **84–86**, 171
    in CASE formulations 382, 410
    in moulded foams 194, 195
    in packaging foams 222
    physical properties 455
    production of 63–64, 65, 78
    reactivity of isocyanate groups
        85–86
    in rigid foams 231
    in wood composites 396, 401, 404
polyols **89–112**
    acid value 98, 447
    analysis of 112
    build-up ratios for 95
    in CASE formulations 349, 350
    in elastomers 286, 292–293, 303,
        332
    equivalent-weight calculations 98,
        448
    in flexible foams 170
        hydrophilic foams 223
        moulded foams 170, 194, 195,
            197
        slabstock foam 170, 206, 207,
            209
    functionality
        factors affecting 109, 110
        meaning of term 98
        typical values 110, 111
    global market 89, 90
    hazards 43
    hydroxyl value
        calculations 447
        meaning of term 98
        typical values 111, 304, 307, 323,
            364
    life-cycle inventory data 25
    molecular-weight calculations 98,
        448
    production of 7–8, **93–100**, 106,
        108–109
    production capacity 14, 15, 16
    purification of 96
    quality 112
    recycled 36, 90, 112, 125, 166
    in rigid foams 232, 250
    storage and transportation of
        47–49
    and surfactants 158, 159
    in TPUs 321–322
    types 90

470

unsaturation value 98
see also polyester polyols; polyether polyols
polyoxyalkylene polysiloxane copolymers
  structures 158
  as surfactants in flexible foams 158–159, 195, 206
polyoxyalkyleneamines 106, 365
polypropylene, properties compared with PU composites 438
poly(propylene oxide/ethylene oxide) copolymers 90
poly(propylene oxide) polyols 90
  production of 93–99
polytetrahydrofuran 90, **106–107**
  applications 107
  benefits conferred on PUs 107
  production of 106
polytetramethylene ether glycol (PTMEG) 106, 322
polytetramethylene glycol (PTMG) 332
polyurea elastomers **344–345**
  applications 344–345
polyurea two-component coatings 370
  applications 370, *371*
  formulations 367
polyurethane-bound concrete 370
polyurethane dispersions (PUDs) **375–376**
  in adhesives 386–387
  application techniques 376
  applications 337, 375, 386, 387
  production of 386
  types 387
polyurethane fibres 6, 315, 338–339
  raw materials and formulations 107, 338–339
  see also elastomeric fibres
polyurethane gels **223**, 332
polyurethanes
  cross-links in 126, 287, 288
  degradation of 123–126
  global market 9, *16*
  phase-separation of 126, 288
  physical chemistry 126
  production of 113, 114
    environmental impacts 28–30
  reaction first discovered 113, 315
poly(vinyl alcohol) (PVA) adhesives, in wood composites 407
population growth *10*
post-treated PU foams 225–226
pot life

two-component adhesives 383, 391
two-component coatings 366, 367, 369
potassium acetate (KA)
  as catalyst *140*
  reaction kinetics *149*
potassium octoate (KO)
  as catalyst *140*, 343
  reaction kinetics *149*
pounding test, flexible foams 182
pouring–extruding process, for TPU soles 313
powder coatings **373**
  applications 17, 18, 373
  formulations 367
PPDI *see p*-phenylene diisocyanate
pre-formulated mixtures, hazards 44
prepolymers 79
  in adhesives 155, 386, 388, 390, 391
  branching agents in 121
  for cast elastomers 332
  catalysts used *140*, 154
  dimerisation in 118
  for flexible foam production *170*, 171
  in moisture-cure coatings 373
  trimerisation in 120
  see also isocyanate derivatives
pressure, conversion factors 453
pressure infuser bags (medical) 319
printing rollers 6
Process Safety code 39
process technology
  fibre-reinforced composites **427–435**
  integral skin foams 341–342
  moulded foams **199–202**
  rigid foams **253–256**
  RIM elastomers 343–344
  slabstock foam **211–216**
  TPUs **327–330**
  wood composites 399–400, 403–404, 406–407
processing advantages of PUs 3
product design and development
  fibre-reinforced composites 421–422
  risk management in 40–42
product stewardship **39–62**
production phase
  environmental impacts
    PU products 28–30
    raw materials 26–27
  production waste 29–30
propenyl ethers

formation from polyols 123, *124*
hydrolysis of 123, *124*
propylene glycol, as chain-extender 164, 448
propylene oxide
  anionic (base-catalysed) polymerisation of
    with caesium hydroxide 94, 97
    with potassium hydroxide 94–95
  cationic (acid-catalysed) polymerisation of 97
  coordination (double metal cyanide catalysed) polymerisation of 94, 97–98
  end uses 91
  global market 89
    split by end-use *91*
  oxidation of 124
  polyether polyols produced from 93–99
  production of 8
    chlorohydrin process 89, 91
    by hydroperoxidation 91–93
propylene oxide/ethylene oxide polyether polyols 99–100, *170*, 171, 322
PUDs *see* polyurethane dispersions
pultrusion **431**
  advantages and limitations 434
  formulations used 433
  properties of composites 433
puncture-resistant tyres 335–336
purchasing parity power per capita 11
  data for various countries *11*
  and flexible foam consumption *20*
purchasing power 10–11
purification
  isocyanates 78–79
  polyols 96
PVC, compared with PU 308, 312

quaternary ammonium carboxylates, as catalysts *140*, 145
quaternary ammonium formate (QAF)
  as catalyst *140*
  reaction kinetics *149*
quinone-imide structure 86

radiation-cured coatings **377**
radiative transfer, in rigid foams 235
radiography, inspection of composites by 425

raw materials
  environmental burdens due to production 26–27
  see also base chemicals; isocyanates; polyols
reaction injection moulding (RIM) 342
  see also low-density structural...; reinforced...; structural reaction injection moulding
reaction injection moulding (RIM) elastomers 285, **342–345**
  applications 342
  catalysts used in 155, 156, 343
  chain-extenders in 163, 343
  market share 12
  process technology 343–344
  properties *343*
  raw materials and formulations 342–343
  reinforced 343, 427–445
reaction kinetics **145–150**
  FTIR experiments 148–149
  ion viscosity studies 150
  model systems 147–148
  traditional experiments 147
reaction moulding, advantages of PUs 3
reaction-to-fire tests 51–52, 55–58
reaction spinning, for elastomeric fibres 339
reactive adhesives 155, 379, **388–393**
  development of strength with time *380*
  effect of temperature on strength *381*
  hot-melt adhesives **393**
  one-component adhesives **388–390**
  two-component adhesives **391–393**
reactive coatings 363, **366–374**
  one-component coatings **373–374**
  oven-curing/stoving systems **371–373**
  two-component coatings **366–370**
reactive extrusion, TPUs produced by 326–327
reactive sealants 410–411
rebonded foams 37, **226–227**
  adhesives used 390
rebound resilience
  test methods 185, 297
  see also resilience
recycled polyols 90, 112, 125, 166
recycling of waste 36–37, 166, 226

refrigerators and freezers
  cold circuit(s) for 246
  compressor(s) for 246–247
  design 246–249
  double-door top-mount 248, *249*
  early models 246
  energy labels for 32, *33*, 252
  multiple-door 245, *246*, 249
  PU foams first used 246
  recycling of 38
  single-door 248
  two-door side-by-side 249
regulations
  chemical control 39–40
  marketing-and-use 40
  on waste management 37–38
reinforced reaction injection moulding (RRIM) materials
  applications 17
  formulations *433*
  process technology 429, **429**
    advantages and limitations *434*
  properties *343*, *433*
renewable materials, polyols made from 23
replacement challenges 21–22
resilience
  test methods 185, 297
  typical values
    elastomers *290*, 331
    technical foams *219*, *221*, *222*
Responsible Care programme 39
reticulated foams **225–226**
reverse heat leakage test, for refrigerators and freezers 252–253
rheology modifiers, in CASE formulations 351
rigid foams **229–284**
  alternative materials 21
  applications 1, 5, 229
  chemical changes during foam formation **230**
  closed cell content
    determination of *241*, 243
    typical values 5, *260*, *269*, *274*, *280*
  compression strength
    test methods *241*, 242
    typical values *251*, *260*, *269*, *274*, *280*
  density
    test methods *241*
    typical values 5, 152, 229, *251*, *260*, *269*, *274*, *280*
  dimensional stability
    test methods *241*, 243

    typical values *280*
  fire retardants in 162
  fire risks 58–61
  formulation design **231–232**
  in integral skin foams *340*
  low-density 4, 5, **284**
  manufacture
    catalysts used *140*, **152–153**
    hazards associated 50
    isocyanate emissions 28, *29*
    surfactants used **159**
  market share 12, *13*, 229
  open cell foams 5, **239–240**
  physical changes during foam formation **230–231**, *232*
  process technology **253–256**
    carousel line 255–256
    drum/tambour chain system 256
    open mould pour technique 256
    top-flow technique 255
  raw materials and formulations 101–102, 229–230, 250, *251*, 258–259, *260*, *269*
  for appliances 250–253
  test methods **241–244**
  thermal conductivity **233–236**
    effect of ageing **236–239**
    effect of blowing agent *251*
    test methods **241–242**
    typical values *251*, *260*, *269*, *274*
  water vapour transmission properties, determination of *241*, 244
RIM *see* reaction injection moulding...
ring opening, of propylene oxide 95
risk, meaning of term 40–42
risks
  in foam manufacture 49–50
  generic mistakes 49
  raw materials 42–44
roller shear test 183
roller skate wheels 331, *332*
roofing panels
  raw materials and formulations 259
  US market 259
  see also boardstock
Ross (flex fatigue) test 295, 297
  typical results *304*, *307*
rotational casting, semi-rigid foams 218
rotational moulding, TPUs 330
Rozenburg MDI plant 27
RRIM *see* reinforced reaction injection moulding...

472

rubber crumb materials, adhesives for 389–390
rupture modulus, wood composites *402, 404*

safety boots *304, 305*
SAG factor, flexible foams 180–181
sample preparation 178, 358
SAN *see* styrene–acrylonitrile...
sandwich panels **266–271**
　adhesives used 390
　advantages in buildings 267–268
　compared with on-site construction 266
　edge profile design 267, 270
　factors affecting use 266
　manufacture
　　compared with boardstock manufacture 270
　　continuous lamination process 269–270
　　discontinuous lamination process 271
　　raw materials and formulations *269*
SATRA (friction) test 298
SBI fire-propagation test 60
scanning electron microscopy (SEM)
　flexible foams 175, *176*
　reticulated foams *225*
　rigid foams *235*
scorch, in curing of foam 50, 215–216, 230, 284
scrap 29–30
sealants **409–416**
　applications 410
　direct-glazing sealants **413–415**
　foam-in-place gaskets **415–416**
　formulations *411*
　historical development 409–410
　market share 409
　one-component sealants **412–413**
　properties *411*
　two-component sealants **410–411**
seals and gaskets 318
second rise, in PIR foams 232
segmented block polymers, elastomers as *287*
semi-rigid foams 5, **217–219**
　applications 17, 171, **217–218**
　in integral skin foams 340
　interaction with PVC 218–219
　manufacture
　　process technology 218
　　raw materials and formulations **218–219**

market share 12, *13*
　properties **219**
S-glass fibre 423
sheet moulding compound (SMC)
　automotive applications 440, 443
　compared with glass fibre reinforced PU composites 444
shelf life, CASE products 357
shoe soles
　adhesives for 384–385, *386*
　advantages of PU 302
　'combi' soles 312, *313*
　component construction 302
　dual-density 305, 310, *311*
　flex fatigue tests 298
　formulation(s)
　　polyester systems *304, 452*
　　polyether systems *307*
　　worked example 451–452
　frictional property tests 298
　manufacture
　　by direct moulding 301, 309, 310, *311*
　　process technology 309–311
　　raw materials and formulations 303–308
　unit soles 301, 310, *311*
　materials 5, 7, 301–302
　TPU used 311–314
　traditional materials 301
Shore A (hardness) scale *296*
Shore D (hardness) scale *296*
silicone surfactants
　in flexible foams 158–159, 195, 206
　in rigid foams 159
　structural parameters 157, *158*
sisal fibre 424
ski-slope, building for artificial 267
slabstock foam 5, **203–216**
　applications 203
　cell-size control 205
　compared with moulded foams 189
　compression cutting of 216
　manufacture
　　batch-block process 215
　　catalysts used *140*, 144
　　CO_2-assisted blowing 214
　　continuous processes 212–214
　　fire risks 49
　　foam curing 214
　　process technology **211–216**
　　scorch during 215–216
　market share *170*, 203
　post-processing of 216

profile cutting of 216
　properties *206, 208, 209*
　raw materials and formulations 103, *170*, **204–211**
slush moulding, semi-rigid foams 218
small angle X-ray spectroscopy (SAXS), flexible foams studied by 176
SMC *see* sheet moulding compound
smoke, effects 61
smoke suppressants *161*
societal pressure, market growth affected by 11
soil stabilisation applications **284**
solid-phase thermal conduction, in rigid foams 235
solid PUs 6
　see also elastomers
soling applications 301–314
　see also shoe soles
solvent-borne adhesives **384–385**
　development of strength with time *380*
　effect of temperature on strength *381*
　production of 385
solvent-borne lacquers **374–375**
　applications 375
solvent-borne oven-curing coatings 372
　applications 372
　formulations *367*
solvent-borne two-component coatings **366–368**
　applications 366, 368
　formulations *367*
solvent (fibre-spinning) process 6
solvent-free two-component coatings **369–370**
　applications 369
　formulations *367*
solvent resistance
　coatings 361
　elastomers 300
　flexible foams 188
solvents
　in CASE formulations *351*
　　adhesives *382, 383, 392*
　　coatings 365–366
sonar sensors, encapsulants in 416
sorbitol polyols 101, 232, 250
sound-insulation foams 192–193
　raw materials and formulations **196**
　testing of 185, *186*

Spandex fibres 338
specific heat, aromatic isocyanates 455
spectroscopy, flexible foams studied by 176
spinning processes, for elastomeric fibres 339
split tear test, microcellular elastomers *290, 296, 297*
sponge-like foams 208
sports shoes 305
    PUs used *304, 307, 308,* 311
sports surface applications 389
spray elastomer coatings 370
spray elastomers 285
spray foams **278–281**
    adhesion to substrates 279, 280–281
    equipment for 281
    for pipe insulation 277–278
    raw materials and formulations *140, 152, 153,* 196, *280, 280*
sprayed chopped fibre process 429–430
    advantages 430, *434,* 445
    applications 430, 444
    formulations used *433*
    limitations *434*
    properties of composites *433*
SRIM *see* structural reaction injection moulding
standard test methods
    CASE products *355–356*
    elastomers *295*
    flexible foams *179*
    rigid foams *241*
stannous octoate (stannous 2-ethylhexanoate)
    as catalyst *140,* 144–145, 151, 174, 197, 207
    production of 139
static comfort
    automotive seating 191
    bedding/seating applications 177
Steiner Tunnel test 59
storage, of base chemicals 47, 48, 82
storage modulus, microcellular elastomers *291*
storage stability, CASE products 357
stoving coatings **371–373**
    see also oven-curing coatings
stress analysis, fibre-reinforced composites 421
stress/strain curves
    composites 424
    flexible foams 180, *181*

rigid foams 242
semi-rigid foams 219
string time 230
    typical values, rigid foams *260, 269, 274*
structural reaction injection moulding (SRIM) 427–428
    advantages and limitations *434*
    formulations used *433*
    properties of composites *433*
styrene–acrylonitrile (SAN) polymer polyols 104, 171, 194
sucrose polyols 101
    applications 232, 250, *433*
super-soft foams 207–208
supply/demand balance 13–16
support factor, flexible foams 180–181
surface interactions, in CASE applications 350–353
surface reactions 352–353
surface wetting and spreading 352
surfactants **156–160**
    in flexible foams 195, 206
    in rigid foams 230
surfboards 283
sustainable development, meaning of term 23
synthetic leathers **336–338**
    manufacture
        one-component processes 337
        two-component processes 336–337
synthetic mortar 370
system houses 8

Taber abrader 360, *361*
tack-free time 230, *231*
    typical values
        sealants *411*
        sprayed rigid foams *280*
talc 204, 423
TBHP *see* butyl hydroperoxide
TDI *see* toluene diisocyanate
TDI-based PUs, growth in market 12
TDI polyester foam
    formulations *206,* 450
    properties *206*
TDI polyether foam, formulation and properties *197, 208*
tear strength
    conversion factors *453*
    elastomers *304, 307,* 312
    flexible foams *195, 197, 206, 208, 209*
    RIM elastomers *343*

sealants *411*
technical foams *219, 221, 222*
test methods 296, *297*
TPUs 323
technical foams **217–228**
    see also packaging...; semi-rigid...; viscoelastic foams
temperature, conversion factors 453
tensile moduli, elastomers, test method 296
tensile strength
    elastomers *290, 296, 304, 307*
    flexible foams *179*–180, *195, 197, 206, 208, 209*
    integral skin foams *341*
    microcellular elastomers *290, 304, 307*
    RIM elastomers *343*
    sealants *411*
    test methods 179–180, 296
    TPUs *323,* 324
tertiary amine catalysts *140, 154,* 155
test methods
    CASE products **353–362**
    composites **424–425**
    elastomers **294–300**
    fire ignitability/propagation 56–59, 60
    flexible foams **177–188**
    rigid foams **241–244**
    see also standard test methods
tetramethylxylene diisocyanate (TMXDI) 63, *449, 456*
thermal analysis
    elastomers 175, **298**
    flexible foams 175
thermal conductivity
    composites *433*
    conversion factors *453*
    rigid foams **233–236**
        effect of ageing **236–239**
        effect of blowing agent *251*
        standards for 239, *241*
        test methods **241–242**
        typical values *251, 260, 269, 274*
    see also lambda value
thermal insulation applications 7, 20–21, 229
    economic advantages 23
    market share 16
thermal resistivity 234
thermography, inspection of composites by 425
thermoplastic polyurethanes 6, 285, **311–314, 315–330**
    in adhesives 384, 387

applications 6, 302, 311, **316–321**
blends with other polymers 327–328
in coatings 374
compared with PVC and thermoplastic rubber *312*
density *312*, 323
expanded/blown TPU 302, 313–314
   advantages 313–314
   properties *312*, 313
hazards 44
process technology 315–316, **327–330**
   blending of polymers 327–328
   compounding 327
   drying 328
   extrusion 329
   injection moulding 328–329
   powder processing 328
   rotational moulding 330
production of **324–327**
   by band casting 325
   batch process 325
   by reactive extrusion 326–327
properties *312*, 323, *324*
raw materials and formulations 321, **321–324**, *322*
   catalysts 155, 156
   polyols 107, 109, *322*
thixotropy control agents 383, 392, 412
threshold limit values (TLVs) 44–45
tin-based catalysts 137, *140*, 144–145, 151, 154, *154*
see also organotin catalysts; stannous octoate
tolidine diisocyanate (TODI), market share 63
toluene, nitration of 64, *66*, 70–71
toluene diamine, production of 64, *66*, 72
ortho-toluene diamines 72
   as initiator in polyol production 101
toluene diisocyanate (TDI) **86**
   in CASE formulations, 349, 364, 382, 391m 412
   chemical structure *66*, 86, *288*, 289
   emissions from foam production 29
   equivalent weight 449
   in flexible foam production 169, *170*, 171
      moulded foam *170*, 189, 193, 194, 195, 197
      slabstock foam *170*, 203, 207, *209*, 210
   foam formulation worked example 450
   functionality 449
   isocyanate value 449
   life-cycle inventory data 25
   market share 63
   molecular weight 449, 455
   physical properties 455
   production of 64, *66*
   purification of *66*, 78
   reactivity 83, 113, *114*, 115
   solvents for 77
   TMP adduct 449
   volatility 43
see also TDI-based PUs
toxic potency 62
training of workers 45
transesterification, of polyols 125
transfer coating process, for synthetic leather production 337
transmission electron microscopy (TEM), flexible foams 175, *176*
transport applications 17, 390, 391
see also automotive applications
transportation, of base chemicals 47–49, 82
triamines
   molecular structure of isomers 75
   production of 73, *74*
trichlorofluoromethane
   as blowing agent 127, 131
   physical and environmental properties 131
trichloropropyl phosphate (TCPP), as fire retardant 161, 162
triethanolamine, as chain extender 448
triethyl phosphate (TEP)
   as catalyst 120, 121
   as fire retardant 162
triethylenediamine (TEDA)
   as catalyst *140*, *154*, 155, 333
      reaction kinetics 147, *148*, *149*, *150*
   reactivity 142
   production of 137, *138*
trifunctional isocyanurates 80
trimethylolpropane, as chain extender 448
tris(N,N-dimethylaminomethyl) phenol, as catalyst 145
N,N'N"-tris(3-dimethylaminopropyl)-hexahydrotriazine, as catalyst *140*, 145

tropical conditions, in testing of flexible foams 187
trouser tear test, microcellular elastomers 296, *297*
tubing applications 319, *320*
two-component adhesives 379, **391–393**
   applications 391, 392–393
   development of strength with time *380*
   effect of temperature on strength *381*
two-component coatings 154, *348*, 363, **366–370**
   polyurea coatings *367*, 370
   solvent-borne coatings 366–368, *367*
   solvent-free coatings *367*, 369–370
   water-borne coatings *367*, 368–369
two-component elastomers
   cast elastomers 6, 285, **331–336**
   elastomeric fibres 6, 286, **338–339**
   footwear applications **303–310**
   integral skin foams 306, **339–342**
   polyurea elastomers 345
   RIM elastomers 285, **342–345**
   synthetic leathers **336–338**
two-component sealants **410–411**
   applications 411, 415
   formulations *411*
   properties *411*
types of PUs 4–6
tyre-fill cast elastomer 335–336

ultrasonic testing, composites 425
ultraviolet (UV) radiation
   effect on PUs 125–126, 167
   urethane acrylate coatings cured with 377
ultraviolet (UV) stabilisers 167–168, 367
United Nations Environment Programme (UNEP), Flexible and Rigid Foams Technical Options Committee 130
upholstery fires 53–55
urea
   formation in isocyanate reactions 115, *116*
   reaction with isocyanates 116–117
   in side reaction of phosgenation 77
urea–formaldehyde (UF) resins, in wood composites 395, 404
urea polymers, formation of 171, *172*, 230
urethane acrylate adhesives 394

urethane acrylate coatings 377
urethane alkyds 376
urethane oils 155, 376
urethane reaction 114
urethanes
  formation in flexible foams 171, *172*
  formation of 114
  thermal decomposition of 88, 114–115
uretidinediones, formation of *118*, *146*
uretonimines 78, **120–121**
USA
  fire loss data *53*, *54*
  fire regulations 59
  flexible foam consumption vs purchasing power *20*
  standard test methods
    elastomers 295
    flexible foams *179*
    rigid foams *241*
  see also ASTM standards/test methods
use phase, environmental impacts 30–34

vacuum forming, semi-rigid foams 218
vacuum insulation technology 21–22
vapour pressure (at various temperatures), isocyanates *455–456*
ventilation systems 44–45
versatility of PUs 21
Vertifoam (slabstock foam production) process 214
vibration energy tests, flexible foams 185–186
viscoelastic foams 196, 205, **220–221**
  applications 220
  formulations 220–221
viscosity
  conversion factors 453
  determination of 81
  isocyanates 81, *455–456*
  polyols 99–100, 109
  sealants *411*
viscosity-reducers 167
viscosity measurements, CASE formulations 354, 356
visual properties
  coatings 360, 374
  composites 444

volatile organic compounds (VOCs) emissions
  from adhesives and coatings 366, 385
  from foam production 28, 29, 261
volume, conversion factors 453

wall stabilisation applications **284**
waste
  amounts *35*
  collection of 36
  composition 36
  recycling of 36–37, 166, 226
  types *35*
waste management of PUs
  by Huntsman Polyurethanes 27
  factors affecting 36–37
  regulations covering 37–38
  technical options 37–38
water, as chain extender *164*, *448*
water-based dispersions 375–376
  applications 337, 375, 386, 387
  see also polyurethane dispersions
water-borne adhesives **386–387**
  applications 386
  development of strength with time *380*
  effect of temperature on strength *381*
water-borne oven-curing coatings **372**
  applications 372
  formulations 367
water-borne two-component coatings 368–369
  applications 17, 18, 368
water-borne two-component PU coatings, formulations 367
water heaters, insulation of **282–283**
water–isocyanate reaction 115
  in blowing agents 127, 135
  in expanded elastomers 306
  in flexible foams 204
  in rigid foams 280, 282, 283
water-scavenging additives *351*, 374, 383
water vapour transmissibility
  determination of
    flexible foams 185
    rigid foams *241*, 244
weathering tests
  CASE products 360
  elastomers 299

weight reduction, in automotive applications 438, 439, 442, *443*
weldable foams 208, 225
wellington boots, insulation properties *308*
wet compression set
  test conditions 187
  typical values
    moulded foams *195*, *197*
    slabstock foam *206*, *208*, *209*
wet spinning, for elastomeric fibres 339
wetting of surfaces 352
wide angle X-ray spectroscopy (WAXS), flexible foams studied by 176
windsails 283
windscreens/windshields (automobiles) 320, 413–415
windsurfing boards 283
winter-conditions applications 187
wire and cable applications 6, 320, 372
wood adhesives 393
wood composites 18–19, **395–408**
  engineered lumber 405–408
  market share *397*
  mechanisms of adhesion 397–399
  medium-density fibreboard *396*, 403–405
  oriented strand board *396*, 399–402
  types *396*
wood products, coatings for 1, 18
wood substrate, properties compared with PU composites *438*
world population growth *10*
woven glass cloth 423
woven glass roving 423

Photographs

3-9		© Corbis
3-10		Edward Milford
8-1		Ozone Processing Team, Goddard Space Flight Center, NASA
12-3		Larry Dale
12-4		Eric Crossan
13-6		Cannon
13-7		Fecken-Kirfel
15	(banner, centre)	Eric Crossan
	(banner, right)	© Corbis
16-1		*id.*
16-2		Schenectady Museum
16-4		Danfoss
16-5		*id.*
16-6		Electrolux
16-7		*id.*
16-8		*id.*
17-2		Siempelkamp
17-3		EcoTherm bv Winterswijk
17-7		Kingspan
17-8		*id.*
17-9		*id.*
18-5		Demilec Inc.
18-6		*id.*
19	(banner, right)	Larry Dale
20-10		APEGO SRL
20-11		*id.*
20-12		Klöckner Desma Schuhmachinen GmbH
21-5		© Corbis
22-6		*id.*
23	(banner, centre)	Stockmeier Kunststoffe GmbH & Co KG
	(banner, left)	APA - The Engineered Wood Association
23-6		Taber® Industries
24-1		© Digital Vision Ltd
24-3		© Corbis
24-4		Rhino Linings USA, Inc.
24-6		© Corbis
25-3/4/5/6		Satra Technology Centre
25-8		Stockmeier Kunststoffe GmbH & Co KG
25-9		Paul Lefton
26-3		APA - The Engineered Wood Association
26-5		Maschinenfabrik J. Dieffenbacher GmbH & Co.
26-6		Siempelkamp (ContiRoll® Press for OSB composite wood panels)
26-7/8/9/10/11		APA - The Engineered Wood Association
28	(banner, left)	© Corbis
	(banner, centre)	© Digital Vision Ltd
29-2		Larry Dale
30-1		*id.*
30-2		*id.*
30-3		*id.*
30-4		*id.*
30-5		*id.*
30-6		Rhino Linings USA, Inc.